Molecular Biology Intelligence Unit

Nitric Oxide Research from Chemistry to Biology: EPR Spectroscopy of Nitrosylated Compounds

Yann A. Henry, Ph.D.

Annie Guissani, Ph.D.

Béatrice Ducastel, M.S.

Unité INSERM 350
Institut Curie
Orsay, France

CHAPMAN & HALL
I(T)P An International Thomson Publishing Company

New York • Albany • Bonn • Boston • Cincinnati • Detroit • London • Madrid • Melbourne •
Mexico City • Pacific Grove • Paris • San Francisco • Singapore • Tokyo • Toronto • Washington

LANDES
BIOSCIENCE

AUSTIN, TEXAS

MOLECULAR BIOLOGY INTELLIGENCE UNIT
NITRIC OXIDE RESEARCH FROM CHEMISTRY TO BIOLOGY:
EPR SPECTROSCOPY OF NITROSYLATED COMPOUNDS

R.G. LANDES COMPANY
Austin, Texas, U.S.A.

Please address all inquiries to the Publishers:
R.G. Landes Company, 810 S. Church Street, Georgetown, Texas, U.S.A. 78626
Phone: 512/ 863 7762; FAX: 512/ 863 0081

North American distributor:
Chapman & Hall, 115 Fifth Avenue, New York, New York, U.S.A. 10003

CHAPMAN & HALL

U.S. and Canada ISBN: 0-412-13561-2

While the authors, editors and publisher believe that drug selection and dosage and the specifications and usage of equipment and devices, as set forth in this book, are in accord with current recommendations and practice at the time of publication, they make no warranty, expressed or implied, with respect to material described in this book. In view of the ongoing research, equipment development, changes in governmental regulations and the rapid accumulation of information relating to the biomedical sciences, the reader is urged to carefully review and evaluate the information provided herein.

Library of Congress Cataloging-in-Publication Data

Henry, Yann A.
    Nitric oxide research from chemistry to biology : EPR spectroscopy of nitrosylated compounds / Yann A. Henry, Annie Guissani & Béatrice Ducastel.
        p. com. -- (Molecular biology intelligence unit)
    Includes bibliographical references and index.
        ISBN 1-57059-390-6 (alk. paper)
        1. Nitric oxide--Metabolism. 2. Metalloproteins--Metabolism. 3. Electron paramagnetic resonance spectroscopy. 4. Organonittrogen compounds--Metabolism. I. Guissani, Annie. II. Ducastel, Béatrice. III. Title. IV. Series.
                [DNLM: 1. Nitrous Oxide. 2. Nitroso Compounds. 3. Metalloproteins. 4. Electron Spin Resonance Spectroscopy.   QD
181.N1 H525n 1996]
QP535.N1H46 1996
574.19'245--dc20
DNLM/DLC
for Library of Congress   96-29465
                                                                                                              CIP

Publisher's Note

R.G. Landes Company publishes six book series: *Medical Intelligence Unit, Molecular Biology Intelligence Unit, Neuroscience Intelligence Unit, Tissue Engineering Intelligence Unit, Biotechnology Intelligence Unit* and *Environmental Intelligence Unit.* The authors of our books are acknowledged leaders in their fields and the topics are unique. Almost without exception, no other similar books exist on these topics.

Our goal is to publish books in important and rapidly changing areas of bioscience and environment for sophisticated researchers and clinicians. To achieve this goal, we have accelerated our publishing program to conform to the fast pace in which information grows in bioscience. Most of our books are published within 90 to 120 days of receipt of the manuscript. We would like to thank our readers for their continuing interest and welcome any comments or suggestions they may have for future books.

Shyamali Ghosh
Publications Director
R.G. Landes Company

DEDICATION

To Maurice, Yves and Jean-François.

We wish to dedicate this work to our parents who tried to make us understand, more or less successfully, that to learn English might become useful some day!

We dedicate this book to those who interested one of us (YH) in EPR spectroscopy during his Ph.D. work many years ago, Dr. Ramaprasad Banerjee, Dr. François Leterrier, Prof. Robert J.P. Williams, Prof. John F. Gibson, and Prof. Jack Peisach, as his post-doctoral fellow.

Yann A. Henry, Annie Guissani and Béatrice Ducastel

CONTENTS

Prologue 1
   Introducing the VIDM Club 1
   A Beautiful Molecule 1

1. Introduction: The Origins of Nitric Oxide 3
   Yann A. Henry
   Nitrogen Cycles 3
   Sources of Nitric Oxide, an Air Pollutant 4
   Use of Nitric Oxide as a Paramagnetic Probe of Metalloproteins 4
   Nitric Oxide in Bacterial Denitrification 5
   Effects of Exogenous Nitric Oxide on Humans 5
   Nitric Oxide Biosynthesis in Mammals 5
   Introductory Chapters 6
   Scaling Problems in the Quantification Methods of Nitric Oxide 7
   A Historical Note 8
   Additional Recent Bibliography 8

2. Basic Chemistry of Nitric Oxide and Related Nitrogen Oxides ... 15
   Yann A. Henry, Béatrice Ducastel and Annie Guissani
   Structure of Nitrogen Oxides 15
   Reactivity of Nitric Oxide and Nitrogen Oxides 22

3. Basic EPR Methodology 47
   Yann A. Henry
   Basic Principles of EPR Spectroscopy 47
   Experimental Conditions 53
   Discussion: Specificity Versus Sensitivity 57

**4. EPR Characterization of Nitric Oxide Binding
   to Hemoglobin** 61
   Yann A. Henry
   Introduction 61
   Nitrosyl Hemoglobin: Kinetic and EPR Data 64
   Reactions of Nitric Oxide with Ferrihemoglobin
     and Oxyhemoglobin 77
   Conclusion 79

5. **Effects of Nitric Oxide on Red Blood Cells** 87
    Yann A. Henry
    Introduction 87
    Effects of Long-Term Exposure of Animals and Humans
      to Nitric Oxide and Nitrogen Dioxide Gas Considered
      as Air Pollutants 89
    Effects of Nitrite and Nitrogenous Vasodilators on Red Blood Cells ... 90
    Effects on Red Blood Cells of Nitric Oxide Inhalation
      in Therapeutic Uses 93
    Conclusions 94

6. **Utilization of Nitric Oxide as a Paramagnetic Probe
    of the Molecular Oxygen Binding Site
    of Metalloenzymes** 99
    Yann A. Henry
    Introduction 99
    Hemoproteins 100
    Cytochrome *c* Oxidase (Cytochrome *aa₃*) 108
    Copper-Containing Proteins 113
    Mononuclear Iron-Containing Dioxygenases 116
    [FeS], Ni and Mo Cluster-Containing Proteins 117
    Binuclear Iron-Containing Proteins 122
    Conclusion 128

7. **Nitric Oxide, an Intermediate in the Denitrification
    Process and Other Bacterial Pathways, as Detected
    by EPR Spectroscopy** 145
    Yann A. Henry
    Introduction: A Historical Background 145
    Respiratory and Assimilatory Nitrite Reductases 149
    Nitric Oxide Reductases 159
    Other Cytochromes Interacting with Nitric Oxide 162
    Nitrous Oxide Reductases 164
    Possible Roles of Nitric Oxide in Nitrification
      and Nitrogen-Fixation 167
    Bacterial Nitric Oxide Synthase 167
    Discussion and Conclusion 167

8. **Nitric Oxide Biosynthesis in Mammals** 175
    Sandrine Vadon-Le Goff and Jean-Pierre Tenu
    Constitutive and Inducible Nitric Oxide Synthases 176
    Enzymatic Mechanism of Nitric Oxide Synthases 176
    Nitric Oxide Synthases' Regulations 179
    Post-Transcriptional Modifications of the Enzyme 180
    Conclusion 184

9. **The Use of EPR Spectroscopy for the Identification of the Nature of Endothelium-Derived Relaxing Factor** 193
Yann A. Henry
The Nature of Endothelium-Derived Relaxing Factor:
A Controversy 193
Nitric Oxide as an EDRF Detected by EPR Spectroscopy 194
Fe(SR)$_2$(NO)$_2$ Ternary Iron Complexes as Plausible EDRF
or Nitric Oxide Transporters 195
Conclusion 199

10. **Enzymatic Targets of Nitric Oxide as Detected by EPR Spectroscopy Within Mammal Cells** 205
Yann A. Henry, Béatrice Ducastel and Annie Guissani
Introduction 205
Soluble Guanylate Cyclase 206
[FeS] Cluster-Containing Enzyme 208
Ribonucleotide Reductase 212
Proteins of the Iron Metabolism 217
Conclusion 223

11. **Overproduction of Nitric Oxide in Physiology and Pathophysiology: EPR Detection** 235
Yann A. Henry
Introduction 235
Cancer and Interleukin-2 Immunotherapy 236
Parasites, Fungi and Bacterial Infections 240
Septic Shock 243
Autoimmune Diseases 248
Organ Transplantations 249
Vascular Diseases 255
Other Physiological and Pathophysiological States 257
Conclusion 258

12. **Palliatives to Underproduction of Nitric Oxide as Assayed by EPR Spectroscopy** 271
Claire Ducrocq and Annie Guissani
Introduction 271
Metal Complexes as Nitric Oxide Donors 273
Nitrite, Organic Nitrites and Nitrates 276
Other Derivatives 280
Inhalation Of Nitric Oxide: NO Gas, a Suppletive Drug 282
Conclusion 284

13. Nitric Oxide-Specific Spin-Trapping EPR Methods 293
    Yann A. Henry
    Use of Hemoglobin as a Nitric Oxide Spin-Trap 293
    Conventional Spin-Trapping 294
    Prospects on Nitric Oxide-Specific Spin-Traps 294
    Scavenging of Nitric Oxide by Reaction with Stable Free Radicals 294
    Line Broadening of Fusinite and Other Free Radicals
        by Nitric Oxide 295
    Use of Fe-(DETC)$_2$ and Similar Complexes
        as Nitric Oxide Spin-Traps 295
    Conclusion 301

**14. General Discussion: Crossregulations of Metalloenzymes
    Triggered by Nitric Oxide** 307
    Yann A. Henry
    Downregulation of Nitric Oxide Synthase by Nitric Oxide 307
    Inhibition of Microsomal Cytochromes P-450 310
    The Urea Cycle 311
    Inhibition or Regulation of Cytochrome *c* Oxidase 312
    Crossregulations of Other Inducible Metalloproteins 313
    Regulation by Nitric Oxide of Other Metalloenzymes
        of the Inflammatory Response 316
    Other Targets 318
    General Conclusion 320

Index 329

EDITORS

Yann A. Henry, Ph.D.
Unité INSERM 350
Institut Curie
Orsay, France
Chapters 1-7, 9-11, 13, 14

Annie Guissani, Ph.D.
Unité INSERM 350
Institut Curie
Orsay, France
Chapters 2, 10, 12

Béatrice Ducastel, M.S.
Unité INSERM 350
Institut Curie
Orsay, France
Chapters 2, 10

CONTRIBUTORS

Sandrine Vadon-Le Goff, Ph.D.
CNRS URA 400
Unitersité René Descartes
Paris, France
Chapter 8

Jean-Pierre Tenu, Ph.D.
CNRS URA 1116
Institut de Biochimie
Orsay, France
Chapter 8

Claire Ducrocq, Ph. D.
CNRS UPR 2301
Institut de Chimie des Substances
   Neturelles
Gif-sur-Yvette, France
Chapter 12

PREFACE

In the rapidly developing field of nitric oxide research, it appears difficult to make a meaningful synthesis, except through the bias of given points of view or of specific analytical methods. This book focuses on the use of electron paramagnetic resonance spectroscopy of nitrosylated metalloprotein complexes in nitric oxide research.

The basic chemical reactivity and physical properties of Nitric Oxide have been summarized. EPR spectroscopy results obtained over 25 years—many of them since 1990—have been comprehensively described, covering several biological fields: hemoglobin allostery, nitric oxide in blood, metalloprotein biochemistry, bacterial denitrification, cellular biochemistry, immunology, transplantation, etc.

Recent use of EPR spectroscopy detection of nitrosylated metalloproteins in cell cultures and in animal models of pathologies, and its potential for understanding health-disease switches in humans, have been discussed.

In this monograph, an attempt has been made to render EPR spectroscopy accessible to biologists interested in nitric oxide research, over a scope as wide as possible, and at the same time to be as comprehensive as possible. The book was written over a period of two and a half years when a flood of articles was pouring in; this might explain some shortcomings.

This book results from recurrent bouts of interest in nitric oxide research, always seen through the specific EPR spectroscopy window frame, three times in YH's career: in 1971-1974: nitric oxide and hemoglobin, in 1979-1984: nitric oxide in denitrification, and since 1988: nitric oxide in mammalian cells.

Orsay, June 11, 1996

ACKNOWLEDGMENTS

We gladly thank our colleagues and friends who corrected the manuscript, Dr. Robert Cassoly, Dr. Jean-Claude Drapier, Dr. Michel Lepoivre, and particularly Dr. Mike C. Marden, who went through the whole book.

This book owes everything to our successive directors, Dr. Ramaprasad Banerjee, Dr. Robert Cassoly, Dr. Jean-Marc Lhoste, and Dr. Daniel Lavalette, who gave us free rein in our scientific choices, for which we are grateful. Our laboratories have been endowed over these years by salaries (to YH and AG) and grants from the Centre National de la Recherche Scientifique, the Institut de Biologie Physico-Chimique, the Institut National de la Santé et de la Recherche Médicale, and the Institut Curie (section de recherche).

Finally we are indebted to our colleagues and friends, Dr. Jean-Claude Drapier, Dr. Claire Ducrocq, Dr. Michel Lepoivre, Dr. Jean-Luc Boucher, Dr. Jean-Pierre Tenu, their students, and many others met at the "Club NO" meetings and forums, for the very rewarding collaborations they have offered us since 1988 and for the time they have spent to share their knowledge with us; these collaborations made this book possible. And not least, we thank Martial Thiébaut who offered a portable Macintosh for work at home!

Yann A. Henry, Annie Guissani and Béatrice Ducastel

ABBREVIATIONS

AABS : A-activator binding site

AIDS : acquired immunodeficiency syndrome

AL : argininosuccinate lyase

AO : ascorbate oxydase

ARDS : adult respiratory distress syndrome

AS : argininosuccinate synthase

$AscH_2$: ascorbic acid

Av1 and Av2 : subunits of nitrogenase from *Azotobacter vinelandii*

BAL : bronchoalveolar cell pellet

BCG: Bacillus Calmette-Guérin *(Myobacterium bouis)*

BFR : bacterioferritin

BH_4 : tetrahydrobiopterin

BSA : bovine serum albumin

BSO : buthionine-SR-sulfoxime

BZF : bezafibrate

CaM : calmodulin

CAT : catalase

CcO : cytochrome aa_3 (cyt aa_3), cytochrome *c* oxidase

CcP : cytochrome *c* peroxidase

2,3CDO : catechol 2,3-dioxygenase

CFA : clofibric acid

cGMP : cyclic guanosine 3',5'-monophosphate

CMV : cytomegalovirus

ABBREVIATIONS

CNS : central nervous system

COX : cyclooxygenase or cytochrome oxidase

CPN : ceruloplasmin

Cp1 and Cp2 : subunits of nitrogenase from *Clostridium pasteurianum*

CPO : chloroperoxidase

CTPO : 3-carbamoyl-2,2,5,5-tetramethyl-3-pyrroline-1-yloxy

CW-EPR : continuous wave EPR

CYP : cytochrome P-450

DBNBS : 3,5-dibromo-4-nitrosobenzene sulphonate

DETC : diethyldithiocarbamate

DMPO : 5,5-dimethyl-1-pyrroline *N*-oxide

DMSO : dimethylsulfoxide

DNIC : dinitrosyl-iron complex

2,3-DPG : 2,3-diphosphoglycerate

DPPH : 1,1-diphenyl-2-picrylhydrazyl stable free radical

DTT : dithiothreitol

EAE : experimental allergic encephalomyelitis

eALAS : erythroid 5-aminolevulinate synthase

EDRF : endothelium-derived relaxing factor

ENDOR : electron nuclear double resonance

EPR : electron paramagnetic resonance

ESR : electron spin resonance

EXAFS : extended X-ray absorption fine structure

ABBREVIATIONS

FAD : flavin adenine dinucleotide

Fd : ferredoxin

FK409 : (E)-4-ethyl-2-[(E)-hydroxyimino]-5-nitro-3-hexenamide

Fld : flavodoxin

FMLP : peptide formylMet-Leu-Phe

FMN : flavin mononucleotide

FT : Fourier transform

FTN : ferritin

G : Gauss

GC, sGC : guanylate cyclase, soluble GC

GC/MS : gas chromatography/mass spectrometry

GHz : gigahertz

GMP, cGMP : guanosine 3',5'-monophosphate, cyclic GMP

GTN : glyceryl trinitrate

GTP : guanosine 5'-triphosphate

GVHD : graft-versus-host disease

Hb : hemoglobin

HbO_2, MetHb, HbNO : oxyhemoglobin, methemoglobin, nitrosyl hemoglobin

Hc : hemocyanin

HETE : hydroxyeicosatetraenoic acid

12-HHT : 12-hydroxyheptadecatrinoic acid

HIV : human immunodeficiency virus

ABBREVIATIONS

HO : heme oxygenase

Hp : haptoglobin

5-HPETE : 5-hydroperoxyeicosatetraenoic acid

Hr : hemerythrin

HRP : horseradish peroxidase

HSP : heat-shock protein

HUVEC : human umbilical vein endothelial cells

IC_{50} : half-inhibitory concentration

IDN : isosorbide dinitrate

IDO : indoleamine 2,3-dioxygenase

IFNγ : γ-interferon

IHP : inositolhexaphosphate

IL : interleukin

IL-1β : interleukin 1β

IL-2R : interleukin-2 receptor

IL-6 : interleukin 6

i.p. : intraperitoneally

IRE : iron responsive element

γ-IRE : IFNγ response element

IRF-1: interferon-response factor 1

IRP : iron regulatory protein or IRE-binding protein

J : Joule

K : Kelvin

ABBREVIATIONS

K : kaiser (1 cm^{-1} = 1 kaiser)

kHz : kilohertz

L-1 : lipoxygenase 1

LegHb : leghemoglobin

LPO : lactoperoxidase

LPS : lipopolysaccharide

LTB4 : leukotriene B4

Mb, metMb : myoglobin and ferric myoglobin

MbNO : nitrosylmyoglobin

MBP : myelin basic protein

MCD : magnetic circular dichroism

MDP : muramyl dipeptide

MDRF : macrophage-derived relaxing factor

MGD : *N*-methyl-D-glucamine dithiocarbamate

MHz : megahertz

MNP : 2-methyl-2-nitrosopropane

MPO : myeloperoxidase

MRE : metal responsive element

MT : metallothionein

mW : milliwatt

NADPH/NADP : nicotinamide adenine dinucleotide phosphate reduced/oxidized

NAME : *N$^\omega$*-nitro-L-arginine methyl ester

ABBREVIATIONS

NANC : nonadrenergic noncholinergic

NaR : nitrate reductase

NFκB : nuclear factor-κB

NF-IL-6 : nuclear factor-interleukin-6

NHE : normal hydrogen electrode

NI : 7-nitroindazole

NiR : nitrite reductase

NMDA : N-methyl-D-aspartate

NMMA : N^ω-monomethyl-L-arginine

NMR : nuclear magnetic resonance

NOCT : nitric oxide cheletropic trap

NOHA : N^ω-hydroxy-L-arginine or N^Ghydroxy-L-arginine

NONOate : diazeniumdiolate, $RR'N[(NO)-N=O]^-$

NoR : nitric oxide reductase

N_2OR : nitrous oxide reductase

NOS : nitric oxide synthase

NOS I (nNOS) : neuronal constitutive NOS

NOS II (iNOS) : inducible NOS

NOS III (eNOS) : endothelial constitutive NOS

mtNOS : mitochondrial NOS

Oe : Oersted

$18\text{-}ONO_2A$: 18-nitro-oxyandrostenedione

P-450 : P-450 monooxygenases, $P\text{-}450_{cam}$ from *Pseudomonas putida*, $P\text{-}450_{scc}$ from bovine adrenal cortex

ABBREVIATIONS

PBN : phenyl *N-tert*-butylnitrone

3,4PDO : protocatechuate 3,4-dioxygenase

4,5PDO : protocatechuate 4,5-dioxygenase

PGE_2 : [15Z,11α,13E,15S]-11,15-dihydroxy-9-oxoprosta-5,13-dienoic acid

PGG_2 : 15-hydroperoxy-9,11-peroxidoprosta-5,13-dienoic acid

PGH_2 : 15-hydroxy-9,11-peroxidoprosta-5,13-dienoic acid

PGHS : prostaglandin H (prostaglandin endoperoxide) synthase

PMA : phorbol 12-myristate 13-acetate

PMN : polymorphonuclear leukocyte, neutrophil

PMS : phenazine methosulfate

POD : post-operation day

ppb : part per billion

PPHN: persistent pulmonary hypertension of the newborn

ppm : part per million

PTI : 2-phenyl-4,4,5,5-tetramethylimidazoline-1-oxyl

PTIO : 2-phenyl-4,4,5,5-tetramethylimidazoline-1-oxyl 3-oxide

R1 and R2 : subunits of ribonucleotide reductase

RBC : red blood cell

REAC : resistance arteriolar endothelial cells

RNR : ribonucleotide reductase

RR : resonance Raman

SDH : succinate dehydrogenase

ABBREVIATIONS

SDS : sodium dodecylsulfate

sGC : soluble guanylate cyclase

SHF : superhyperfine (structure)

SIN-1 : 3-morpholinosydnonimine *N*-ethylcarbamide

SIN-1A : N-morpholinoiminoacetonitrate

SNAP : *S*-nitroso-*N*-acetylpenicillamine

SNP : sodium nitroprusside

SOD : superoxide dismutase

sox : oxidative stress

SR : thiol residue

sTF, oTF, lTF : serum transferrin, ovotransferrin, lactoferrin

T : Tesla

t-BuOOH : *tert*-butylhydroperoxide

TCT : tracheal cytotoxin

TGFβ : transforming growth factor β

TNF : tumor necrosis factor

TNFα : tumor necrosis factor α

TPO : L-tryptophan 2,3-dioxygenase

TSST-1 : toxic shock syndrome toxin 1

U : unit

XDH : xanthine dehydrogenase

XO : xanthine oxidase

PROLOGUE

INTRODUCING THE VIDM CLUB

Nitric oxide has recently joined a very select club, that of the "Very Important Diatomic Molecules," together with three old members elected in the XVIIIth century, O_2, H_2, and N_2. Other candidates such as CO, ClO⁻, or CN⁻ are still in a position of begging for admittance, being rather murderous crooks, while $O_2^{-\bullet}$ and OH$^\bullet$ make pretense to belonging to the club; some say these last two are either too flimsy for the former or really too boisterous for the latter to carry real weight in the club's decisions. There are two others which could enter the club very soon, being NO kins, NO⁺ and NO⁻ ; but their precedence quarrel over NO could well indispose the old electors. When one remembers the long life of NO itself in the underground world of killers, for so many years, one understands the refusal for its first membership request after acquiring a beautiful mansion in "Bacteria Land" in the late 1970s with money borrowed from N_2 but never returned. Then more recently, being offered a real palace in "Mammalia" (1987) and having received massive popular acclaim,[1,2] NO could enter the VIDM club.

A BEAUTIFUL MOLECULE

We shall recount here one chapter of a long tale. There are many others to bind into "A Thousand and One Nights" book of tales. It is better anyway to tell the story right away while we have it fresh in mind, and before the beautiful palace walls start peeling and crumbling. In all fairy tales there is a question of taste and fashion. We hope to interest you readers before the beautiful NO molecule becomes too well known, or even appears old and ugly. Nobody knows whether the strong but brutal personality shown by NO in previous years will prevail over its present amenity and grace channeled to help the poor and the sick. We shall tell how we have been able to keep track of NO, through a peculiar looking glass called electron paramagnetic resonance spectroscopy, for many years in its bad and good deeds and to explain in latter years that NO could be a rather decent person. Any resemblance between the life of NO that we shall relate and a short novel about a schizophrenic Dr. Jekyll is simply fortuitous.[3]

REFERENCES
1. Koshland DE Jr. The molecule of the year 1992. Science 1992; 258:1861.
2. Gibaldi M. What is nitric oxide and why are so many people studying it? J Clin Pharmacol 1993;33:488-496.
3. Stevenson RL. The Strange Case of Dr. Jekyll and Mr. Hyde, 1886. New York: Dover Thrift Editions, 1991.

CHAPTER 1

INTRODUCTION: THE ORIGINS OF NITRIC OXIDE

Yann A. Henry

NITROGEN CYCLES

Nitric oxide NO[•] is part of several interconnected nitrogen cycles of varying global importance. The geochemical nitrogen cycle relates to atmospheric layers[1] such as the thermosphere (> 110 km),[2] the airglow layer (85-110 km),[3] the troposphere (2-15 km) and stratosphere (12-35 km).[4-6]

The biological nitrogen cycle spreads over seas, soils and the whole biosphere.[7-10] The biological nitrogen cycle was first understood in the restricted domain of microbiology through the interpretation of six processes having global impacts, atmospheric nitrogen fixation, the ammonia-amine loop, nitrification, assimilatory and dissimilatory reductions of nitrate and nitrite, and finally denitrification (See chapter 7).[11,12]

From the newly discovered NO[•] synthesis from L-arginine, the nitrogen cycle has now been understood to include mammals (largely covered in chapters 8 to 12) and a number of animals in various phyla of invertebrates and vertebrates, molluscs (gastropods, snails *Lymnaea stagnalis*,[13] *Limax maximus*,[14] a very old "conserved" arthropod horse-shoe crab *Limulus polyphemus*),[15] echinoderms (e.g. starfish *Marthasterias glacialis*, also a very conserved species),[16] insects (e.g. bug *Rhodnius proxilus*,[17] fly *Drosophila melanogaster*,[18] silkworm *Bombyx mori*),[19] amphibians (toads, *Bufo marinus*),[20] fishes (Atlantic salmon *Salmo salar*),[21] birds (quail *Coturnix coturnix*), etc. Recently NO[•] synthesis from L-arginine was discovered for the first time in a microorganism, a *Nocardia* species which is an obligate aerobe, a filamentous actinomycetes bacteria found in soil.[22] An acellular macroplasmodial slime mold of the class Myxomycota, *Physarum polycephalum*, contains also a NO synthase, with an activity depending on the organism cycle.[23] These data indicate that the biosynthesis of NO[•] from L-arginine is a pathway of very early evolutionary origin.[15]

Nitric Oxide Research from Chemistry to Biology: EPR Spectroscopy of Nitrosylated Compounds, edited by Yann A. Henry, Annie Guissani and Béatrice Ducastel. © 1997 R.G. Landes Company.

SOURCES OF NITRIC OXIDE, AN AIR POLLUTANT

The origins of $NO^{\bullet}$, $NO_2^{\bullet}$, of other nitrogen oxides sometimes labeled NO_x ($NO^{\bullet} + NO_2^{\bullet}$) and NO_y (total reactive nitrogen) in the various atmospheric layers are partially anthropogenic—i.e., arising from human activities-related pollution—from car and aircraft exhaust emissions and industrial fumes.[4,24] They arise also massively from lightning discharges[25] and volcanic activities.[26,27] The importance for global climate of the existence of nitrogen oxides in the atmosphere is related to their inverse contributions to the photochemical production of the protective ozone layer contained in the stratosphere and to its destruction by the reaction, catalyzed by $NO^{\bullet}$ of O_3 with monoatomic oxygen.[6,24,28,29] Due to the short lifetime of NO_x, its concentration can vary by several orders of magnitude with latitude and longitude, altitude and time.

Other important sources of $NO^{\bullet}$, with great impact on human health, are industrial smogs and accidental building fires, producing wood and plastics smokes, and plain cigarette smoking.[30-32] Several free radicals are produced at high (300-1000°C) temperatures and, depending on the two smoke phases (gas and tar suspension), can exist for long periods at concentrations increasing with time, due to chain reactions and steady-state mechanisms. As we shall see in the following chapters nitric oxide toxic effects to humans are not only related to its own reactions with cellular targets (chapters 5, 7-12 and 14) but to the formation; by reaction with oxygen, superoxide anion, peroxyl radicals, etc.; of more highly toxic species such as nitrogen dioxide $NO_2^{\bullet}$, peroxynitrites $ROONO$, and peroxynitrates $ROONO_2$ (chapter 2).[31,33-38] Severe smoke inhalation may cause death after a delay of hours or days even in the absence of burns. On the contrary the effects of cigarette smoking may be delayed over twenty years or more producing serious diseases such as emphysema through the inactivation of an antiprotease, α-1-proteinase inhibitor. Nitrogen oxides,

in particular $NO_2^{\bullet}$ and nitrite, are certainly genotoxic and carcinogenic through several different mechanisms involving DNA base deamination, nitrosation of primary amines, and formation of N-nitroso compounds from secondary amines, etc.[39] However, the effects of $NO^{\bullet}$ itself seem to be rather slight, and those of nitrate appear null. Several types of human cancer are also related to production of nitrogen oxides at specific sites, i.e., lungs, throat or jaw.

The toxicity of nitric oxide was long attributed mostly to its strong, nearly irreversible binding to deoxyhemoglobin. In fact this property was diverted to the use of $NO^{\bullet}$ as a paramagnetic probe in studies relating hemoglobin structure to its function. These studies by electron paramagnetic resonance (EPR) spectroscopy led for 50 years to important academic contributions in the general understanding of hemoglobin considered by biochemists and biophysicists as a prototype of allosteric proteins (an "honorary enzyme", to quote Jeffries Wyman (1901-1995), a brilliant scientist and humanist I had the honor to know while working on my Ph.D. thesis under Dr. Ramaprasad Banerjee's guidance). This source of knowledge seems inexhaustible, as interesting later developments are still published (chapter 4).

Also due to its noxious effects as an air pollutant, nitric oxide effects on red blood cells were carefully studied (chapter 5). These results have had powerful and renewed impacts since $NO^{\bullet}$ synthesis was discovered in mammals (chapters 8-12).

USE OF NITRIC OXIDE AS A PARAMAGNETIC PROBE OF METALLOPROTEINS

Nitric oxide has a variable affinity for transition metals (chapter 2) and binds to many metalloproteins. In studies analogous to those performed on hemoglobin it was used as a protein inhibitor and a paramagnetic probe in studies relating their structure to their function as oxygen carriers, oxidases and oxygenases, etc.[40,41] Chapter 6 shall be devoted to the use of electron paramagnetic resonance spectroscopy in these diverse academic studies.

The first and the only remaining undisputed example in which nitric oxide was discovered to activate rather than inhibit an enzyme—guanylate cyclase—was discovered early.[42-46] The importance of this activation remained confined for several years to a limited scientific circle until the simultaneous discoveries (1986-1988) of the identification of nitric oxide as an endothelium-derived relaxing factor (EDRF)[47-57] and of the synthesis of nitrate and nitrite in mammals upon nonspecific immune response to bacterial infection,[58-67] which focused on the general importance of NO$^•$ in previously unconnected fields.

NITRIC OXIDE IN BACTERIAL DENITRIFICATION

Another important step in the history of nitric oxide was reached when it was recognized (in the early 1980s)[68-71] as a somewhat controversial intermediate in the bacterial denitrification process, that is the dissimilatory respiration of anaerobic bacteria over nitrate and nitrite.[12,72,73] These biogenic emissions of NO$^•$ in the atmosphere of rural areas[8] could have rates similar to those of anthropogenic emissions in urban areas.[9] The role of bacterial denitrification also has large implications for human health since many anaerobic species are hosts to humans.[74] NO$^•$ production and metabolism in denitrifiers shall be dealt with specifically in chapter 7. Nitric oxide bacterial reduction yields N_2O, another serious pollutant, implicated in the greenhouse effect and ozone depletion.

EFFECTS OF EXOGENOUS NITRIC OXIDE ON HUMANS

Nitric oxide is also known as a side product of nitrite addition in meat curing.[75] Nitrite has been used since the end of the 19th century to give cooked meat a bright red color similar to that of fresh meat. In fact our grandmothers' dry sausages and smoked ham had a rapid tendency to turn dark brown. This "fresh meat" red color obtained upon nitrite addition corresponds to nitrosylated myoglobin MbNO$^•$ and protein-dissociated heme-NO$^•$ (pentacoordinated Fe) which have visible absorption bands similar to those of MbO_2. If a gross excess of nitrite is added to meat, nitrite will react with the hemin ring itself, and the meat will turn green! The main advantage of nitrite meat curing apart from the psychological color effect is to inhibit bacterial proliferation, especially that of deadly *Clostridium botulinum*,[76] which can occur under mixed anaerobic conditions such as those existing in sealed packages. Another possible advantage of nitrite addition could be related to its chemical conversion to NO$^•$ at low pH values. This could act as a defense mechanism against ingested pathogenic microorganisms in the saliva and on the surface of the tongue and in the stomach at acidic pH levels.[77]

However, the use of excessive concentrations of added nitrite, converted to NO$^•$ and $NO_2^•$, could have long-term serious implications in N-nitrosation reactions leading to carcinogens.[58,60,78-84] Whatever their origin, chronic repeated bacteria or parasite infections, or pollution of water and food, the importance of nitrosating nitrogen oxides in the ethiology of certain types of human cancer of specific organs, i.e., bladder, stomach, colon, etc., was recognized many years ago and is still a hot topic in cancer research.[85-87]

NITRIC OXIDE BIOSYNTHESIS IN MAMMALS

The identification within a short period in 1986-1988[48-52] between nitric oxide and the earlier defined EDRF[47] in the mammalian cardiovascular system,[53-57] together with the simultaneous recognition of its roles in host defense mechanisms[58-67] and in the central and peripheral nervous systems,[88-93] provided suddenly unifying views over previously separated scientific fields: chemistry, biochemistry, microbiology, pollution and toxicology, cancerology, cardiovascular research, immunology, neurology, medicine, and surgery, etc.

This can be asserted from the incredible increase of the annual rate of scientific articles, as detected in a given data bank (Medline), which maintains an exponential pace, more than seven years after the discovery of NO synthesis in mammals (Fig. 1.1).

Chapters 8-14 shall be devoted firstly to a short introduction of NO• synthesis in mammals (chapter 8) and a specific chapter (chapter 9) on the use of EPR spectroscopy for the identification of EDRF. The three following chapters describe the determination, through the use of EPR spectroscopy, of some molecular targets in mammal cells (chapter 10) and in animals and humans following overproduction of NO• in pathophysiological cases (chapter 11). The palliatives to underproduction of NO• are chemically diverse; EPR spectroscopy has also been useful in studying the metabolisms of various NO• donors (chapter 12). In the final chapters (13 and 14) we shall introduce several discussions on the methodologies which should be adapted in detecting such targets in vitro and in vivo and on other still hypothetical cellular targets that could be of some biological importance.

INTRODUCTORY CHAPTERS

Due to the considerable importance of the discovery of NO• biosynthesis in mammals, many forgotten data have to be brought under a new light. In the field covered by the present book, we have attempted three such reviews, confronting

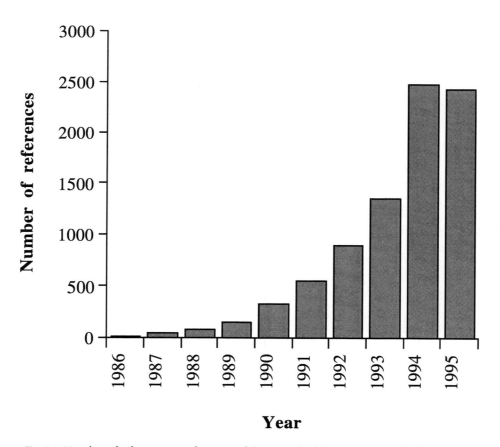

Fig. 1.1. Number of references as a function of the year of publication as found by bibliographical research using Medline as the data base (year 1995 number, when referenced in March 1996, could be somewhat underestimated). Key words used are nitric oxide or EDRF.

old data with new ones.[40,94,95] The purpose of the present review is also to bear in mind some forgotten or ill-understood basic facts about $NO^\bullet$ intrinsic paramagnetism and some of its complexes with transition metals, as studied by EPR spectroscopy. We shall also warn against some pitfalls of the method, as compared to other analytical methods in the detection of nitric oxide.[96] Due to its capacity to be either electron acceptor "electrophilic" ($NO^\bullet + e^- \rightarrow NO^-$ or N_2O) or electron donor "nucleophilic" ($NO^\bullet \rightarrow e^- + NO^+$ or NO_2^-) (see a discussion on this in chapter 2),[93,97-103] $NO^\bullet$ reactivity depends on the redox potential of its partners. In order to help make correct appraisals of the existence of $NO^\bullet$ in biological systems a short introduction to the basic chemistry of nitric oxide and "related" nitrogen oxides shall be first presented (chapter 2), followed by an introduction to basic EPR methodology (chapter 3).

SCALING PROBLEMS IN THE QUANTIFICATION METHODS OF NITRIC OXIDE

Estimates of quantities of $NO^\bullet$, which could be small in a given biological microcompartment, have been found to be very difficult and often nonspecific. Available analytical methods have been thoroughly reviewed by Archer.[96] It is interesting at this point to mention a few basic data in order to put the problem in perspective.

The nitrogen dioxide $NO_2^\bullet$ legally-accepted upper limit as mandated by the Clean Air Act (USA) is 0.053 ppm or 100 $\mu g/m^3$ (10^{15} molecules l) according to Pryor.[31] This is to be related to a well-documented example of $NO_2^\bullet$ air pollution in Paris, France; from June 15-August 15, 1994, a particularly hot period, the mean concentration of $NO_2^\bullet$ detected by three captors on the Eiffel Tower was 122 $\mu g/m^3$ (80 ppb) with several daily peaks above 200 $\mu g/m^3$, leading to public warnings of pollution. Recently it reached 490 $\mu g/m^3$ on a general transport strike day in Paris, with lovely autumn weather (Oct. 10, 1995). Worse situations may happen in

long tunnels like the Saint-Gothard (17 km) or Holland and Lincoln tunnels under the Hudson River. Another recommendation put the upper limit for $NO_2^\bullet$ inhalations to 5 ppm (cited in refs. 104-107). For nitric oxide there is no evidence of toxicity if its level is below 50 ppm and the legally defined 8-hour maximal working-exposure level is fixed at 25 ppm (see ref. 108 and references quoted therein).

Nitric oxide (detected by the chemiluminescence method) is found in the expired gas of rabbits (15 ppb) and humans (mean 9 to 23 ppb during oral or nasal breathing).[109-111] The amount of $NO^\bullet$ exhaled by asthmatics is increased 2- to 3- fold.[110] There is much more NO detected in nasal breathing than in mouth breathing, and very little is detected following a tracheostomy. Thus most NO is synthesized in the nasopharynx. There are enormous individual variations, ranging up to 10-fold within each subset: healthy controls, smokers or asthmatic subjects. A rabbit breathing artificial air (20% O_2 in N_2) expirates $NO^\bullet$ at a rate of 12 pmol/min (determined as nitrite by the diazotation method). In "king-size" cigarette smoke inspiration, Norman and Keith[30] measured $NO^\bullet$ levels as high as 400 to 1000 ppm per puff (ten 35 ml puffs of gas-phase smoke per cigarette) but no $NO_2^\bullet$. Using Fourier transform infrared spectroscopy Cueto and Pryor[112] were able to detect both $NO^\bullet$ and $NO_2^\bullet$ and their respective kinetics of appearance and decay.

In the vascular system, Malinski et al[113] using porphyrinic electrochemical microsensors, measured $NO^\bullet$ concentration at the immediate vicinity of the endothelial wall. A bradykinin injection (10 nmols) induces a rapid $NO^\bullet$ release on the endothelial cell membranes of the rabbit aorta at an initial rate of 0.4 $\mu mol/l$ sec; after 13 sec a maximal concentration of 1.3 $\mu mol/l$ is recorded. A second sensor placed in the muscle cell layer, 100 μm apart from the endothelial cell layer, measured a maximal concentration of 0.85 $\mu mol/l$, 20 sec after the bradykinin injection. These authors estimate that about 40% of the $NO^\bullet$ produced is consumed in chemical reactions in the aorta in the course of its otherwise free diffusion.[113] NO$^\bullet$ concentration on the surface of cultured rat endothelial cells stimulated by 0.5 µmol/l bradykinin was measured at 5 µmol/l five minutes after stimulation.[114] Similarly the induction of NOS of cultured rat vascular smooth muscle cells by interleukin (IL-1β) (100 U/ml) led to a NO$^\bullet$ concentration of 0.5 µmol/l over the cell layer sixty minutes after treatment.

Such rather high local rates and concentrations result from multiple fluxes, very dependent upon intracellular compartmentation and intercellular exchanges. The synthesis, transport and storage, the reversible inhibition and activation of enzymes by NO$^\bullet$, and the irreversible oxidation reactions of NO$^\bullet$ are under intensive study, often with contradictory results, ever since the physiologic and pathophysiologic roles of nitric oxide have been discovered in mammals. As we shall see, EPR spectroscopy has the advantage over other methods to give usually unambiguous proof of NO$^\bullet$ presence in a given system; but, having a poor sensitivity, it has been seldom used to quantitate NO$^\bullet$-bound target concentrations and even more rarely to indirectly assess local free NO$^\bullet$ concentrations.

A HISTORICAL NOTE

At the end of this overview of the multiple origins of nitric oxide, written in order to point out its importance in physiology and physiopathology, let us mention the now historical role which EPR spectroscopy played 30 years ago, yielding correctly interpreted results published in well- known journals, that went however mostly unnoticed. In fact the pioneering simultaneous works by groups in the USA,[115-117] the then USSR[118,119] and in Japan,[120] in which EPR signals of paramagnetic nitrosylated complexes were detected in mammal cells, were the first clues (1965-1971) of NO synthesis in mammals. Unfortunately the scientific community did not pay enough attention to it until another generation was ready to accept a biological role for nitric oxide.

ADDITIONAL RECENT BIBLIOGRAPHY

Hemocytes of the freshwater snail *Viviparus ater* contain a NOS activity.[121,122] It is inhibited by mammalian NOS inhibitor, N^ω-monomethyl-L-arginine and seems Ca^{2+}-independent. The treatment of hemocytes, the cells responsible for defense mechanisms, by LPS increased total NOS activity 2.4-fold. Immunoreactivity for NOS and calmodulin was also found in plant embryonic tissues (pea embryonic axes and wheat germs).[123] Further work; however, needed to ascertain the existence in plants of a NOS activity similar to that found in mammals.

REFERENCES

1. Bates DR. Cause of terrestrial nightglow continuum. Proc R Soc Lond A 1993; 443:227-237.

2. Sharma RD, Sun Y, Dalgarno A. Highly rotationally excited nitric oxide in the terrestrial thermosphere. Geophys Res Letters 1993; 20:2043-2045.

3. Mende SB, Swenson GR, Geller SP et al. Limb view spectrum of the earth's airglow. J Geophys Res 1993; 98:117-125.

4. Johnson C, Henshaw J, McInnes G. Impact of aircraft and surface emissions of nitrogen oxides on tropospheric ozone and global warming. Nature 1992; 355:69-71.

5. Kondo Y, Kitada T, Koike M et al. Nitric oxide and ozone in the free troposphere over the western pacific ocean. J Geophys Res 1993; 98:527-535.

6. Lelieveld J, Crutzen PJ. Role of deep cloud convection in the ozone budget of the troposphere. Science 1994;264:1759-1761.

7. Remde A, Ludwig J, Meixner FX et al. A study to explain the emission of nitric oxide from a marsh soil. J Atmosph Chem 1993; 17:249-275.

8. Rondon A, Johansson C, Sanhueza E. Emission of nitric oxide from soils and termite nests in a trachypogon savanna of the Orinoco basin. J Atmosph Chem 1993; 17:293-306.

9. Valente RJ, Thornton FC. Emissions of NO from soil at a rural site in central Tennessee. J Geophys Res 1993; 98:745-753.

10. Williams EJ, Davidson EA. An intercomparison of two chamber methods for the determination of emission of nitric oxide from soil. Atmosph Environ 1993; 27A:2107-2113.

11. Payne WJ. Denitrification. John Wiley & Sons, New York. 1981.

12. Zumft WG. The biological role of nitric oxide in bacteria. Arch Microbiol 1993; 160:253-264.

13. Elofsson R, Carlberg M, Moroz L et al. Is nitric oxide (NO) produced by invertebrate neurones? NeuroReports 1993; 4:279-282.

14. Gelperin A. Nitric oxide mediates network oscillations of olfactory interneurons in a terrestrial mollusc. Nature 1994; 369:61-63.

15. Radomski MW, Martin JF, Moncada S. Synthesis of nitric oxide by the haemocytes of the american horseshoe crab (*Limulus polyphemus*). Phil Trans Royal Soc Lond B 1991; 334:129-133.

16. Martinez A, Riveros-Moreno V, Polak JM et al. Nitric oxide (NO) synthase immunoreactivity in the starfish *Marthasterias glacialis*. Cell Tissue Res 1994; 275:599-603.

17. Ribeiro JMC, Hazzard JMH, Nussenzveig RH et al. Reversible binding of nitric oxide by a salivary heme protein from a blood-sucking insect. Science 1993; 260:539-541.

18. Müller U, Buchner E. Histochemical localization of NADPH-diaphorase in the adult *Drosophila* brain. Is nitric oxide a neuronal messenger also in insects? Naturwissenschaften 1993; 80:524-526.

19. Choi SK, Choi HK, Kadono-Okuda K et al. Occurence of novel types of nitric oxide synthase in the silkworm, *Bombyx mori*. Biochem Biophys Res Commun 1995; 207:452-459.

20. Li ZS, Furness JB, Young HM et al. Nitric oxide synthase immunoreactivity and NADPH diaphorase enzyme activity in neurones of the gastrointestinal tract of the toad, *Bufo marinus*. Arch Histol Cytol 1992; 55:333-350.

21. Holmqvist BI, Östholm T, Alm P, Ekström P. Nitric oxide synthase in the brain of a teleost. Neuroscience Letters 1994; 171:205-208.

22. Chen Y, Rosazza JPN. A bacterial nitric oxide synthase from a *Nocardia* species. Biochem Biophys Res Commun 1994; 203:1251-1258.

23. Werner-Felmayer G, Golderer G, Werner ER et al. Pteridine biosynthesis and nitric oxide synthase in *Physarum polycephalum*. Biochem J 1994; 304:105-111.

24. Brooks SB, Lewis MJ, Dickerson RR. Nitric oxide emissions from the high-temperature viscous boundary layers of hypersonic aircraft within the stratosphere. J Geophys Res 1993; 98:755-760.

25. Goldenbaum GC, Dickerson RR. Nitric oxide production by lightning discharges. J Geophys Res 1993; 98:333-338.

26. Hoffman DJ, Solomon S. Ozone destruction through heterogeneous chemistry following the eruption of El Chicon. J Geophys Res 1989; 94:5029-5041.

27. Brasseur GP, Granier C, Walters S. Future changes in stratospheric ozone and the role of heterogeneous chemistry. Nature 1990; 348:626-628.

28. Rodriguez JM, Ko MKW, Sze ND. Role of heterogeneous conversion of N_2O_5 on sulphate aerosols in global ozone losses. Nature 1991; 352:134-137.

29. Torres AL, Thompson AM. Nitric oxide in the equatorial pacific boundary layer: SAGA3 measurements. J Geophys Res 1993; 98:949-954.

30. Norman V, Keith CH. Nitrogen oxides in tobacco smoke. Nature 1965; 205:915-916.

31. Pryor WA. Biological effects of cigarette smoke, wood smoke, and the smoke from plastics: the use of electron spin resonance. Free Rad Biol Med 1992; 13:659-676.

32. Eiserich JP, Vossen V, O'Neill CA et al. Molecular mechanisms of damage by excess nitrogen oxides: nitration of tyrosine by gas-phase cigarette smoke. FEBS Lett 1994; 353:53-56.

33. Blough NV, Zafiriou OC. Reaction of superoxide with nitric oxide to form peroxonitrite in alkaline solution. Inorg Chem 1985; 24:3502-3504.

34. Wink DA, Darbyshire JF, Nims RW et al. Reactions of the bioregulatory agent nitric oxide in oxygenated aqueous media: determination of the kinetics for oxidation and nitrosation by intermediates

generated in the NO/O_2 reaction. Chem Res Toxicol 1993; 6:23-27.

35. Ford PC, Wink DA, Stanbury DM. Autooxidation of aqueous nitric oxide. FEBS Lett 1993; 326:1-3.

36. Huie RE, Padmaja S. The reaction of NO with superoxide. Free Rad Res Comms 1993; 18:195-199.

37. Padmaja S, Huie RE. The reaction of nitric oxide with organic peroxyl radicals. Biochem Biophys Res Commun 1993; 195:539-544.

38. Kharitonov VG, Sundquist AR, Sharma VS. Kinetics of nitric oxide autoxidation in aqueous solution. J Biol Chem 1994; 269:5881-5883.

39. Victorin K. Review of the genotoxicity of nitrogen oxides. Mutation Res 1994; 317:43-55.

40. Henry Y, Ducrocq C, Drapier J-C et al. Nitric oxide, a biological effector. Electron paramagnetic resonance detection of nitrosyl-iron-protein complexes in whole cells. Eur Biophys J 1991; 20:1-15.

41. Ruggiero CE, Carrier SM, Antholine WE et al. Synthesis and structural and spectroscopic characterization of mononuclear copper nitrosyl complexes: models for nitric oxide adducts of copper proteins and copper-exchanged zeolites. J Am Chem Soc 1993; 115:11285-11298.

42. Arnold WP, Mittal CK, Katsuki S et al. Nitric oxide activates guanylate cyclase and increases guanosine 3'-5'-cyclic monophosphate levels in various tissue preparations. Proc Natl Acad Sci USA 1977; 74:3203-3207.

43. Murad F, Mittal CK, Arnold WP et al. Guanylate cyclase: activation by azide, nitro compounds, nitric oxide and hydroxyl radical and inhibition by hemoglobin and myoglobin. Adv Cyclic Nucleotides Res 1978; 9:145-158.

44. Craven PA, DeRubertis FR. Restoration of the responsiveness of purified guanylate cyclase to nitrosoguanidine, nitric oxide, and related activators by heme and hemoproteins. Evidence for involvement of the paramagnetic nitrosyl-heme complex in enzyme activation. J Biol Chem 1978; 253:8433-8443.

45. Craven PA, DeRubertis FR, Pratt DW. Electron spin resonance study of the role of NO·catalase in the activation of guanylate cyclase by NaN$_3$ and NH$_2$OH. Modulation of enzyme responses by heme proteins and their nitrosyl derivatives. J Biol Chem 1979; 254:8213-8222.

46. Ignarro LJ, Edwards JC, Gruetter DY et al. Possible involvement of S-nitrosothiols in the activation of guanylate cyclase by nitroso compounds. FEBS Lett 1980; 110: 275-278.

47. Furchgott RF, Zawadzki JV. The obligatory role of endothelial cells in the relaxation of arterial smooth muscle by acetylcholine. Nature 1980; 288:373-376.

48. Ignarro LJ, Buga GM, Wood KS et al. Endothelium-derived relaxing factor produced and released from artery and vein is nitric oxide. Proc Natl Acad Sci USA 1987; 84:9265-9269.

49. Palmer RMJ, Ferrige AG, Moncada S. Nitric oxide release accounts for the biological activity of endothelium-derived relaxing factor. Nature 1987; 327:524-526.

50. Radomski MW, Palmer RMJ, Moncada S. The role of nitric oxide and cGMP in platelets adhesion to vascular endothelium. Biochem Biophys Res Commun 1987; 148:1482-1489.

51. Moncada S, Radomski MW, Palmer RMJ. Endothelium-derived relaxing factor. Identification as nitric oxide and role in the control of vascular tone and platelet function. Biochem Pharmacol 1988; 37:2495-2501.

52. Palmer RMJ, Rees DD, Ashton DS et al. L-arginine is the physiological precursor for the formation of nitric oxide in endothelium dependent relaxation. Biochem Biophys Res Commun 1988; 153:1251-1256.

53. Stuehr DJ, Gross SS, Sakuma I et al. Activated murine macrophages secrete a metabolite of arginine with the bioactivity of endothelium-derived relaxing factor and the chemical reactivity of nitric oxide. J Exp Med 1989; 169:1011-1020.

54. Ignarro LJ. Biosynthesis and metabolism of endothelium-derived nitric oxide. Annu Rev Pharmacol Toxicol 1990; 30:535-560.

55. Ignarro LJ. Signal transduction mechanisms involving nitric oxide. Biochem Pharmacol

1991; 41:485-490.

56. Moncada M, Palmer RMJ, Higgs EA. Biosynthesis of nitric oxide from L-arginine. A pathway for the regulation of cell function and communication. Biochem Pharmacol 1989; 38:1709-1715.

57. Moncada S, Palmer RMK, Higgs EA. Nitric oxide: physiology, pathophysiology, and pharmacology. Pharmacol Rev 1991; 43:109-142.

58. Green LC, Ruiz de Luzuriaja K, Wagner DA et al. Nitrate biosynthesis in man. Proc Natl Acad Sci USA 1981; 78:7764-7768.

59. Wagner DA, Young VR, Tannenbaum SR. Mammalian nitrate biosynthesis: incorporation of $^{15}NH_3$ into nitrate is enhanced by endotoxin treatment. Proc Natl Acad Sci USA 1983; 80:4518-4531.

60. Miwa M, Stuehr DJ, Marletta et al. Nitrosation of amines by stimulated macrophages. Carcinogenesis 1987; 7:955-958.

61. Hibbs JB, Taintor RR, Vavrin Z et al. Nitric oxide: a cytotoxic activated macrophage effector molecule. Biochem Biophys Res Commun 1988; 157:87-94.

62. Marletta MA. Mammalian synthesis of nitrite, nitrate, nitric oxide, and N-nitrosating agents. Chem Res Toxicol 1988; 1:249-257.

63. Marletta MA, Yoon PS, Iyengar R et al. Macrophage oxidation of L-arginine to nitrite and nitrate: nitric oxide is an intermediate. Biochemistry 1988; 27:8706-8711.

64. Stuehr DJ, Nathan CF. Nitric oxide. A macrophage product responsible for cytostasis and respiratory inhibition in tumor target cells. J Exp Med 1989; 169:1543-1555.

65. Hibbs JB, Taintor RR, Vavrin Z et al. Synthesis of nitric oxide from a terminal guanidino nitrogen atom of L-arginine: a molecular mechanism regulating cellular proliferation that targets intracellular iron. In: Moncada S, Higgs EA, eds. Nitric oxide from L-arginine: a bioregulatory system. Amsterdam: Elsevier Science Publishers B.V., 1990; 189-223.

66. Nathan C. Nitric oxide as a secretory product of mammalian cells. FASEB J 1992; 6:3051-3064.

67. Stuehr DJ, Griffith OW. Mammalian nitric oxide synthases. Adv Enzymol 1992; 65:287-346.

68. Averill BA, Tiedje JM. The chemical mechanism of microbial denitrification—a hypothesis. FEBS Lett 1982; 261:8-11.

69. Kim CH, Hollocher TC. ^{15}N tracer studies on the reduction of nitrite by purified dissimilatory nitrite reductase of *Pseudomonas aeruginosa*. Evidence for direct production of N_2O without free NO as an intermediate. J Biol Chem 1983; 258:4861-4863.

70. Bessières P, Henry Y. Stoichiometry of nitrite reduction catalyzed by *Pseudomonas aeruginosa* nitrite-reductase. Biochimie 1984; 66:313-318.

71. Henry Y, Bessières P. Denitrification and nitrite reduction: *Pseudomonas aeruginosa* nitrite-reductase. Biochimie 1984; 66:259-289.

72. Brittain T, Blackmore R, Greenwood C et al. Bacterial nitrite-reducing enzymes. Eur J Biochem 1992; 209:793-802.

73. Ye RW, Averill BA, Tiedje JM. Denitrification: production and consumption of nitric oxide. Applied Environm microbiol 1994; 60:1053-1058.

74. Luckey TD. Introduction to intestinal microecology. Am J Clin Nutr 1972; 25:1292-1294.

75. Bonnett R, Chandra S, Charalambides AA et al. Nitrosation and nitrosylation of haemoproteins and related compounds. Part 4. Pentaco-ordinate nitrosylprotohaem as the pigment of cooked cured meat. Direct evidence from ESR spectroscopy. J Chem Soc Perkin I 1980; 1980:1706-1710.

76. Reddy D, Lancaster JR, Cornforth DP. Nitrite inhibition of *Clostridium botulinum*: electron spin resonance detection of iron-nitric oxide complexes. Science 1983; 221:769-770.

77. Benjamin N, O'Driscoll F, Dougall H et al. Stomach NO synthesis. Nature 1994; 368:502.

78. Iqbal ZM, Dahl K, Epstein SS. Role of nitrogen dioxide in the biosynthesis of nitrosamines in mice. Science 1980; 207:1475-1477.

79. Iyengar R, Stuehr DJ, Marletta MA. Macrophages synthesis of nitrite, nitrate, and N-nitrosamines: precursors and role of the respiratory burst. Proc Natl Acad Sci USA 1987; 84:6369-6373.

80. Challis BC, Fernandes MHR, Glover BR et al. Formation of diazopeptides by nitrogen oxides. In: Bartsch H, O'Neill IR, Schulte-Hermann R, eds. Relevance of *N*-nitroso compounds to human cancer: exposures and mechanisms. IARC Scientific Publications No 84, Lyon, France, 1987; 308-314.

81. Mirvish SS, Ramm MD, Babcook DM et al. Lipidic nitrosating agents produced from atmospheric nitrogen dioxide and a nitrosamine produced in vivo from amyl nitrite. In: Bartsche H, O'Neill IK, Schulte-Hermann R, eds. Relevance of *N*-nitroso Compounds to Human Cancer: Exposures and Mechanisms. IARC Scientific Publications No 84, Lyon, France, 1987; 315-318.

82. Tannenbaum SR. Endogenous formation of *N*-nitroso compounds: a current perspective. In: Bartsch H, O'Neill IK, Schulte-Hermann R, eds. Relevance of *N*-nitroso Compounds to Human Cancer: Exposures and Mechanisms. IARC Scientific Publications No 84, Lyon, France, 1987; pp 292-296.

83. Bartsch H, Ohshima H, Pignatelli B. Inhibitors of endogenous nitrosation mechanisms and implications in human cancer prevention. Mutation Res 1988; 202:307-324.

84. Ohshima H, Tsuda M, Adachi H et al. L-arginine-dependent formation of *N*-nitrosamines by the cytosol of macrophages activated with lipopolysaccharide and interferon-γ. Carcinogenesis 1991; 12:1217-1220.

85. Esumi H, Tannenbaum SR. U.S.-Japan cooperative cancer research program: seminar on nitric oxide synthase and carcinogenesis. Cancer Res 1994; 54:297-301.

86. Haswell-Elkins MR, Satarug S, Tsuda M et al. Liver fluke infection and cholangiocarcinoma: model of endogenous nitric oxide and extragastric nitrosation in human carcinogenesis. Mutation Res 1994; 305:241-252.

87. Ohshima H, Bartsch H. Chronic infections and inflammatory processes as cancer risk factors: possible role of nitric oxide in carcinogenesis. Mutation Res 1994; 305:253-264.

88. Garthwaite J, Charles SL, Chess-Williams R. Endothelium-derived relaxing factor release on activation of NMDA receptors suggests role as intercellular messenger in the brain. Nature 1988; 336:385-388.

89. Knowles RG, Palacios M, Palmer RMJ et al. Formation of nitric oxide from L-arginine in the central nervous system: a transduction mechanism for stimulation of the soluble guanylate cyclase. Proc Natl Acad Sci USA 1989; 86:5159-5162.

90. Gally JA, Montague PR, Reeke GN et al. The NO hypothesis: Possible effects of a short-lived, rapidly diffusible signal in the development and function of the nervous system. Proc Natl Acad Sci USA 1990; 87:3547-3551.

91. Snyder SH. Nitric oxide and neurons. Current Opinion in Neurobiol 1992; 2:323-327.

92. Snyder SH. Nitric oxide: first in a new class of neurotransmitters? Science 1992; 257:494-496.

93. Lipton SA, Choi Y-B, Pan Z-H et al. A redox-based mechanism for the neuroprotective and neurodestructive effects of nitric oxide and related nitroso-compounds. Nature 1993; 364:626-632.

94. Henry Y, Lepoivre M, Drapier J-C et al. EPR characterization of molecular targets for NO in mammalian cells and organelles. FASEB J 1993; 7:1124-1134.

95. Henry YA, Singel DJ. Metal-nitrosyl interactions in nitric oxide biology probed by electron paramagnetic resonance spectroscopy. In: Feelisch M, Stamler JS, eds. Methods in Nitric Oxide Research. John Wiley and Sons, 1996:357-372.

96. Archer S. Measurement of nitric oxide in biological models. FASEB J 1993; 7:349-360.

97. Enemark JH, Feltham RD. Stereochemical control of valence and its application to the reduction of coordinated NO and N_2. Proc Natl Acad Sci USA 1972; 69:3534-3536.

98. Bottomley F, Brooks WVF, Clarkson SG et al. Electrophilic behaviour of the coordinated nitrosyl cation. J Chem Soc Comm 1973; 1973:919-920.

99. McCleverty JA. Reactions of nitric oxide coordinated to transition metals. Chem Rev 1979; 79:53-76.

100. Koppenol WH, Moreno JJ, Pryor WA et al. Peroxynitrite, a cloaked oxidant formed by nitric oxide and superoxide. Chem Res

Toxicol 1992; 5:834-842.

101. Stamler JS, Singel DJ, Loscalzo J. Biochemistry of nitric oxide and its redox-activated forms. Science 1992; 258:1898-1902.

102. Richter-Addo GB, Legzdins P. Metal nitrosyls. Oxford University Press, Oxford, UK, 1992.

103. Bonner FT, Hughes MN. No lack of NO activity. Science 1993; 260:145-146.

104. Foubert L, Fleming B, Latimer R et al. Safety guidelines for use of nitric oxide. Lancet 1992; 339:1615-1616.

105. Bouchet M, Renaudin M-H, Raveau C et al. Safety requirement for use of inhaled nitric oxide in neonates. Lancet 1993; 341:968-969.

106. Laguenie G, Berg A, Saint-Maurice J-P et al. Measurement of nitrogen dioxide formation from nitric oxide by chemiluminescence in ventilated children. Lancet 1993; 341:969.

107. Miller OI, Celermajer DS, Deanfield JE et al. Guidelines for the safe administration of inhaled nitric oxide. Arch Disease Childhood 1994; 70:F47-F49.

108. Rossaint R, Falke KJ, Lopez F et al. Inhaled nitric oxide for the adult respiratory distress syndrome. N Eng J Med 1993; 328:399-405.

109. Gustafsson LE, Leone AM, Persson MG et al. Endogenous nitric oxide is present in the exhaled air of rabbits, guinea pigs and humans. Biochem Biophys Res Commun 1991; 181:852-857.

110. Alving K, Weitzberg E, Lundberg JM. Increased amount of nitric oxide in exhaled air of asthmatics. Eur Respir J 1993; 6:1368-1370.

111. Leone AM, Gustafsson LE, Francis PL et al. Nitric oxide is present in exhaled breath in humans: direct GC-MS confirmation. Biochem Biophys Res Commun 1994; 201:883-887.

112. Cueto R, Pryor WA. Cigarette smoke chemistry: conversion of nitric oxide to nitrogen dioxide and reactions of nitrogen oxides with other smoke components as studied by Fourier transform infrared spectroscopy. Vibrational Spectr 1994; 7:97-111.

113. Malinski T, Taha Z, Grunfeld S et al. Diffusion of nitric oxide in the aorta wall monotored in situ by porphyrinic microsensors. Biochem Biophys Res Commun 1993; 193:1076-1082.

114. Malinski T, Kapturczak M, Dayharsh J et al. Nitric oxide synthase activity in genetic hypertension. Biochem Biophys Res Commun 1993; 194:654-658.

115. Vithayathil AJ, Ternberg JL, Commoner B. Changes in electron spin resonance signals of rat liver during chemical carcinogenesis. Nature 1965; 207:1246-1249.

116. Commoner B, Woolum JC, Senturia BH et al. The effects of 2-acetylaminofluorene and nitrite on free radicals and carcinogenesis in rat liver. Cancer Res 1970; 30:2091-2097.

117. Woolum JC, Commoner B. Isolation and identification of a paramagnetic complex from the livers of carcinogen-treated rats. Biochim Biophys Acta 1970; 201:131-140.

118. Emanuel NM, Saprin AN, Shabalkin VA et al. Detection and investigation of a new type of ESR signal characteristic of some tumour tissues. Nature 1969; 222:165-167.

119. Vanin AF, Vakhnina LV, Chetverikov AG. Nature of the EPR signals of a new type found in cancer tissues. Biofizika 1970; 15:1044-1051 (English translation 1082-1089).

120. Maruyama T, Kataoka N, Nagase S et al. Identification of three-line electron spin resonance signal and its relationship to ascites tumors. Cancer Res 1971; 31:179-184.

121. Conte A, Ottaviani E. Nitric oxide activity in molluscan hemocytes. FEBS Lett 1995; 365:120-124.

122. Franchini A, Conte A, Ottaviani E. Nitric oxide: an ancestral immunocyte effector molecule. Adv Neuroimmunol 1995; 5:463478.

123. Sen S, Cheema IR. Nitric oxide synthase and calmodulin immunoreactivity in plant embryonic tissue. Biochem Arch 1995; 11:221-227.

CHAPTER 2

BASIC CHEMISTRY OF NITRIC OXIDE AND RELATED NITROGEN OXIDES

Yann A. Henry, Béatrice Ducastel and Annie Guissani

The general purpose of this chapter is to briefly recall the basic chemistry of NO•, especially its own paramagnetism and that of some of the complexes it forms with transition metals. As a full review is out of our scope, we shall only mention data that could be useful to understand the following chapters. In particular we shall attempt to compile some kinetic data which can enable us to propose whether a given reaction between NO• and a molecular target could be relevant to the biology of a cell or an animal.

A very comprehensive review of the chemistry of NO has appeared which offers most interesting mechanistic discussions of recent results.[1]

STRUCTURE OF NITROGEN OXIDES

NITROGEN OXIDES, REDUCTION POTENTIALS

Nitrogen being less electronegative than oxygen forms oxides and oxidized compounds with an oxidation number n between +5 and +1 (Table 2.1).

It also forms compounds with negative oxidation numbers: -1 (hydroxylamine NH_2OH), -2 (hydrazine N_2H_2), and -3 (ammonia NH_3) (Fig. 2.1).[2,3]

This diagram of the relative energy, ΔE, Δn times the reduction potential E° in volts, versus the oxidation number n, allows us to predict on equilibrium thermodynamic grounds, whether a given compound is stable or not and whether a dismutation (disproportionation) reaction may occur, independently of kinetic considerations. It explains why N_2O_4 dismutates in water to NO_3^- and NO_2^-. The dismutation of NO• to NO_2^- and N_2O is in principle thermodynamically favorable. However it does

Nitric Oxide Research from Chemistry to Biology: EPR Spectroscopy of Nitrosylated Compounds, edited by Yann A. Henry, Annie Guissani and Béatrice Ducastel.
© 1997 R.G. Landes Company.

Table 2.1. Summary of the redox properties of nitrogen oxides, acids and ions.

Oxidation Number	Oxide	Acid	Ions	Names	Redox Properties
+5		HNO_3	NO_3^-	nitric anhydride, nitrate	oxidant
+5		$ONOOH$	$ONOO^-$	peroxynitrite	powerful oxidant
+4				nitrogen dioxide	
				nitrogen tetroxide	dismutates in solution
+3		$HO-N=O$	NO_2^-	nitrous anhydride, nitrite	oxidant and reducer, cannot dismutate except in acidic media
			NO^+	nitrosonium	

Table 2.1. (continued)

	Structure	Name	Property
+2	$\overset{\bullet}{N}=O \quad \overset{\bullet}{N}=O$	nitric oxide or nitrogen monoxide	oxidant and reducer, reacts rapidly with O_2
+1	$N^-=N^+=O \qquad N\equiv N^+-O^-$	nitrous oxide	neither oxidant nor reducer
	$HO-N=N-OH \qquad O^{\ominus}-N=N-O^{\ominus}$	cis and trans hyponitrite	does not dismutate
	$HO-N=N-OH \qquad O^{\ominus}-N=N-O^{\ominus}$		
	$H-N=O \qquad NO^-$	nitroxyl	
	$O^{\ominus}-N(=O)-N(O^{\ominus})$	oxyhyponitrite or trioxodinitrate	

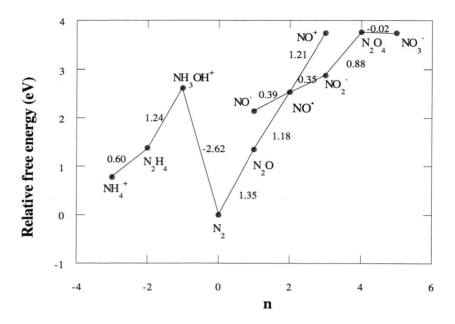

Fig. 2.1. Diagram at pH 7 of the free energy (eV), relative to the N_2 molecule, of various N-derivatives as a function of their oxidation number **n**. The redox potentials are given in volts relative to the normal hydrogen electrode.

not occur at pH 7, for kinetic reasons which transition metal catalysis can overcome.

In compounds such as oxyhyponitrite (or trioxodinitrate in the recommended nomenclature) $^-O-N-NO-O^-$ (Table 2.1) the nitrogen atoms can assume inequivalent oxidation numbers (0 and +4). In many instances two species can have the same oxidation number with very different reduction potentials: nitrite NO_2^- and NO^+ with **n** +3, $E^{l°}(NO_2^-/NO^•) = 0.24$ V and $E^{l°}(NO^+/NO^•) = 1.21$ V; NO^- and N_2O with **n** +1, $E^{l°}(NO^•/NO^-) = 0.39$ V and $E^{l°}(NO^•/N_2O) = 1.28$ V. $NO^•$ can therefore be either an oxidant or a reducer. Peroxynitrite or peroxonitrite (oxoperoxonitrate), $ONOO^-$, has the same oxidation number (+5) as nitrate NO_3^- but is a much stronger oxidant with $E^{l°} = 1.4$ V as compared to -0.15 V. In a similar and more elaborate diagram Koppenol et al have assigned an oxidation number +6 to the nitrosyldioxyl radical $ONOO^•$ based on the standard -2 oxidation number of oxygen.[3]

ELECTRONIC STRUCTURE AND PARAMAGNETISM OF NITRIC OXIDE

Nitric oxide or, in a more precise nomenclature, nitrogen monoxide $NO^•$ (usually written simply as NO and wholly different from nitrosonium or nitrosyl cation NO^+ and from nitroxyl anion or oxonitrate NO^-, as we shall see below) is an uncharged lipophilic paramagnetic gas. As with O_2 and CO it is fairly soluble in water (4.7 ml per 100 ml under 1 atm, that is 1.95 mmol/l, at 20°C according to the Merck Index, 3.27 mmol/l at 0°C and 0.08 mmol/l at 60°C) (Fig. 2.2).[4] It has no acid-base property.

Its diffusion coefficient in aqueous solution was measured by a porphyrinic sensor at 37°C as 3.30×10^{-5} cm^2/s,[5-7] close to that of O_2, $HO_2^•$ and $O_2^{-•}$.[8] More recent measurements by phosphorescence from Ru and Pd chelates of mesoporphyrin IX yielded much smaller values, by two or three orders of magnitude (5×10^{-7} to 9×10^{-8} cm^2/s).[9] These odd data would predict a mean path

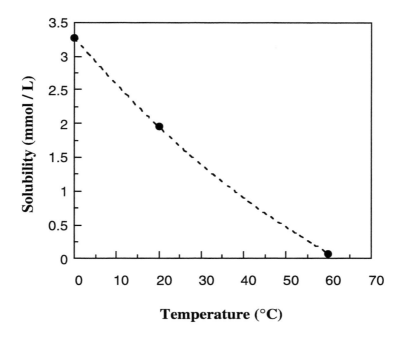

Fig. 2.2. Water solubility (mmol/l or mM) of nitric oxide as a function of temperature (°C).

of NO⁺ within 10 seconds of only 10 μm, that is of the order of one cell diameter, which is contradictory to all the known physiological properties of NO⁺. It is admitted that its charge neutrality allows easy diffusability across cell membranes and through several cell layers.[6,7]

Its N-oxidation number is +2. It is the most stable of "odd molecules" as[10]—contrarily to most free radicals such as superoxide anion $O_2^{-\bullet}$, ascorbyl, and their respective acid forms—it does not dimerize (except as a solid below -164°C),[11,12] nor dismutate. It carries a single electron delocalized in the π^* orbital, as explained in the following energy diagram (Fig. 2.3).

One can write two resonance structures, $:\overset{\bullet}{N} = \overset{\bullet\bullet}{O}:$ and $:\overset{\bullet\bullet}{N} = \overset{\bullet}{O}:^+$, or a so-called three-electron bond, $:N \doteq \overset{\bullet\bullet}{O}:$ (bond order 2.5, bond length 1.151 Å, bond stretch 1840 cm⁻¹). It carries an electron doublet on the N atom, on which the odd electron is also somewhat (60%) delocalized.[10,13] NO⁺ is isoelectronic with O_2^+.

This structure should be compared to those of two diamagnetic species with a different N-oxidation number, often mentioned as "related" to nitric oxide,[14] or "redox-activated" forms[15]—nitrosonium or nitrosyl cation NO⁺ (N-oxidation number +3) (isoelectronic to N_2) and nitroxyl anion or oxonitrate (in the recommended nomenclature) NO⁻ (N-oxidation number +1) (isoelectronic to O_2). These can be written respectively as: $:N \equiv O:^+$ (bond order 3, bond length 1.062 Å, bond stretch 2300 cm⁻¹) with an empty π^* orbital and $:\overset{\bullet\bullet}{N} = \overset{\bullet\bullet}{O}:$ (bond order 2, bond length 1.26 Å, bond stretch 1290 cm⁻¹) with two electrons in the π^* orbital. As shown by these bond lengths and bond strengths the three species have different structures, and therefore wholly different chemical properties are expected.

It is also interesting to compare these electronic structures to those of diamagnetic carbon monoxide, paramagnetic (fundamental triplet state) dioxygen and para-

NO

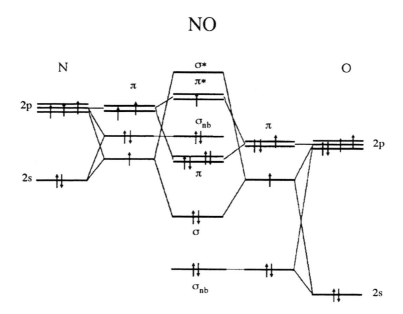

Fig. 2.3. Molecular orbital diagram of the NO· molecule.

magnetic superoxide anion, respectively, :C=Ö: or :C≡Ö: for CO, :O⊥O: for O_2, and [:O---O:]⁻ for $O_2^{-\cdot}$.[10] All these diatomic molecules carry (as do NO·, NO⁺ and NO⁻) a free doublet, thus allowing a bonding with transition metals.

EPR SPECTROSCOPY OF NITRIC OXIDE

Nitric oxide was one of the first paramagnetic species submitted to the newly invented electron paramagnetic (or spin) resonance spectroscopy (EPR or ESR) (circa 1950). In $^{14}N^{16}O^{\cdot}$, the natural isotopic abundant NO· molecule, the electronic orbital motion and spin are strongly coupled. In the ground electronic state, $\Lambda = 1$ and $\Sigma = 1/2$. The resulting doublet has a large (~ 120 cm^{-1}) separation between the $^2\Pi_{1/2}$ and the $^2\Pi_{3/2}$ doublets. The upper or $^2\Pi_{3/2}$ doublet component possesses an electronic magnetic moment and is responsible for the paramagnetism of NO· (Fig. 2.4).[16,17]

A nine-line ($\Delta M_J = \pm 1$, $\Delta M_I = 0$) magnetic resonance spectrum of NO·, as a gas at low pressure (1 mmHg) and room-temperature, is observed at magnetic fields in the range of 8600 gauss

(0.86 T) at a frequency of 9.36 GHz, corresponding to a g-value of $g_J = 0.774$ (see chapter 3 for the basic principles of EPR spectroscopy and the definition of g-value) (Fig. 2.5).

This spectrum arises from magnetic-dipole transitions between Zeeman levels in the $J = 3/2$ state of the spin component. The magnetic IJ coupling constant due to $I = 1$ of the ^{14}N atom is 30 MHz (27.5 ± 1 gauss at 9.36 GHz or 1x10^{-3} cm^{-1}; see chapter 3).

Another rotational ($J = 1/2 \rightarrow 3/2$) transition, consisting of ten lines split into two groups of five each, of the $^2\Pi_{1/2}$ state occurs in the 150 GHz microwave range (2 mm wavelength) (Fig. 2.6).[18]

Several fully detailed theoretical papers have attempted to account for the electronic structure of NO·.[13,17,19,20] EPR spectroscopy measurements have been used to quantitate other radicals in the gas phase.[21,22]

What should be remembered by the biochemist or biologist NO-user is that, probably due to its very short relaxation time, NO· in room temperature solutions or in frozen solutions in liquid nitrogen

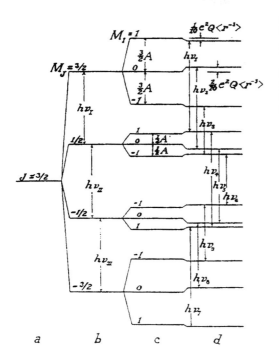

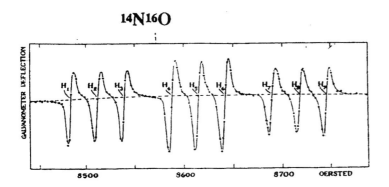

Fig. 2.4. Energy level diagram of the J = 3/2 level of the $^2\Pi_{3/2}$ state of $^{14}N^{16}O^{\cdot}$ (a) in the absence of magnetic field, (b) Zeeman effect, (c) including the IJ coupling. The nine transitions $\Delta M_J = \pm 1$, $\Delta M_I = 0$ are shown in (d). (Reproduced with permission from: Beringer R, Castle JG. Magnetic resonance absorption in nitric oxide. Phys Rev 1950; 78:581-586). Copyright © The American Physical Society, College Park, MD, USA.

Fig. 2.5. Observed EPR spectrum of $^{14}N^{16}O^{\cdot}$ at room temperature and at the pressure of 1.0 mm Hg. (Reproduced with permission from: Beringer R, Castle JG. Magnetic resonance absorption in nitric oxide. Phys Rev 1950; 78:581-586). Copyright © The American Physical Society, College Park, MD, USA.

(77 K), does not usually give rise per se to any detectable EPR signal in usual experimental conditions of frequency and magnetic fields (see chapter 3). However an EPR signal detected at 77 K with a g-value of 1.97, tentatively assigned to "matrix-bound" NO·—the nature of the matrix being unknown but certainly not of proteic origin—has been reported by many independent groups.[23-30] From the detected g-value close to 2.0, the orbital motion appears to be quenched, a spin-only motion remaining. A definite assignment of this matrix-bound NO· is required (Ducastel B, unpublished results). The gas should be in principle detectable at room temperature, however the poor sensitivity of the method would render it totally impractical. As we shall see at length throughout this book, NO· is actually detectable when forming paramagnetic complexes with metal cations or stable paramagnetic adducts with compounds of higher molecular weight.

REACTIVITY OF NITRIC OXIDE AND NITROGEN OXIDES

REACTIONS OF NITRIC OXIDE WITH OTHER N DERIVATIVES

Although nitric oxide does not dismutate it reacts with several other N de-

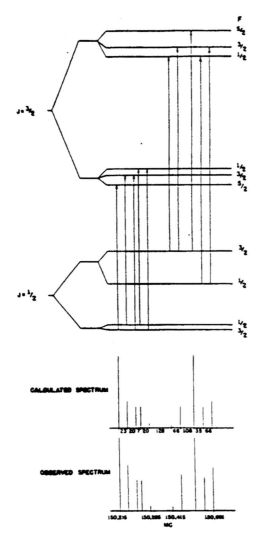

Fig. 2.6. Energy level diagram and spectrum of the $J = 1/2$ to $J = 3/2$ rotatioanl transition of the $^2\Pi_{1/2}$ state of $^{14}N^{16}O^{\cdot}$ (Reproduced with permission from: Gallagher JJ, Bedard FD, Johnson CM. Microwave spectrum of $^{14}N^{16}O$. Phys Rev 1954; 93:729-733). Copyright © The American Physical Society, College Park, MD, USA.

rivatives, for instance hyponitrous acid:[31,32]

$$NO^{\cdot} + HO\text{-}N = N\text{-}OH \rightarrow N_2O + OH^{\cdot} + HNO$$

It is reduced by hydroxylamine NH_2OH ($pK_a \sim 14$):[32-34]

$$2\,NO^{\cdot} + 2\,NH_2OH \rightarrow N_2 + N_2O + 3\,H_2O$$

through the following mechanism:

$$NO^{\cdot} + NH_2OH \rightarrow HNO + {}^{\cdot}NHOH$$
$$NO^{\cdot} + {}^{\cdot}NHOH \rightarrow [ON\text{-}NHOH] \rightarrow N_2O + H_2O$$
$$HNO + NH_2OH \rightarrow N_2 + 2H_2O$$
$$HNO + HNO \rightarrow N_2O + H_2O.$$

Nitric oxide reacts also with nitroxyl by a probably slow net electron exchange reaction to nitrous oxide and nitrite:[32,35,36]

$$2NO^{\cdot} + NO^- \rightarrow N_2O + NO_2^-$$

Finally and probably the most important, $NO^{\cdot}$ reacts with nitrogen dioxide $NO_2^{\cdot}$ to produce the anhydride nitrogen trioxide N_2O_3, which hydrates into nitrous acid HNO_2 ($pK = 3.2$) and nitrite NO_2^-:

$$NO^{\cdot} + NO_2^{\cdot} \rightarrow N_2O_3$$

This reaction is concurrent in the gas phase with the dimerization of $NO_2^{\cdot}$ into N_2O_4 which in turn dismutates into N_2O_3 and N_2O_5 and hydrates yielding equimolar amounts of nitrite and nitrate in water solution. These reactions are of importance in the oxidation of $NO^{\cdot}$ by O_2 (see below the reactions with molecular oxygen). Nitrogen dioxide is a free radical with powerful oxidative and nitrosating reactivity,[37-40] and an extreme pulmonary toxicity.[41,42]

PRODUCTION OF NITRIC OXIDE BY REDUCTION OF NITRITE OR OXIDATION OF NITROXYL

Nitric oxide is produced by the reduction of nitrite in two reactions which could occur in some biological acidic media (stomach, lysozomes). At acidic pH values nitrite dismutates to $NO^{\cdot}$ and nitrate by the following reaction:

$$3HNO_2 \rightarrow 2NO^{\cdot} + HNO_3 + H_2O$$

which may have some importance in the stomach and the mouth,[43,44] concurrently with the NOS II (iNOS)-catalyzed production of $NO^{\cdot}$ in the airways.[45]

Between pH 3 and 6 nitrite is reduced by ascorbic acid ($AscH_2$) to produce the semihydroascorbyl radical ($AscH^{\cdot}$) and dehydroascorbate (Asc):

$AscH_2 + H^+ + NO_2^- \rightarrow AscH^\bullet + NO^\bullet + H_2O$

$AscH^\bullet + NO_2^- \rightarrow Asc + NO^\bullet + H_2O$

Nitroxyl (HNO/NO$^-$, pK_a = 4.7) is difficult to study, especially in biological systems, as it is metastable, undergoing dimerization into hyponitrous acid and dehydration (overall rate 1.8 x 10^9 mol/l·sec) into nitrous oxide N_2O:[36]

$HNO + HNO \rightarrow HON=NOH \rightarrow N_2O + H_2O$

HNO can be readily oxidized by O_2, superoxide, metal cations or metalloenzymes,[46,47] properties which might account for the biological (EDRF-like) activities of HNO:

$HNO + O_2 \rightarrow NO^\bullet + HO_2^\bullet$

$HNO + HO_2^\bullet \rightarrow NO^\bullet + H_2O_2$

$NO^- + SOD(Cu^{II}) \leftrightarrow NO^\bullet + SOD(Cu^I)$

Nitroxyl and nitrite can form an equilibrium with trioxodinitrate (also called oxyhyponitrite or Angeli's salt $Na_2N_2O_3$) (Table 2.1):[32,36,48-50]

$NO^- + NO_2^- \leftrightarrow {}^-O\text{-}N\text{-}NO\text{-}O^-$

REACTIONS WITH MOLECULAR OXYGEN

Although NO$^\bullet$ oxidation by oxygen is well-known in textbooks as yielding highly toxic red vapors of $NO_2^\bullet$ which dimerize into uncolored N_2O_4, the stoichiometry and rate law have been controversial until recent concurrent reports on the reaction in aqueous solutions.[51-53]

Aqueous solution reaction

Nitric oxide autoxidation yields NO_2^- (and not NO_3^-) and is given by the stoichiometry:

$4NO^\bullet + O_2 + 2H_2O \rightarrow 4NO_2^- + 4H^+$

It has a third order kinetic law (pH independent):[52]

$-d[NO^\bullet]/dt = 4k_{aq}[O_2][NO^\bullet]^2$,

with a fairly good agreement among different measurements for the value of the overall rate constant: $4k_{aq}$ = (8.8 ± 0.4) x 10^6 mol^{-2} l^2 sec^{-1} at 25°C with an activation energy of (2.8 ± 0.2) kcal mol^{-1},[54] (6 ± 1.5) x 10^6 mol^{-2} l^2 sec^{-1} at 22°C and (3.5 ± 0.7) x 10^6 mol^{-2} l^2 sec^{-1} at 37°C,[51] (8-9) x 10^6 mol^{-2} l^2 sec^{-1} at 25°C,[52] and (6.3 ± 0.4) x 10^6 mol^{-2} l^2 sec^{-1} at 25°C.[55] Similar values of 8.4 x

10^6 mol^{-2} l^2 sec^{-1} at 23°C and 9.6 x 10^6 mol^{-2} l^2 sec^{-1} at 37°C were found to be independent of pH between 4.9 and 7.4.[56,57] The rate of formation of NO_2^- equaled that of reaction of NO$^\bullet$, and there was no detectable formation of NO_3^-. The finding by Taha et al,[58] using an electrochemical method that the reaction rate law was zero order with regard to NO$^\bullet$ concentration, has been repeatedly disproven. The generally accepted mechanism is as follows:

$2NO^\bullet + O_2 \rightarrow 2NO_2^\bullet$

$NO_2^\bullet + NO^\bullet \rightarrow N_2O_3$

$N_2O_3 + H_2O \rightarrow 2NO_2^- + 2H^+$

The third order rate law and more precisely the second order in NO$^\bullet$ concentration is of prime importance to understand the highly variable stability and availability of NO$^\bullet$ in different biological systems. At low "constitutive" NO$^\bullet$ concentrations which one can imagine as being at less than 1 µmol/l levels, NO$^\bullet$ could have a half-life of minutes to hours and therefore could diffuse over several cell layers or circulate over a great distance, in order to reach its "protracted" target and act its role of mediator, EDRF or neurotransmitter (Fig. 2.7).

At higher "induced" concentrations, NO$^\bullet$ would have much shorter half-lives on the order of seconds. Reactions rates with other targets, i.e., metalloenzymes, hemoglobin, etc, would have to be very fast and/or indiscriminate before NO$^\bullet$ oxidation occurs.

Gas phase reaction

Nitric oxide oxidation in the gas phase is also a thermolecular but much slower reaction than in the aqueous phase following the rate law:

$-d/dt [NO^\bullet] = 2k_g [O_2] [NO^\bullet]^2$

with k_g = (6 - 7) x 10^3 mol^{-2} l^2 sec^{-1} at 20°C.[59] This author compiled other rate constant values covering one order of magnitude from 3 to 30 x 10^3 mol^{-2} l^2 sec^{-1} at 20°C.

In view of the novel importance of NO$^\bullet$ inhalation in clinical use on adult and neonatal humans, as a pulmonary vasodilator not affecting the systemic circulation

Fig. 2.7a.

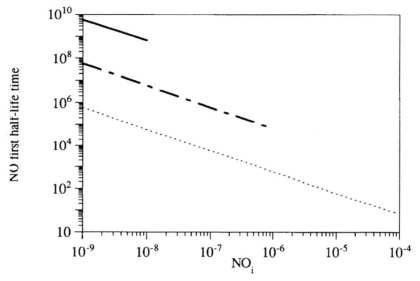

Fig. 2.7b.

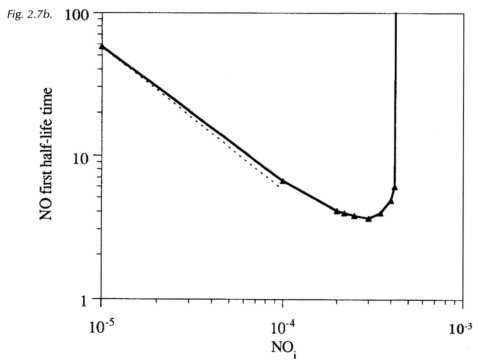

Fig. 2.7. First half-life time (s) versus [NO·]$_i$ initial concentration (mol/l) at various O$_2$ concentrations (····220 μmol/l, –·–· 2.2 μmol/l, and —— 22 nmol/l) using a third order kinetic law with a rate constant of $4k = 8 \times 10^6$ mol^2 l^{2}s^{-1} and assuming a constant O$_2$ bulk concentration. Straight lines are obtained when initial O$_2$ concentration is greater than that of NO (a). Curve (b) shows the first half-life of NO· at higher initial NO· concentrations, between 0.1 to 1 mmol/l in the presence of air (O$_2$, 220 μmol/l), calculated by the second integration method of Runge-Kutta. The deviation from the straight line only above 100 μmol/l therefore justifies the assumption of a constant O$_2$ concentration over the reaction course.

and of the extreme toxicity of nitrogen dioxide, the reaction rate was recently remeasured by several groups desiring to provide safety guidelines for $NO^\bullet$ use (see also chapters 5 and 12). The following rate constants were estimated to calculate the time required to yield 5 ppm $NO_2^\bullet$, the estimated upper limit set as a safety standard, at varying concentrations of $NO^\bullet$ and O_2: 1.93×10^{-38} cm^6 molecule^{-2} sec^{-1} at 27°C.[60] It was remeasured by Bouchet et al[61] who calculated the following k values (expressed in 10^{-38} cm^6 molecule^{-2} sec^{-1}) : 5.22 at 20°C and 5.39 at 37°C, that is 2.5-fold faster than those given by Foubert et al.[60] Converted into molarity, the value at 20°C is 14×10^3 mol^{-2} l^2 sec^{-1}, that is twice that measured by Olbregts.[59] More recent measurements yielded (expressed in 10^{-38} cm^6 molecule^{-2} sec^{-1}) 1.89 at 37°C and 2.27 at 25°C, in agreement with Foubert.[62] The rates were found to be faster as temperature is lower and independent of the gas humidity. The reaction proceeds through the following mechanism:[59]

$$NO^\bullet + O_2 \leftrightarrows OONO$$
$$OONO + NO^\bullet \leftrightarrows ONOONO \leftrightarrows 2\,NO_2^\bullet \leftrightarrows N_2O_4$$
$$2NO^\bullet + O_2 \leftrightarrows 2NO_2^\bullet$$

As seen above, oxidation of $NO^\bullet$ by oxygen in aqueous solution yields only nitrite and no nitrate.[51-56] This is not true however if oxidation occurs in the presence of an oxygenated hemoprotein, oxyhemoglobin, oxymyoglobin or NO-synthase contained in unpurified cytosolic fractions.[53] $NO^\bullet$ and nitrite are then partially converted to nitrate through several possible reactions:

$$Fe^{2+}O_2 + NO^\bullet \rightarrow Fe^{3+} + NO_3^-$$
$$2Fe^{2+}O_2 + 3NO_2^- + 2H^+ \rightarrow 2Fe^{3+} + 3NO_3^- + H_2O$$

or

$$4Fe^{2+}O_2 + 4NO_2^- + 4H^+ \rightarrow 4Fe^{3+} + 4NO_3^- + O_2 + 2H_2O$$

See the additional bibliographical note at the end of this chapter for a more complete analysis and discussion of NO autooxidation.

REACTIONS WITH SUPEROXIDE ANION AND HYDROGEN PEROXIDE, FORMATION OF PEROXYNITRITE

Peroxynitrite (also commonly called peroxonitrite or in the adequate international nomenclature, oxoperoxonitrate) ^-O-O-N=O and its acidic form H-O-O-N=O,[3] are two very potent oxidants formed by several reactions involving $NO^\bullet$ and its related "brothers" NO^- and NO^+, and on the other hand $O_2^{-\bullet}$ and its "parent" compounds O_2 and H_2O_2:[50,63]

$$\left.\begin{array}{l} NO^\bullet \\ NO_2^- + H_2O_2 \\ NO^+ + O_2 \end{array}\right\} \rightarrow O=N-O-O^- \rightarrow NO_3^-$$

As well as for the reaction of $NO^\bullet$ with O_2, a precise knowledge of the various rates is of great importance to ascertain the half-life period of each compound and their relevance in terms of yield, chemical reactivity and by way of consequence, biological activities as an EDRF, cytotoxin, nitrosating agent, etc. Further emphasis derives from the fact that some specialized cells such as neutrophils, polymorphonuclear leukocytes or macrophages, can produce "reactive oxygen" and "reactive oxynitrogen" when triggered or activated.[64] However the production could not be simultaneous; while NO biosynthesis requires several hours induction in the macrophages, the triggering of the respiratory burst induces immediate and transient release of reactive oxygen intermediates by NADPH oxidase (see below).

In fact the interaction of EDRF with $O_2^\bullet$ was historically important in the reasoning that led to the discovery of $NO^\bullet$ as an EDRF.[65-67]

The rate of the first reaction at physiological pH (7.2-7.4), which should be written, contrarily to what is usually done, as between $NO^\bullet + HO_2^\bullet$, since the pKa of $HO_2^\bullet/O_2^{-\bullet}$ is 4.8:

$$NO^\bullet + HO_2^\bullet \rightarrow H\text{-}O\text{-}O\text{-}N=O$$

was determined by pulse-radiolysis as being equal to $(3.7 \pm 1.1) \times 10^7$ mol^{-1} l sec^{-1}.[68] The decay of peroxynitrite follows a first order law (1.4 sec^{-1}) into the two strong oxidants, HO$^\bullet$ and NO$_2$$^\bullet$, and nitrate.[3,68-70] The yield of hydroxyl radical and nitrogen dioxide in peroxynitrite decomposition is very pH-dependent:[71]

$$H-O-O-N=O \rightarrow HO^\bullet + NO_2^\bullet$$
or $$H-O-O-N=O \rightarrow NO_2^+ + OH^-$$
and $$H-O-O-N=O \rightarrow H^+ + NO_3^-$$

In fact peroxynitrous acid has two distinct pK values at 7.5-8,[69] and 6.8,[3,72] which could be related to the existence of two *cis* and *trans* conformations having different oxidative properties.[72-74]

Recent reports brought an important controversy to this "hot" topic.[75,76] The rate constant for the reaction of NO$^\bullet$ with O$_2$$^\bullet$$^-$ was determined, by laser-flash photolysis, to be $(6.7 \pm 0.9) \times 10^9$ mol^{-1} l sec^{-1} at pH 7.5, close to the diffusion rate constant;[75] that is 200-fold faster than that reported by Saran et al.[8,68,77] Pulse radiolysis experiments yielded similar values $(4.3 \pm 0.5) \times 10^9$ mol^{-1} l sec^{-1}, independent of ionic strength and pH value in the range of 6.1-10.0.[76] All groups nearly agreed over the rates of peroxynitrite first order decay. Saran and colleagues' results were interpreted by Huie and Padmaja and by Goldstein and Czapski as resulting from the other possible peroxynitrite build-up reaction:[50]

$$NO^- + O_2 \rightarrow {}^-OONO$$

Another report on NO$^\bullet$ reactivity with a simple short-lived radical such as CO$_2$$^-$$^\bullet$, (SCN)$_2$$^-$$^\bullet$, or Br$_2$$^-$$^\bullet$, yielded rate constants on the order of 3 to 5×10^9 mol^{-1} l sec^{-1},[78] quite similar to that determined by Huie and Padmaja and by Goldstein and Czapski,[75,76] bringing another argument in favor of these authors data. See the additional bibliographical note at the end of this chapter for more recent evidence.

A further alternative, with no rate constant given, was that NO$^\bullet$ reacts with hydrogen peroxide to yield singlet oxygen, a highly reactive species which might heavily damage cellular components, instead of peroxynitrite:[79]

$$2NO^\bullet + H_2O_2 \rightarrow {}^1O_2 + N_2O + H_2O$$

Neutrophils and macrophages produce another microbicidal agent, hypochlorite HClO, from the reaction of chloride with H$_2$O$_2$ catalyzed by myeloperoxidase. Hypochlorite was predicted to react with NO$^\bullet$ or with peroxynitrite to yield nitrosyl chloride ClNO, a strongly oxidizing agent.[80]

Finally NO$^\bullet$ reacts with ozone. It is the basis of the chemiluminescence method of detection of NO$^\bullet$:[4,81]

$$NO^\bullet + O_3 \rightarrow O_2 + NO_2^* \rightarrow NO_2^\bullet + hv$$

ESTIMATION OF THE FREE NITRIC OXIDE AVERAGE PATHLENGTHS

Due to the importance of NO$^\bullet$ as a cellular messenger and as a potentially toxic compound, not much by itself but through the generation of very reactive (oxidative) species: $^-$OONO, HO$^\bullet$ + NO$_2$$^\bullet$, and ^{1}O$_2$, the knowledge of the average pathlengths of NO$^\bullet$ and O$_2$$^-$$^\bullet$, etc, in a given medium is of great interest. It depends not only on the rate constants but equally on the local concentrations of its targets, proteins, hemoproteins, transferrin, ceruloplasmin, bound iron, bound copper, cysteine, ascorbic acid, oxygen and H$_2$O$_2$, etc. The half-life of a given species, say NO$^\bullet$, is $1/\Sigma k_i C_i$, where k_i is the second-order "on" rate-constants of NO$^\bullet$ with species i, and C_i their concentrations.

Whatever the solution to the dilemma between the two proposed values of the rate constant of the reaction between NO$^\bullet$ and superoxide anion O$_2$$^-$$^\bullet$ or its acid perhydroxyl radical HO$_2$$^\bullet$, one must recall that the lifetime of HO$_2$$^\bullet$ at physiological pH values is extremely short due to its spontaneous and strongly pH-dependent dismutation reactions:[82]

$$HO_2^\bullet + HO_2^\bullet \rightarrow O_2 + H_2O_2$$
$$k = (8.6 \pm 0.6) \times 10^5 \text{ mol}^{-1} \text{ l sec}^{-1}$$
$$O_2^-{}^\bullet + HO_2^\bullet + H^+ \rightarrow O_2 + H_2O_2$$
$$k = (1.0 \pm 0.5) \times 10^8 \text{ mol}^{-1} \text{ l sec}^{-1}$$
and
$$O_2^-{}^\bullet + O_2^-{}^\bullet + 2H^+ \rightarrow O_2 + H_2O_2$$

$k < 3.5 \times 10^{-1}$ mol^{-1} l sec^{-1} with an apparent dismutation rate constant of $k = 3 \times 10^5$ mol^{-1} l sec^{-1} at pH 7.4. Including the other fast (also in the range 1 to 10×10^5 mol^{-1} l sec^{-1}) reactions of HO$_2$$^\bullet$/O$_2$$^-$$^\bullet$ with species occuring at high concentrations in cells or in plasma, such as ascorbic acid (50 μmol/l), transferrin (40 μmol/l) or ceruloplasmin (2 μmol/l), one can estimate that with a local concentration of 10 pmol/l of $HO_2^\bullet/O_2^{-\bullet}$, its lifetime would be less than 50 ms and its mean pathlength shorter than 40 μm or a few (< 10) cell diameters. The same reasoning can be made with NO$^\bullet$ which could be produced at higher levels than $HO_2^\bullet/O_2^{-\bullet}$, for instance 100 nmol/l. The actual level of peroxynitrite produced should be very low, except in instances when both $HO_2^\bullet/O_2^{-\bullet}$ and NO$^\bullet$ are produced in large quantities over a large time period. One example could be that of phagocytes or macrophages following their activation; steady-state concentration as high as 100 nmol/l could be expected and lead to effective bactericidal properties.[83,84] An acute question remains: are $HO_2^\bullet/O_2^{-\bullet}$ and NO$^\bullet$ produced simultaneously or at least are there meaningful concentrations of both reactants at a given time to produce peroxynitrite? It is "a question of balance".[85]

Answers are apparently contradictory or at least seem to depend on the system under study. Evidences of production of NO and $O_2^{-\bullet}$ was found in rat peritoneal neutrophils (PMN) stimulated by the peptide formylMet-Leu-Phe (FMLP), by leukotriene B4 (LTB4),[86] or by opsonized zymosan.[87] Rat alveolar macrophages activated by phorbol-myristate-acetate (PMA) were also found to produce a nitrosating compound as assayed by nitration of 4-hydroxyphenylacetate in the presence of superoxide dismutase.[88] The release of peroxynitrite was supposed to be responsible for the nitration reaction. In human neutrophils stimulated by PMA to produce the respiratory burst, the yields of nitric oxide and hydrogen peroxide production show a similar (within a factor of ten) PMA dose-dependence and a similar time-dependence favoring the formation of peroxynitrite.[89] Other recent results seem to give a negative answer in two systems, activated macrophages in the presence of *Leishmania*

major,[90] or stimulated polymorphonuclear leukocytes.[64]

Interesting attempts at a similar compilation of kinetic data and modelization of the diffusion of NO$^\bullet$ have recently been made by several authors.[6-8,91-93] See a very recent kinetic analysis at the end of this chapter.

REACTIONS OF NITRIC OXIDE WITH ORGANIC FREE RADICALS

Like O_2, NO$^\bullet$ and $NO_2^\bullet$ react with alkyl R$^\bullet$, peroxyl $RO_2^\bullet$ and alcoxyl RO$^\bullet$ radicals at various rates, mostly through addition reactions:[94,95]

$R^\bullet + O_2 \rightarrow RO_2^\bullet$
$R^\bullet + NO^\bullet \rightarrow R\text{-}NO$ (nitroso)
$R\text{-}NO + R^\bullet \rightarrow R_2\text{-}NO^\bullet$ (nitroxide)
$R^\bullet + NO_2^\bullet \rightarrow R\text{-}NO_2$ (nitro)
$RO_2^\bullet + NO^\bullet \rightarrow RO^\bullet + NO_2^\bullet$
and
$RO_2^\bullet + NO^\bullet \rightarrow ROONO$ (peroxynitrite)
$RO_2^\bullet + NO_2^\bullet \leftrightarrow RO_2\text{-}NO_2$ (peroxynitrate)
$RO^\bullet + NO^\bullet \rightarrow RO\text{-}NO$ (nitrite)
$RO^\bullet + NO_2^\bullet \rightarrow RO\text{-}NO_2$ (nitrate)

Depending on the rates of decomposition of the organic peroxynitrite and peroxynitrate, these reactions could either lead to branched chain reactions as in the case of organic peroxidation or on the contrary quench the formation of peroxides; in the last case NO$^\bullet$ would act as an antioxidant.[94]

Some of these reactions are also used in spin-trapping experiments of NO$^\bullet$ when R$^\bullet$ is a nitronyl nitroxide,[96] or other stable radical, imidazolineoxyl N-oxide,[97,98] or carbon-centered alkyl radical.[99] See chapter 13 for more details.

Other radical reactions could be important, for instance with thiyl RS$^\bullet$:[100]
$RS^\bullet + NO^\bullet \leftrightarrow RS\text{-}NO$ (thionitrite)
or with the tyrosyl stable radical TyrO$^\bullet$ found in several proteins (ribonucleotide reductase, prostaglandin H synthase, galactose oxidase, photosystem II, etc.). This reversible reaction has been for the first time hypothesized by use of NO$^\bullet$-releasing thionitrites on ribonucleotide reductase R2 subunit (Scheme 2.1):[101]

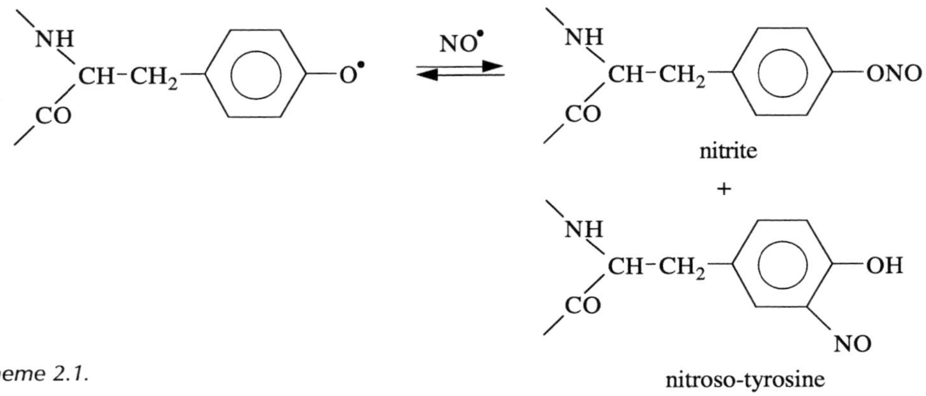

Scheme 2.1.

nitrite

+

nitroso-tyrosine

or more generally with phenoxyl radicals:[102]

PheO• + NO• ↔ Phe-O-NO (nitrite) and Phe(=O)=NO (2- or 4-nitroso-cyclohexadienone).

By utilizing a pulse-radiolytic technique, the rate constants for the reaction of NO• with TyrO• and TrpO• have been determined to be near diffusion-controlled rates on the order of $(1-2) \times 10^9$ mol^{-1} l sec^{-1}.[103] Being reversible these reactions could contribute to the regulatory properties of NO•.

ELECTROPHILIC NITROSATION REACTIONS AND NITRATION REACTIONS

The electrophilic or nucleophilic character of NO• is not clear cut; this has led to many contradictory reports in the interpretation of the reactions of NO• with thiols, amines and phenols. The ambiguity is probably related to the fact that NO• has a *doublet* on the N atom which can be given to a nucleophile and at the same time can undergo exchange of *its lone electron* to yield NO$^+$ (one electron loss leaving only a doublet on the N atom) or NO$^-$ (one electron capture to form a second doublet on the N atom). It is altogether an ambiguity in terms over the actual meaning of "electrophilic" or "nucleophilic", considered on the one hand by the organic chemist as related to the existence of an electronic doublet which can be donated (nucleophilic) or accepted (electrophilic),

and on the other hand by the coordination chemist as related to an electron donor (reducer: NO• - e$^-$ → NO$^+$) or acceptor (oxidant: NO• + e$^-$ → NO$^-$) character.[104-109] This ambiguity in terms has to be related to the prodigious ambivalence or polyvalence of NO• chemistry and therefore to its unique biological versatility ("Dr. Jekyll and Mr. Hyde", "Janus face", "foe or friend", "double-edged sword", etc.).

It is now quite clear that NO• itself is not an electrophilic nitrosating agent. It has to be oxidized into NO$^+$, NO$_2$•, N$_2$O$_3$, peroxynitrite or other unknown species.[51] As we have seen before, oxidizing agents could be oxygen, superoxide, transition metal cations or metalloproteins (see below), etc. NO$^+$ gives additional reactions with nucleophiles and aromatic compounds:

ArH + NO$^+$ → ArNO + H$^+$

RSH + NO$^+$ → RSNO + H$^+$

ROH + NO$^+$ → RONO + H$^+$

RR'NH + NO$^+$ → RR'N-NO + H$^+$

Similar reactions could be written with nitrite in acidic media, since it generates NO$^+$ or N$_2$O$_3$:

NO$_2^-$ + 2H$^+$ → [H$_2$NO$_2^+$] → NO$^+$ + H$_2$O

On the contrary in the absence of air or other oxidant NO• would react with thiols in the presence of bases to yield disulfide and N$_2$O:

RSH + B$^-$ → RS$^-$ + BH

RS$^-$ + NO• ↔ [RS-NO]$^-$ → 1/2RSSR + NO$^-$

NO$^-$ + BH → B$^-$ + 1/2N$_2$O + 1/2H$_2$O

An interesting instance of the difference of reactivity of NO• and oxidized NO$^+$

or $NO_2^•$ is that of the reaction with secondary amine yielding $RR'N[(NO)-N=O]^-$ (diazeniumdiolate, NONOate) in the first case (in the absence of oxygen) (Scheme 2.2) and nitrosamine RR'N-NO in the second:

$$RR'NH + NO^+ \rightarrow RR'N-N=O + H^+$$

The chemistry of $NO^•$ complexes with secondary amines and polyamines (NONOates) is widely studied,[110,111] since these compounds are excellent $NO^•$ donors and currently sought after for their platelet antiaggregate and vasodilator properties.[112-114]

As already mentioned, formation of *N*-nitroso amines (RR'N-NO) from secondary or cyclic amines such as morpholine (tetrahydro-1,4-oxazine) or proline by NO-derived NO^+ or NO_2, N_2O_3 or N_2O_4 has been evoked as a cause of human carcinogenesis following chronic infections and inflammatory processes.[37,115-127]

Nitric oxide can react with aromatic amines leading to deamination reactions. An important instance, which could perhaps be significant at high "disruptive" $NO^•$ concentrations, is the irreversible deamination of deoxynucleotides, deoxynucleosides and bases in DNA in oxygenated solutions,[128,129] such as the following (Scheme 2.3) leading to base alterations:

guanine → xanthine
adenine → hypoxanthine
cytosine → uracyl
5-methylcytosine → thymine

reactions which would greatly contribute to genome alterations by point mutations (G·C → T·A; A·T → T·A; G·C → A·T).

Another nitration reaction which could be of great physiological importance is that of catecholamines, norepinephrine, epinephrine or dopamine, yielding the corresponding 6-nitro-derivatives.[130] The reaction occurs in mild conditions, with nitrite at slightly acidic pH values (3 < pH < 6) or $NO^•$ gas in non-deaerated phosphate buffer (pH 7.4).

Several amino acid residues can be nitrosated and nitrated. For instance methionine is oxidized to ethylene and to methionine sulfoxide by peroxynitrite.[131] Nitrosation reactions of tyrosine by nitrogen oxides $NO_2^•$ or peroxynitrite have been proposed in several cases. For instance nitrotyrosine is slowly (1 mol^{-1} l sec^{-1}) formed in Cu-Zn, Mn and Fe superoxide dismutase by an autocatalyzed reaction.[72,84] 3-Nitrotyrosine (3-NO_2-Tyr) is formed from cigarette smoke.[132] Phenylalanine is simultaneously hydroxylated, more evidence for hydroxyl radical production from peroxynitrite.[133] Phenols are similarly hydroxylated.[102] Extensive nitration of protein

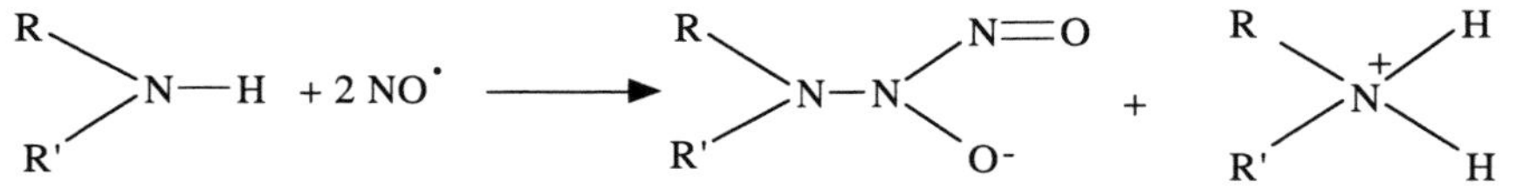

Scheme 2.2.

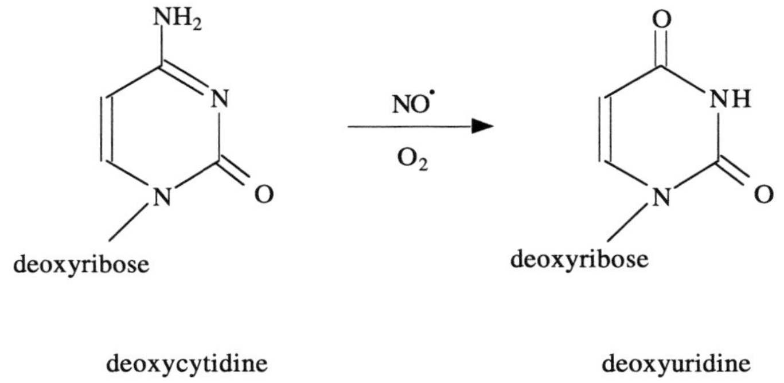

Scheme 2.3.

tyrosines in human atherosclerosis and other diseases has actually been detected by immunochemistry and other analytical methods,[134-137] through the proposed nitration mechanism:

$$Tyr\text{-}H + NO_2^{\bullet} \rightarrow Tyr^{\bullet} + NO_2^{-} + H^{+}$$
$$Tyr^{\bullet} + NO_2^{\bullet} \rightarrow Tyr\text{-}NO_2$$

A similar finding of free nitrotyrosine in serum and synovial fluid from rheumatoid patients,[138] or patients with acute lung injury,[139] presents another evidence for NO$^{\bullet}$-mediated oxidative damage in inflammatory diseases.

Similarly thiols can be oxidized with the formation of S-nitrosothiol (thionitrite, RSNO) intermediates of variable stability, often obtained pure, yielding finally disulfide and NO$^{\bullet}$:

$$RSH + NO^{+} \rightarrow RSNO + H^{+}$$
or
$$2RSH + NO_2^{\bullet} \rightarrow 2RSNO \rightarrow RSSR + 2NO^{\bullet}$$
or
$$RSH + R'ONO \rightarrow ROH + R'SNO$$

These reactions can occur on free cysteine, glutathione and many proteins.[51,101,140-150] Exchange equilibria between S-nitrosothiol and thiol have been characterized kinetically (rate constants from 1 to 300 mol^{-1} l sec^{-1}):[148,151]

$$RSH + R'SNO \leftrightarrow RSNO + R'SH$$
and
$$RSH + R'SNO \rightarrow R'SSR + NO^{\bullet} + H^{+}$$

Whenever reversible, these reactions may contribute to several regulatory processes assigned to NO$^{\bullet}$, such as the regulation of some enzymes[101,146,152] or the "incognito" transport of NO$^{\bullet}$, in a form protected from its oxidation leading to toxic products,[140-144,152] etc. A most interesting hypothesis stemmed from the vasorelaxant properties of S-nitrosothiols, that S-nitrosocysteine in particular would be closer to EDRF than NO$^{\bullet}$ itself;[140] this point shall be discussed in chapter 9. The reaction yielding nitrosothiol led recently to a spectrofluorimetric method for NO$^{\bullet}$ determination in the 0.5 to 100 µmol/l range.[153] Some thionitrites, such as the universally used S-nitroso-N-acetyl-D,L-penicillamine (SNAP) (Scheme 2.4), are good NO$^{\bullet}$-donors (see below).

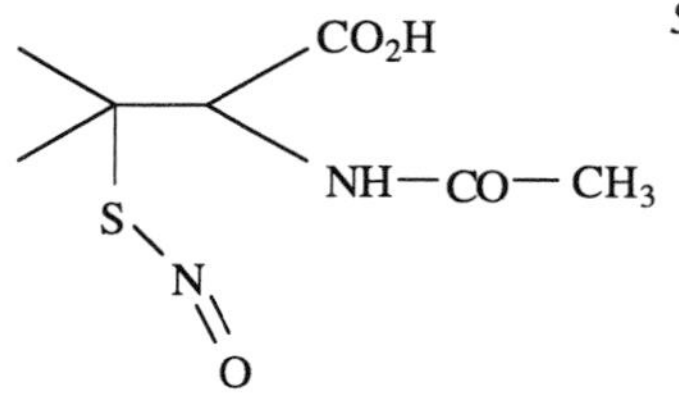

Scheme 2.4.

S-nitroso-N-acetyl penicillamine (SNAP

NITRIC OXIDE DONORS

This subject would deserve a chapter by itself. We shall propose a mere list of some NO$^{\bullet}$-donors, not attempting to be exhaustive. Some NO$^{\bullet}$-donors were actually used long before anything was known about EDRF; they certainly contributed massively, together with NO-synthase inhibitors, to the discovery of the role of endogenous nitric oxide in mammals. The best known are nitroprusside and nitroglycerin.

Used as a vasodilator, sodium nitroprusside (SNP, orange-red pentacyanonitrosylferrate) [Fe^{III}(CN)$_5$NO]$^{2-}$ does not however spontaneously release NO$^{\bullet}$ except when reduced into the paramagnetic green compound [Fe^{II}(CN)$_5$NO]$^{3-}$ or activated with light. It should be recalled that it releases intrinsically active (toxic) Fe^{2+} cations at the same time, and whenever there are some acceptors (ferric hemoproteins for instance), toxic cyanide anions thus induce gross experimental artifacts in complex systems.[154,155] The mechanism of NO$^{\bullet}$ release is the following:[155-160]

$$[Fe(CN)_5NO]^{2-} + X^{-} \rightarrow [Fe(CN)_5N{<}^{O}_{X}]^{3-}$$
$$\rightarrow [Fe(CN)_5NO]^{3-} + X^{\bullet}$$

X^{-} being OH^{-}, SH^{-}, NH$_2$R, NHR^{-}, NHR$_2$, CHR^{2-} or SR^{-}:

$$[Fe(CN)_5NO]^{3-} \rightarrow [Fe(CN)_4NO]^{2-} + CN^{-}$$
$$[Fe(CN)_4NO]^{2-} \rightarrow [Fe(CN)_4]^{2-} + NO^{\bullet}$$
$$6\,[Fe(CN)_4]^{2-} + 6CN^{-} \rightarrow 5\,[Fe(CN)_6]^{4-} + Fe^{2+}$$

SNAP can also act as a NO^{+} donor, provided that a nucleophilic species is present.

Many complexes of NO$^{\bullet}$ with transition metal cations are excellent NO$^{\bullet}$-donors. Among them iron-sulfur cluster nitrosyls are studied intensively for their vasodilator activity.[158-162]

Nitroglycerin (glyceryl trinitrate) is also commonly used as a vasodilator. It is metabolized within the endothelium by various enzymatic systems like cytochrome P-450 and glutathione S-transferase, and its product acts as an EDRF (see chapter 12). Many other organic nitrates and nitrites are also used as NO$^{\bullet}$-donors, for instance isosorbide dinitrate or amyl nitrite (commonly known as "poppers").

Another class of NO$^{\bullet}$-donors is sydnonimines derived from morpholine, which decompose spontaneously in the presence of oxygen into stoichiometric amounts of NO$^{\bullet}$ and superoxide (which could produce peroxynitrite and in turn hydroxyl radical), such as in the case of 3-morpholinosydnonimine N-ethylcarbamide (SIN-1) through the intermediate SIN-1-A, to yield in the presence of oxygen, the superoxide anion, O$_2^{-\bullet}$, SIN-1-C and NO$^{\bullet}$ (Scheme 2.5).[163]

Other NO-donors are C-nitroso compounds which are cleaved by light (Scheme 2.6).[164,165]

Complexes of NO$^{\bullet}$ with secondary amines and polyamines (diazeniumdiolates, NONOates) are widely studied (i.e. DEA-NO with diethylamine, SPER-NO with spermine, etc.) as:

$$2\,RR'NH + 2\,NO^{\bullet} \rightarrow RR'N[(NO)-N=O]^- + RR'NH_2^+$$

which decompose spontaneously (Scheme 2.7).

These compounds are excellent NO$^{\bullet}$-donors, being stable as solids, quite soluble in water and releasing NO$^{\bullet}$ at known rates.[111-114] A simulation of NO fluxes generated by SPER-NO and DEA-NO has been computed recently, which takes into account the autoxidation of NO by air (Fig. 2.8.).[114b]

We have already mentioned S-nitrosothiols (thionitrites) as NO$^{\bullet}$-donors.[166]

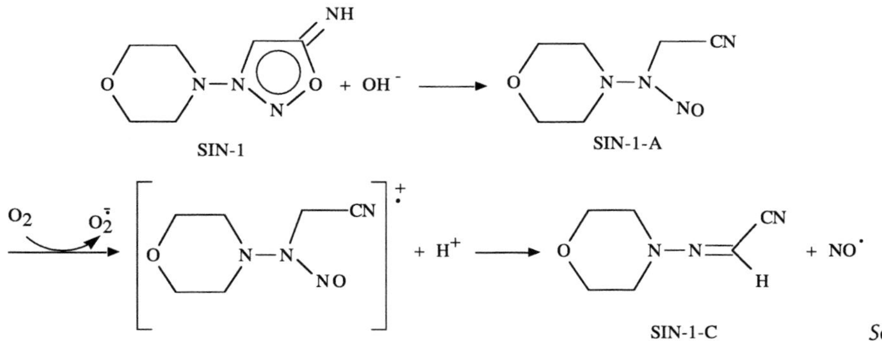

Scheme 2.5.

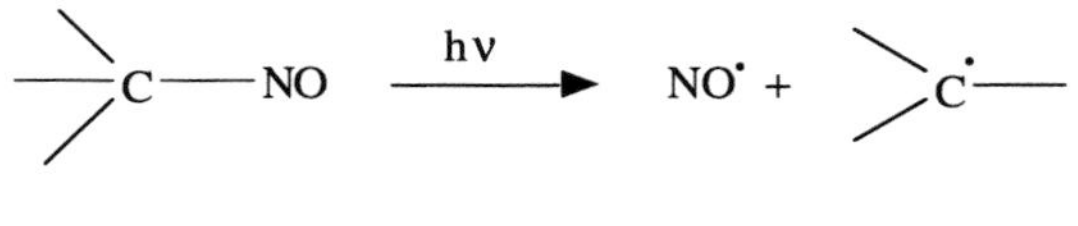

Scheme 2.6.

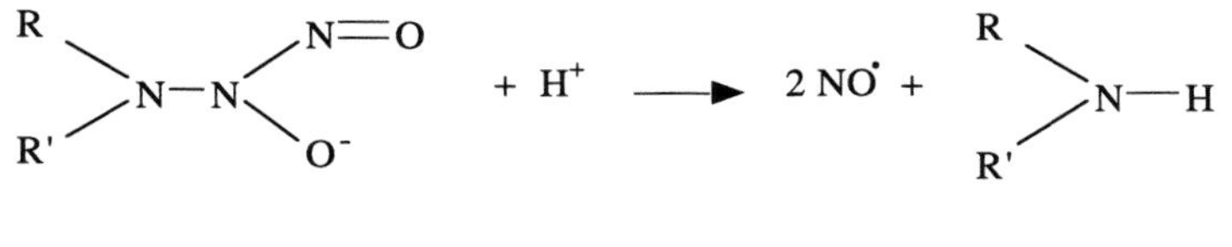

Scheme 2.7.

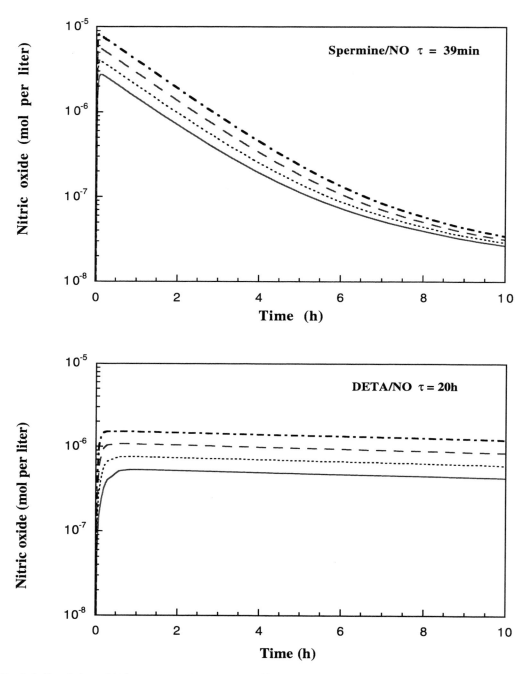

Fig. 2.8. Simulation of NO concentrations generated by two NO-donors. The molar concentration of NO generated by the spontaneous decomposition of spermine-NO ($\tau_{1/2}$ = 39 min at 37°C) -upper panel - or diethylenetriamine-NO (DETA-NO; $\tau_{1/2}$ = 20 h at 37°C) - lower panel - was simulated/calculated by the second integration method of Runge-Kutta, as a function of time for increasing initial concentrations of the NO-donating compounds ; the concentration of air was assumed to be constant(220 μmol/l). (initial NO-donor concentration in μmol/l: —75 ; ···· 150 ; —300 ; –·–· 600.)

Depending upon the pH value and their redox partner RSH/RSSR, they can release NO• or NO+:[148,167]

$$RSNO \rightarrow RS^• + NO^•$$
$$2RS^• \rightarrow RSSR$$

S-nitrosothiols are able to nitrosate proteins on various amino acid residues as exemplified on the NMDA receptor,[14] or ribonucleotide reductase R1 and R2 sub-

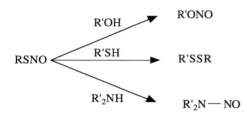

Scheme 2.8.

units (Scheme 2.8).[101]

Their decomposition is catalyzed by Cu^{2+}, Fe^{3+} or Fe^{2+}, with apparently highly variable rates.[168,169] Most *S*-nitrosothiols are photo-sensitive and release NO• under visible light, a property which has potential photochemotherapeutic applications.[162,170,171]

A great deal of research is currently performed to find new NO•-donors, stable enough, biologically inert until NO• is released, and safe by themselves and by their secondary products (for instance iron-sulphur cluster nitrosyls,[162] or so-called "caged nitric oxide").[172]

REACTIVITY OF NITRIC OXIDE WITH TRANSITION METALS

As mentioned at the end of chapter 1, metalloproteins were historically the first recognized NO• targets in mammal cells, long before the biosynthesis of NO• in mammals could be imagined (see chapter 8). The earliest examples were detected by Commoner and colleagues using EPR spectroscopy in chemically induced carcinogenesis of rat liver,[173] and correctly interpreted as iron-NO• complexes with specific thiol-containing proteins.[174,175] Similar pioneering discoveries were made simultaneously by Vanin and colleagues,[176,177] and by Maruyama et al.[178]

We have also mentioned in the introductory chapter that the turning point in the NO• story was the demonstration of guanylate cyclase activation by NO• and nitrogenous compounds,[166,179,180] before anybody knew about the existence of EDRF/NO•, and that guanylate cyclase was a hemoprotein, susceptible to NO• binding![66,67,181-187] EPR rapidly brought evidence that a nitrosyl-heme complex was involved in the enzyme activation.[188,189] This has received enormous recent developments by Stone and Marletta.[190-193] See chapters 10 and 14 for more details.

In fact this binding to some metalloproteins is quite a general phenomenon as NO• binds to nearly all transition metals (Mo, Mn, Fe, Co, Ni, Cu, to mention those found in active sites of proteins), in the corresponding columns (groups IV to VIII) in the periodic table of Dimitri Ivanovitch Mendeliev, including those of Ti and V, and other "cornered" elements such as Sn and Pb, with some possible exceptions such as Zn^{2+} which has an alkaline-earth character like Mg^{2+}. The binding occurs through their *d* or *f* electron orbitals. A thorough review on metal nitrosyls appeared in 1992 in the form of a 369-page book by Richter-Addo and Legzdins.[193] We shall only recall here some very basic data necessary to understand the following chapters on nitrosylated metalloproteins.

The coordination chemistry of NO• was studied very early, like that of metal carbonyls or cyanide complexes. The 18th and 19th centuries discovered many highly colored salts of Fe, Co, Mn, etc., such as Prussian blue (ferric ferrocyanide$[Fe(CN)_6]_3Fe_4$), Turnbull blue (ferric and potassic double salt of ferrocyanide$[Fe(CN)_6]^{4-}Fe^3,K^+$), orange-red nitroprusside (sodium pentacyanonitrosylferrate $[Fe^{III}(CN)_5NO]^{2-}$, $2Na^+$), black Roussin salt $[Fe_4S_3(NO)_7]^-$, etc. It has still, at the end of the 20th century, a considerable industrial importance through the necessary search for the catalytic destruction of atmospheric pollutants such as the species we are interested in: CO, NO• and NO_2• and *S*-oxides, etc.[195-198]

Reasonable structures of metal-nitrosyl complexes are $M^- = N^+ = \ddot{O}:$ and $M - \ddot{N} = \ddot{O}:$.[10] Unlike CO which has ten outershell electrons and can be a two-electron donor, $NO^\bullet$ with eleven outershell electrons, posseses a doublet and a lone electron on its N atom; it acts as a three-electron donor, leading to very strong $M-NO^\bullet$ bonds and sometimes to a weakened bond in *trans* (see below). This is similar to the binding of O_2 (twelve outershell electrons) (triplet) which takes an $O_2^{-\bullet}$ character. Upon binding, some internal electron transfers between the metal and $NO^\bullet$ can take place leading to a NO^+ or NO^- character. To describe the electron density on the M-NO bonds one can write three limiting forms in equilibrium:[105,199-201]

$$[M^{(n+1)+}\text{-}NO^-] \leftrightarrow [M^{n+}\text{-}NO^\bullet] \leftrightarrow [M^{(n-1)+}\text{-}NO^+]$$
$$\quad\quad A \quad\quad\quad\quad\quad B \quad\quad\quad\quad\quad C$$

form A being characterized by a long M-N bond (>1.80 Å), a low N-O frequency (1525-1590 cm^{-1}), a rather bent M-N-O angle (120-140°) and a nucleophilic character; while form C on the contrary has a shorter M-N bond (~1.6 Å), a larger N-O frequency (1650-1985 cm^{-1}), a linear form (165-180°) and an electrophilic character.[106-109] Depending upon stereochemistry of the complexes, $NO^\bullet$ may have *within one given complex* either a NO^+ or a NO^- character; the bending of the M-N-O group is indicative of the withdrawal of an electron pair from the metal to the N atom, for instance in $[Co^{III}(N = O^\bullet)]^{2+}$ (angle Co-N-O is 135°) or $[Co^I(N \equiv O^+)]^{2+}$ (angle Co-N-O is 179°).[105,193,201-203] One should be cautious however in generalization: the NO^+ or NO^- "character" of NO and the linear or bent angle in a complex do not necessarily mean that NO would be released from the complex in the free form $NO^\bullet$, or as NO^+ or NO^-.

This electron transfer could intervene in redox reactions with some enzymes, such as SOD, copper enzymes or heme-proteins:[47,203-206]

$$NO^\bullet + Fe^{II} \rightarrow NO^- + Fe^{III}$$
$$NO^- + Cu^{II}(SOD) \rightarrow NO^\bullet + Cu^I(SOD)$$
$$2NO^\bullet + Fe^{III}(heme) \rightarrow Fe^{II}(heme)\text{-}$$
$$NO^\bullet + NO^+ \ (or\ NO_2^-)$$

In general $NO^\bullet$ may act as a reducer of metal cations, more rarely as an oxidant.

These complexes' formations explain why metals are able to catalyze nitrosation reactions, which $NO^\bullet$ itself does not perform (for instance the *S*-nitrosothiols formation) (Scheme 2.9).

In mononuclear complexes $NO^\bullet$ binds to transition metals through their *d* electrons' orbitals, forming a d_{z^2}-σ^* orbital, $NO^\bullet$ being an σ-donor ligand. There is also a strong $d\pi$-$p\pi$ interaction (π-back-donation) of their $d\pi$ (d_{xz}, d_{yz}) electrons to the antibonding π^* orbital of $NO^\bullet$,[105,201,207-210] as shown in Fig. 2.9.

Apart from the mononuclear complexes mentioned, $NO^\bullet$, like NH_2, can be a bridging ligand between two (often written as μ_2-NO) or even three (μ_3-NO) metal atoms.[197,206] As we shall see below $NO^\bullet$ binds also to metals in clusters.

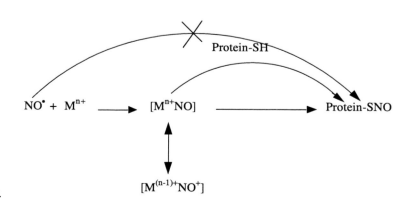

Scheme 2.9.

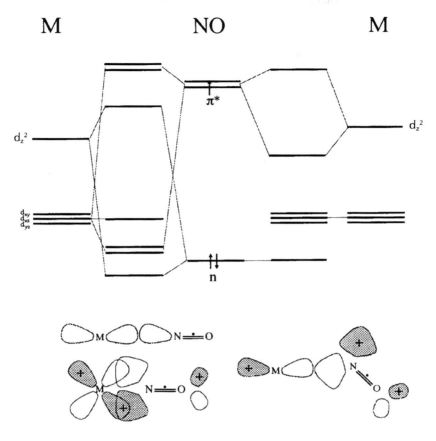

Fig. 2.9. Molecular orbital diagram of M-NO complex with a linear M-N-O (left) or a bent M-NO bond(right).

NO• generally has a higher affinity for a given metal than O_2 and CO.[194,206] The NO• affinity is usually higher for Fe^{2+} than for Fe^{3+}; it binds to Mo and Co in various oxidation numbers and to Mn^{2+}, Ni^{2+}, Cu^+ and Cu^{2+}. It creates a medium-to-strong ligand field in compounds such as diamagnetic (orange-red) oxidized nitroprusside $[Fe^{III}(CN)_5NO]^{2-}$, paramagnetic (green) reduced nitroprusside $[Fe^{II}(CN)_5NO]^{3-}$ or $[Fe^{II}(H_2O)_5NO]^{2+}$.[199,212,213]

PARAMAGNETISM AND EPR SPECTROSCOPY OF METAL NITROSYL COMPLEXES

Since nitric oxide has a strong-to-medium ligand field character and carries a spin angular momentum S of 1/2, its complexes with metals will be paramagnetic whenever the S of the metal is S $\neq$ 1/2, that is the resulting total $S_T \neq 0$. As explained in chapter 3, conventional EPR spectroscopy usually applies to systems with S_T = 1/2, 3/2, 5/2, but most often does not apply to systems with integer total electronic spin S_T = 1, 2, 3 etc.

The usual formalism of the MNO complexes' electron configuration was proposed by Enemark and Feltham as $\{MNO\}^n$, in which n is the number of *d* type electrons in the system when the nitrosyl ligand is formally considered as NO^+.[109,151,201] Complexes such as $[Cr(CN)_5(NO)]^{3-}$ or $[Fe(CN)_5(NO)]^{3-}$ with the respective electronic configuration $\{CrNO\}^5$ and $\{FeNO\}^7$ are paramagnetic and EPR detectable with g-values close to 2.0. On the contrary complexes

$[Fe(CN)_5(NO)]^{2-}$ and $[Mn(CN)_5(NO)]^{3-}$ with the respective electron configuration of $\{FeNO\}^6$ and $\{MnNO\}^6$ are diamagnetic.

Similarly, binding of $NO^\bullet$ to high-spin (S=2) deoxyhemoglobin (Fe^{II}, d^6 electronic configuration) results in a paramagnetic (S = 1/2) EPR detectable species (chapter 4), while binding to methemoglobin (Fe^{III}, d^5 electronic configuration) (S = 1/2 or 5/2) results in a diamagnetic complex (S = 0). Similarly Mn^{II}-substituted hemoglobin, when nitrosylated, is EPR-silent being either diamagnetic (S = 0) or a spin integer (S = 1 or 2).[208,209]

As we shall see in following chapters $NO^\bullet$ binds to [FeS] clusters, to Fe-thiol and to Fe-amino acid complexes to form $Fe(NO)_2L_2$ ternary compounds.[158,160,176,214] These compounds are EPR-detectable with features compatible with a $S_T = 1/2$. The electronic configuration cannot therefore be formally Fe^{II}, as often described, as the combination of S = 2 for Fe^{II} with two S = 1/2 for each NO ligand would result in an S_T integer, thus EPR-silent in usual experimental conditions. We must therefore propose a more unusual Fe^I formal configuration (S = 1/2 or 3/2), and the complex should be written as $Fe^I(NO)_2L_2$.

No more instances are required here as they are going to constitute the bulk of this book, particularly in chapter 6. Some have already been reviewed.[215-218]

ADDITIONAL BIBLIOGRAPHICAL NOTE

Two recent papers by Gidon Czapski and Sara Goldstein have been published since the bulk of this chapter was written.[219,220] In a very complete kinetic analysis they discussed the reactions of $NO^\bullet$ with $HO_2^\bullet/O_2^{-\bullet}$ and O_2, and the possible role of intermediate products of these reactions, $ONOO^-$, $ONOO^\bullet$, $ONOONO$, $NO_2^\bullet$, N_2O_3, etc. in their oxidizing properties, and in determining the life-span of NO in biological media.[219,220]

REFERENCES

1. Fontecave M, Pierre J-L. The basic chemistry of nitric oxide and its possible biological reactions. Bull Soc Chim Fr 1994; 131:620-631.
2. Henry Y, Bessières P. Denitrification and nitrite reduction: *Pseudomonas aeruginosa* nitrite-reductase. Biochimie 1984; 66:259-289.
3. Koppenol WH, Moreno JJ, Pryor WA et al. Peroxynitrite, a cloaked oxidant formed by nitric oxide and superoxide. Chem Res Toxicol 1992; 5:834-842.
4. Archer S. Measurement of nitric oxide in biological models. FASEB J 1993; 7:349-360.
5. Malinski T, Taha Z, Grunfeld S et al. Diffusion of nitric oxide in the aorta wall monitored in situ by porphyrinic microsensors. Biochem Biophys Res Commun 1993; 193:1076-1082.
6. Lancaster JR. Simulation of the diffusion and reaction of endogenously produced nitric oxide. Proc Natl Acad Sci USA 1994; 91:8137-8141.
7. Wood J, Garthwaite J. Models of the diffusional spread of nitric oxide: implications for neural nitric oxide signalling and its pharmacological properties. Neuropharmacol 1994; 33:1235-1244.
8. Saran M, Bors W. Signalling by $O_2^{-\bullet}$ and $NO^\bullet$: how far can either radical, or any specific reaction product, transmit a message under in vivo conditions? Chem-Biol Interact 1994; 90:35-45.
9. Vanderkooi JM, Wright WW, Erecinska M. Nitric oxide diffusion coefficients in solutions, proteins and membranes determined by phosphorescence. Biochim Biophys Acta 1994; 1207:249-254.
10. Pauling L. The one-electron bond and the three-electron bond; electron-deficient substances. In: Pauling L ed. The Nature of the Chemical Bond and the Structure of Molecules and Crystals. third edition, Cornell University Press, Ithaca, NY, USA, 1960:340-363.
11. Dulmage WJ, Meyers EA, Lipscomb WN. On the crystal and molecular structure of N_2O_2. Acta Cryst 1953; 6:760-764.
12. Sluyts EJ, Van der Veken BJ. On the behaviour of nitrogen oxides in liquefied argon and krypton. Dimerisation of nitric oxide. J Mol Struct 1994; 320:249-267.
13. Dousmanis GC. Magnetic hyperfine effects

and electronic structure of NO. Phys Rev 1955; 97:967-970.

14. Lipton SA, Choi Y-B, Pan Z-H et al. A redox-based mechanism for the neuro-protective and neurodestructive effects of nitric oxide and related nitroso-compounds. Nature 1993; 364:626-632.

15. Stamler JS, Singel DJ, Loscalzo J. Biochemistry of nitric oxide and its redox-activated forms. Science 1992; 258:1898-1902.

16. Beringer R, Castle JG. Magnetic resonance absorption in nitric oxide. Phys Rev 1950; 78:581-586.

17. Whittaker JW. Molecular paramagnetic resonance of gas-phase nitric oxide. J Chem Educ 1991; 68:421-423.

18. Gallagher JJ, Bedard FD, Johnson CM. Microwave spectrum of $N^{14}O^{16}$. Phys Rev 1954; 93:729-733.

19. Gallagher JJ, Johnson CM. Uncoupling effects in the microwave spectrum of nitric oxide. Phys Rev 1956; 103:1727-1737.

20. Mizushima M. Theory of the hyperfine structure of NO molecule. Electronic structure. Phys Rev 1957; 105:1262-1270.

21. Westenberger AA. Intensity relations for determining gas-phase OH, Cl, Br, I, and free-electron cooncentrations by quantitative ESR. J Chem Phys 1965; 43:1544-1549.

22. Westenberger AA, de Haas N. Quantitative ESR measurements of gas-phase H and OH concentrations in the $H-NO_2$ reaction. J Chem Phys 1965; 43:1550-1556.

23. Galpin JR, Veldink GA, Vliegenthart JFG et al. The interaction of nitric oxide with soybean lipoxygenase-1. Biochim Biophys Acta 1978; 536:356-362.

24. Stevens TH, Brudvig GW, Bocian FP et al. Structure of cytochrome a_3-Cua_3 couple in cytochrome c oxidase as revealed by nitric oxide binding studies. Proc Natl Acad Sci USA 1979; 76:3320-3324.

25. Brudvig GW, Stevens TH, Chan SI. Reactions of nitric oxide with cytochrome c oxidase. Biochemistry 1980; 19:5275-5285.

26. Martin CT, Morse RH, Kanne RM et al. Reactions of nitric oxide with tree and fungal laccase. Biochemistry 1981; 20:5147-5155.

27. Arciero DM, Lipscomb JD, Huynh BH et al. EPR and Mössbauer studies of proto-catechuate 4,5-dioxygenase. Characterization of a new Fe^{2+} environment. J Biol Chem 1983; 258:14981-14991.

28. Nocek JM, Kurtz DM, Pickering RA et al. Oxidation of deoxyhemerythrin to semi-methemoglobin by nitrite. J Biol Chem 1984; 259:12334-12338.

29. Nelson MJ. The nitric oxide complex of ferrous soybean lipoxygenase-1. Substrate, pH and ethanol effects on the active site iron. J Biol Chem 1987; 262:12137-12142.

30. Musci G, Di Marco S, Bonaccorsi di Patti M et al. Interaction of nitric oxide with ceruloplasmin lacking an EPR-detectable type 2 copper. Biochemistry 1991; 30:9866-9872.

31. Akhtar MJ, Bonner FT, Hughes M. Reaction of nitric oxide with hyponitrous acid: a hydrogen atom abstraction reaction. Inorg Chem 1985; 24:1934-1935.

32. Bonner FT, Hughes MN. No lack of NO activity. Science 1993; 260:145-146.

33. Bonner FT, Dzelzkalns LS, Bonucci JA. Properties of nitroxyl as intermediate in the nitric oxide-hydroxylamine reaction and in trioxodinitrate decomposition. Inorg Chem 1978; 17:2487-2494.

34. Bonner FT, Wang N-Y. Reduction of nitric oxide by hydroxylamine. 1. Kinetics and mechanism. Inorg Chem 1986; 25:1858-1862.

35. Grätzel M, Tanigushi S, Henglein A. Pulsradiolytische untersuchung kurzlebiger zwischenprodukte der NO-reduction in wässriger lösung. Ber Bunsen-Ges Phys Chem 1970; 10:1003-1010.

36. Bazylinski D, Hollocher TC. Evidence from the reaction between trioxodinitrate(II) and ^{15}NO that trioxodinitrate(II) decomposes into nitrosyl hydride and nitrite in neutral aqueous solution. Inorg Chem 1985; 24:4285-4288.

37. Iqbal ZM, Dahl K, Epstein SS. Role of nitrogen dioxide in the biosynthesis of nitrosamines in mice. Science 1980; 207:1475-1477.

38. Halliwell B, Gutteridge JMC. Free radicals in biology and medicine. Clarendon Press, Oxford, 1985.

39. Kosaka H, Uozumi M. Inhibition by amines indicates involvement of nitrogen dioxide

in autocatalytic oxidation of oxyhemoglobin by nitrite. Biochim Biophys Acta 1986; 871:14-18.

40. Huie RE. The reaction kinetics of NO_2. Toxicology 1994; 89:193-216.

41. Stephens RJ, Freeman G, Evans MJ. Early response of lungs to low levels of nitrogen dioxide: light and electron microscopy. Arch Environ Health 1972; 24:160-179.

42. Halliwell B, Hu M-L, Louie S et al. Interaction of nitrogen dioxide with human plasma: antioxidant depletion and oxidative damage. FEBS Lett 1992; 313:62-66.

43. Benjamin N, O'Driscoll F, Dougall H et al. Stomach NO synthesis. Nature 1994; 368:502.

44. Duncan C, Dougall H, Johnston P et al. Chemical generation of nitric oxide in the mouth from the enterosalivary circulation of dietary nitrate. Nature Medicine 1995; 1:546-551.

45. Lundberg JON, Farkas-Szallasi T, Weitzberg E et al. High nitric oxide production in human paranasal sinuses. Nature Medicine 1995; 1:370-373.

46. Murphy ME, Sies H. Reversible conversion of nitroxyl anion to nitric oxide by superoxide dismutase. Proc Natl Acad Sci USA 1991;88:10860-10864.

47. Fukuto JM, Hobbs AJ, Ignarro LJ. Conversion of nitroxyl (HNO) to nitric oxide (NO) in biological systems: the role of physiological oxidants and relevance to the biological activity of HNO. Biochem Biophys Res Commun 1993;196:707-713.

48. Akhtar MJ, Lutz CA, Bonner FT. Decomposition of sodium trioxodinitrate ($Na_2N_2O_3$) in the presence of added nitrite in aqueous solution. Inorg Chem 1979; 18:2369-2375.

49. Doyle MP, Mahapatro SN. Nitric oxide dissociation from trioxodinitrate (II) in aqueous solution. J Am Chem Soc 1984; 106:3678-3679.

50. Donald CE, Hughes MN, Thompson JM et al. Photolysis of the N=N bond in trioxonitrate: reaction between triplet NO^- and O_2 to form peroxonitrite. Inorg Chem 1986; 25:2676-2677.

51. Wink DA, Darbyshire JF, Nims RW et al. Reactions of the bioregulatory agent nitric oxide in oxygenated aqueous media: determination of the kinetics for oxidation and nitrosation by intermediates generated in the NO/O_2 reaction. Chem Res Toxicol 1993; 6:23-27.

52. Ford PC, Wink DA, Stanbury DM. Autoxidation kinetics of aqueous nitric oxide. FEBS Lett 1993; 326:1-3.

53. Ignarro LJ, Fukuto JM, Griscavage JM et al. Oxidation of nitric oxide in aqueous solution to nitrite but not nitrate: Comparison with enzymatically formed nitric oxide from L-arginine. Proc Natl Acad Sci USA 1993; 90:8103-8107.

54. Pogrebnaya VL, Usov AP, Baranov AV et al. Oxidation of nitric oxide by oxygen in the liquid phase. J Applied Chem USSR 1975; 48:1004-1007.

55. Kharitonov VG, Sundquist AR, Sharma VS. Kinetics of nitric oxide autoxidation in aqueous solution. J Biol Chem 1994; 269:5881-5883.

56. Lewis RS, Deen WM. Kinetics of the reaction of nitric oxide with oxygen in aqueous solutions. Chem Res Toxicol 1994; 7:568-574.

57. Mayer B, Klatt P, Werner ER et al. Kinetics and mechanism of tetrahydrobiopterin-induced oxidation of nitric oxide. J Biol Chem 1995; 270:655-659.

58. Taha Z, Kiechle F, Malinski T. Oxidation of nitric oxide by oxygen in biological systems monitored by porphyrin sensor. Biochem Biophys Res Commun 1992; 188:734-739.

59. Olbregts J. Thermolecular reaction of nitrogen monoxide and oxygen: a still unsolved problem. Int J Chem Kinetics 1985; 17:835-848.

60. Foubert L, Fleming B, Latimer R et al. Safety guidelines for use of nitric oxide. Lancet 1992; 339:1615-1616.

61. Bouchet M, Renaudin M-H, Raveau C et al. Safety requirement for use of inhaled nitric oxide in neonates. Lancet 1993; 341:968-969.

62. Miyamoto K, Aida A, Nishimura M et al. Effects of humidity and temperature on nitrogen dioxide formation from nitric oxide. Lancet 1994; 343:1099-1100.

63. Blough NV, Zafiriou OC. Reaction of superoxide with nitric oxide to form per-

oxonitrite in alkaline solution. Inorg Chem 1985; 24:3502-3504.

64. Bastian NR, Hibbs JB. Assembly and regulation of NADPH oxidase and nitric oxide synthase. Current Opinion in Immunol 1994; 6:131-139.

65. Gryglewski RJ, Palmer RMJ, Moncada S. Superoxide anion is involved in the breakdown of endothelium-derived vascular relaxing factor. Nature 1986; 320:454-456.

66. Palmer RMJ, Ferrige AG, Moncada S. Nitric oxide release accounts for the biological activity of endothelium-derived relaxing factor. Nature 1987; 327:524-526.

67. Palmer RMJ, Rees DD, Ashton DS et al. L-arginine is the physiological precursor for the formation of nitric oxide in endothelium dependent relaxation. Biochem Biophys Res Commun 1988; 153:1251-1256.

68. Saran M, Michel C, Bors W. Reaction of NO with O_2^{-} Implications for the action of endothelium-derived relaxing factor (EDRF). Free Rad Res Comms 1990; 10:221-226.

69. Beckman JS, Beckman TW, Chen J et al. Apparent hydroxyl radical production by peroxynitrite: implications for endothelial injury from nitric oxide and superoxide. Proc Natl Acad Sci USA 1990; 87:1620-1624.

70. Yang G, Candy TEG, Boaro M et al. Free radical yields from the homolysis of peroxynitrous acid. Free Rad Biol Med 1992; 12:327-330.

71. Crow JP, Spruell C, Chen J et al. On the pH-dependent yield of hydroxyl radical products from peroxynitrite. Free Rad Biol Med 1994; 16:331-338.

72. Beckman JS, Ischiropoulos H, Zhu L et al. Kinetics of superoxide dismutase- and iron-catalyzed nitration of phenolics by peroxynitrite. Arch Biochem Biophys 1992; 298:438-445.

73. Tsai JM, Harrison JG, Martin JC et al. Role of conformation of peroxynitrite anion (ONOO⁻) in its stability and toxicity. J Am Chem Soc 1994; 116:4115-4116.

74. Beckman JS, Chen J, Ischiropoulos H et al. Oxidative chemistry of peroxynitrite. Methods Enzymol 1994; 233:229-240.

75. Huie RE, Padmaja S. The reaction of NO with superoxide. Free Rad Res Comms 1993; 18:195-199.

76. Goldstein S, Czapski G. The reaction of NO· with $O_2^{-•}$ and $HO_2^•$: a pulse radiolysis study. Free Rad Biol Med 1995; 19:505-510.

77. Saran M, Bors W. Pulse radiolysis for investigation of nitric oxide-related reactions. In: Packer L, ed. Oxygen Radicals in Biological systems, Pt C. Methods Enzymol Academic Press, 1994; 233:20-34.

78. Czapski G, Holcman J, Bielski BHJ. Reactivity of nitric oxide with simple short-lived radicals in aqueous solutions. J Am Chem Soc 1994; 116:11465-11469.

79. Noronha-Dutra AA, Epperlein MM, Woolf N. Reaction of nitric oxide with hydrogen peroxide to produce potentially cytotoxic singlet oxygen as a model for nitric oxide-mediated killing. FEBS Lett 1993; 321:59-62.

80. Koppenol WH. Thermodynamic considerations on the formation of reactive species from, hypochlorite, superoxide and nitrogen monoxide. Could nitrosyl chloride be produced by neutrophils and macrophages? FEBS Lett 1994; 347:5-8.

81. Dunham AJ, Barkley RM, Sievers RE. Aqueous nitrite ion determination by selective reduction and gas phase nitric oxide chemiluminescence. Anal Chem 1995; 34:220-224.

82. Bielski BHJ, Cabelli DE. Highlights of current research involving superoxide and perhydroxyl radicals in aqueous solutions. Int J Radiat Biol 1991; 59:291-319.

83. Rubanyi GM, Ho EH, Cantor EH et al. Cytoprotective function of nitric oxide: inactivation of superoxide radicals produced by human leukocytes. Biochem Biophys Res Commun 1991; 181:1392-1397.

84. Ischiropoulos H, Zhu L, Beckman JS. Peroxynitrite formation from macrophage-derived nitric oxide. Arch Biochem Biophys 1992; 298:446-451.

85. Darley-Usmar V, Wiseman H, Halliwell B. Nitric oxide and oxygen radicals: a question of balance. FEBS Lett 1995; 369:131-135.

86. McCall TB, Boughton-Smith NK, Palmer RMJ et al. Synthesis of nitric oxide from L-arginine by neutrophils. Release and interaction with superoxide anion. Biochem J 1989; 261:293-296.

87. Rodenas J, Mitjavila MT, Carbonell T. Simultaneous generation of nitric oxide and superoxide by inflammatory cells in rats. Free Rad Biol Med 1995; 18:869-875.

88. Ischiropoulos H, Zhu L, Chen J et al. Peroxynitrite-mediated tyrosine nitration catalyzed by superoxide dismutase. Arch Biochem Biophys 1992; 298:431-437.

89. Carreras MC, Pargament GA, Catz SD et al. Kinetics of nitric oxide and hydrogen peroxide production and formation of peroxynitrite during the respiratory burst of human neutrophils. FEBS Lett 1994; 341:65-68.

90. Assreuy J, Cunha FQ, Epperlein M et al. Production of nitric oxide and superoxide by activated macrophages and killing of *Leishmania Major*. Eur J Immunol 1994; 24:672-676.

91. Squadrito GL, Pryor WA. The formation of peroxynitrite in vivo from nitric oxide and superoxide. Chem-Biol Interac 1995; 96:203-206.

92. Garthwaite J, Boulton CL. Nitric oxide signaling in the central nervous system. Annu Rev Physiol 1995; 57:683-706.

93. Laurent M, Lepoivre M, Tenu J-P. Kinetic modelling of the nitric oxide gradient generated in vitro by adherent cells expressing inducible nitric oxide synthase. Biochem J 1996; 314:109-113.

94. Padmaja S, Huie RE. The reaction of nitric oxide with organic peroxyl radicals. Biochem Biophys Res Commun 1993; 195:539-544.

95. Cueto R, Pryor WA. Cigarette smoke chemistry: conversion of nitric oxide to nitrogen dioxide and reactions of nitrogen oxides with other smoke components as studied by Fourier transform infrared spectroscopy. Vibrational Spectroscopy 1994; 7:97-111.

96. Joseph J, Kalyanaraman B, Hyde JS. Trapping of nitric oxide by nitronyl nitroxides: an electron spin resonance investigation. Biochem Biophys Res Commun 1993; 192:926-934.

97. Akaike T, Yoshida M, Miyamoto Y et al. Antagonistic action of imidazolineoxyl N-oxides against endothelium-derived relaxing factor/·NO through a radical reaction. Biochemistry 1993; 32:827-832.

98. Yoshida M, Akaike T, Wada Y et al. Therapeutic effects of imidazolineoxyl N-oxide against endotoxin shock through its direct nitric oxide-scavenging activity. Biochem Biophys Res Commun 1994; 202:923-930.

99. Lagercrantz C. Spin trapping of nitric oxide as aminoxyl radicals by its reaction with two species of short-lived radicals derived from azo compounds such as 2,2'-azobisisobutyronitrile and some aliphatic alcohols. Free Rad Res Comms 1993; 19:387-395.

100. Gatti RM, Radi R, Augusto O. Peroxynitrite-mediated oxidation of albumin to the protein-thiyl free radical. FEBS Lett 1994; 348:287-290.

101. Roy B, Lepoivre M, Henry Y et al. Inhibition of ribonucleotide reductase by nitric oxide derived from thionitrites: reversible modifications of both subunits. Biochemistry 1995; 34:5411-5418.

102. Janzen EG, Wilcox AL, Manoharan V. Reactions of nitric oxide with phenolic antioxidants and phenoxyl radicals. J Org Chem 1993; 58:3597-3599.

103. Eiserich JP, Butler J, Van Der Vliet A et al. Nitric oxide rapidly scavenges tyrosine and tryptophan radicals. Biochem J 1995; 310:745-749.

104. Bottomley F, Brooks WVF, Clarkson SG et al. Electrophilic behaviour of the coordinated nitrosyl cation. J Chem Soc Comm 1973; 1973:919-920.

105. Enemark JH, Feltham RD. Principles of structure, bonding, and reactivity for metal nitrosyl complexes. Coord Chem Reviews 1974; 13:339-406.

106. Huheey JE. Inorganic Chemistry. Harper and Row Pub, 1983, 610-615.

107. Lipscomb LA, Lee BS, Yu NT. Resonance Raman investigation of nitric oxide bonding in iron porphyrins: detection of the Fe-NO stretching vibration. Inorg Chem 1993; 32:281-286.

108. Pombeiro AJL. Coordination chemistry of small unsaturated-N molecules at electron-rich mononuclear centers: cyanamide, organonitriles, nitric oxide and related species. New J Chem 1994; 18:163-174.

109. Westre TE, Di Cicco A, Filipponi A, et al. Determination of the Fe-N-O angle in [FeNO]^7 complexes using multiple-scattering EXAFS analysis by GNXAS. J

Am Chem Soc 1994; 116:6757-6768.

110. Saavedra JE, Dumans TM, Flippen-Anderson JL et al. Secondary amine/nitric oxide complex ions, R₂N[N(O)NO]⁻. O-functionalization chemistry. J Org Chem 1992; 57:6134-6138.

111. Hrabie JA, Klose JR, Wink DA et al. New nitric oxide-releasing zwitterions derived from polyamines. J Org Chem 1993; 58:1472-1476.

112. Diodati JG, Quyyumi AA, Hussain N et al. Complexes of nitric oxide with nucleophiles as agents for the controlled biological release of nitric oxide: antiplatelets effect. Thrombosis and Haemostasis 1993; 70:654-658.

113. Morley D, Keefer LK. Nitric oxide/nucleophile complexes: a unique class of nitric oxide-based vasodilators. J Cardiovasc Pharmacol 1993; 22:S3-S9.

114a. Vanderford PA, Wong J, Chang R et al. Diethylamine/nitric oxide (NO) adduct, an NO donor, produces potent pulmonary and systemic vasodilation in intact newborn lambs. J Cardiovasc Pharmacol 1994; 23:113-119.

114b. Petit J-F, Nicaise N, Lepoivre M et al. Protection of glutathione against the antiproliferative effects of nitric oxide: dependence on kinetics of NO release. Biochem Pharmacol 1996; in press.

115. Wagner DA, Young VR, Tannenbaum SR. Mammalian nitrate biosynthesis: incorporation of ¹⁵NH₃ into nitrate is enhanced by endotoxin treatment. Proc Natl Acad Sci USA 1983; 80:4518-4531.

116. Challis BC, Fernandes MHR, Glover BR et al. Formation of diazopeptides by nitrogen oxides. In: Bartsch H, O'Neill IR, Schulte-Hermann R, eds. Relevance of N-nitroso Compounds to Human Cancer: Exposures and Mechanisms. IARC Scientific Publications No 84, Lyon, France, 1987; 308-314.

117. Iyengar R, Stuehr DJ, Marletta MA. Macrophages synthesis of nitrite, nitrate, and N-nitrosamines: precursors and role of the respiratory burst. Proc Natl Acad Sci USA 1987; 84:6369-6373.

118. Mirvish SS, Ramm MD, Babcook DM et al. Lipidic nitrosating agents produced from atmospheric nitrogen dioxide and a nitrosamine produced in vivo from amyl nitrite. In: Bartsche H, O'Neill IK, Schulte-Hermann R, eds. Relevance of N-nitroso Compounds to Human Cancer: Exposures and Mechanisms. IARC Scientific Publications No 84, Lyon, France, 1987; 315-318.

119. Miwa M, Stuehr DJ, Marletta MA et al. Nitrosation of amines by stimulated macrophages. Carcinogenesis 1987; 8:955-958.

120. Tannenbaum SR. Endogenous formation of N-nitroso compounds: a current perspective. In: Bartsch H, O'Neill IK, Schulte-Hermann R, eds. Relevance of N-nitroso Compounds to Human Cancer: Exposures and Mechanisms. IARC Scientific Publications No 84, Lyon, France, 1987; 292-296.

121. Bartsch H, Ohshima H, Pignatelli B. Inhibitors of endogenous nitrosation mechanisms and implications in human cancer prevention. Mutation Res 1988; 202:307-324.

122. Marletta MA. Mammalian synthesis of nitrite, nitrate, nitric oxide, and N-nitrosating agents. Chem Res Toxicol 1988; 1:249-257.

123. Ohshima H, Tsuda M, Adachi H et al. L-arginine-dependent formation of N-nitrosamines by the cytosol of macrophages activated with lipopolysaccharide and interferon-γ. Carcinogenesis 1991; 12:1217-1220.

124. Ohshima H, Bartsch H. Chronic infections and inflammatory processes as cancer risk factors: possible role of nitric oxide in carcinogenesis. Mutation Res 1994; 305:253-264.

125. Haswell-Elkins MR, Satarug S, Tsuda M et al. Liver fluke infection and cholangiocarcinoma: model of endogenous nitric oxide and extragastric nitrosation in human carcinogenesis. Mutation Res 1994; 305:241-252.

126. Lewis RS, Tannenbaum SR, Deen WM. Kinetics of N-nitrosation in oxygenated nitric oxide solutions at physiological pH: role of nitrous anhydride and effects of phosphate and chloride. J Am Chem Soc 1995; 117:3933-3939.

127. Liu RH, Hotchkiss JH. Potential genotoxicity of chronically elevated nitric oxide: a review. Mutation Res 1995; 339:73-89.

128. Wink DA, Kasprzak KS, Maragos CM et al. DNA deaminating ability and geno-

toxicity of nitric oxide and its progenitors. Science 1991; 254:1001-1003.

129. Nguyen T, Brunson D, Crespi CL et al. DNA damage and mutation in human cells exposed to nitric oxide *in vitro*. Proc Natl Acad Sci USA 1992; 89:3030-3034.

130. De La Bretèche ML, Servy C, Lenfant M et al. Nitration of cathecolamines with nitrogen oxides in mild conditions: a hypothesis for the reactivity of NO in physiological systems. Tetrahedron Let 1994; 35:7231-7232.

131. Pryor WA, Jin X, Squadrito GL. One- and two-electron oxidations of methionine by peroxynitrite. Proc Natl Acad Sci USA 1994; 91:11173-11177.

132. Eiserich JP, Vossen V, O'Neill CA et al. Molecular mechanisms of damage by excess nitrogen oxides: nitration of tyrosine by gas-phase cigarette smoke. FEBS Lett 1994; 353:53-56.

133. Van der Vliet A, O'Neill CA, Halliwell B et al. Aromatic hydroxylation and nitration of phenylalanine and tyrosine by per-oxynitrite. Evidence for hydroxyl radical production from peroxynitrite. FEBS Lett 1994; 339:89-92.

134. Beckman JS, Carson M, Smith CD et al. ALS, SOD and peroxynitrite. Nature 1994; 364:584.

135. Beckman JS, Ye YZ, Anderson PG et al. Extensive nitration of protein tyrosines in human atherosclerosis detected by immu-nohistochemistry. Biol Chem Hoppe-Seyler 1994; 375:81-88.

136. Szabo C, Salzman AL, Ischiropoulos H. Endotoxin triggers the expression of an in-ducible isoform of nitric oxide synthase and the formation of peroxynitrite in the rat aorta in vivo. FEBS Lett 1995; 363:235-238.

137. Salman-Tabcheh S, Guérin MC, Torreilles J. Nitration of tyrosyl-residues from ex-tra- and intracellular proteins in human whole blood. Free Rad Biol Med 1995; 19:695-698.

138. Kaur H, Halliwell B. Evidence for nitric oxide-mediated oxidative damage in chronic inflammation. Nitrotyrosine in serum and synovial fluid from rheumatoid patients. FEBS Lett 1994; 350:9-12.

139. Kooy NW, Royall JA, Ye YZ et al. Evi-dence for in vivo peroxynitrite production in human acute lung injury. Am J Resp Crit Care Med 1995; 151:1250-1254.

140. Myers PR, Minor RL, Guerra R et al. Vasorelaxant properties of the endothelium-derived relaxing factor more closely resemble S-nitrosocyteine than nitric oxide. Nature 1990; 345:161-163.

141. Stamler JS, Jaraki O, Osborne J et al. Ni-tric oxide circulates in mammalian plasma primarily as an *S*-nitroso adduct of serum albumin. Proc Natl Acad Sci USA 1992; 89:7674-7677.

142. Stamler JS, Simon DI, Jaraki O et al. *S*-nitrosylation of tissue-type plasminogen activator confers vasodilatory and antiplatelet properties on the enzyme. Proc Natl Acad Sci USA 1992; 89:8087-8091.

143. Stamler JS, Simon DI, Osborne JA et al. *S*-nitrosylation of proteins with nitric oxide: synthesis and characterization of biologically active compounds. Proc Natl Acad Sci USA 1992; 89:444-448.

144. Gaston B, Reilly J, Drazen JM et al. En-dogenous nitrogen oxides and bronchodilator *S*-nitrosothiols in human airways. Proc Natl Acad Sci USA 1993; 90:10957-10961.

145. Wink DA, Nims RW, Darbyshire JF et al. Reaction kinetics for nitrosation of cysteine and glutathione in aerobic nitric oxide so-lutions at neutral pH. Insights into the fate and physiological effects of intermediates generated in the NO/O_2 reaction. Chem Res Toxicol 1994; 7:519-525.

146. Mohr S, Stamler JS, Brüne B. Mechanism of covalent modification of glyceraldehyde-3-phosphate dehydrogenase at its active site thiol by nitric oxide, peroxynitrite and re-lated nitrosating agents. FEBS Lett 1994; 348:223-227.

147. Moro MA, Darley-Usmar VM, Goodwin DA et al. Paradoxical fate and biological action of peroxynitrite on human platelets. Proc Natl Acad Sci USA 1994; 91:6702-6706.

148. Roy B, Du Moulinet d'Hardemare A, Fontecave M. New thionitrites: synthesis, stability, and nitric oxide generation. J Org Chem 1994; 59:7019-7026.

149. DeMaster EG, Quast BJ, Redfern B et al. Reaction of nitric oxide with the free sulf-hydryl group of human serum albumin

yields a sulfenic acid and nitrous oxide. Biochemistry 1995; 34:11494-11499.

150. Kharitonov VG, Sundquist AR, Sharma VS. Kinetics of nitrosation of thiols by nitric oxide in the presence of oxygen. J Biol Chem 1995; 270:28158-28164.

151. Meyer DJ, Kramer H, Özer N et al. Kinetics and equilibria of *S*-nitrosothiol-thiol exchange between glutathione, cysteine, penicillamines and serum albumin. FEBS Lett 1994; 345:177-180.

152. Clancy RM, Levartovsky D, Leszczynska-Piziak J et al. Nitric oxide reacts with intracellular glutathione and activates the hexose monophosphate shunt in human neutrophils: evidence for *S*-nitrosoglutathione as a bioactive intermediary. Proc Natl Acad Sci USA 1994; 91:3680-3684.

153. Gabor G, Allon N. Spectrofluorimetric method for NO determination. Anal Biochem 1994; 220:16-19.

154. Smith RP, Kruszyna H. Nitroprusside produces cyanide poisoning via a reaction with hemoglobin. J Pharm Exp Ther 1974; 191:557-563.

155. Wilcox DE, Kruszyna H, Kruszyna R et al. Effect of cyanide on the reaction of nitroprusside with hemoglobin: relevance to cyanide inference with biological activity of nitroprusside. Chem Res Toxicol 1990; 3:71-76.

156. Morando PJ, Borgy EB, de Schteingart LM. The reaction of cysteine with pentacyanonitrosylferrate (2-) ion. J Chem Soc Dalton Trans 1981; 435-440.

157. Butler AR, Calsy-Harrison AM, Glidewell C et al. The pentacyanonitrosylferrate ion - V. The course of the reactions of nitroprusside with a range of thiols. Polyhedron 1988; 7:1197-1202.

158. Butler AR, Glidewell C, Li M-H. Nitrosyl complexes of iron-sulfur clusters. Adv Inorg Chem 1988; 32:335-393.

159. Butler AR, Calsy AM, Johnson IL. Enzyme inhibition by sodium nitroprusside. Polyhedron 1990; 9:913-919.

160. Butler AR, Williams DLH. The physiological role of nitric oxide. Chem Soc Rev 1993:233-241.

161. Flitney FW, Megson IL, Clough T et al. Nitrosylated iron-suphur clusters, a novel class of nitrovasodilator: studies on the rat isolated tail artery. J Physiol 1990; 430:42P.

162. Flitney FW, Megson IL, Flitney DE et al. Iron-sulphur cluster nitrosyls, a novel class of nitric oxide generator: mechanism of vasodilator action on rat isolated tail artery. Br J Pharmacol 1992; 107:842-848.

163. Hogg N, Darkley-Usmar VM, Wilson MT et al. Production of hydroxyl radicals from the simultaneous generation of superoxide and nitric oxide. Biochem J 1992; 281:419-424.

164. Chamulitrat W, Jordan SJ, Mason RP et al. Nitric oxide formation during light-induced decomposition of phenol *N-tert*-butylnitrone. J Biol Chem 1993; 268:11520-11527.

165. Pou S, Anderson DE, Surichamorn W et al. Biological studies of a nitroso compound that release nitric oxide upon illumination. Mol Pharmacol 1994; 46:709-715.

166. Ignarro LJ, Edwards JC, Gruetter DY et al. Possible involvement of *S*-nitrosothiols in the activation of guanylate cyclase by nitroso compounds. FEBS Lett 1980; 110: 275-278.

167. Mathews WR, Kerr SW. Biological activity of nitrosothiols: the role of nitric oxide. J Pharmacol Exp Ther 1993; 267:1529-1537.

168. McAninly J, Williams DLH, Askew SC et al. Metal ion catalysis in nitrosothiols (RSNO) decomposition. J Chem Soc Chem Commun 1993:1758-1759.

169. Askew SC, Barnett DJ, McAninly J et al. Catalysis by Cu^{2+} of nitric oxide release from *S*-nitrosothiols (RSNO) J Chem Soc Perkin Trans 2 1995:741-745.

170. Sexton DJ, Muruganandam A, McKenney DJ et al. Visible light photochemical release of nitric oxide from *S*-nitrosoglutathione: potential photochemotherapeutic applications. Photochem Photobiol 1994; 59:463-467.

171. Singh RJ, Hogg N, Joseph J et al. Photosensitized decomposition of *S*-nitrosothiols and 2-methyl-2-nitrosopropane. Possible use for site-directed nitric oxide production. FEBS Lett 1995; 360:47-51.

172. Makings LR, Tsien RY. Caged nitric oxide. Stable organic molecules from which nitric oxide can be photoreleased. J Biol Chem 1994; 269:6282-6285.

173. Vithayathil AJ, Ternberg JL, Commoner B. Changes in electron spin resonance signals of rat liver during chemical carcinogenesis. Nature 1965; 207:1246-1249.

174. Woolum JC, Tiezzi E, Commoner B. Electron spin resonance of iron-nitric oxide complexes with amino acids, peptides and proteins. Biochim Biophys Acta 1968; 160:311-320.

175. Woolum JC, Commoner B. Isolation and identification of a paramagnetic complex from the livers of carcinogen-treated rats. Biochim Biophys Acta 1970; 201:131-140.

176. Vanin AF. Identification of divalent iron complexes with cysteine in biological systems by the EPR method. Biokhimiya 1967; 32:277-282 (English translation 228-232).

177. Vanin AF, Vakhnina LV, Chetverikov AG. Nature of the EPR signals of a new type found in cancer tissues. Biofizika 1970; 15:1044-1051 (English translation 1082-1089).

178. Maruyama T, Kataoka N, Nagase S et al. Identification of three-line electron spin resonance signal and its relationship to ascites tumors. Cancer Res 1971; 31:179-184.

179. Arnold WP, Mittal CK, Katsuki S et al. Nitric oxide activates guanylate cyclase and increases guanosine 3':5'-cyclic monophosphate levels in various tissue preparations. Proc Natl Acad Sci USA 1977; 74:3203-3207.

180. Murad F, Mittal CK, Arnold WP et al. Guanylate cyclase: activation by azide, nitro compounds, nitric oxide and hydroxyl radical and inhibition by hemoglobin and myoglobin. Adv Cyclic Nucleotides Res 1978; 9:145-158.

181. Ignarro LJ, Buga GM, Wood KS et al. Endothelium-derived relaxing factor produced and released from artery and vein is nitric oxide. Proc Natl Acad Sci USA 1987; 84:9265-9269.

182. Radomski MW, Palmer RMJ, Moncada S. The role of nitric oxide and cGMP in platelets adhesion to vascular endothelium. Biochem Biophys Res Commun 1987; 148:1482-1489.

183. Moncada S, Radomski MW, Palmer RMJ. Endothelium-derived relaxing factor. Identification as nitric oxide and role in the control of vascular tone and platelet function. Biochem Pharmacol 1988; 37:2495-2501.

184. Moncada M, Palmer RMJ, Higgs EA. Biosynthesis of nitric oxide from L-arginine. A pathway for the regulation of cell function and communication. Biochem Pharmacol 1989; 38:1709-1715.

185. Ignarro LJ. Biosynthesis and metabolism of endothelium-derived nitric oxide. Ann Rev Pharmacol Toxicol 1990; 30:535-560.

186. Ignarro LJ. Signal transduction mechanisms involving nitric oxide. Biochem Pharmacol 1991; 41:485-490.

187. Moncada S, Palmer RMJ, Higgs EA. Nitric oxide: physiology, pathophysiology, and pharmacology. Pharmacol Rev 1991; 43:109-142.

188. Craven PA, DeRubertis FR. Restoration of the responsiveness of purified guanylate cyclase to nitrosoguanidine, nitric oxide, and related activators by heme and hemoproteins. Evidence for involvement of the paramagnetic nitrosyl-heme complex in enzyme activation. J Biol Chem 1978; 253:8433-8443.

189. Craven PA, DeRubertis FR, Pratt DW. Electron spin resonance study of the role of NO·catalase in the activation of guanylate cyclase by NaN_3 and NH_2OH. Modulation of enzyme responses by heme proteins and their nitrosyl derivatives. J Biol Chem 1979; 254:8213-8222.

190. Stone JR, Marletta MA. Soluble guanylate cyclase from bovine lung: activation with nitric oxide and carbon monoxide and spectral characterization of the ferrous and ferric states. Biochemistry 1994; 33:5636-5640.

191. Stone JR, Marletta MA. Heme stoichiometry of heterodimeric soluble guanylate cyclase. Biochemistry 1995; 34:14668-14674.

192. Stone JR, Marletta MA. The ferrous heme of soluble guanylate cyclase: formation of hexacoordinate complexes with carbon monoxide and nitrosomethane. Biochemistry 1995; 34:16397-16403.

193. Stone JR, Sands RH, Dunham WR et al. Electron paramagnetic resonance spectral evidence for the formation of a pentaco-

ordinated nitrosyl-heme complex on soluble guanylate cyclase. Biochem Biophys Res Commun 1995; 207:572-577.

194. Richter-Addo GB, Legzdins P. Metal nitrosyls. Oxford University Press, Oxford, UK, 1992.

195. Taylor KC. Nitric oxide catalysis in automobile exhaust systems. Catal Rev-Sci Eng 1993; 35:457-481.

196. Coichev N, Van Eldik R. Metal catalyzed atmospheric oxidation reactions. A challenge to coordination chemistry. New J Chem 1994; 18:123-131.

197. Proust A, Gouzerh P, Robert F. Reactivity of acetone oxime towards oxomolybdenum(VI) complexes. Part 2. Synthesis, crystal structures and reactivity of Molybdenum nitrosyl complexes. J Chem Soc Dalton Trans 1994:825-833.

198. Qiu S, Ohnishi R, Ichikawa M. Formation and interaction of carbonyls and nitrosyls on Gold(I) in ZSM-5 zeolite catalytically active in NO reduction with CO. J Phys Chem 1994; 98:2719-2721.

199. Beinert H, Dervartanian DV, Hemmerich P et al. On the ligand field of redox active non-heme iron in proteins. Biochim Biophys Acta 1965; 96:530-533.

200. Gans P. Reaction of nitric oxide with cobalt (II) ammine complexes and other reducing agents. J Chem Soc 1967; 1967 A:943-946.

201. Enemark JH, Feltham RD. Stereochemical control of valence and its application to the reduction of coordinated NO and N_2. Proc Natl Acad Sci USA 1972; 69:3534-3536.

202. McCleverty JA. Reactions of nitric oxide coordinated to transition metals. Chem Rev 1979; 79:53-76.

203. Wade RS, Castro CE. Redox reactivity of iron(III) porphyrins and heme proteins with nitric oxide. Nitrosyl transfer to carbon, oxygen, nitrogen, and sulfur. Chem Res Toxicol 1990; 3:289-291.

204. Doyle MP, Mahapatro SN, Broene RD et al. Oxidation and reduction of hemoproteins by trioxodinitrate(II). The role of nitrosyl hydride and nitrite. J Am Chem Soc 1988; 110:593-599.

205. Castro CE, Bartnicki EW. The interconversion of nucleic acid bases by iron(III) porphyrins and nitric oxide. J Org Chem

1994; 59:4051-4052.

206. Ruggiero CE, Carrier SM, Tolman WB. Reductive disproportionation of NO mediated by copper complexes: modeling N_2O generation by copper proteins and heterogenous catalysts. Angew Chem Int Ed Engl 1994; 33:895-897.

207. McNeil DAC, Raynor JB, Symons MCR. Structure and reactivity of transition-metal complexes with polyatomic ligands. Part 1. Electron spin resonance spectra of $[Mn(CN)_5NO]^{2-}$ and $[Fe(CN)_5NO]^{3-}$. J Chem Soc 1965; 1965:410-415.

208. Kon H, Kataoka N. Electron paramagnetic resonance of nitric oxide-protoheme with some nitrogenous base. Model systems of nitric oxide hemoproteins. Biochemistry 1969; 8:4757-4762.

209. Yonetani T, Yamamoto H, Erman JE et al. Electron properties of hemoproteins. V. Optical and electron paramagnetic characteristics of nitric oxide derivatives of metalloporphyrin-apo-hemoprotein complexes. J Biol Chem 1972; 247:2447-2455.

210. Tsai A-l. How does NO activates hemeproteins? FEBS Lett 1994; 341:141-145.

211. Traylor TG, Sharma VS. Why NO? Biochemistry 1992; 31:2847-2849.

212. Van Voorst JDW, Hemmerich P. Electron spin resonance of $Fe(CN)_5NO^{3-}$ and $Fe(CN)_5NOH^{2-}$. J Chem Phys 1966; 45:3914-3913.

213. Swinehart JH. The nitroprusside ion. Coord Chem Rev 1967; 2:385-402.

214. McDonald CC, Phillips WD, Mower HF. An electron spin resonance of some complexes of iron, nitric oxide, and anionic ligands. J Am Chem Soc 1965; 87:3319-3326.

215. Henry Y, Ducrocq C, Drapier J-C et al. Nitric oxide, a biological effector. Electron paramagnetic resonance detection of nitrosyl-iron-protein complexes in whole cells. Eur Biophys J. 1991; 20:1-15.

216. Henry Y, Lepoivre M, Drapier J-C et al. EPR characterization of molecular targets for NO in mammalian cells and organelles. FASEB J. 1993; 7:1124-1134.

217. Henry YA, Singel DJ. Metal-nitrosyl interactions in nitric oxide biology probed by electron paramagnetic resonance spectros-

copy. In: Feelisch M, Stamler J, eds. Methods in Nitric Oxide Research. John Wiley and Sons, 1996:357-372.

218. Singel SJ, Lancaster JR. Electron paramagnetic resonance spectroscopy and nitric oxide biology. In: Feelisch M, Stamler J, eds. Methods in Nitric Oxide Research. John Wiley and Sons, 1996:341-356.

219. Czapski G, Goldstein S. The role of the reactions of NO with superoxide and oxygen in biological systems: a kinetic approach. Free Rad Biol Med 1995; 19:785-794.

220. Goldstein S, Czapski G. Kinetics of nitric oxide autoxidation in aqueous solution in the absence and presence of various reductants. The nature of oxidizing intermediates. J Am Chem Soc 1995; 117:12078-12084.

CHAPTER 3

BASIC EPR METHODOLOGY

Yann A. Henry

The present chapter presents the basic methodology of electron paramagnetic (spin) resonance (EPR or ESR), illustrated by instances developed in the following chapters. Although it is meant to be read on the present spot, it will be more fully understood on a second reading. In our opinion, a full understanding of EPR—including quantum mechanics, klystron, wave guides and all that—is probably not necessary to collaborate with an EPR spectroscopist and to record significant and biologically relevant spectra.[1-3]

BASIC PRINCIPLES OF EPR SPECTROSCOPY

CW-EPR (CW, continuous wave as opposed to pulsed) is an absorption spectroscopy in the microwave range, the most familiar being around ν = 9 GHz ($\lambda \sim 3$ cm). The special feature of EPR spectroscopy as compared to optical transitions is that absorption of photons occurs when the sample is subjected to an external magnetic field H_0 (Zeeman effect)[†] and is thus connected to the presence of magnetic dipoles in the sample. Electrons are ubiquitous and possess spin; however most materials are diamagnetic because electrons go in pairs so that there is no net resulting magnetic moment. Diamagnetic materials will not give rise to any EPR spectrum, and this is most convenient since it means that extensive isolation and purification of the sample under study, which often implies alterations and loss, is often unnecessary. However paramagnetic impurities are also numerous which can contaminate a sample: so are rubber, ashes, dusts, some plastic materials, some modeling clays and texture agents, transition metal cations in solvent, etc. Furthermore, several paramagnetic species, contained in a given sample, say living cells, can have overlapping spectra.

[†] One should in fact use the magnetic induction or flux density B_0, measured in tesla (T) as NMR spectroscopists do (1 T = 10^4 G), rather than gauss (G); traditionally EPR spectroscopists continue to use the magnetic field strength H_0, measured in oersted (Oe) or in ampere per meter or in gauss (1 Oe = 1 G).

Nitric Oxide Research from Chemistry to Biology: EPR Spectroscopy of Nitrosylated Compounds, edited by Yann A. Henry, Annie Guissani and Béatrice Ducastel.
© 1997 R.G. Landes Company.

There must be unpaired electrons in a material to give an EPR spectrum; however as we shall see below, this condition is necessary but not sufficient for a spectrum to be observed. There are two main classes of such materials: first, paramagnetic ions from the transition groups of the periodic table, which contain partly filled d (or f for lanthanides) electron shells, such as Fe, Cu, Ni, Mn, Mo and Co, encountered in biological samples; and second, free radicals and radical ions, such as $NO^\bullet$, O_2, $O_2^{-\bullet}$, organic radicals such as ascorbyl radical or the tyrosyl radical of ribonucleotide reductase, to name species of interest in the present book. *S*-nitrosothiols, *N*-nitrosamines, other nitroso derivatives, organic nitrites and nitrates, peroxynitrite, etc., are all diamagnetic.

MAGNETIC MOMENT, g-VALUES AND ZEEMAN EFFECT

An electron has a spin angular momentum $\overline{S}$, an orbital angular momentum $\overline{L}$ and a total angular momentum $\overline{J} = \overline{S} + \overline{L}$. Due to the electron charge the spinning motion of an electron generates a magnetic moment $\overline{\mu}$. Moments $\overline{\mu}$ and $\overline{J}$ are related by:

$$\overline{\mu} = -g\beta\overline{J}$$

where g is a dimensionless constant called g-factor, equal to 2.00232 for the "free" electron; β is the Bohr magneton (4.66858 cm^{-1} G^{-1} or 9.2741×10^{-24} J T^{-1}), a constant of the electron related to its charge and mass, the velocity of light and the Planck constant h (3.33564×10^{-11} cm^{-1} s or 6.6262×10^{-34} J s).

The orientation of the electron magnetic moment $\overline{\mu}$ in a static external magnetic field $\overline{H}_o$ (that of an electromagnet) is characterized by the electron spin quantum number $M_S = \pm 1/2$. Transitions are possible between the two $M_S = 1/2$ and $-1/2$ sublevels (Zeeman effect) for which M_S changes by ± 1 (Fig. 3.1).

The resonance condition, corresponding to the hv photon resonant absorption, is expressed by the following relationship between the microwave frequency v and the magnetic field H_0:

$$\Delta E = h\nu = g\beta H_0$$

corresponding to the spin Zeeman Hamiltonian :

$$\mathcal{H} = -\overline{\mu} \cdot \overline{H}_o = \beta\overline{S} \cdot \overline{\overline{g}} \cdot \overline{H}_o,$$

where $\overline{\overline{g}}$ is a tensor, with principal g-values, g_X, g_Y and g_Z, depending on the complex symmetry, characteristic of the paramagnetic sample under study. The symmetry is termed rhombic when the principal g-values are distinct, axial when two of them are equal and isotropic when the three values are identical. As a function of the environment offered to the unpaired electron in the complex under study, that is the atoms "visited" by the electron, the g-values may be more or less different from g = 2.0023. For instance in transition elements, g-values varying between 0.6 and 10 are found. As we have seen in the preceding chapter, in the $NO^\bullet$ free molecule the electronic orbital motion and spin are strongly coupled, and $NO^\bullet$ has a $J = 3/2$ excited state in which is observed the Zeeman effect with a $g_J = 0.774$. In paramagnetic nitrosylated metal complexes, $NO^\bullet$ loses its symmetry; the odd electron is highly delocalized over the metal orbitals, so that its orbital angular momentum $\overline{L}$ is largely quenched, and the complexes have a pure S = 1/2 character, with g-values different from g = 2.0023, depending on their symmetry.

As stated above, a complex is paramagnetic when it contains at least one unpaired electron. For instance, the Fe^{III} cation ($3d^5$ configuration), depending on the strength of the field produced by its four to six ligands, can exist under the so-called high-spin (S = 5/2) and low-spin (S = 1/2) states. Such complexes in both states are usually observable by EPR spectroscopy (see an instance in ref. 4) (Fig. 3.2).

Their reduced Fe^{II} ($3d^6$ configuration) counterparts (S = 2) and (S = 0) are not EPR-detectable—the low-spin complexes because they are diamagnetic, the high-spin ones for theoretical reasons (Kramer's theorem) that we are unable to explain here. Indeed the latter fact is quite general; paramagnetic species with an even number of unpaired electrons (S = 1, 2, 3, etc.) are

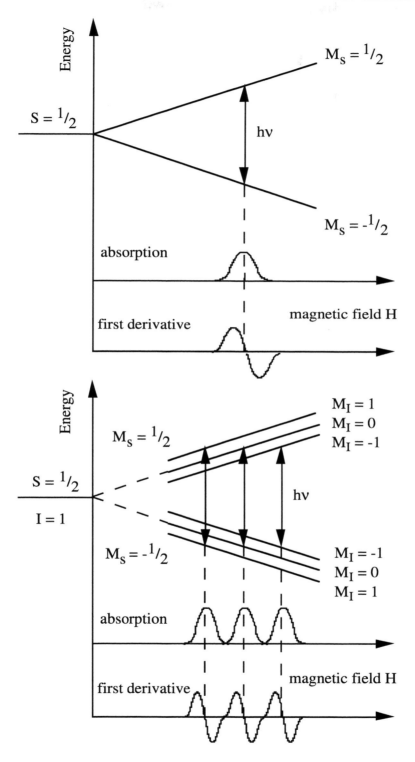

Fig. 3.1. Electronic energy levels as a function of the magnetic field strength (Zeeman effect) (a) with S = 1/2 and hyperfine coupling (b) with S = 1/2 and I = 1.

Fig. 3.2. EPR spectra at 77 K of isolated hemoglobin α-subunit in the oxidized state heme concentation (1.2 mmol/l) with various axial Fe^{III} "sixth" ligands: weak ligand field ligands, H_2O (pH 6), F^-, giving high-spin axial complexes with g-values at 6 and 2 and strong ligand field ligands, N_3^- and OH^- (pH 9), giving low-spin rhombic complexes with three g-values around 2. (Reproduced with permission from: Henry Y, Banerjee R. Electron spin resonance spectra of isolated ferrihemoglobin (α, β and γ) chains. An attempted correlation with optical absorption spectra. J Mol Biol 1970; 50:99-110) Copyright © Academic Press Ltd, London, UK.

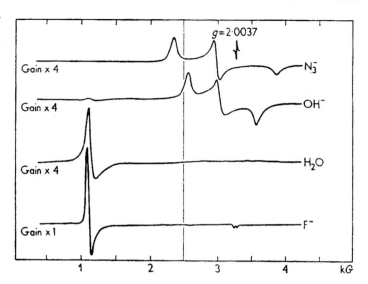

usually EPR-silent, at least under standard frequency and temperature experimental conditions. Only species with an odd number of unpaired electrons (S = 1/2, 3/2, 5/2, etc.) give rise to EPR spectra.

HYPERFINE INTERACTIONS

EPR spectra often present hyperfine structures when μ interacts with magnetic moments of neighboring nuclei, characterized by a quantum number I different from zero. It is the case for ^1H (I = 1/2), ^{14}N (I = 1), ^{15}N (I = 1/2), ^{57}Fe (I = 1/2), natural abundance $^{63,65}Cu$ (I = 3/2), ^{55}Mn (I = 5/2), etc., but not for natural abundance ^{12}C, ^{16}O, ^{56}Fe, etc. (I = 0). The corresponding Hamiltonian, added to the Zeeman term, is:

$$\mathcal{H} = \beta \bar{S} \cdot \bar{\bar{g}} \cdot \bar{H}_o + \bar{S} \cdot \bar{\bar{A}} \cdot \bar{I},$$

where $\bar{\bar{A}}$ is also a tensor with principal values A_X, A_Y and A_Z, characteristic of the observed species.[‡] In general, a nucleus with a spin I splits an absorption line into (2I + 1) equally spaced components (Fig. 3.1b). For instance, the nitrogen

nucleus of $^{14}NO^{\bullet}$ gives a triplet for the hyperfine structure which splits further into nine lines due to the proximal histidine ^{14}N atom, as detected in Hb(Fe^{II})NO (Fig. 3.3).[5-9]

In another instance no hyperfine structure of $^{14}NO^{\bullet}$ is resolved in EPR spectra at 77 K of $Fe^I(SR)_2(NO)_2$ complexes, yet substitution of ^{15}N for ^{14}N causes a line shape modification (Fig. 3.4).[10]

TEMPERATURE, MICROWAVE POWER SATURATION AND LINE-WIDTHS

The third feature of a spectrum, once g-values and A-values have been measured, is the shape and width of the absorption lines, which do not derive from the spin Hamiltonian. The information derived from the line shape analysis is often nonspecific and related to several overlapping phenomena, the most important being the relaxation processes. However when this difficult analysis is properly carried out, the reward in terms of structural information can be important.

[‡] The hyperfine splitting **a** (measured in gauss) is related to the A-value (expressed in cm^{-1} or kaiser K, or in MHz) by the following relationships :
$A (cm^{-1}) = g \cdot (4.6686 \times 10^{-5} cm^{-1} G^{-1}) \cdot \mathbf{a} (G)$
$A (MHz) = g \cdot (1.3996 MHz G^{-1}) \cdot \mathbf{a} (G)$

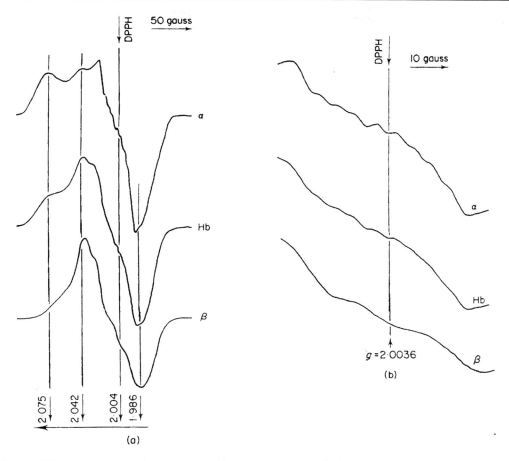

Fig. 3.3. EPR spectra at 77 K of various nitrosyl hemoproteins: α and β hemoglobin subunits as compared to hemoglobin tetramer Hb (heme concentrations 0.5 to 1 mmol/l) (a); (b) same as (a) with 5-fold expanded field scale around g = 2.0, to show the hyperfine structures, well resolved as nine-lines for α-subunits and poorly resolved as three-lines for β-subunits; HbNO spectrum is the sum of both subunits contributions. (Reproduced with permission from: Henry Y, Banerjee R. Electron paramagnetic studies of nitric oxide haemoglobin derivatives: isolated subunits' and nitric oxide hybrids. J Mol Biol 1973; 73:469-482) Copyright © Academic Press Ltd, London, UK.

The fractional population of spins in the two Zeeman energy sublevels is expressed as a Boltzmann distribution, which states that the probability to find an electron at a given energy level E is proportional to exp(-E/kT), where k is the Boltzmann constant and T the absolute temperature. At elevated temperature, both population fractions N_- and N_+ approach 1/2, and only at low temperatures is the fractional population of the lower state M_S = -1/2 appreciably greater than that of the upper state. At any particular ν and T the EPR absorption is proportional to the number of paramagnetic centers (that is the spin concentration in a given sample volume) and to the fractional population difference $n = N_- - N_+$ between the two Zeeman sublevels. The total number of unpaired electrons being $N = N_- + N_+$, the ratio n/N called the "polarization" is an index of the population dependence with T. One can demonstrate that the n to N relationship is :

$$n = N \cdot h\nu/2kT$$

Therefore, other things remaining equal, the absorption can be increased, and the sensitivity improved, by working at a

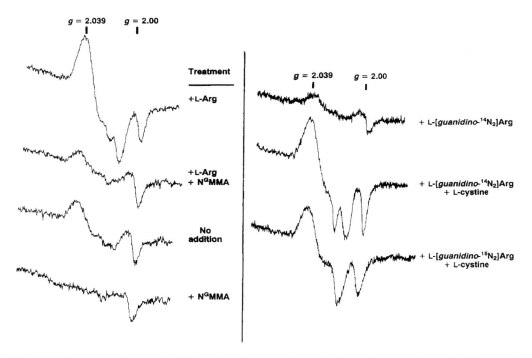

Fig. 3.4. EPR spectra of cytotoxic BCG-activated macrophages. Left panel: comparison of cells treated with L-arginine or with N ^{G}MMA, an analogue of L-arginine inhibitor of NO synthase. Right panel: effects of L-cystine and of L-arginine guanidino ^{14}N to ^{15}N isotopic substitution, showing a modification of hyperfine structure from three-lines to two-lines around g = 2.015. (Reproduced with permission from the authors: Lancaster JR, Hibbs JB. EPR demonstration of iron-nitrosyl complex formation by cytotoxic activated macrophages. Proc Natl Acad Sci USA 1990; 87:1223-1227L.)

higher frequency or higher magnetic field H_0 and by lowering the sample temperature. There are however other considerations for selecting the best temperature.

If the electron spins present in the sample were "free" in the sense that they interact only with H_0 and hv, the populations of the two Zeeman sublevels would equalize immediately. In fact this is not the case, because relaxation processes force the spin populations to return to their state of thermal equilibrium after being disturbed by the hv photon. One of these processes, called spin-lattice relaxation (T_1), that is the rate at which the spins can deliver the microwave power to the bulk (lattice) of the sample, is related to the microwave power saturation behavior of the sample. Measurements of T_1 or measurements of the saturation behavior may help in identifying the species which contains

the unpaired electron. A theoretical development of this phenomenon is difficult and out of the scope of the present book. Let us simply state two facts. Firstly the spin-lattice relaxation time T_1 increases as the temperature of the sample is lowered and results in a line sharpening. Secondly, at a given temperature, the integrated area (or more simply the amplitude, see below) of the EPR signal of a sample is proportional to the square-root of the microwave power, $P^{1/2}$, if T_1 is short enough. If not, there is a saturation behavior. The normalized curve $A/P^{1/2}$ versus P or logP is a horizontal line in nonsaturating conditions and a downward curve when saturation occurs. The shape of this curve at a given temperature is related to the $T_1.T_2$ product (see below). See instances of such curves in the case of the tyrosyl free radical of the R2 (B2 or M2) subunit of ribonucleotide reductase.[11]

Another relaxation phenomenon is the spin-spin relaxation. The interaction of the microwave photon hv and its associated magnetic field H_1 with the spin system under study, oriented in the static magnetic field $\overline{H}_0$, requires a certain time to reach a new equilibrium state; it is related to the spin-spin relaxation time T_2, usually close to T_1. In a first approximation a simple absorption line has a line-width proportional to $1/T_2$.

EXPERIMENTAL CONDITIONS

MICROWAVE FREQUENCY AND MAGNETIC FIELD

Due to the large difference in the electron magnetic moment as compared to that of a proton (600-fold), the experimental setups for EPR and NMR experiments are quite different. EPR spectrometers usually operate in the X-microwave-band (9.0-9.6 GHz). The free electron resonance condition occurs in magnetic fields H_0 around 3200-3600 G, that is 320-360 mT. There exist EPR spectrometers operating in other microwave-bands, such as S (3.5 GHz),[12] K (23 GHz),[6] Q (35 GHz), etc. (Fig. 3.5).

The absorption lines have Gaussian or Lorenzian line shapes. However due to low-frequency fluctuations of the microwave frequency v, the resonance signal has to be modulated (the same phenomenon occurs for HF or TV frequencies, which requires frequency modulation). In EPR spectroscopy the simplest way is to modulate $\overline{H}_0$ by a parallel magnetic field $\overline{H}_m$ varying at high frequency (100 kHz). The signal detection, synchronous at 100 kHz, yields the derivative of the absorption curve affecting a sinusoidal shape. In practice many EPR spectra are not well resolved. Therefore the description of a spectrum is somewhat factual and empirical. Depending on the signal shape, the g-values are measured at maximal, minimal or crossing points of the derivative. The signal line-width is either defined as the half-maximum width

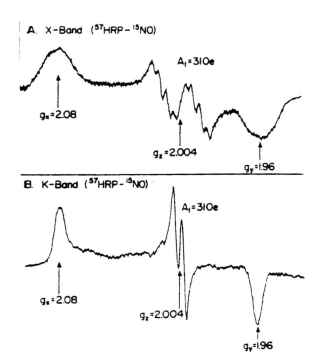

Fig. 3.5. X-band and K-band EPR spectra at 77 K of of [15]NO-ferrous horseradish peroxidase (HRP) containing [57]Fe substituted heme ([57]HRP-[15]NO). The magnetic field is expressed in the unconventional unit Oersted (Oe) instead of Gauss. The comparison shows the same g-values and A-values and the effect of microwave frequency on the apparent line-widths. (Reproduced with permission from: Yonetani T, Yamamoto H, Erman JE et al. Electron properties of hemoproteins. V. Optical and electron paramagnetic characteristics of nitric oxide derivatives of metalloporphyrin-apohemoprotein complexes. J Biol Chem 1972; 247:2447-2455.) Copyright © The American Society for Biochemistry & Molecular Biology, Bethesda, MD, USA.

or the peak-to-peak line-width. In practice the exact determination of the g-values requires simulations of experimental spectra, which suppose a correct estimate of the line-widths.

When the three principal g-values are resolved and different, the complex is said to have a rhombic symmetry; that is the case of most nitrosylated hemoproteins, such as P-450(Fe^{II})-NO or HRP(Fe^{II})-NO (Fig. 3.6).

If only two g-values are measured, often noted $g_{//}$ and $g_{\perp}$, the complex has an axial symmetry, which is the case of the $Fe^I(SR)_2(NO)_2$ complexes (Fig. 3.4) with g = 2.04 and g = 2.015, the directions // and $\perp$ being arbitrary. Hyperfine interactions are sometimes easy to measure when regular features are detected; for instance in Hb(Fe^{II})-NO a hyperfine structure of nine lines (a triplet of a triplet) is detected in the $g_z = 2.005$ region. This hyperfine structure is attributed to the interaction of the electron spin with the nuclear spin of ^{14}N of $NO^{\bullet}$ (A_z = 21 G) on the z-axis perpendicular to the heme plane, and the ^{14}N of the proximal histidine on the other side of that plane (A_z = 7 G)[5-7,9] (Figs. 3.3 and 3.6). Another example of interest for the present chapter is that of the tyrosyl radical of the R2 subunit of RNR, for which the hyperfine structures due to the protons on C-3, C-5 and C-β carbons are nearly fully resolved at 30 K, but only two broad lines are detected at 77 K or at room temperature (Fig. 3.7).[11,13]

TEMPERATURE

As mentioned in the previous section, lowering the temperature increases

Fig. 3.6. Effects of isotopic substitution of ^{14}NO to ^{15}NO and ^{56}Fe to ^{57}Fe on X-band EPR spectra of nitric oxide ferrous horseradish peroxidase (HRP-NO). As in Fig. 3.5., the magnetic field is expressed in kOe instead of kG (1 kG = 0.1 T). (Reproduced with permission from: Yonetani T, Yamamoto H, Erman JE et al. Electron properties of hemoproteins. V. Optical and electron paramagnetic characteristics of nitric oxide derivatives of metalloporphyrin-apo-hemoprotein complexes. J Biol Chem 1972; 247:2447-2455.) Copyright © The American Society for Biochemistry & Molecular Biology, Bethesda, MD, USA.

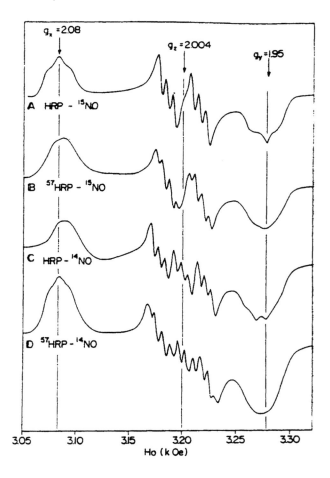

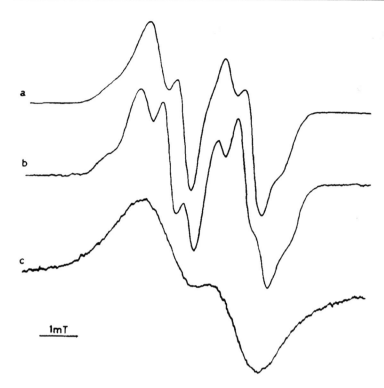

a

b

c

1mT

the signal intensity and decreases the linewidths, thus increasing the overall resolution. But lowering the temperature too much may create microwave saturation problems. So a compromise in the choice of the right temperature is necessary. Usually convenience of the experimental setup and cost are overwhelming over other reasons. Experiments with liquid nitrogen (77 K) are extremely easy and uncostly. The only equipment is a small quartz dewar in the shape of a finger tip containing liquid nitrogen into which a cylindrical (inner diameter = 2-3 mm) EPR tube is freely inserted and removed. This small dewar is easily positioned in the EPR detection cavity at X-band (9-9.6 GHz). This kind of tube (Φ_i = 3 mm) requires a sample volume of 250 μl, larger volumes being unnecessary. A scan over 400 G for instance would usually require only 4

to 16 minutes. These experimental setups are easy to maintain in quasi routine conditions. To give an idea of the time and effort required, results such as those described by Lepoivre et al[13] were derived from approximately 25 independent experiments with over 100 sample tubes. Another research study published by others in a "highly rated" journal resulted from one experiment and four sample tubes. No need to give away the reference!

In many instances the chemical nature of the sample requires one to record spectra at much lower temperatures, 30 K down to 4 K (liquid helium) or even 1.4 K (pumped helium). Not many EPR laboratories have the necessary dewar equipment; each experiment takes more time, and helium is costly.

Due to the bulk of proteins, EPR work at room temperature yields broad unresolved spectra and requires huge protein concentration. However whenever the protein carries a free radical species which has itself some degree of movement, recording EPR spectra at room temperature gives interesting information on these movements (see refs. 11 and 14).

Finally small molecular complexes, due to their more rapid movement or short correlation time, give well resolved spectra at room temperature.[15,16] A good example is that of the ternary complex $Fe^{II}(DETC)_2$-NO (DETC: diethyldithiocarbamate) formed in NO• trapping experiments (see chapter 13).[16] The great advantage of these experiments at room temperature is to test a fluid medium containing NO•, which can be a biological fluid.

As already mentioned, recording EPR spectra at X-band and 77 K requires 3 mm inner diameter quartz tubes containing an optimal volume of 250 µl. Freezing the sample requires some care as an upward meniscus forms in the tube and, as the aqueous volume is increased upon freezing, forms a fragile contact with the tube inner surface. Usually a slow and progressive freezing avoids sample tube breakage. On the contrary sample thawing should be rapid, by immersion in warm water. For experiments at room temperature, quartz "flat" capillary cells are necessary, containing 250 µl (nonthermostated) or 50 µl (thermostated) sample volume. In order to make quantitations at room temperature, a regulated temperature around the sample is necessary.

SATURATION CURVES

A rather slight theoretical indication of its interest has already been given. Measuring the signal amplitude rather than performing a double integration, which gives a number really proportional to the signal intensity, is often a sufficient first-order approximation. Practically the determination of saturation curves over four to five decades (200 mW to 20 µW or less) for different sets of g-values allows to discriminate two different paramagnetic species in the same sample giving overlapping

signals (see refs. 13,17 and 18). In Drapier and colleagues' work it allowed one to distinguish the EPR signals of nitrosylated [Fe-S]-containing proteins from those of nitrosylated ferritin. Such curves discriminated nitrosylated [Fe-S] species from tyrosyl radicals species contained in the same activated tumor cells,[13] the first being saturated above 2 mW, the second being unsaturable (Fig. 3.8.).

As a rule nitrosyl-hemoproteins' EPR signals are not saturable at 77 K. A more precise determination of saturation curves using the double-integrated signal intensity is necessary to have precise determination of T_1 and T_2 and to derive precise structural data.[11]

SAMPLING AND CONCENTRATION

Cultured cell sampling

Due to poor sensitivity, samples containing 30 to 50 x 10^6 cells are necessary to record good, easily quantified, EPR spectra.[13,18] The cultured cell suspensions (around 600 µl) transferred in the EPR tubes are packed into a final volume of 250 µl by spinning the tubes at 500 to 2000 g for 15-20 minutes. The supernatant is removed, and the cells are immediately frozen and stored in liquid nitrogen. Storage at -80°C should be avoided as slow electron transfer could occur as is the case in frozen alcoholic glasses.

Artifacts may happen through the use of inadequate materials to scrap cells in culture dishes. For instance the so-called "rubber policeman" can produce paramagnetic bits contaminating the sample. Any copper-containing material, and contamination by iron, manganese or copper cations should be carefully avoided. Teflon and most plastics are good materials.

Blood sampling

Rigorous conditions have to prevail in blood sampling, as the stability of nitrosylated hemoglobin is very temperature-dependent.[19] At all temperatures the EPR signal decays with first-order kinetics; the half-life is 40 min at 37°C, 3.5 hrs at

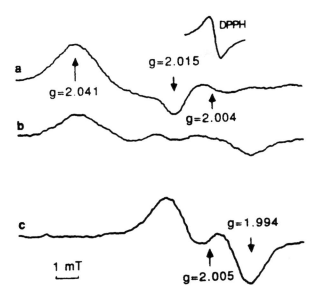

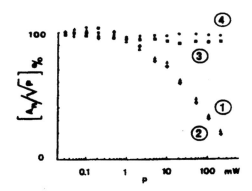

room-temperature (20°C), and 6.6 days at refrigerator temperature (about 4°C). No decay was observed over 30 days at -20°C or lower temperatures. The temperature of blood sampling is therefore of great importance in any attempt at detection and quantification of NO• in humans and animals.

Packing of RBC by low speed centrifugation (1000-2000 g for 5-10 min) before sampling increases the EPR detection sensitivity provided the initial volume is large enough.[20,21] Pooling of the RBC of two or three animals can be necessary.[22] Subtracting the EPR signal of control animals greatly increases the signal-to-noise ratio.[22] Furthermore this separation from plasma allows one to avoid the ceruloplasmin Cu^{2+} (types 1 and 2) EPR signals,[19,23] with a large trough at g = 2.05, which presents individual variations in intensity.[24-27]

DISCUSSION: SPECIFICITY VERSUS SENSITIVITY

EPR spectroscopy is per se not sensitive, as compared to visible-UV spectroscopy for instance. It requires large amounts of biological material, though less than NMR, X-ray or EXAFS.[28] That being given, some compromises in experimental settings are often necessary to record spectra, even at the cost of some signal distortion. Usually the owner of the EPR spectrometer is well aware of these necessary compromises, but often a chemist used to working with mmol/l samples does not understand the agony of a biochemist working at µmol/l levels or worse that of a biologist using nmol/l or pmol/l samples.

Microwave power should be set at the highest value (on the order of 1 to 10 mW for metal-nitrosyl complexes) where saturation is minimal. Performing a rapid test of the power saturation, over several decibel scales, is always necessary in order to avoid wild errors. The EPR signal is proportional to the amplitude of the magnetic field modulation $\overline{H}_m$, as far as its amplitude is smaller than one fifth of the signal line width. When the amplitude is on the order of (or greater than) the width some signal distortion occurs. It is advantageous to increase $\overline{H}_m$ as much as possible, as the noise brought about by $\overline{H}_m$ is much smaller than that brought by the microwave power or the apparatus gain.

The sensitivity of EPR spectroscopy is very dependent on the signal line-width, as the detection is that of an absorption derivative. Nitrosylated complexes of metal cations or metalloproteins have comparable line-widths on the order of 10 G (1 mT). The lower limit of EPR detection of HbNO in frozen state (77 to 90 K) is approximately 1 µmol/l,[29-31] as compared to 1 nmol/l by the spectrophotometric method.[32] By use of agarose-bound hemoglobin as a trap (in the presence of dithionite), a threshold limit of 1 to 10 nmol/l NO or nitrite in plasma was reached.[30,33] The sensitivity problem of EPR spectroscopy shall more fully discussed in chapters 13 and 14.

References

1. Henry YA, Singel DJ. Metal-nitrosyl interactions in nitric oxide biology probed by electron paramagnetic resonance spectroscopy. In: Feelisch M, Stamler J, eds. Methods in Nitric Oxide Research. John Wiley and Sons, 1996:357-372.

2. Singel DJ, Lancaster JR. Electron paramagnetic resonance spectroscopy and nitric oxide biology. In: Feelisch M, Stamler J, eds. Methods in Nitric Oxide Research. John Wiley and Sons. 1996: 341-356.

3. Wilcox DE, Smith RP. Detection and quantification of nitric oxide using electron magnetic resonance spectroscopy. Methods: a companion to methods in enzymology 1995; 7:59-70.

4. Henry Y, Banerjee R. Electron spin resonance spectra of isolated ferrihemoglobin (α, β and γ) chains. An attempted correlation with optical absorption spectra. J Mol Biol 1970; 50:99-110.

5. Kon H, Kataoka N. Electron paramagnetic resonance of nitric oxide-protoheme with some nitrogenous base. Model systems of nitric oxide hemoproteins. Biochemistry 1969; 8:4757-4762.

6. Yonetani T, Yamamoto H, Erman JE et al. Electron properties of hemoproteins. V. Optical and electron paramagnetic characteristics of nitric oxide derivatives of metalloporphyrin-apo-hemoprotein complexes. J Biol Chem 1972; 247:2447-2455.

7. Henry Y, Banerjee R. Electron paramagnetic studies of nitric oxide haemoglobin derivatives: isolated subunits and nitric oxide hybrids. J Mol Biol 1973; 73:469-482.

8. Henry Y, Cassoly R. Chain non-equivalence in nitric oxide binding to hemoglobin. Biochem Biophys Res Commun 1973; 51:659-665.

9. Henry Y, Ducrocq C, Drapier J-C et al. Nitric oxide, a biological effector. Electron paramagnetic resonance detection of nitrosyl-iron-protein complexes in whole cells. Eur Biophys J 1991; 20:1-15.

10. Lancaster JR, Hibbs JB. EPR demonstration of iron-nitrosyl complex formation by cytotoxic activated macrophages. Proc Natl Acad Sci USA 1990; 87:1223-1227.

11. Sahlin M, Petersson L, Gräslund A et al. Magnetic interaction between the tyrosyl free radical and the antiferromagnetically coupled iron center in ribonucleotide reductase. Biochemistry 1987; 26:5541-5548.

12. Komarov A, Mattson D, Jones MM et al. In vivo spin trapping of nitric oxide in mice.

Biochem Biophys Res Commun 1993; 195:1191-1198.

13. Lepoivre M, Flaman J-M, Henry Y. Early loss of the tyrosyl radical in ribonucleotide reductase of adenocarcinoma cells producing nitric oxide. J Biol Chem 1992; 267:22994-23000.

14. Henry Y, Peisach J. Photoreduction of copper chromophores in blue oxidases. J Biol Chem 1978; 253:7751-7756.

15. McDonald CC, Phillips WD, Mower HF. An electron spin resonance of some complexes of iron, nitric oxide, and anionic ligands. J Am Chem Soc 1965; 87:3319-3326.

16. Mordvintcev P, Mülsch A, Busse R et al. On-line detection of nitric oxide formation in liquid aqueous phase by electron paramagnetic resonance spectroscopy. Anal Biochem 1991; 199:142-146.

17. Pellat C, Henry Y, Drapier J-C. IFN-γ-activated macrophages: detection by electron paramagnetic resonance of complexes between L-arginine-derived nitric oxide and non-heme iron proteins. Biochem Biophys Res Commun 1990; 166:119-125.

18. Drapier J-C, Pellat C, Henry Y. Generation of EPR-detectable nitrosyl-iron complexes in tumor target cells cocultured with activated macrophages. J Biol Chem 1991; 266:10162-10167.

19. Cantilena LR, Smith RP, Frazur S et al. Nitric oxide hemoglobin in patients receiving nitroglycerin as detected by elctron paramagnetic resonance spectroscopy. J Lab Clin Med 1992; 120:902-907.

20. Langrehr JM, Müller AR, Bergonia HA et al. Detection of nitric oxide by electron paramagnetic resonance spectroscopy during rejection and graft-versus-host disease after small-bowel transplantation in the rat. Surgery 1992; 112:395-402.

21. Lancaster JR, Langrehr JM, Bergonia HA et al. EPR detection of heme and nonheme iron-containing protein nitrosylation by nitric oxide during rejection of rat heart allograft. J Biol Chem 1992; 267:10994-10998.

22. Wang Q, Jacobs J, DeLeo J et al. Nitric oxide hemoglobin in mice and rats in endotoxic shock. Life Sci 1991; 49:PL55-PL60.

23. Wennmalm Å, Benthin G, Petersson A-S. Dependence of the metabolism of nitric oxide (NO) in healthy human whole blood on the oxygenation of its red cell haemoglobin. Br J Pharmacol 1992; 106:507-508.

24. Dodd NJF. Electron spin resonance study of changes during the development of a mouse myeloid leukaemia. I. Paramagnetic metal ions. Br J Cancer 1975; 32:108-120.

25. Bomba M, Camagna A, Cannistraro S et al. EPR study of serum ceruloplasmin and iron transferrin in myocardial infarction. Physiol Chem Physics 1977; 9:175-180.

26. Pocklington T, Foster MA. Electron spin resonance of caeruloplasmin and iron transferrin in blood of patients with various malignant diseases. Br J Cancer 1977; 36:369-374.

27. Horn RA, Friesen EJ, Stephens RL et al. Electron spin resonance studies of properties of ceruloplasmin and transferrin in blood from normal human subjects and cancer patients. Cancer 1979; 43:2392-2398.

28. Archer S. Measurement of nitric oxide in biological models. FASEB J 1993; 7:349-360.

29. Westenberger U, Thanner S, Ruf HH et al. Formation of free radicals and nitric oxide derivative of hemoglobin in rats during shock syndrome. Free Rad Res Comms 1990; 11:167-178.

30. Wennmalm Å, Lanne B, Petersson A-S. Detection of endothelium-derived factor in human plasma in the basal state and following ischemia using electron paramagnetic resonance spectrometry. Anal Biochem 1990; 187:359-363.

31. Wennmalm Å, Benthin G, Edlund L et al. Metabolism and excretion of nitric oxide in humans. An experimental and clinical study. Circ Res 1993; 73:1121-1127.

32. Kelm M, Feelisch M, Spahr R et al. Quantitative and kinetic characterisation of nitric oxide and EDRF released from cultured endothelial cells. Biochem Biophys Res Commun 1988; 154:236-244.

33. Greenberg SS, Wilcox DE, Rubanyi GM. Endothelium-derived relaxing factor released from canine femoral artery by acetylcholine cannot be identified as free nitric oxide by electron paramagnetic resonance spectroscopy. Circ Res 1990; 67:1446-1452.

EPR CHARACTERIZATION OF NITRIC OXIDE BINDING TO HEMOGLOBIN

Yann A. Henry

INTRODUCTION

Since the early work of Hermann (1865) followed by that of Haurowitz (1924), Keilin and Hartree (1937) and Gibson and Roughton (1957)—to point out only a few landmarks—nitric oxide is like O_2 and CO, a well-known ligand of deoxygenated reduced hemoglobin.[1,2] The complex was first characterized by its red color with large absorbances of the a and b bands of the heme at 574.5 and 536 nm and secondly by a magnetic susceptibility of 3.07 Bohr magnetons,[3-5] and by an electron paramagnetic resonance with a g-value close to 2.0.[6] The first property was used to reveal the cooperative binding of NO to ferrous iron in an analogous manner to that of O_2 and CO;[2] the second property, paramagnetism, made NO a choice probe of the physiological O_2 binding site. This paramagnetic probe was of great importance at a period (late 60s-early 70s) when X-ray crystallographic data were still at limited resolutions (2.8-3.5 Å) and referred only to the quaternary structure of methemoglobin—instead of oxyhemoglobin due to autoxidation under X-ray beams, as compared to that of deoxyhemoglobin. Other spectroscopic methods such as far-infrared or resonance Raman were then pioneer instruments in the hands of very few scientists, while EPR was more readily accessible. X-ray structural analysis of nitric oxide hemoglobin was only performed later which demonstrated a close analogy between HbNO and HbO_2.[7]

Although a historical description of the specific impacts of NO binding to hemoglobin on the knowledge of hemoglobin allosteric equilibria could be of some interest, we shall focus here on the peculiar

Nitric Oxide Research from Chemistry to Biology: EPR Spectroscopy of Nitrosylated Compounds, edited by Yann A. Henry, Annie Guissani and Béatrice Ducastel.

contribution of the studies of the paramagnetism and EPR properties of the HbNO complex, keeping in mind the specific applications of NO detection in human red blood cells described in the following chapters (5, 9, 11 and 12).

The paramagnetic complex Hb(Fe^{II})NO$^\bullet$ was one of the first "organic" derivatives to be probed by the new spectroscopic method, "paramagnetic resonance", ten years after the first experiments in the USSR, UK and USA,[8-11] in the 1944-1946 period which followed the development of radar during World War II.[6] It is easily detectable by use of a conventional (X-band, 9 to 10 GHz) EPR spectrometer at 77 K.[12] Kon was the first to describe an EPR spectrum of properly prepared (i.e. nondenaturated) nitrosyl-hemoglobin at 77 K and to offer some structural explanation of its features.[13] Although not clearly resolved, three g-factors at 2.060, 2.023 and 1.986 indicated a rhombic symmetry (Fig. 4.1).

The superhyperfine (SHF) structures due to N atoms (I = 1) were found to be poorly resolved (physicist's nomenclature refers to "hyperfine" when the nuclear-electron spin coupling occurs with the metal nucleus and to "superhyperfine" when ligand atom nuclei are involved; see chapter 3). Let us recall that the iron atom has six N ligands, four of the porphyrin ring, the N_ε of "proximal" histidine (F8) and that of NO. The splitting of the three lines expected for each of the six N atoms is proportional to the probability distribution of the unpaired electron on the N atoms; by use of ^{15}NO (I = 1/2) as compared to natural ^{14}NO (I = 1) the unpaired electron was found to be little associated with NO nitrogen, but to be delocalized on iron orbitals and on the $2p\pi^*$ orbital of the N_ε atom of the proximal histidine ligand. There is little or no spin distribution over the N atoms of the porphyrin ring pyrroles. This was in accordance with an earlier Mössbauer absorption spectrum interpreta-

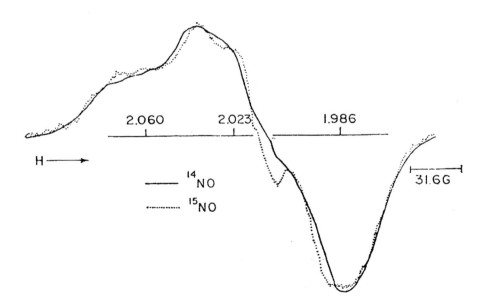

Fig. 4.1. EPR spectra of ^{14}NO (—) and ^{15}NO (⋯) hemoglobin solutions at 77 K. Both spectra were normalized to the same intensity. The numbers on the base line are the g-values (Reproduced with permission from: Kon H. Paramagnetic resonance study of nitric oxide hemoglobin. J Biol Chem 1968; 243:4350-4357.) Copyright © The American Society for Biochemistry & Molecular Biology, Bethesda, MD, USA.

tion which implied firstly a strong covalent bonding and a large spin transfer to the iron and secondly a tilted angle of the axis of the NO molecule with respect to the porphyrin perpendicular.[14] Partial denaturation by sodium dodecyl sulfate or salicylate and dehydration led to large modifications in EPR spectra with a well resolved three line SHF structure due to ^{14}N of the NO group,[13] implying large conformational changes.

Kon's pioneering contribution was substantiated by several fundamental results. In order to provide an explanation for EPR spectra of nitrosyl-hemoglobin[13] and cytochrome c,[15] several model systems of

nitric oxide-protoheme complexed with a nitrogenous base in the iron sixth coordination position (pyridine, quinoline, imidazole derivatives, etc.) were studied.[16-19] Four classes could be empirically defined according to g-values and to superhyperfine A-values of ^{14}N of NO or of the nitrogenous base in the "sixth" coordination position (Fig. 4.2, Table 4.1).

The delocalization of the unpaired electron to the iron d_{z^2}-σ^* orbital was demonstrated and roughly quantified to 50% in two of the model systems by [57]Fe (I = 1/2) substitution (natural [56]Fe has I = 0) experiments. Furthermore the importance of the $d\pi$-$p\pi$ interaction between the

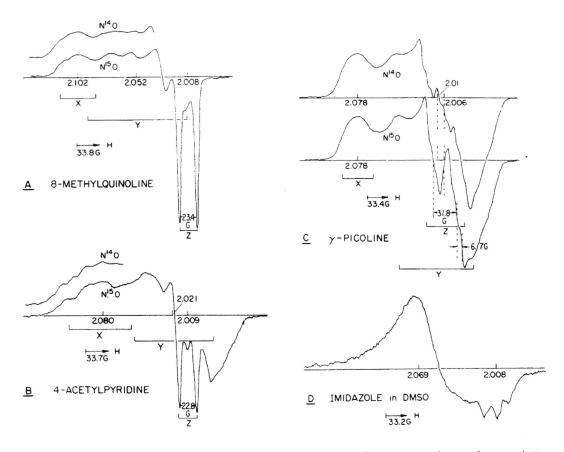

Fig. 4.2. Representative EPR spectra of NO-heme(Fe[II]) complexes with nitrogenous base in frozen solution at 77 K. (A) 8-methylquinoline, (B) 4-acetylpyridine, (C) γ-picoline, (D) Me$_2$SO saturated with imidazole. The g$_y$-values were calculated from the g$_x$, g$_z$ at 77 K and the g$_{soln}$ (measured in solution at room temperature). (Reproduced with permission from: Kon H, Kataoka N. Electron paramagnetic resonance of nitric oxide-protoheme with some nitrogenous base. Model systems of nitric oxide hemoproteins. Biochemistry 1969; 8: 4757-4762.) Copyright © The American Chemical Society, Columbus, OH, USA.

iron d_{xz} or d_{yz} and the π orbitals of the base N atom in determining the electronic structure of the four different classes of complexes was underlined. Although not fully recognized by the authors, the possibility that a pentacoordinated nitrosyl-heme could exist even in the presence of base (imidazole in DMSO, triethylamine, or Type D complexes) was expressed. This kind of complex was later recognized to be of great importance in the hemoglobin allosteric equilibrium and was only correctly interpreted by correlation with infrared spectroscopy (see below).[20]

A more complete description was provided in a beautiful piece of work by Yonetani et al[21] by use of nitrosylated ([14]NO and [15]NO) complexes of Fe^{II}, (natural [56]Fe-heme and [57]Fe-enriched-heme), Fe^{III}, Mn^{II}, Co^{II}-protoporphyrin-IX with several apohemoproteins, horseradish peroxidase, cytochrome c peroxidase, lactoperoxidase, cytochrome c oxidase, catalase, hemoglobin and myoglobin (Figs. 4.3-4.4).[21]

Mn^{II}, Fe^{III} and Co^{II}-nitrosylated heme complexes were EPR silent, through electron transfer from nitric oxide to the metal.

All the Fe^{III}-hemoprotein NO complexes appeared to be reversible, but while those of hemoglobin and myoglobin were reduced by NO, those of ferric peroxidases were not reduced under comparable conditions, in relation to their oxidation-reduction potentials. Furthermore Mn^{III} and Co^{III}-hemes did not react with NO. Finally a complete interpretation of the SHF structure of nitrosylated Fe^{II}-heme complexes was provided (Figs. 4.5-4.6).[21]

All these results provided a sound interpretative basis of hemoglobin and more generally of hemoprotein EPR spectra, which allowed the development of studies on the structure to function relationships within the hemoglobin tetramer or in heme-containing oxidases and oxygenases (chapters 5 and 6).

NITROSYL HEMOGLOBIN: KINETIC AND EPR DATA

Nitric Oxide Binding to Ferrous Hemoglobin

Nitric oxide is certainly a potential poison as it binds with an enormous

---

Table 4.1. Electron paramagnetic resonance parameters of NO-heme-nitrogenous base complexes.

Type	Base Molecule	g_{soln}	g_x	$g_y{}^a$	g_z	$A_N{}^b$ (G)
A	Quinoline	2.053	2.100	(2.051)	2.008	16.0
	8-Methylquinoline	2.054	2.102	(2.052)	2.009	16.7
B	3-Acetylpyridine	2.036	2.081	(2.017)	2.009	16.5
	4-Acetylpyridine	2.037	2.080	(2.021)	2.009	16.4
C	Pyridine	2.032	2.082	(2.007)	2.006	21.4
	γ-Picoline	2.030	2.080	(2.007)	2.005	23.5
D	Imidazole^c	2.049	(2.069)		2.008	16.6
	Triethylamine^d	2.060				

a: calculated from g_{soln}, g_x and g_z
b: SHF splitting of base ^{14}N observed
c: in DMSO
d: in H_2O
Reproduced with permission from Kon H, Kataoka N. Electron paramagnetic resonance of nitric oxide-photoheme complexes with some nitrogenous base. Model systems of nitric oxide hemoproteins. Biochemistry 1969; 8:4757-4762. © The American Chemical Society, Columbus, Ohio, U.S.A.

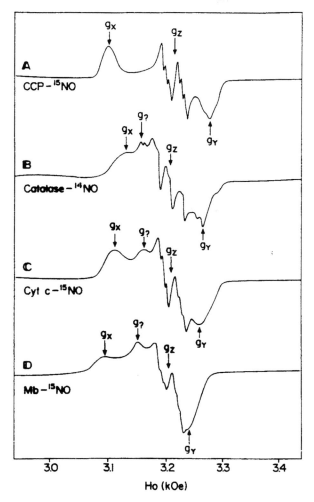

Fig. 4.3. X-band EPR spectra at 77 K of nitric oxide ferrous hemoproteins. (A) ^{15}NO-ferrous cytochrome c peroxidase (CCP-^{15}NO), (B) ^{14}NO-ferrous catalase (Catalase-^{14}NO), (C) ^{15}NO-ferrous cytochrome c (Cyt c-^{15}NO), (D) ^{15}NO-ferrous myoglobin (Mb-^{15}NO). (Reproduced with permission from: Yonetani T, Yamamoto H, Erman JE et al. Electron properties of hemoproteins. V. Optical and electron paramagnetic resonance characteristics of nitric oxide derivatives of metalloporphyrin-apohemoprotein complexes. J Biol Chem 1972; 247:2447-2455.) Copyright © The American Society for Biochemistry & Molecular Biology, Bethesda, MD, USA.

affinity to hemoglobin. This was established before and during the Second World War by FJW Roughton and QH Gibson following JS Haldane's laws (1895-1897),[22] using the Hartridge reversion spectroscope, Gibson's newly invented "stopped-flow" apparatus and the blood from a lamb kept on Cambridge (UK) lawns. The apparent binding constants at mean saturation of the hemoglobin tetramer are 5×10^4 mol^{-1} l for O_2, 2×10^7 mol^{-1} l for CO and 3×10^{10} mol^{-1} l for NO. While the "mean" binding rates are similar within one or two orders of magnitude (5×10^6 mol^{-1} l s^{-1} for O_2, 2×10^5 mol^{-1} l s^{-1} for CO and 2×10^7 mol^{-1} l s^{-1} for NO), the "mean" ligand dissociation rates are widely different (35

s^{-1} for O_2, 2×10^{-2} s^{-1} for CO and 5×10^{-5} s^{-1} for NO). This can conveniently be expressed as follows: the "mean" residence time ($\tau_{1/2}$) of O_2 on hemoglobin is 20 ms, that of CO is 35 s and that of NO is ~4 hours.[2, 22-26]

This large residence time should be a good reason for NO to be poisonous, though probably it is not, as inhaled NO is perhaps more or less irreversibly reacted with other targets before it could reach deoxygenated hemoglobin sites (only ~1/4 at venous blood pressure). It all depends on NO modes of delivery in inhalation as exemplified by its use as a nonsystemic and specific lung capillary vasodilator. It explains also in retrospect the danger of

Fig. 4.4. Effects of substitution of nitric oxide and iron with ^{15}NO and ^{57}Fe on X-band EPR spectra of nitric oxide ferrous horseradish peroxidase (HRP-NO) at 77 K. (Reproduced with permission from: Yonetani T, Yamamoto H, Erman JE et al. Electron properties of hemoproteins. V. Optical and electron paramagnetic resonance characteristics of nitric oxide derivatives of metalloporphyrin-apohemoprotein complexes. J Biol Chem 1972; 247:2447-2455.) Copyright © The American Society for Biochemistry & Molecular Biology, Bethesda, MD, USA.

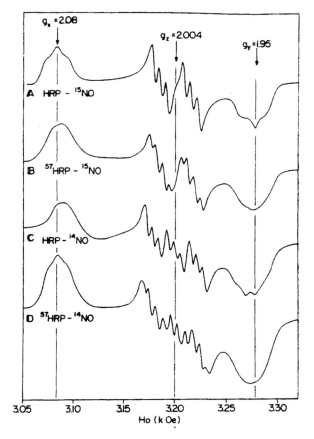

Fig. 4.5. Interactions of NO with (A) ferric, (B) manganous, (C) cobaltous and (D) ferrous protoporphyrin IX-containing hemoproteins. (Reproduced with permission from: Yonetani T, Yamamoto H, Erman JE et al. Electron properties of hemoproteins. V. Optical and electron paramagnetic resonance characteristics of nitric oxide derivatives of metalloporphyrin-apo-hemoprotein complexes. J Biol Chem 1972; 247:2447-2455.) Copyright © The American Society for Biochemistry & Molecular Biology, Bethesda, MD, USA.

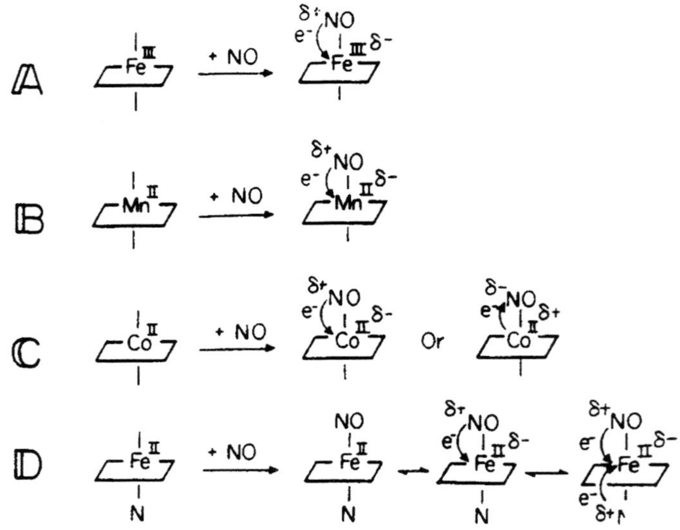

having cell-free circulating hemoglobin in plasma under pathophysiological conditions associated with hemolysis; in this case, it is hemoglobin which is poisonous towards the NO function as an endothelial-derived relaxing factor (EDRF).[27,28] Finally it explains the failure of "hemoglobin experts" for so many years in trying to use hemoglobin as a blood substitute in human trauma treatments (ressucitation) following hemorraghic shock, for instance in intensive care units, as human recipients had unexpected and unwarranted reactions such as severe systemic and pulmonary hypertension or endothelial oxidative stress and cytotoxicity which offsets the potential advantage of the increased blood oxygen content.[29-32]

Another interesting experimental point to note is a large effect of temperature on NO dissociation from NO saturated hemoglobin (Q_{10} = 5.5),[2] which has been recently confirmed on human blood by Cantilena et al.[33] If the mean half-dissociation rate of NO is 212 min (3.5 hr) at 20°C, it is 40 min at 37°C and 6.6 days at 4°C. This is an important parameter in blood collecting for nitrosyl hemoglobin quantitation in animal or human pathological studies (see chapter 11). It could well have some incidence in fever conditions.

By taking advantage of the very slow dissociation of NO from nitrosylated isolated α and β subunits (4 x 10^{-5} s^{-1} at 20°C) and by performing the mixing

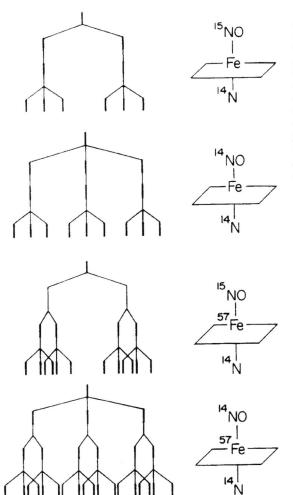

Fig. 4.6. Interpretation of hyperfine and superhyperfine structure in the z-EPR absorption of nitric oxide ferrous peroxidases. ^{15}N and ^{57}Fe give a doublet splitting each, whereas ^{14}N induces a triplet splitting. (Reproduced with permission from: Yonetani T, Yamamoto H, Erman JE et al. Electron properties of hemoproteins. V. Optical and electron paramagnetic resonance characteristics of nitric oxide derivatives of metalloporphyrin-apo-hemoprotein complexes. J Biol Chem 1972; 247:2447-2455.) Copyright © The American Society for Biochemistry & Molecular Biology, Bethesda, MD, USA.

experiments at 5°C, hybrid molecules of the types $(\alpha NO\text{-}\beta_{deoxy})_2$ and $(\alpha_{deoxy}\text{-}\beta NO)_2$ were prepared and studied for their CO binding kinetic properties.[23] Both systems were found to behave similarly to fully deoxygenated hemoglobin. The same property was later used to prepare and study other nitrosylated hemoglobin hybrids in order to study the allosteric equilibrium (see below).[34-38]

ALLOSTERIC EQUILIBRIUM, SUBUNIT INEQUIVALENCE IN NITRIC OXIDE BINDING

The notion of the slow NO dissociation has to be refined depending on the R or T structure of hemoglobin, that is on the degree of saturation of the hemoglobin tetramer and the presence of allosteric effectors. The four association rate constants (j'_1 to j'_4) of NO to deoxyhemoglobin were found to differ by only a factor of four (2.6 x 10^7 to 1.02 x 10^8 mol^{-1} l s^{-1}).[39] On the contrary all the cooperativity in the NO + Hb equilibrium appears to be related to its dissociation rates differing by 100-fold.[40] In the T state—also called in the hemoglobin literature "unliganded", "deoxy-like" or "low-affinity"—the NO dissociation rate constant was measured as $j_T = 10^{-3}$ s^{-1} ($\tau_{1/2}$ = 12 min).[25,40] This represents the dissociation of NO from tetrameric hemoglobin molecules such as $\alpha NO\alpha_{deoxy}\beta NO\beta_{deoxy}$ or $\alpha NO\alpha_{deoxy}\beta_{deoxy}\beta_{deoxy}$ or from fully saturated nitrosyl hemoglobin in the presence of allosteric effectors 2,3-diphosphoglycerate (2,3-DPG) normally found in red blood cells or its artificial analogue inositolhexaphosphate (IHP). Conversely in the R state ("liganded", "oxy-like" or "high-affinity"), the NO dissociation rate constant is two orders of magnitude smaller (j_R = 10^{-5} s^{-1}, that is $\tau_{1/2}$ = 20 h). This would apply to asymmetrical hybrid hemoglobin molecules such as $\alpha NO\alpha NO\beta_{deoxy}\beta_{deoxy}$, $\alpha NO\alpha NO\beta O_2\beta O_2$ or to fully nitrosylated hemoglobin $\alpha NO\alpha NO\beta NO\beta NO$ in the absence of any allosteric effector.[25,40] Thus these rate constant data predicted

that nitrosylated hemoglobin hybrids could exist as rather stable components and should have different behavior in venous and in arterial blood (see below).

At the same period EPR spectroscopy provided interesting evidence which led to refinements in the allosteric model. EPR spectra of nitrosyl hemoglobin single crystals at -195°C revealed four nonequivalent paramagnetic centers which fell into two sets, corresponding respectively to α and β subunits within the Hb tetramer, though having the same principal g-values: g_{xx} = 2.0820, g_{yy} = 2.0254 and g_{zz} = 1.9909, and the same principal values for hyperfine A-tensor: 77 MHz, 54 MHz and 18 MHz (See chapter 3 for the equivalence in gauss or in cm^{-1}).[41] These results confirmed a bent Fe-N-O bond with an angle of 110° ± 2.5°. However EPR studies at room temperature and in frozen solutions (ca. -100°C) of isolated hemoglobin subunits showed that α and β chains' spectra are dissimilar and that their contribution in HbNO spectrum is simply arithmetic.[42] This was later confirmed at 77 K (Fig. 4.7).[34] The symmetry of αNO is rhombic with a well resolved nine-line SHF structure, that of βNO being more axial with a poorly resolved hyperfine structure. The spectrum of nitrosyl hemoglobin is the sum of those of its two types of subunits.

In a second step we demonstrated by two independent techniques of quick freeze and EPR on the one hand, and stopped-flow kinetics on the other hand,[43] a preferential binding of NO to α over β subunits of deoxyhemoglobin at partial saturation of the heme sites (Fig. 4.8).

The ratio of the association rate constants ($j'_1\alpha/j'_1\beta$) of the first NO molecule to deoxyhemoglobin was found to be 4- to 10-fold in favor of binding to α subunits within the tetramer. A similar result obtained on fully ligated nitrosyl hemoglobin showed that the first NO ligand dissociated faster ($j_4\beta/j_4\alpha$ = 2 to 3) from β subunits than from α ones.[43] The structural and functional inequivalence of α and β subunits within the hemoglobin tetramer

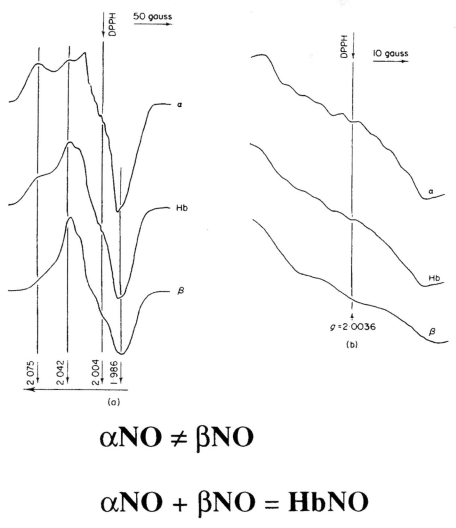

$$\alpha NO \neq \beta NO$$

$$\alpha NO + \beta NO = HbNO$$

Fig. 4.7. Frozen solution EPR spectra at 77 K of HbNO and its isolated α and β subunits. The spectra were normalized to the same integrated intensity. (b) same (a) with 5-fold expanded field scale around g = 2.00. DPPH, 1,1-diphenyl-2-picrylhydrazyl stable radical. (Reproduced with permission from: Henry Y & Banerjee R. Electron paramagnetic studies of nitric oxide haemoglobin derivatives: isolated subunits and nitric oxide hybrids. J Mol Biol, 1973; 73:469-482.) © Academic Press, Ltd, London, UK.

with respect to the binding of NO, CO or *n*-butylisocyanide was confirmed by many others,[25,35-37,39,44-50] sometimes with contradictory results.[51]

Complementary information on subunits' inequivalence and R–T allosteric shift in nitrosyl hemoglobin was gathered in photodissociation and NO geminate recombination experiments together with experiments on nitrosylmyoglobin and model complexes.[38,52-60]

As a result of these kinetic data (independently of the existence of the allosteric equilibrium displacement mentioned above), *hybrid species* such as $\alpha NO\alpha O_2\beta O_2\beta O_2$, $\alpha NO\alpha_{deoxy}\beta_{deoxy}\beta_{deoxy}$ or $\alpha NO\alpha NO\beta_{deoxy}\beta_{deoxy}$ were expected to occur whenever NO quantity was limiting, rather than statistically distributed NO on either subunits resulting in mixtures of $\alpha NO\alpha NO\beta NO\beta NO$, $\alpha NO\alpha_{deoxy}\beta NO\beta_{deoxy}$, etc. (see below).

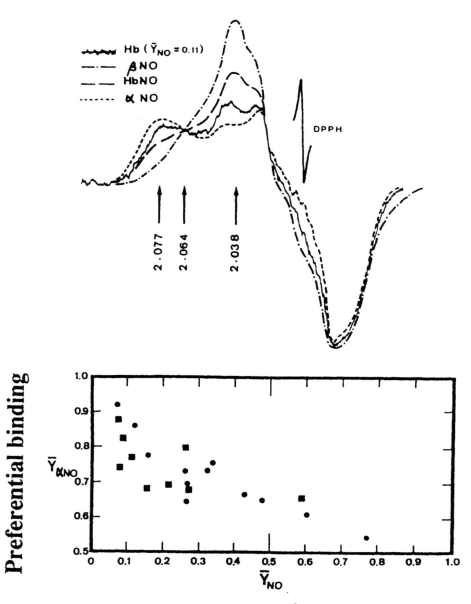

Partial saturation of Hb

Fig. 4.8. Upper panel. EPR spectra at 77 K of hemoglobin partially ligated with NO (Y = 0.11) as compared to the spectra of aNO, bNO and HbNO computed for the same nitrosyl-heme concentration and apparatus settings. Lower panel. Preferential binding of NO to a subunit of hemoglobin. $\bar{Y} \alpha_{NO}$ is plotted as a function of the partial saturation $\bar{Y}_{NO}$ of hemoglobin. Filled squares refer to EPR experiments and filled circles to independent stopped-flow experiments (Reproduced with permission from: Henry Y, Cassoly R. Chain non-equivalence in nitric oxide binding to hemoglobin. Biochem Biophys Res Commun 1973; 51:659-665) Copyright © Academic Press, Inc, Orlando, Florida, USA.

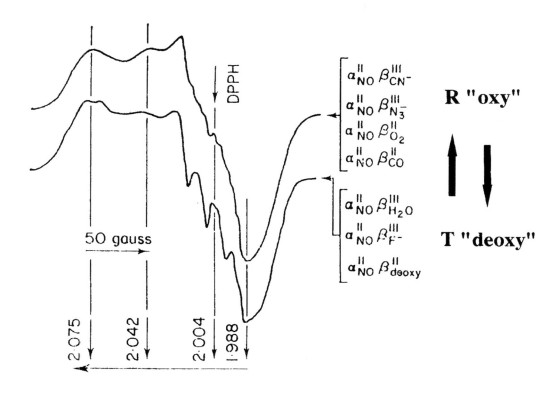

Fig. 4.9. EPR spectra at 77 K of various derivatives of αNOβ hybrid recorded at the same heme concentrations and the same apparatus settings. (Reproduced with permission from: Henry Y, Banerjee R. Electron paramagnetic studies of nitric oxide haemoglobin derivatives: isolated subunits and nitric oxide hybrids. J Mol Biol 1973; 73:469-482.) Copyright © Academic Press, Ltd, London, UK.

EPR Spectroscopy as a Probe of the Allosteric Equilibrium

In the same line, a detailed study of artificial αNO hemoglobin hybrids carrying the β subunits' partner within the hemoglobin tetramer, in the deoxy state or in ferric states, demonstrated that the contribution of αNO subunits in the EPR spectra was different whether the β hemes were deoxygenated or oxygenated, or more generally whether hemoglobin was in the R "oxy-like" or the T "deoxy-like" structure (Fig. 4.9).[34] The R to T transition was mostly characterized by a change of the hyperfine structure from a nine-line pattern to a three-line pattern.

Similar results had been independently obtained by Trittelvitz et al[61] and by Rein

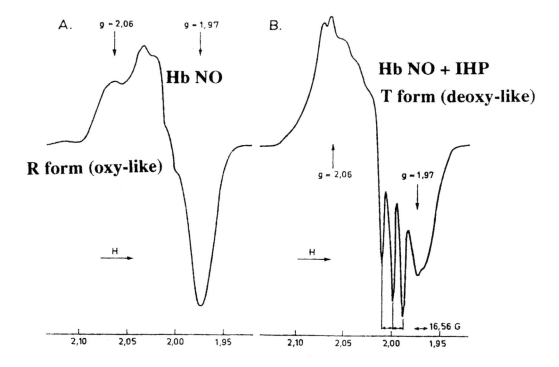

Fig. 4.10. EPR spectra of HbNO in 0.2 M Tris buffer, pH 6.9 in the absence (A) and in the presence (B) of 0.5 mM IHP, pH 6.9. (Reproduced with permission from: Rein H, Ristau O, Scheler W. On the influence of allosteric effectors on the electron paramagnetic spectrum of nitric oxide hemoglobin. FEBS Lett 1972; 24:29-26) Copyright © Elsevier Science, Sara Burgerhartstraat 25, 1055 KV Amsterdam, The Netherlands.

et al[62] which shifted the R to T allosteric equilibrium by the addition of 2,3-DPG or IHP (Fig. 4.10). NO as a probe of the R to T shift can be used in many instances, such as that of the effects of bezafibrate (BZF) and clofibric acid (CFA), two drugs displaying antigelling and antihyperlipoproteinemia activities.[63,64]

In fact the shift of the pH dependence of the SHF structure upon 2,3-DPG or IHP binding reflected not only the R to T shift but also an ionization of the αNO subunit (Fig. 4.11).[34,49,61,62]

The final clue to this hyperfine structure modification was brought by Maxwell and Caughey making use of both EPR and infrared spectroscopy. (Fig. 4.12).[20]

It confirmed the bent-end-on Fe-N-O bonding with an appreciable double bond (covalent) character of the Fe-N bond, with

net donation of electron density from Fe[II] to NO. They suggested also that the bond between iron and the proximal His of the R form (nine-line SHF structure) was cleaved in the T form (three-line SHF structure), leaving a *pentacoordinated nitrosyl ferrous heme* in two subunits of hemoglobin tetramer (Fig. 4.13).[20]

This was confirmed and correctly explained by infrared, resonance Raman and EPR spectroscopic studies of hemoglobin as compared to model complexes,[18,65-67] and by theoretical investigations of EPR hyperfine interaction.[68,69] The pentacoordinated nitrosyl ferrous heme responsible for the three-line SHF structure is exclusively confined to the two α subunits of the tetramer, the β hemes remaining hexacoordinated independently of the quaternary structure shift.[34,37,48] This was confirmed in

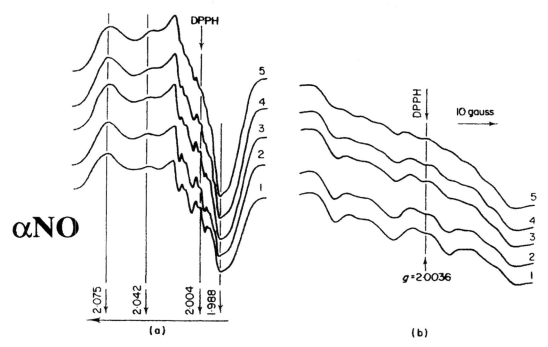

Fig. 4.11. pH dependence of the EPR spectrum at 77 K of isolated αNO hemoglobin subunit (1: pH = 5.95; 2: 6.20; 3: 6.40; 4: 6.85; 5: 7.20) (b) same as (a) with a 5-fold expanded field scale around g=2.00. (Reproduced with permission from: Henry Y, Banerjee R. Electron paramagnetic studies of nitric oxide haemoglobin derivatives: isolated subunits and nitric oxide hybrids. J Mol Biol 1973; 73:469-482.) Copyright © Academic Press, Ltd, London, UK.

abnormal mutant hemoglobin studies (see below). Nitric oxide accepts electron density from Fe^{II} in bent-end-on Fe^{2+}-^{14}N-^{16}O (v_{NO} = 1616.5 cm^{-1}) in Hb(Fe^{II})-NO.[67] On the contrary NO donates electron density to Fe^{III} in linear Fe^{3+}-^{14}N-^{16}O (v_{NO} = 1925 cm^{-1}) in Hb(Fe^{III})-NO.

These findings allowed one to interpret correctly the occurrence of NO in a pentacoordinate nitrosyl heme complex in cooked nitrite cured meat![70] Incidentally a similar pentacoordinated nitrosyl heme structure has been assumed to occur during the activation of guanylate cyclase,[26,71-77] and has been recently detected by EPR (See chapter 10).[78]

Another instance of use of NO as an EPR probe of hemoglobin structure is the investigation of its interaction with haptoglobin (Hp). Haptoglobin is a plasma α_2-globulin which combines with hemoglobin in an almost irreversible manner to form a 1:1 complex. The Hp-Hb complex has an

increased oxygen affinity and an increased peroxidatic activity. The Hp-HbNO complex is maintained in the R state and is markedly protected from denaturation.[79,80] This is also true of the αNOβ$_{deoxy}$-Hp hybrid complex.[81]

As mentioned above, hybrids of nitrosylated hemoglobin in R or T structure are expected to occur and have distinct EPR spectra. Such molecules have effectively been detected in RBC of animal models under pathological conditions as shall be detailed in chapter 11.[82-86]

The earliest example was given by Maruyama et al in mammal ascites tumors also containing RBC (Fig. 4.14).[87] The top spectrum is that of hemoglobin in the T structure. When oxygen is added in order to saturate the hemoglobin tetramer it shifts to the R "oxy-like" structure, NO being bound only to the α subunits.

A complete interpretation of the EPR spectral transition along the arteriovenous

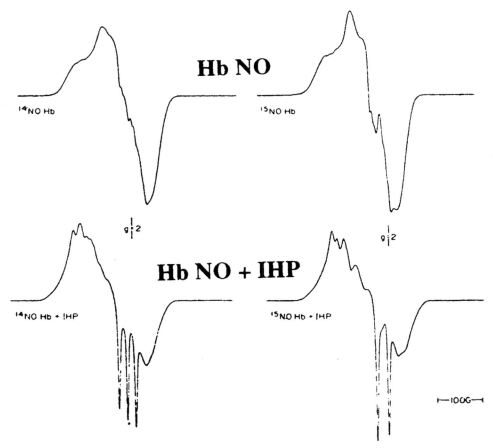

Fig. 4.12. EPR spectra at 77 K of ^{14}NO and ^{15}NO hemoglobin complexes without and with IHP (2 mol per Hb tetramer). (Reproduced with permission from: Maxwell JC, Caughey W. An infrared study of NO bonding to heme B and hemoglobin A. Evidence for inositol hexaphosphate induced cleavage of proximal histidine to iron bonds. Biochemistry 1976; 15:388-396.) Copyright © The American Chemical Society, Columbus, OH, USA.

cycle of nitric oxide hemoglobin in the blood of cytokine-treated rats has been correctly proposed recently.[88] It includes both the displacement of the R–T allosteric equilibrium and the subunits' kinetic inequivalence. Thus academic in vitro results obtained more than twenty years ago have eventually received interesting in vivo applications in pathophysiological cases resulting in overproduction of nitric oxide (chapter 11) or in the use of palliative medication (chapter 12).

ABNORMAL HUMAN HEMOGLOBINS

A number of abnormal mutant hemoglobins have been probed using EPR spec-troscopy of their nitrosyl ferrous complexes and compared to normal adult (HbA) and fetal (HbF) hemoglobins.[89]

Some of these mutant hemoglobins, called HbM, result from point mutations in the heme pocket of the proximal histidine (M-Iwate His(F8)87α → Tyr) or of the distal histidine (M-Zürich His(E7)63β → Arg, M-Saskatoon His(E7)63β → Tyr, M-Boston His(E7)58α → Tyr), or close by on the distal side (M-Milwaukee Val(E11)67β → Glu).[90-93] These hemoglobins are naturally occurring valency hybrids in which the iron in the abnormal subunit has become oxidized as a result of the amino acid replacement in the heme pocket. Thus the

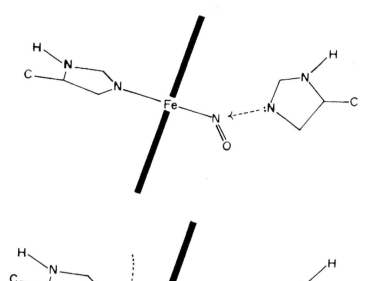

Fig. 4.13. Structures of HbNO in the R state and T state as derived from EPR and infrared spectroscopy. (Reproduced with permission from: Maxwell JC, Caughey W. An infrared study of NO bonding to heme B and hemoglobin A. Evidence for inositol hexaphosphate induced cleavage of proximal histidine to iron bonds. Biochemistry 1976; 15:388-396.) Copyright © The American Chemical Society, Columbus, OH, USA.

oxidized subunit is no longer able to bind O_2 and CO. HbMs have altered oxygen affinity (as compared to HbA) and no heme-heme interaction. When reduced by NO itself or by dithionite, both the normal and the abnormal subunits bind NO. The EPR spectroscopy results were all interpreted by alterations of the R–T allosteric equilibrium; for instance HbM-Saskatoon is frozen in the R state,[93] while on the contrary HbM-Boston and HbM-Iwate are shifted totally to the T state independently of the presence of organic phosphates 2,3-DPG or IHP.[90-92]

Substitutions of residues in contact at the α1-β2 interface were probed by the same method. Hb-Wood (His(FG4)97β → Leu) presents a high affinity for oxygen even when deoxygenated. EPR spectroscopy of Hb-Wood-NO shows that it remains in the R state even in the presence of IHP or at low pH.[94] Another substitution at the

α1-β2 interface far removed from the heme in Hb-Kansas (Asn(G4)102β → Thr) induces on the contrary a low O_2 affinity and a decreased cooperativity. EPR spectroscopy of Hb-Kansas-NO demonstrated that it stays in the T state, even in the absence of organic phophates.[41,92,95]

Thus NO appears an excellent probe of the R–T allosteric equilibrium and its shifts induced by natural mutations. The probe is sensitive, and interpretations of EPR results appear to be reliable, as the correlations with the O_2 affinity and heme-heme interaction properties are reciprocal.

NITROSYL-FE$^{(II)}$ HEME FINE-STRUCTURE, RECENT DEVELOPMENTS

Hemoglobin has been considered a choice macromolecule to test theories such as allostery (see refs. 96-98) and to test new techniques such as subpicosecond laser photolysis.[53,54,57-60] Biophysicists feel safe to use

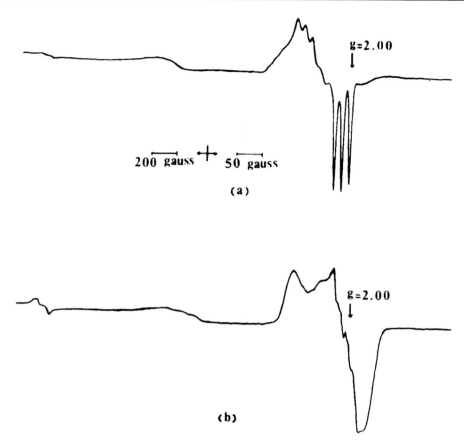

Fig. 4.14. EPR spectra at 77 K of ascites tumors. (a) whole ascites hepatoma (AH 173), (b) after bubbling oxygen gas to whole ascite hepatoma. The low-field part of the spectra shows the near absence of methemoglobin at g = 6. (Reproduced with permission from: Maruyama T, Kataoka N, Nagase S et al. Identification of three-line electron spin resonance signal and its relationship to ascites tumors. Cancer Res 1971; 31:179-184.) Copyright © The American Association for Cancer Research, Inc, Philadelphia, PA, USA.

it, since it is cheap, stable on a laboratory bench, and one can easily argue and imagine mechanisms based on the description of the X-ray crystal structures at high resolution of deoxyhemoglobin,[99] HbO_2,[100] $HbCO$[101] and $HbNO$.[7]

The paramagnetic properties of HbNO have attracted more attention than that of EPR-silent deoxyhemoglobin (S = 2) or of high spin methemoglobin (S = 5/2) because of the well resolved EPR spectra and the richness imparted by the SHF structure due to N ligands or to experimentally heme-incorporated ^{57}Fe. The main technical approaches were: the use of low temperature

EPR down to liquid helium (4.2 K),[48,102-105] use of single crystals,[102,106] use of ^1H and ^{14}N ENDOR,[107] and use of photolysis combined to EPR.[48,103,104]

Low temperature photodissociation, EPR studies, ^1H and ^{14}N ENDOR spectroscopic studies confirmed with many details the subunit inequivalence and the influence of the quaternary structure on the coordination and the fine structures of the α and β subunits within nitrosyl hemoglobin and its hybrids.[48,107] Other low temperature experiments indicated phonon-induced changes in the orientation of NO with respect to the heme plane, explaining the change of the Fe-N-O bond angle from 145° detected in X-ray diffraction at room temperature to 110° estimated at 77 K, and the large modifications of EPR spectral line-shapes of both α and β subunits with temperature.[48,103-105] A quantum mechanics study of the electronic structure of nitrosyl heme complexes demonstrated the existence of two distinct, low-energy electronic states corresponding to the formal assignment of the unpaired electron to either NO or iron atomic orbitals.[69] These two states are nearly degenerate at axial ligand distances experimentally observed. It was further shown that both the extent of the degeneracy and the relative energy order of the two electronic states are sensitive to minor changes in the axial ligand bond distances and angles and can be thermally or electronically mixed.[69,108,109] This theoretical study provided a coherent interpretation of the various EPR spectra of nitrosyl hemoglobin observed in various experimental conditions.

REACTIONS OF NITRIC OXIDE WITH FERRIHEMOGLOBIN AND OXYHEMOGLOBIN

REACTION OF NITRIC OXIDE WITH FERRIC HEMOGLOBIN AND HEMOPROTEINS

Some other details of the multiple reactions of NO with hemoglobin in its various forms (oxy, ferric) as compared to that with deoxyhemoglobin are given in the following paragraphs, in order to warn against possible artifacts in a qualitative or quantitative use of hemoglobin as a NO trap[110,111] (Table 4.2).

Contrary to gaseous ligands, O_2 and CO, binding only to pentacoordinate deoxyhemoglobin; and anions such as F^- or N_3^- binding only to methemoglobin ($Hb(Fe^{III})$ or MetHb) (replacing a weak field distal water molecule); NO is the only ligand (with the exception of CN^-) of hemoproteins able to react both with ferrous and ferric forms. NO binds slowly ($k_{on} = 10^3$-10^4 mol^{-1} l s^{-1}) to MetHb, forming a reversible ($k_{off} = 1$ s^{-1}) complex $Hb(Fe^{III})$-NO with absorption bands at 568 and 531 nm as compared to 574.5 and 536 nm for the ferrous complex $Hb(Fe^{II})$-NO.[1,112] The complex ground state is, depending on the ligand field

Table 4.2. Summary of the equilibrium and rate constants of hemoglobin reactions with O_2, NO and NO_2^-

	Reactions		$K_{1/2}$ (mol^{-1})	k_{on} (mol^{-1}ls^{-1})	k_{off} (s^{-1})	$\tau_{1/2off}$
1	$Hb(Fe^{II}) + O_2 \Leftrightarrow Hb(Fe^{II})O_2$		$5\ 10^4$	$5\ 10^6$	35	20 ms
2	$Hb(Fe^{II}) + NO \Leftrightarrow Hb(Fe^{II})NO$	mean	$3\ 10^{10}$	$2\ 10^7$	$5\ 10^{-5}$	4 hr
2'		R-state	–	$1\ 10^8$	10^{-5}	20 hr
2"		T-state	–	$2.6\ 10^7$	10^{-3}	12 min
3	$Hb(Fe^{II})O_2 + NO \rightarrow Hb(Fe^{III}) + NO_3^-$		–	$3.7\ 10^7$	–	–
4	$Hb(Fe^{III}) + NO \Leftrightarrow Hb(Fe^{III})NO$	mean	$4\ 10^3$	$4\ 10^3$	1	0.7s
5	$Hb(Fe^{III})NO + NO \rightarrow Hb(Fe^{II})NO$		–	–	10^{-3}	12 min
6	$Hb(Fe^{II})O_2 + NO_2^- \Leftrightarrow Hb(Fe^{III}) + O_2^{2-} + NO_2^{\cdot}$		slow with a			
6'	$Hb(Fe^{II})O_2 + NO_2^- \Leftrightarrow Hb(Fe^{III}) + O_2 + NO_3^-$		lag phase			
7	$Hb(Fe^{III}) + NO_2^- \Leftrightarrow Hb(Fe^{III})NO_2^-$		30-100	2-700	n.d.	n.d.

n.d. not determined

strength imparted to heme(Fe^{III}) by NO, either diamagnetic (strong field) or spin-even and EPR-silent S = 2 (weak field).[112,113] The important difference between Hb(Fe^{III})-NO and Hb(Fe^{II})-NO is reflected in the above mentioned difference in the N-O frequency in the infrared range, 1925 cm^{-1} and 1616.5 cm^{-1} respectively. It corresponds to a linear versus a bent Fe-N-O angle and conversely to an electron density donation to the iron atom from NO in Hb(Fe^{III})-NO and an electron density capture by NO in Hb(Fe^{II})-NO.[67]

In fact, hemoglobin subunits are also inequivalent in NO binding kinetics: NO combines to and dissociates from MetHb β subunits (k_{on} = 6.4 x 10^3 mol^{-1} l s^{-1} and k_{off} = 1.5 s^{-1}) 3-4 times faster than with α subunits (k_{on} = 1.71 x 10^3 mol^{-1} l s^{-1} and k_{off} = 0.65 s^{-1} respectively).[114] The presence of the so-called distal histidine (E7) is of prime importance to modulate the NO-combination rate, probably through the interaction between the distal histidine and the ferric iron.[114,115] In the absence of the distal histidine, the rates are much faster and similar to those observed for the pentacoordinate heme in microperoxidase (k_{on} = 1.1 x 10^6 mol^{-1} l s^{-1} and k_{off} = 3.4 s^{-1}).

Another important feature of nitrosyl ferrihemoglobin is its slow autoreduction, in the absence of oxygen into Hb(Fe^{II})-NO, already noticed by Keilin and Hartree and by Chien.[1,112] The reaction kinetics are first order in Hb(Fe^{III})-NO (k' = 10^{-3} s^{-1}; $\tau_{1/2}$ = 12 min).[116] Other ferrihemoproteins forming stable adducts with NO are not further reduced by it, such as P-450,[117] cytochrome c oxidase,[118] cytochrome c peroxidase, catalase, HRP[18] or cytochrome c.[113]

Similar phenomena in NO binding to ferric heme and ferric heme autoreduction could be extremely important to impart hemoproteins with widely different functions such as, to choose the most relevant instances, (i) guanylate cyclase activated by NO in the Fe^{II} form,[26,71-78] (ii) NO synthase which binds its own product in both the Fe^{II} and the Fe^{III} forms[119,120] or (iii) NO transport by salivary ferrihemoproteins called nitrophores in the Fe^{III} form.[121-123] See discussions in chapters 10 and 14.

Finally, MetHb when oxidized to its oxyferryl free radical species by *tert*-butyl hydroperoxide (t-BuOOH), is reduced by NO.[124] This reaction has been followed by several spectroscopic methods including EPR. It results in the formation of an oxyferryl center (Hb(Fe^{IV})=O) and an EPR detectable tyrosyl free radical, similar to that found in ribonucleotide reductase;[125] NO reacts through a two-electron reduction yielding first MetHb and then eliminating the tyrosyl radical. Nitric oxide could thus have a protective role against ferryl hemoprotein-induced oxidations,[124] and generally act as an antioxidant.[126]

REACTIONS OF NITRIC OXIDE AND OTHER NITROGEN OXIDES WITH OXYHEMOGLOBIN

Nitric oxide very quickly and irreversibly oxidizes HbO_2 to MetHb and nitrate. The overall reaction is the base for the well-known quantification method of NO,[110,111,127]

$$Hb(Fe^{II})\text{-}O_2 + NO \rightarrow Hb(Fe^{III}) + NO_3^-$$

Although the apparent binding constant has not to our knowledge been reported, a calculated estimate of the reaction rate of NO with oxymyoglobin was given as 3.7 x 10^7 mol^{-1} l s^{-1},[128] which is faster than the rates of binding O_2 and NO to deoxyhemoglobin. It is obviously an extreme limit to an experimental determination of a rate constant by conventional methods. The production of nitrate but not nitrite in the overall reaction was confirmed by Ignarro et al;[129] this was to be expected as any nitrite formed would oxidize HbO_2. The very fast rate of MetHb formation from HbO_2 allowed real-time, continuous assay of NO release by human PMN leukocytes.[130]

Nitrite rather slowly oxidizes HbO_2 to MetHb by an autocatalytic reaction with a lag phase according to the stoichiometry:[131]

$$4HbO_2 + 4NO_2^- + 4H^+ \rightarrow 4Hb^{3+} + 4NO_3^- + 2H_2O + O_2$$

Several intermediates in the complex chain reaction were detected: oxyferryl Hb (symbolized below as [Hb^{4+}=O]) and an

EPR-detectable radical which is most probably a tyrosyl radical with a loss of a hydrogen atom (symbolized as $[\text{'}Hb^{4+}{=}O]$ in the following equations).[125,131,132] However the actual rates of the intermediate reactions are not all known:

$$3HbO_2 + 3NO_2^- + 6H^+ \rightarrow 3Hb^{3+} + 3NO_2^{\text{'}} + 3H_2O_2 \tag{1}$$

$$3Hb^{3+} + 3H_2O_2 \rightarrow 3[\text{'}Hb^{4+}{=}O] + 3H_2O \tag{2}$$

$$3[\text{'}Hb^{4+}{=}O] + 3NO_2^- \rightarrow 3[Hb^{4+}{=}O] + 3NO_2^{\text{'}} \tag{3}$$

$$3[Hb^{4+}{=}O] + 3NO_2^- + 6H^+ \rightarrow 3Hb^{3+} + 3NO_2^{\text{'}} + 3H_2O \tag{4}$$

$$HbO_2 + NO_2^{\text{'}} \rightarrow Hb^{3+} + O_2 + NO_2^- \tag{5}$$

and the termination reaction:

$$8NO_2^{\text{'}} + 4H_2O \rightarrow 4NO_2^- + 4NO_3^- + 8H^+ \tag{6}$$

— — — — — — — — — — — — —

$$4HbO_2 + 4NO_2^- + 4H^+ \rightarrow 4Hb^{3+} + 4NO_3^- + 2H_2O + O_2$$

The half-reaction time of a stoichiometric mixture of HbO_2 and nitrite, at 1 mmol l^{-1} each, is given as 5 minutes.[131] As nitrite is the cause of acute methemoglobinemia,[133] the variation of the time course of the overall reaction with the reactants concentration and the inhibition of the catalytic stage of the reaction was further studied, demonstrating more complications such as dependence on the R–T allosteric equilibrium.[134,135]

The method of quantification of NO by reaction with HbO_2 could in some cases be further complicated by other adventitious reactions such as the following. Peroxynitrite reacts with HbO_2, inducing spectral changes of HbO_2 identical with those elicited by NO. It could thus interfere with the photometrical determination of NO.[136]

Nitrite binds weakly and reversibly to MetHb to yield a mixture of EPR-detectable $S = 5/2$ and $S = 1/2$ complexes, with optical absorptions very different from that of acidic ($S = 5/2$) MetHb. Therefore it is safe to check for absorbance linearity versus concentration and for the existence of isosbestic points in the titration of HbO_2 by NO in order to avoid titrating contaminant nitrite.

According to Doyle and Hoekstra dinitrogen trioxide N_2O_3 also rapidly oxidizes HbO_2 to MetHb, while oxidation by N_2O_4 is slow.[128]

Finally several other complex reactions occur between trioxodinitrate $HN_2O_3^-$, its decomposition products HNO or NO^-, NO and NO_2^- and hemoglobin in its various forms, deoxy, oxy and met; they were studied as tools in the detection of nitrite reduction products in the anaerobic denitrification process.[137-141]

CONCLUSION

The reactions of hemoglobin with NO and its oxidation products are certainly the best documented reactions, in terms of stoichiometry, structure and kinetic analysis, of a metalloprotein with NO. It indicates several nearly irreversible reactions together with a few rather slow redox loops. It also leads to several predictions as to the existence of rather stable and specific hybrid hemoglobin molecules.

In the following chapter we shall focus on the effects of NO on hemoglobin within circulating red blood cells, where the multiple in vitro reactions of purified hemoglobin described above are regulated by many other enzymatically catalyzed reactions. The effects are different whether NO is an exogenous gas (following inhalation for instance), or results from the cellular metabolization of nitrogenous compounds (e.g. organic nitrates, see also chapter 12), or is endogenous, functioning as EDRF.[142] Many instances of NO captured by hemoglobin in RBC in pathophysiological cases shall be given in subsequent chapters (chapters 11 and 12).

ADDITIONAL BIBLIOGRAPHICAL NOTE

Another review on the subject has recently appeared after this chapter was completed.[143] Very recently two interesting articles were published in which the authors proposed new physiological roles for NO in oxygen transport by RBC.[144,145] In blood containing NO bound to one α subunit of hemoglobin, the oxygen affinity is significantly decreased.[144] According to Kosaka and Seiyama, *S*-nitrosothiols would serve as a NO reservoir, and releasing NO would help O_2 delivery from RBC to peripheral tissues. For Jia et al, hemoglobin is itself *S*-nitrosylated on residue Cysβ–93 at different rates along the arteriovenous cycle: *S*-nitrosylation is much faster on oxyhemoglobin (R-state) than on deoxyhemoglobin (T-state).[145] Thus SNO-Hb might act as a NO donor in the systemic circulation and would participate in the regulation of normal blood flow and delivery of O_2 to tissues.[145] It is a new role for the EDRF.

REFERENCES

1. Keilin D, Hartree EF. Reaction of nitric oxide with haemoglobin and methaemoglobin. Nature 1937; 139:548.

2. Gibson QH, Roughton FJW. The kinetic and equilibria of the reactions of NO with sheep haemoglobin. J Physiol 1957; 136:507-526.

3. Coryell CD, Pauling L, Dodson RW. The magnetic properties of intermediates in the reactions of hemoglobin. J Phys Chem 1939; 43:825-839.

4. Griffith JS. On the magnetic properties of some haemoglobin complexes. Proc Roy Soc A 1956; 235:23-36.

5. Chien JCW. Electron paramagnetic resonance study of the stereochemistry of nitrosylhemoglobin. J Chem Phys 1969; 51:4220-4227.

6. Ingram DJE, Bennett JE. Paramagnetic resonance in phtalocyanine, haemoglobin, and other organic derivatives. Discussion Faraday Soc 1955; 19:140-146.

7. Deatherage JF, Moffat K. Structure of nitric oxide hemoglobin. J Mol Biol 1979; 134:401-417.

8. Zavoisky E. Paramagnetic relaxation of liquid solutions for perpendicular fields. Journal of Physics (USSR) 1945; 9:211-216.

9. Griffiths JHE. Anomalous high-frequency resistance of ferromagnetic metals. Nature 1946; 158:670-671.

10. Bloch F, Hansen WW, Packard M. The nuclear induction experiment. Phys Rev 1946; 70:474-485.

11. Cummerov RL, Halliday D. Paramagnetic losses in two manganous salts. Phys Rev 1946; 70:433.

12. Sancier K, Freeman G, Mills I. Electron spin resonance of nitric oxide-hemoglobin complexes in solution. Science 1962; 137:752-754.

13. Kon H. Paramagnetic resonance study of nitric oxide hemoglobin. J Biol Chem 1968; 243:4350-4357.

14. Lang G, Marshall W. Mössbauer effect in some haemoglobin compounds. J Mol Biol 1966; 18:385-404.

15. Kon H. Electron paramagnetic resonance of nitric oxide cytochrome *c*. Biochem Biophys Res Commun 1969; 35:423-427.

16. Kon H, Kataoka N. Electron paramagnetic resonance of nitric oxide-protoheme with some nitrogenous base. Model systems of nitric oxide hemoproteins. Biochemistry 1969; 8:4757-4762.

17. Piciulo PL, Rupprecht G, Scheidt WR. Stereochemistry of nitrosylmetalloproteins. Nitrosyl-α,β,γ,δ-tetraphenylporphynato(1-methylimidazole)iron and nitrosyl-α,β,γ,δ-tetraphenyl-porphynato(4-methylpiperidine) manganese. J Am Chem Soc 1974; 96:5293-5295.

18. Wayland BB, Olson LW. Spectroscopic studies and bonding model for nitric oxide complexes of iron porphyrins. J Am Chem Soc 1974; 96:6037-6041.

19. Kon H. An interpretation of the three line EPR spectrum of nitric oxide hemoproteins and related model systems: the effect of the heme environment. Biochim Biophys Acta 1975; 379:103-113.

20. Maxwell JC, Caughey W. An infrared study of NO bonding to heme B and hemoglobin A. Evidence for inositol hexaphosphate induced cleavage of proximal histidine to iron bonds. Biochemistry 1976; 15:388-396.

21. Yonetani T, Yamamoto H, Erman JE et al.

Electron properties of hemoproteins. V. Optical and electron paramagnetic resonance characteristics of nitric oxide derivatives of metalloporphyrin-apo-hemoprotein complexes. J Biol Chem 1972; 247:2447-2455.

22. Antonini E, Brunori M. Hemoglobin and myoglobin in their reactions with ligands. North-Holland Publishing Company, Amsterdam, London. 1971.

23. Antonini E, Brunori M, Wyman J et al. Preparation and kinetic properties of intermediates in the reaction of hemoglobin with ligands. J Biol Chem 1966; 241:3236-3238.

24. Gray RD, Gibson QH. The effect of inositol hexaphosphate on the kinetics of CO and O_2 binding by human hemoglobin. J Biol Chem 1971; 246:7168-7174.

25. Sharma VS, Ranney HM. The dissociation of NO from nitrosylhemoglobin. J Biol Chem 1978; 253:6467-6472.

26. Traylor TG, Sharma VS. Why NO? Biochemistry 1992; 31:2847-2849.

27. Motterlini R, Macdonald VW. Cell-free hemoglobin potentiates acetylcholine-induced coronary vasoconstriction in rabbit hearts. J Appl Physiol 1993; 75:2224-2233.

28. Voelkel NF, Lobel K, Westcott JY et al. Nitric oxide-related vasoconstriction in lung perfused with red cell lysate. FASEB J 1995; 9:379-386.

29. Winslow RM. Hemoglobin-based red cell substitutes. The Johns Hopkins University Press. 1992.

30. Hess JR, Macdonald VW, Brinkley WW. Systemic and pulmonary hypertension after resuscitation with cell-free hemoglobin. J Appl Physiol 1993; 74:1769-1778.

31. Winslow RM. Vasoconstriction and the efficacy of hemoglobin-based blood substitutes. Transfusion Clin Biol 1994; 1:9-14.

32. Motterlini R, Foresti R, Vandegriff K et al. Oxidative-stress response in vascular endothelial cells exposed to acellular hemoglobin solutions. Am J Physiol 1995; 269:H648-H655.

33. Cantilena LR, Smith RP, Frasur S et al. Nitric oxide hemoglobin in patients receiving nitroglycerin as detected by electron paramagnetic resonance spectroscopy. J Lab Clin Med 1992; 120:902-907.

34. Henry Y, Banerjee R. Electron paramag-

netic studies of nitric oxide haemoglobin derivatives: isolated subunits and nitric oxide hybrids. J Mol Biol 1973; 73:469-482.

35. Cassoly R. Relations between optical spectrum and structure in nitrosylhemoglobin and hybrids. J Mol Biol 1975; 98:581-595.

36. Cassoly R. Use of nitric oxide as a probe for assessing the formation of asymmetrical hemoglobin hybrids. An attempted comparison between $\alpha^{NO}\beta^{NO}\alpha^{deoxy}\beta^{deoxy}$, $\alpha2^{NO}\beta2^{deoxy}$, and $\alpha2^{deoxy}\beta2^{NO}$ hybrids. J Biol Chem 1978; 253:3602-3066.

37. Sugita Y. Differences in spectra of α and β chains of hemoglobin between isolated state and tetramer. J Biol Chem 1975; 250:1251-1256.

38. Kiger L, Poyart C, Marden MC. Oxygen and CO binding to triply NO and asymmetric NO/CO hemoglobin hybrids. Biophys J 1993; 65:1050-1058.

39. Cassoly R, Gibson QH. Conformation, cooperativity and ligand binding in human hemoglobin. J Mol Biol 1975; 91:301-313.

40. Moore EG, Gibson QH. Cooperativity in the dissociation of nitric oxide from hemoglobin. J Biol Chem 1976; 251:2788-2794.

41. Chien JCW, Dickinson LC. Nonequivalence of subunits in [^{15}N]nitrosylhemoglobin Kansas. A single crystal electron paramagnetic resonance investigation. J Biol Chem 1977; 252:1331-1335.

42. Shiga T, Hwang R-J, Tyuma I. Electron paramagnetic resonance studies of nitric oxide hemoglobin derivatives. I. Human hemoglobin subunits. Biochemistry 1969; 8:378-383.

43. Henry Y, Cassoly R. Chain non-equivalence in nitric oxide binding to hemoglobin. Biochem Biophys Res Commun 1973; 51:659-665.

44. Cassoly R. Relation entre spectre d'absorption optique et structure de la nitrosyl hémoglobine. C R Acad Sc Paris 1974; 278:1417-1420.

45. Salhany JM, Ogawa S, Shulman RG. Spectral-kinetic heterogeneity in reactions of nitrosyl hemoglobin. Proc Natl Acad Sci USA 1974; 71:3359-3362.

46. Salhany JM, Ogawa S, Shulman RG. Correlation between quaternary structure and ligand dissociation kinetics for fully

liganded hemoglobin. Biochemistry 1975; 14:2180-2190.

47. Reisberg P, Olson JS, Palmer G. Kinetic resolution of ligand binding to the α and β chains within human hemoglobin. J Biol Chem 1976; 251:4379-4383.

48. Nagai K, Hori H, Yoshida S et al. The effect of quaternary structure on the state of the α and β subunits within nitrosyl haemoglobin. Low temperature photodissociation and the ESR spectra. Biochim Biophys Acta 1978; 532:17-28.

49. Taketa F, Antholine WE, Chen JY. Chain non equivalence in binding of nitric oxide to hemoglobin. J Biol Chem 1978; 253:5448-5451.

50. Louro SRW, Ribeiro PC, Bemski G. EPR spectral changes of nitrosyl hemes and their relation to the hemoglobin T-R transition. Biochim Biophys Acta 1981; 670:56-63.

51. Hille R, Palmer G, Olson JS. Chain equivalence in reaction of nitric oxide with hemoglobin. J Biol Chem 1977; 252:403-405.

52. Rose EJ, Hoffman BM. Nitric oxide ferrohemes: kinetics of formation and photodissociation quantum yields. J Am Chem Soc 1983; 105:2866-2873.

53. Martin JL, Migus A, Poyart C et al. Femtosecond photolysis of CO-ligated protoheme and hemoproteins: appearance of deoxy species with a 350-fsec time constant. Proc Natl Acad Sci USA 1983; 80:173-177.

54. Petrich JW, Lambry J-C, Kuczera K et al. Ligand binding and protein relaxation in heme proteins: a room temperature analysis of NO geminate recombination. Biochemistry 1991; 30:3975-3987.

55. Das TK, Mazumdar S, Mitra S. Micelle-induced release of haem-NO from nitric oxide complex of myoglobin. J Chem Soc Chem Commun 1993; 18:1447-1448.

56. Traylor TG, Magde D, Marsters J et al. Geminate processes in the reaction of nitric oxide with 1-methylimidazole-iron(II) porphyrin complexes. Steric, solvent polarity, and viscosity effects. J Am Chem Soc 1993; 115:4808-4813.

57. Carlson ML, Regan R, Elber R et al. Nitric oxide recombination to double mutants of myoglobin: role of ligand diffusion in a fluc-tuating heme pocket. Biochemistry 1994; 33:10597-10606.

58. De Sanctis G, Falcioni G, Polizio F et al. Mini-myoglobin: native-like folding of the NO-derivative. Biochim Biophys Acta 1994; 1204:28-32.

59. Walda KN, Liu XY, Sharma VS et al. Geminate recombination of diatomic ligands CO, O_2, NO with myoglobin. Biochemistry 1994; 33:2198-2209.

60. Duprat AF, Traylor TG, Wu G-Z et al. Myoglobin-NO at low pH: free four-coordinated heme in the protein pocket. Biochemistry 1995; 35:2634-2644.

61. Trittelvitz E, Sick H, Gersonde K. Conformational isomers of nitrosyl-haemoglobin. An electron-spin resonance study. Eur J Biochem 1972; 31:578-584.

62. Rein H, Ristau O, Scheler W. On the influence of allosteric effectors on the electron paramagnetic spectrum of nitric oxide hemoglobin. FEBS Lett 1972; 24:29-26.

63. Lalezari I, Lalezari P, Poyart C et al. New effectors of human hemoglobin: structure and function. Biochemistry 1990; 29:1515-1523.

64. Ascenzi P, Coleta M, Desideri A et al. Effect of bezafibrate and clofibric acid on the spectroscopic properties of the nitric oxide derivative of ferrous human hemoglobin. J Inorg Biochem 1992; 48:47-53.

65. Szabo A, Perutz MF. Equilibrium between six- and five-coordinated hemes in nitrosyl-hemoglobin: interpretation of electron spin resonance spectra. Biochemistry 1976; 15:4427-4428.

66. Scholler DM, Wang MYR, Hoffman BM. Resonance Raman and EPR of nitrosyl human hemoglobin and chains, carp hemoglobin, and model compounds. Implications for the nitrosyl heme coordination state. J Biol Chem 1979; 254:4072-4078.

67. Sampath V, Zhao X-J, Caughey WS. Characterization of interactions of nitric oxide with human hemoglobin A by infrared spectroscopy. Biochem Biophys Res Commun 1994; 198:281-287.

68. Mun SK, Chang JC, Das TP. Origin of observed changes in ^{14}N hyperfine interaction accompanying R → T transition in nitrosylhemoglobin. Proc Natl Acad Sci USA 1979; 76:4842-4846.

69. Waleh A, Ho N, Chantranupong L et al. Electronic structure of nitrosyl ferrous heme complexes. J Am Chem Soc 1989; 111:2767-2772.

70. Bonnett R, Chandra S, Charalambides AA et al. Nitrosation and nitrosylation of haemoproteins and related compounds. Part 4. Pentaco-ordinate nitrosylprotohaem as the pigment of cooked cured meat. Direct evidence from ESR spectroscopy. J Chem Soc Perkin I 1980; 1706-1710.

71. Craven PA, DeRubertis FR. Restoration of the responsiveness of purified guanylate cyclase to nitrosoguanidine, nitric oxide and realated activators by heme and hemoproteins. Evidence for involvement of the paramagnetic nitrosyl-heme complex in enzyme activation. J Biol Chem 1978; 253:8433-8443.

72. Craven PA, DeRubertis FR, Pratt DW. Electron spin resonance study of the role of NO-catalase in the activation of guanylate cyclase by NaN_3 and NH_2OH. Modulation of enzyme responses by heme proteins and their nitrosyl derivatives. J Biol Chem 1979; 254:8213-8222.

73. Ignarro LJ, Adams JB, Horwitz PM et al. Activation of soluble guanylate cyclase by NO-hemoproteins involves NO-heme exchange. Comparison of heme-containing and heme-deficient enzyme forms. J Biol Chem 1986; 261:4997-5002.

74. Traylor TG, Duprat AF, Sharma VS. Nitric oxide-triggered heme-mediated hydrolysis: a possible model for biological reactions of NO. J Am Chem Soc 1993; 115:810-811.

75. Tsai A-L. How does NO activates hemeproteins? FEBS Lett 1994; 341:141-145.

76. Stone JR, Marletta MA. Soluble guanylate cyclase from bovine lung: activation with nitric oxide and carbon monoxide and spectral characterization of the ferrous and ferric states. Biochemistry 1994; 33:5636-5640.

77. Yu AE, Hu S, Spiro TG et al. Resonance Raman spectroscopy of soluble guanylate cyclase reveals displacement of distal and proximal heme ligands by NO. J Am Chem Soc 1994; 116:4117-4118.

78. Stone JR, Sands RH, Dunham WR et al. Electron paramagnetic spectral evidence for the formation of a pentacoordinate nitrosyl-

heme complex on soluble guanylate cyclase. Biochem Biophys Res Commun 1995; 207:572-577.

79. Makinen MW, Kon H. Circular dichroism and electron paramagnetic resonance of the haptoglobin-hemoglobin complex. Biochemistry 1971; 10:43-52.

80. Makinen MW, Milstein JB, Kon H. Specificity of interaction with mammalian hemoglobin. Biochemistry 1972; 11:3851-3860.

81. Bannai S, Sugita Y. Absorption spectra and reaction with haptoglobin of hemoglobin (α-NO, β-unliganded). J Biol Chem 1973; 248:7527-7529.

82. Henry Y, Ducrocq C, Drapier J-C et al. Nitric oxide, a biological effector. Electron paramagnetic resonance detection of nitrosyl-iron-protein complexes in whole cells. Eur Biophys J 1991; 20:1-15.

83. Henry Y, Lepoivre M, Drapier J-C et al. EPR characterization of molecular targets for NO in mammalian cells and organelles. FASEB J 1993; 7:1124-1134.

84. Henry YA, Singel DJ. Metal-nitrosyl interactions in nitric oxide biology probed by electron paramagnetic resonance spectroscopy. In: Feelisch M, Stamler J, eds. Methods in nitric oxide research. John Wiley and Sons, 1996:357-372.

85. Singel DJ, Lancaster JR. Electron paramagnetic resonance spectroscopy and nitric oxide biology. In: Feelisch M, Stamler J, eds. Methods in nitric oxide research. John Wiley and Sons, 1996:341-356.

86. Wilcox DE, Smith RP. Detection and quantification of nitric oxide using electron magnetic resonance spectroscopy. Methods: a Companion to Methods Enzymol 1995; 7:59-70.

87. Maruyama T, Kataoka N, Nagase S et al. Identification of three-line electron spin resonance signal and its relationship to ascites tumors. Cancer Res 1971; 31:179-184.

88. Kosaka H, Sawai Y, Sakagushi H et al. ESR spectral transition by arteriovenous cycle in nitric oxide hemoglobin of cytokine-treated rats. Am J Physiol 1994; 266:C1400-1405.

89. Chevion M, Stern A, Peisach J et al. Analoguous effect of protons and inositol hexaphosphate on the alteration of structure

of nitrosyl fetal human hemoglobin. Biochemistry 1978; 17:1745-1750.

90. Trittelvitz E, Gersonde K, Winterhalter KH. Electron spin resonance of nitrosyl haemoglobins: normal α and β chains and mutants Hb M Iwate and Hb Zürich. Eur J Biochem 1975; 51:33-42.

91. Chevion M, Blumberg WE, Peisach J. EPR studies of human nitrosylhemoglobins and their relation to molecular function. In: Pullman B, Goldblum N, eds. Metal-ligand interactions in organic chemistry and biochemistry. Part 2. D Reidel Pub Company, Dordrecht-Holland, 1977:153-162.

92. Nagai K, Hori H, Morimoto H et al. Influence of amino acid replacements in the heme pocket on the electron paramagnetic resonance spectra and absorption spectra of nitrosylhemoglobin M Iwate, M Boston, and M Milwaukee. Biochemistry 1979; 18:1304-1308.

93. John ME, Waterman MR. Structural basis for the conformational states of nitrosyl-hemoglobins M Saskatoon and M Milwaukee. Influence of distal histidine residues on proximal histidine bonds. J Biol Chem 1980; 255:4501-4506.

94. Taketa F, Antholine WE, Mauk AB et al. Nitrosylhemoglobin Wood: effects of inositol hexaphosphate on thiol reactivity and electron paramagnetic resonance spectrum. Biochemistry 1975; 14:3229-3233.

95. Chevion M, Salhany JM, Peisach J et al. Iron-nitrosyl bond configuration in nitrosyl-hemoproteins: a comparative EPR study of hemoglobin A and hemoglobin Kansas. Israel J Chem 1976; 15:311-317.

96. Brzozowski A, Derewenda Z, Dodson E, et al. Bonding of molecular oxygen to T state human hemoglobin. Nature 1984; 307:74-76.

97. Marden MC, Kister J, Bohn B et al. T-state hemoglobin with four ligands bound. Biochemistry 1988; 27:1659-1664.

98. Mozzarelli A, Rivetti C, Rossi GL et al. Crystals of haemoglobin with the T quaternary structure bind oxygen noncooperatively with no Bohr effect. Nature 1991; 351:416-419.

99. Fermi G, Perutz MF, Shaanan B et al. The crystal structure of human deoxyhemoglobin at 1.74 Å resolution. J Mol Biol 1984; 175:159-174.

100. Shaanan B. Structure of human oxyhemoglobin at 2.1 Å resolution. J Mol Biol 1983; 171:31-59.

101. Silva MM, Rogers PH, Arnone A. A third quaternary structure of human hemoglobin A at 1.7-Å resolution. J Biol Chem 1992; 267:17248-17256.

102. Doetschmann DC, Utterback SG. Electron paramagnetic resonance study of nitrosyl-hemoglobin and its chemistry in single crystals. J Am Chem Soc 1981; 103:2847-2852.

103. Linhares MP, El-Jaick LJ, Bemski G et al. EPR studies of photolysis of nitrosyl haemoglobin at low temperatures. Int J Biol Macromol 1990; 12:59-63.

104. El-Jaick LJ, Wajnberg E, Linhares MP. EPR studies of photolysis of nitrosyl haemoglobin at low temperature: effects of quaternary structure. Int J Biol Macromol 1991; 13:289-294.

105. Wajnberg E, Linhares MP, El-Jaick LJ et al. Nitrosyl hemoglobin: EPR components at low temperatures. Eur Biophys J 1992; 21:57-61.

106. Doetschmann DC, Rizos AK, Szumowski J. The cooperativity of nitric oxide hemoglobin ligand exchange in single crystals as studied by EPR. J Chem Phys 1984; 81:1185-1191.

107. Höhn M, Hüttermann J, Chien JCW et al. ^{14}N and ^{1}H ENDOR of nitrosylhemoglobin. J Am Chem Soc 1983; 105:109-115.

108. Morse RH, Chan SI. Electron paramagnetic resonance studies of nitrosyl ferrous heme complexes. Determination of an equilibrium between two conformations. J Biol Chem 1980; 255:7876-7882.

109. Hori H, Ikeda-Saito M, Yonetani T. Single crystal EPR of myoglobin nitroxide. Freezing-induced reversible changes in the molecular orientation of the ligand. J Biol Chem 1981; 256:7849-7855.

110. Feelisch M, Noack EA. Correlation between nitric oxide formation during degradation of organic nitrates and activation of guanylate cyclase. Eur J Pharmacol 1987; 139:19-30.

111. Murphy ME, Noack E. Nitric oxide assay using hemoglobin method. Methods Enzymol 1994; 233:240-250.

112. Chien JCW. Reactions of nitric oxide with methemoglobin. J Am Chem Soc 1969; 91:2166-2168.

113. Ehrenberg A, Szczepkowski TW. Properties and structure of the compounds formed between cytochrome c and nitric oxide. Acta Chem Scand 1960; 14:1684-1692.

114. Sharma VS, Traylor TG, Gardiner R et al. Reaction of nitric oxide with heme proteins and model compounds of hemoglobin. Biochemistry 1987; 26:3837-3843.

115. Sharma VS, Isaacson RA, John ME et al. Reaction of nitric oxide with heme proteins: studies on metmyoglobin, opossum methemoglobin, and microperoxidase. Biochemistry 1983; 22:3897-3902.

116. Addison AN, Stephanos JJ. Nitrosyl (III) hemoglobin: autoreduction and spectroscopy. Biochemistry 1986; 25:4104-4113.

117. O'Keeffe DH, Ebel RE, Peterson JA. Studies of the oxygen binding site of cytochrome P-450. Nitric oxide as a spin-label probe. J Biol Chem 1978; 253:3509-3516.

118. Brudvig GW, Stevens TH, Chan SI. Reactions of nitric oxide with cytochrome c oxidase. Biochemistry 1980; 19:5275-5285.

119. Wang J, Rousseau DL, Abu-Soud HM et al. Heme coordination of NO in NO synthase. Proc Natl Acad Sci USA 1994; 91:10512-10516.

120. Hurshman AR, Marletta MA. Nitric oxide complexes of inducible nitric oxide synthase: spectral characterization and effect on catalytic activity. Biochemistry 1995; 34:5627-5634.

121. Ribeiro JMC, Hazzard JMH, Nussenzveig RH et al. Reversible binding of nitric oxide by a salivary heme protein from a blood-sucking insect. Science 1993; 260:539-541.

122. Ribeiro JMC, Nussenzveig RH. Nitric oxide synthase from a hematophagous insect salivary gland. FEBS Lett 1993; 330:165-168.

123. Ribeiro JMC, Walker FA. High affinity histamine-binding and antihistaminic activity of the salivary nitric oxide-carrying hemeprotein (nitrophorin) of *Rhodnius proxilus*. J Exp Med 1994; 180:2251-2257.

124. Gorbunov NV, Osipov AN, Day BW, et al. Reduction of ferrylmyoglobin and ferryl-hemoglobin by nitric oxide: a protective mechanism against ferryl hemoprotein-induced oxidations. Biochemistry 1995; 34:6689-6699.

125. Giulivi C, Cadenas E. Ferrylmyoglobin: formation and chemical reactivity toward electron-donating compounds. Methods Enzymol 1994; 233:189-202.

126. Kanner J, Harel S, Granit R. Nitric oxide as an antioxidant. Arch Biochem Biophys 1991; 289:130-136.

127. Kelm M, Schrader J. Control of coronary vascular tone by nitric oxide. Circ Res 1990; 66:1561-1575.

128. Doyle MP, Hoekstra JW. Oxidation of nitrogen oxides by bound dioxygen in hemoproteins. J Inorg Biochem 1981; 14:351-358.

129. Ignarro LJ, Fukuto JM, Griscavage JM et al. Oxidation of nitric oxide in aqueous solution to nitrite but not nitrate: comparison with enzymatically formed nitric oxide from L-arginine. Proc Natl Acad Sci USA 1993; 90:8103-8107.

130. Lärfars G, Gyllenhammar H. Measurement of methemoglobin formation from oxyhemoglobin. A real-time, continuous assay of nitric oxide release by human polymorphonuclear leucocytes. J Immunol Methods 1995; 184:53-62.

131. Kosaka H, Imaizumi K, Tyuma I. Mechanism of autocatalytic oxidation of oxyhemoglobin by nitrite, an intermediate detected by electron spin resonance. Biochim Biophys Acta 1982; 702:237-241.

132. Kosaka H, Uozumi M. Inhibition of amines indicates involvement of nitrogen dioxide in autocatalytic oxidation of oxyhemoglobin by nitrite. Biochim Biophys Acta 1986; 871:14-18.

133. Hegesh E, Shiloah J. Blood nitrates and infantile methemoglobinemia. Clin Chim Acta 1982; 125:107-115.

134. Doyle MP, Herman JG, Dykstra RL. Autocatalytic oxidation of hemoglobin induced by nitrite: activation and chemical inhibition. J Free Rad Biol Med 1985; 1:145-154.

135. Spagnuolo C, Rinelli P, Coletta M et al. Oxidation reaction of human oxyhemoglobin with nitrite: a reexamination. Biochim Biophys Acta 1987; 911:59-65.

136. Schmidt K, Klatt P, Mayer B. Reaction of peroxynitrite with oxyhemoglobin: interference with photometrical determination of nitric oxide. Biochem J 1994; 301:645-647.

137. Doyle MP, Mahapatro SN. Nitric oxide dissociation from trioxodinitrate (II) in aqueous solution. J Am Chem Soc 1984; 106:3678-3679.

138. Bazylinski DA, Hollocher TC. Metmyoglobin and methemoglobin as efficient traps for nitrosyl hydride (nitroxyl) in neutral aqueous solution. J Am Chem Soc 1985; 107:7982-7986.

139. Bazylinski DA, Hollocher TC. Evidence from the reaction between troxodinitrate(II) and [15]NO that triooxdinitrate(II) decomposes into nitrosyl hydride and nitrite in neutral aqueous solution. Inorg Chem 1985; 24:4285-4288.

140. Bazylinski DA, Goretski J, Hollocher TC. On the reaction of trioxodinitrate (II) with hemoglobin and myoglobin. J Am Chem Soc 1985; 107:7986-7989.

141. Doyle MP, Mahapatro SN, Broene RD et al. Oxidation and reduction of hemoproteins by trioxodinitrate (II). The role of nitrosyl hydride and nitrite. J Am Chem Soc 1988; 110:593-599.

142. Kosaka H, Uozumi M, Tyuma I. The interaction between nitrogen oxides and hemoglobin and endothelium-derived relaxing factor. Free Rad Biol Med 1989; 7:653-658.

143. Kosaka H, Shiga T. Detection of nitric oxide by electrom paramagnetic resonance using hemoglobin. In: Feelisch M, Stamler JS. Methods in nitric oxide research. John Wiley & Sons, 1996:373-381.

144. Kosaka H, Seiyama A. Physiological role of nitric oxide as an enhancer of oxygen transfer from erythrocytes to tissues. Biochem Biophys Res Commun 1996; 218:749-752.

145. Jia L, Bonaventura C, Bonaventura J et al. S-nitrosohaemoglobin: a dynamic activity of blood involved in vascular control. Nature 1996; 380:221-226.

CHAPTER 5

EFFECTS OF NITRIC OXIDE ON RED BLOOD CELLS

Yann A. Henry

INTRODUCTION

The present chapter shall be divided in an apparently artificial way into three parts. In fact they represent three stages in the understanding of the effects of NO on animals and humans. We shall summarize the effects of NO on hemoglobin within red blood cells (RBC) circulating in the arteriovenous cycle, cells with high metabolic and redox capability. RBC contain hemoglobin at 5 mmol/l concentration with its allosteric effector 2,3-DPG (also 5 mmol/l), a high reducing pool with GSH (2 mmol/l), methemoglobin (normal concentration being less than 0.1 mmol/l) and the enzymatic system cytochrome b_5 (0.2-0.8 mmol/l) plus methemoglobin reductase (70 nmol/l) (Scheme 5.1).[1]

We have summarized in this scheme several cycles of reaction of hemoglobin in its various forms, deoxyHb, HbO_2, HbNO and MetHb (Hb(Fe^{III}) ferric hemoglobin) with NO, which were analyzed in detail in chapter 4.

The first instances of such studies on RBC dealt with NO and other oxides considered as air pollutants.[2,3] The main conclusion appears to be that, due to several metabolic pathways, NO reacts with hemoglobin within RBC with a much lower yield than expected from in vitro kinetic data. Furthermore NO does not appear to be so highly toxic as its enormous affinity for deoxyhemoglobin and its very rapid reaction with oxyhemoglobin would have induced one to think. On the contrary the toxicity of NO reaction products with air, especially $NO_2^{\bullet}$ is fearsome, although the molecular and cellular mechanisms are still largely unknown.[4]

These fundamental experiments were followed by studies of the metabolism of nitrogenous vasodilators,[5-7] especially organic nitrates and nitrites. The main questions at that time, when the idea was just simmering

Nitric Oxide Research from Chemistry to Biology: EPR Spectroscopy of Nitrosylated Compounds, edited by Yann A. Henry, Annie Guissani and Béatrice Ducastel.

that NO was the EDRF was just simmering, were to decide which between nitrite and NO, was the metabolite and the active vasodilating component, and whether RBC were efficient in organic nitrate metabolism.

A new perspective was taken in studies of the interaction of hemoglobin and EDRF,[8-10] rapidly shifted by the purpose to use NO inhalation in the clinic for specific pulmonary diseases.

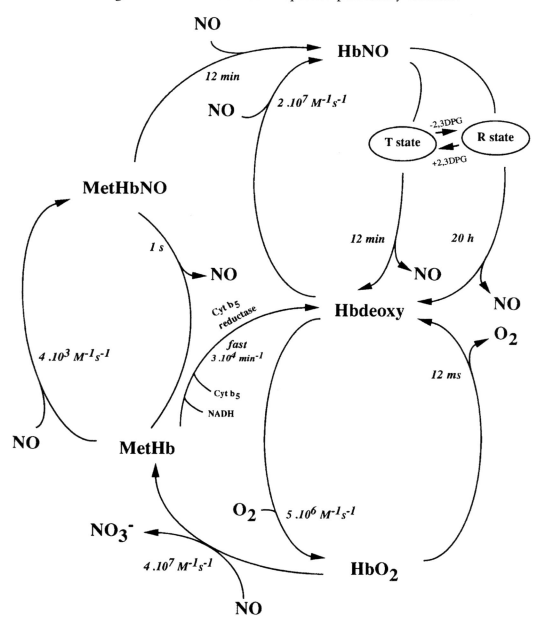

Scheme 5.1.

EFFECTS OF LONG-TERM EXPOSURE OF ANIMALS AND HUMANS TO NITRIC OXIDE AND NITROGEN DIOXIDE GAS CONSIDERED AS AIR POLLUTANTS

The effects of nitric oxide, considered as a pollutant (see introductory chapter 1), upon RBC have been thoroughly studied. Due to its third-order reaction with O_2 to yield the highly toxic nitrogen dioxide $NO_2^{\bullet}$, the overall $NO^{\bullet}$ effects depend largely on the way through which it is inhaled.

By use of a special exposure chamber to small quantities of NO (<10 ppm), nearly excluding $NO_2^{\bullet}$ (<0.8 ppm), the blood of exposed animals (rats and mice) was shown by EPR spectroscopy to contain $Hb(Fe^{II})NO^{\bullet}$.[2] In control experiments a linear correlation was established between the EPR signal intensity and $Hb(Fe^{II})NO^{\bullet}$ concentration, the lower detection limit being around 2 µmol/l. By exposure to 10 ppm NO the $Hb(Fe^{II})NO^{\bullet}$ content in blood reached a plateau (0.13% of total hemoglobin, i.e. ~ 6 µmol/l) after 20 minutes; after one hour of NO exposure it fully disappeared after 30 minutes of recovery in clean air. There was also a dose-response relationship between the maximal $Hb(Fe^{II})NO^{\bullet}$ content in blood of mice exposed for one hour and the concentration of NO. Rabbit blood showed $Hb(Fe^{II})NO^{\bullet}$ only when dithionite was added to the blood sample. There could be a species specificity in the redox balance between NO and nitrite/nitrate on the one hand and $Hb(Fe^{II})NO^{\bullet}$ and methemoglobin $Hb(Fe^{III})$ on the other hand.[2]

Exposure of mice to higher doses of NO (40 ppm) resulted in larger $Hb(Fe^{II})NO^{\bullet}$ (0.7%, ~ 35 µmol/l) and $Hb(Fe^{III})$ (5 %, ~ 0.25 mmol/l) formations, both species being detected by EPR spectroscopy, around g = 2 for $Hb(Fe^{II})NO^{\bullet}$ and around g = 6 for high spin (S = 5/2) $Hb(Fe^{III})$ (Fig. 5.1).[11] Strangely enough, exposure of mice to $NO_2^{\bullet}$ resulted in less $Hb(Fe^{II})NO^{\bullet}$ and no $Hb(Fe^{III})$ production.

The linear dose-response data of mice exposed to 2-80 ppm NO showed that the amount of $Hb(Fe^{II})NO^{\bullet}$ in the blood is much smaller than expected from in vitro kinetic data.[8,11] $Hb(Fe^{II})NO^{\bullet}$ is certainly partially converted to $Hb(Fe^{III})$ in the arteriovenous cycle, and $Hb(Fe^{III})$ is in turn reduced to Hb through RBC enzymatic systems, essentially methemoglobin reductase (Scheme 5.1).[1] Most of inhaled NO is excreted in the urine as nitrate, only a part is reduced by bacteria to NH_3 excreted in the feces or to N_2 in the mouth and the stomach.[3] Methemoglobin reductase certainly plays a protective role for NO intoxication, and this explains the rapid recovery of all RBC functions after ending the exposure to NO.[12] It would be of interest to study the in vitro effect of NO on the system cytochrome b_5-methemoglobin reductase (NADH-cytochrome b_5 reductase) from RBC and from liver microsomes,[1,13] as the occurrence of nitrite and nitrate in methemoglobinemia (over 2 % $Hb(Fe^{III}$, normal level being 0.5-0.7 %) in blood) is well documented.[8,14]

Lifetime exposure of mice to 2.4 ppm NO did not result in a difference in survival rate or in spontaneous disease appearance over 29 months normal lifespan.[15] The level of $Hb(Fe^{II})NO^{\bullet}$ was only 0.01% and that of $Hb(Fe^{III})$ (< 0.3%) did not increase. Hematological examination also did not show any difference of exposed versus normal groups. The same was true upon lung examination. Thus the effects of NO are considered to be milder than that of $NO_2^{\bullet}$ with respect to the effects on the respiratory organs.[4,8,11,16]

Exposure of animals and humans to nitrogen dioxide resulted also in EPR-detectable NO-heme complexes in various organs (liver, lung, bronchoalveolar (BAL) cell pellet) and in the blood (Fig. 5.2).[17] The signal found in tissue cells could arise from P-450-NO complex. The EPR relative intensity of rat or human BAL cell pellets was linearly related to $NO_2^{\bullet}$ concentration (a single 6 hr exposure up to 30 ppm for rats and six 20 minute exposures every other day to 1.5 or 4 ppm for humans).[17]

Fig. 5.1. EPR spectra at -140°C of HbNO (upper) and MetHb (lower) in the blood of mice exposed to 40 ppm NO for 1 hour. (Reproduced with permission from: Oda H, Nogami H, Nakajima T. Reaction of hemoglobin with nitric oxide and nitrogen dioxide in mice. J Toxicol Environ Health 1980; 6:673-678.) Copyright © Taylor & Francis, Inc. 1101 Vermont Ave, N.W. Ste 200, Washington, DC, USA.

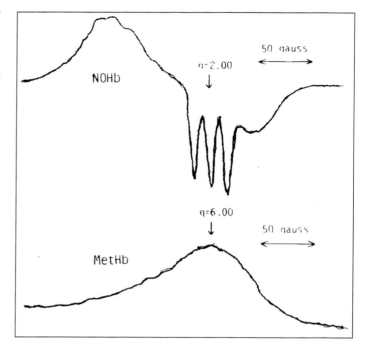

In vitro studies of the effects of NO on rats and human erythrocytes showed modifications of several RBC properties. The oxygen transport is altered through an increase of affinity for oxygen and a decrease of heme-heme cooperativity as the percentage of NO ligation increases.[18] A large MetHb formation is also observed in these in vitro experiments, up to 30% $Hb(Fe^{III})$ for 50% NO ligation. These overall effects probably reflect the formation of hybrid hemoglobin molecules (see below). Changes in blood rheology, as measured by viscosity, erythrocyte morphology (echinocyte formation) and deformability were detected in NO-exposed human or rat erythrocytes.[12] They were accompanied by changes of ghost-membrane proteins, a large decrease of monomeric spectrin, diminution and alteration of band 3 protein and a large increase of globin quantity, as measured by densitography of polyacrylamide gel electrophoresis.[12] These effects were reversed upon addition of reducer mercaptoethanol to ghost-membrane preparation, implying an oxidative crosslinking of the membrane proteins and hemoglobin.

EFFECTS OF NITRITE AND NITROGENOUS VASODILATORS ON RED BLOOD CELLS

After the experiments defining EDRF by Furchgott and Zawadzki,[19] one seminal discovery leading to the identification of NO as the EDRF was the study of the relaxation of rabbit aorta by acidified sodium nitrite.[20]

As mentioned in chapter 4, nitrite oxidizes HbO_2 in aqueous solution. In RBC, addition of nitrite also results in oxidation of several ferric complexes and to reduction of nitrite into NO.[5,21] The α and β subunits in hemoglobin are inequivalent in their redox properties as they are in the ligand binding properties. This is reflected in the oxidation of hemoglobin by ferricytochrome c,[22] the reduction of MetHb by ascorbate,[23,24] and in the MetHb reduction by NADH catalyzed by RBC-NADH-cytochrome b_5 reductase.[1,13] All these reactions yield stable intermediates, the tetrameric valency hybrids $(\alpha^{3+}\beta^{2+})_2$ and $(\alpha^{2+}\beta^{3+})_2$ in variable proportions depending mostly on the presence of allosteric effectors, 2,3-DPG or IHP. Similarly RBC exposure to nitrite yields mixtures of

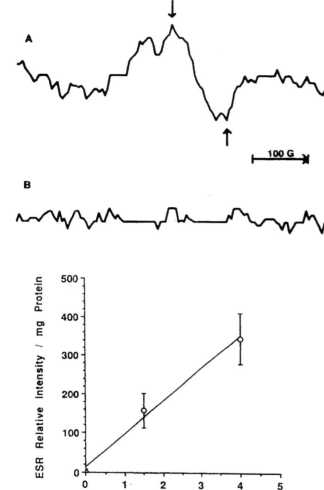

Fig. 5.2. Upper panel: EPR spectra at room temperature of human broncho-alveolar (BAL) cell pellet samples. (A) BAL cell pellet obtained 24 hr after exposure of a human volunteer to 4 ppm NO_2 every other day for six exposures. (B) BAL cell pellet obtained 3 weeks before exposure to NO_2 from the same individual in (A). Lower panel: Human dose-response relationship. (Reproduced with permission from: Maples KR, Sandström T, Su Y-F et al. The nitric oxide/heme protein complex as a biological marker of exposure to nitrogen dioxide in humans, rats, and in vitro models. Am J Respir Cell Mol Biol 1991; 4:538-543.) Copyright © American Lung Association, New York, NY, USA.

fully oxidized $Hb(Fe^{III})$, fully oxygenated HbO_2, and nitrosylated valency hybrids $(\alpha^{3+}\beta^{2+}-O_2)_2$ and $(\alpha^{2+}-NO\beta^{3+})_2$ as revealed by EPR spectroscopy (Fig. 5.3).[5]

Similar nitrosylated valency hybrids were detected by EPR spectroscopy in rat, mice or human RBC suspensions under the action of nitrogenous vasodilators such as nitroglycerin (glyceryl trinitrate, GTN), isosorbide dinitrate (IDN) or sodium nitroprusside (SNP) (Fig. 5.4).[6,21,25]

They were also detected in the blood of nitrogenous vasodilator-treated mice[7]

and of GTN-treated patients.[26] EPR spectroscopy was also used to unravel the reaction mechanisms of hemoglobin with GTN, SNP and 3-morpholinosydnonimine (SIN-1), in particular the influence of reduced thiols.[27-29] It pointed to cyanide poisoning and Fe^{2+} release with its interference with the biological activity of nitroprusside as detailed in chapter 2.[27,28]

As we shall see later such hybrid hemoglobin molecules were also detected in the blood of animals in pathological states (see chapters 11 and 12).

Fig. 5.3. EPR spectra at 77 K for various valency species of hemoglobin when all reduced subunits are fully nitrosylated: full line: $(\alpha^{2+}\beta^{2+})_2$; dashed line: $(\alpha^{2+}\beta^{3+})_2$ and spotted line: $(\alpha^{3+}\beta^{2+})_2$; in 0.1 M bis-Tris buffer, pH 7.0, 0.1 M NaCl. (Reproduced with permission from: Kruszyna R, Kruszyna H, Smith RP et al. Red blood cell generate nitric oxide from directly acting, nitrogenous vasodilators. Toxicol Applied Pharmacol 1987; 91:429-438.) Copyright © Academic Press, Ltd, London, UK.

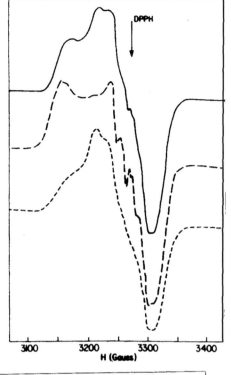

Fig. 5.4. EPR spectra at 110 K of ^{15}N-HbNO production versus time, after anaerobic addition of ^{15}N-isosorbide dinitrate (1 mmol/l) to fresh human whole blood (A) and to washed erythrocytes (B), normalized to a similar relative signal intensity; the increased signal-to-noise ratio indicates an increasing absolute HbNO concentration with time. (Reproduced with permission from: Kosaka H, Tanaka S, Yoshii T et al Direct proof of nitric oxide formation from a nitrovasodilator metabolised by erythrocytes. Biochem Biophys Res Commun 1994; 204:1055-1060.) Copyright © Academic Press, Inc, Orlando, Florida, USA.

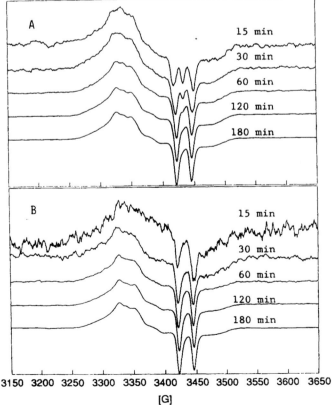

EFFECTS ON RED BLOOD CELLS OF NITRIC OXIDE INHALATION IN THERAPEUTIC USES

The discovery in 1987 that NO was the, or at least one, EDRF led rapidly to the subsequent use of NO inhalation to selectively reverse pulmonary vasoconstriction on animal models.[30,31] Soon afterwards "compassional" clinical trials were performed on newborn patients with severe persistent pulmonary hypertension of the newborn (PPHN)[32-34] and adult patients, particularly those with adult respiratory distress syndrome (ARDS) (see refs. 35-41). These trials have completely modified the perspective as to the relevancy of NO breathing versus its potential toxicity as well as to the estimation of the danger of concomittant $NO_2^\bullet$ production.[42-44] The clinical incidence led to experiments with drawn human blood, with sheep, healthy human subjects and patients (reviewed in ref. 4).

As it seems obvious, the proportion of oxidized and nitrosylated reduced hemes within hemoglobin depends on the oxygenation state of the RBC when incubated with NO.[46] Using EPR spectroscopy to quantify both paramagnetic species in experiments mimicking venous or arterialized blood in vitro, Wennmalm et al showed that NO and $Hb(Fe^{II})O_2$ are mostly converted to $Hb(Fe^{III})$ (MetHb) and NO_3^- in arterial blood conditions (94-99% O_2 saturation) with little $Hb(Fe^{II})NO^\bullet$ and no nitrite. On the contrary the conversion was mostly to $Hb(Fe^{II})NO^\bullet$ and a little NO_3^- in conditions mimicking venous blood (36-85% O_2 saturation). This is probably the way EDRF is eliminated via the kidney.[46,47] In another somewhat contradictory report, deoxyHb in fresh sheep blood also appeared to be oxidized into MetHb by NO.[48]

Intact deoxygenated human RBC have been used to monitor NO by EPR spectroscopy at room temperature (293 K).[49] However this study does not take into account the full metabolic processes occurring in vivo (Scheme 5.1).

A detailed experimental and clinical study of the metabolism and excretion of NO has been performed on human subjects by the same Swedish group, Wennmalm et al.[47] Plasma levels of nitrate and nitrite and EPR detection of HbNO and MetHb were followed in eight healthy volunteers inhaling 25 ppm NO for 60 minutes. While the basal nitrate level before inhalation was 26 ± 2 μmol/l it reached 38 ± 2 μmol/l after 60 min, parallel to an increase in MetHb from 7 ± 1 to 13 ± 1 μmol/l, a level remaining low at 0.64%. The basal level of plasma nitrite did not change significantly (1.3 ± 0.15 μmol/l) after inhalation, and no HbNO was ever detected by EPR before or during inhalation, within the EPR method detection limit (2 μmol/l).[47] In the same study, eight patients with severe heart failure were examined before cardiac transplantation and submitted to NO inhalation (20 to 800 ppm) for 10 minute periods. Plasma nitrate, nitrite, MetHb and HbNO were measured in arterial and pulmonary arterial blood. Before inhalation patients had basal levels of plasma nitrate and MetHb higher than healthy volunteers, and both components increased significantly and concurrently following NO inhalation. The basal nitrite level was smaller than normal but did not increase after inhalation. A significant HbNO level was detectable by EPR (12 ± 3 μmol/l) in the basal state, but it did not increase significantly (14 ± 1 μmol/l) following 80 ppm NO inhalation.

The two main conclusions of this fundamental clinical study are the following. The uptake within RBC of NO, either endogenously formed in endothelial cells and released luminally or inhaled, followed by conversion to nitrate and MetHb, is the major metabolic pathway before elimination of plasma nitrate via the kidney. The formation of HbNO in vivo probably indicates NO binding to partially deoxygenated blood. The low levels of nitrite and HbNO are explained by the existence of highly efficient and fast metabolic loops, such as nitrite reduction into NO (perhaps by methemoglobin reductase itself), oxidation of HbNO by O_2 especially in the lung alveolar capillaries, oxidation of HbO_2 by NO and methemoglobin reductase (Scheme 5.1). The higher basal levels of plasma nitrate and blood HbNO in heart failure patients, as compared to healthy controls, may have several reasons: previous intravenous doses of nitrovasodilators, facilitated vascular formation of NO or cytokine-induced macrophage activation.[46,47]

Methemoglobin level was assessed during low-dose NO inhalation therapy in 15 newborns with PPHN. MetHb level (basal 0.8%) reached 1.4% following 4 hr inhalation at 20 ppm then decreased to 1% in the subsequent period of 6 ppm inhalation during 20 hr.[50]

Nitric oxide inhalation was tested from the initial view that it had selective pulmonary vasodilating effects without affecting the systemic arterial blood pressure. It is in fact the mere presence of hemoglobin which inactivates vasodilation by NO and thus restricts its effect to lung circulation.[10,51]

CONCLUSIONS

In experiments involving NO interaction with RBC, NO inhalation or NO-donors, the various derivatives of hemoglobin, in particular nitrosyl hybrids and valency hybrids, are effectively detected by EPR spectroscopy but at much smaller levels than predicted from kinetic data (see Scheme 5.1 and chapter 4).

Many other instances of NO capture by hemoglobin in physiological and pathophysiological cases, such as pregnancy,[52] immune diseases, shock situations, etc., shall be given in subsequent chapters (chapters 11 and 12).

The subtle interaction of NO and hemoglobin and the subsequent metabolic processes have other physiological consequences, not always predictable by its presently known biochemistry. Hemoglobin in several instances appears to modulate the physiological or pathophysiological functions of NO. The actual function of NO as an EDRF appears to be regulated by the degree of hemoglobin saturation by O_2; the ability of blood to inhibit the vasodilation caused by NO varies inversely with the saturation of hemoglobin by oxygen.[53] Cell-free hemoglobin also potentiates acetylcholine-induced coronary vasoconstriction.[9] Another interesting example is the augmentation of interleukin-1β-induced production of NO in smooth-muscle cells caused by hemoglobin released from extravasated RBC.[54] The respective effects of NO and Hb, free or within RBC, are therefore extremely intertwined and not easy to predict (Scheme 5.1).[55]

Two recent articles provide new data in this prospect which leads us to assign new physiological roles to NO in interactions with hemoglobin within RBC. In blood containing the hybrid hemoglobin derivative $\alpha NO\alpha_{deoxy}\beta_{deoxy}\beta_{deoxy}$ the oxygen affinity is significantly decreased as compared to normal hemoglobin.[56] NO release from S-nitrosothiols would help O_2 release from RBC to peripheral tissues.[56] A similar conclusion was derived by Jia et al. Hemoglobin was found to be S-nitrosylated onto residue Cysβ-93 at a faster rate on oxyhemoglobin than on deoxyhemoglobin.[57] Thus SNO-Hb might act as an endogenous NO donor to the systemic circulation and participate in the regulation for normal blood flow and delivery of O_2 to tissues.[57]

REFERENCES

1. Kuma F. Properties of methemoglobin reductase and kinetic study of methemoglobin reduction. J Biol Chem 1981; 256:5518-5523.
2. Oda H, Kusumoto S, Nakajima T. Nitrosyl-hemoglobin formation in the blood of animals exposed to nitric oxide. Arch Environ Health 1975; 30:453-456.
3. Yoshida K, Kasama K. Biotransformation of nitric oxide. Environ Health Perspect 1987; 73:201-206.
4. Gaston B, Drazen JM, Loscalzo J et al. The biology of nitrogen oxides in the airways. Am J Respir Crit Care Med 1994; 149:538-551.
5. Kruszyna H, Kruszyna R, Smith RP et al. Red blood cells generate nitric oxide

from directly acting, nitrogenous vasodilators. Toxicol Applied Pharmacol 1987; 91:429-438.

6. Kruszyna R, Kruszyna H, Smith RP et al. Nitrite conversion to nitric oxide in red cells and its stabilization as a nitrosylated valency hybrid of hemoglobin. J Pharmacol Exp Therap 1987; 241:307-313.

7. Kruszyna R, Kruszyna H, Smith RP et al. Generation of valency hybrids and nitrosylated species of hemoglobin in mice by nitric oxide vasodilators. Toxicol Applied Pharmacol 1988; 94:458-465.

8. Kosaka H, Uozumi M, Tyuma I. The interaction between nitrogen oxides and haemoglobin and endothelium-derived relaxing factor. Free Rad Biol Med 1989; 7:653-658.

9. Motterlini R, Macdonald VW. Cell-free hemoglobin potentiates acetylcholine-induced coronary vasoconstriction in rabbit hearts. J Appl Physiol 1993; 75:2224-2233.

10. Rimar S, Gillis CN. Selective pulmonary vasodilation by inhaled nitric oxide is due to hemoglobin inactivation. Circulation 1993; 88:2884-2884.

11. Oda H, Nogami H, Kusumoto S et al. Lifetime exposure to 2.4 ppm nitric oxide in mice. Environ Res 1980; 22:245-263.

12. Maeda N, Imaizumi K, Kon K et al. A kinetic study on functional impairment of nitric oxide-exposed rat erythrocytes. Environ Health Perspect 1987; 73:171-177.

13. Tomoda A, Yubisui T, Tsuji A et al. Kinetic studies on methemoglobin reduction by human red cell NADH cytochrome b_5 reductase. J Biol Chem 1979; 254:3119-3123.

14. Hegesh E, Shiloah J. Blood nitrates and infantile methemoglobinemia. Clin Chim Acta 1982; 125:107-115.

15. Oda H, Nogami H, Nakajima T. Reaction of hemoglobin with nitric oxide and nitrogen dioxide in mice. J Toxicol Environ Health 1980; 6:673-678.

16. Stephens RJ, Freeman G, Evans MJ. Early response of lungs to low levels of nitrogen dioxide. Arch Environ Health 1972; 24:160-179.

17. Maples KR, Sandström T, Su Y-F et al. The nitric oxide/heme protein complex as a biological marker of exposure to nitrogen dioxide in humans, rats, and in vitro models. Am J Respir Cell Mol Biol 1991; 4:538-543.

18. Kon K, Maeda N, Shiga T. Effect of nitric oxide on the oxygen transport of human erythrocytes. J Toxicol Environ Health 1977; 2:1109-1113.

19. Furchgott RF, Zawadzki JV. The obligatory role of endothelial cells in the relaxation of arterial of arterial smooth muscle by acetylcholine. Nature 1980; 288:373-376.

20. Furchgott RF. Studies on relaxation of rabbit aorta by sodium nitrite: the basis for the proposal that the acid-activatable inhibitory factor from bovine retractor penis is inorganic nitrite and that the endothelial-derived relaxing factor is nitric oxide. In: Vanhoutte PM, ed. Vasodilatation: vascular smooth muscle, peptides, autonomic nerves, and endothelium. New York, Raven Press, 1988:401-414.

21. Kohno M, Masumizu T, Mori A. ESR demonstration of nitric oxide production from nitroglycerin and sodium nitrite in the blood of rats. Free Rad Biol Med 1995; 18:451-457.

22. Tomoda A, Tsuji A, Yoneyama Y. Mechanism of hemoglobin oxidation by ferricytochrome *c* under aerobic and anaerobic conditions. J Biol Chem 1980; 255:7978-7983.

23. Tomoda A, Takeshita M, Yoneyama Y. Characterization of intermediate hemoglobin produced during methemoglobin reduction by ascorbic acid. J Biol Chem 1978; 253:7415-7419.

24. Tomoda A, Tsuji A, Matsukawa S et al. Mechanism of methemoglobin reduction by ascorbic acid under anaerobic conditions. J Biol Chem 1978; 253:7420-7423.

25. Kosaka H, Tanaka S, Yoshii T et al. Direct proof of nitric oxide formation from a nitrovasodilator metabolised by erythrocytes. Biochem Biophys Res Commun 1994; 204:1055-1060.

26. Cantilena LR, Smith RP, Frazur S et al. Nitric oxide hemoglobin in patients receiving nitroglycerin as detected by electron paramagnetic resonance spectroscopy. J Lab Clin Med 1992; 120:902-907.

27. Smith RP, Kruszyna H. Nitroprusside pro-

duces cyanide poisoning via a reaction with hemoglobin. J Pharm Exp Ther 1974; 191:557-563.

28. Wilcox DE, Kruszyna H, Kruszyna R et al. Effect of cyanide on the reaction of nitroprusside with hemoglobin: relevance to cyanide interference with the biological activity of nitroprusside. Chem Res Toxicol 1990; 3:71-76.

29. Kruszyna H, Kruszyna R, Rochelle LG et al. Effects of temperature, oxygen, heme ligands and sufhydryl alkylation on the reactions of nitroprusside and nitroglycerin with hemoglobin. Biochem Pharmacol 1993; 46:95-102.

30. Fratacci MD, Frostell CG, Chen TY et al. Inhaled nitric oxide. A selective pulmonary vasodilator of heparin-protamine vasoconstriction in sheep. Anesthesiol 1991; 75:990-999.

31. Frostell C, Fratacci MD, Wain JC et al. Inhaled nitric oxide. A selective pulmonary vasodilator reversing hypoxic pulmonary vasoconstriction. Circulation 1991; 83:2038-2047.

32. Roberts JD, Polaner DM, Lang P et al. Inhaled nitric oxide in persistent pulmonary hypertension of the newborn. Lancet 1992; 340:818-819.

33. Kinsella JP, Neish SR, Shaffer E et al. Low-dose inhalational nitric oxide in persistent pulmonary hypertension of the newborn. Lancet 1992; 340:819-820.

34. Kinsella JP, Neish SR, Dunbar D et al. Clinical responses to prolonged treatment of persistent pulmonary hypertension of the newborn with low doses of inhaled nitric oxide. J Pediatr 1993; 123:103-108.

35. Higenbottam T. Inhaled nitric oxide: a magic bullet? Quarterly J Med 1993; 86:555-558.

36. Pearl RG. Inhaled nitric oxide. The past, the present, and the future. Anesthesiol 1993; 78:413-416.

37. Adnot S, Kouyoumdjian C, Defouilloy C et al. Hemodynamic and gas exchange responses to infusion of acetylcholine and inhalation of nitric oxide in patients with chronic obstructive lung disease and pulmonary hypertension. Am Rev Resp Dis 1993; 148:310-316.

38. Frostell CG, Blomqvist H, Hedenstierna G et al. Inhaled nitric oxide selectively reverses human hypoxic pulmonary vasoconstriction without causing systemic vasodilation. Anesthesiol 1993; 78:427-435.

39. Gerlach H, Rossaint R, Pappert D et al. Time-course and dose-response of nitric oxide inhalation for systemic oxygenation and pulmonary hypertension in patients with adult respiratory distress syndrome. Eur J Clin Invest 1993; 23:499-502.

40. Rossaint R, Falke KJ, Lopez F et al. Inhaled nitric oxide for the adult respiratory distress syndrome. N Eng J Med 1993; 328:399-405.

41. Bigatello LM, Hurford WE, Kacmarek RM et al. Prolonged inhalation of low concentrations of nitric oxide in patients with severe adult respiratory distress syndrome. Anesthesiol 1994; 80:761-770.

42. Foubert L, Fleming B, Latimer R et al. Safety guidelines for use of nitric oxide. Lancet 1992; 339: 1615-1616.

43. Bouchet M, Renaudin MH, Raveau C et al. Safety requirements for use of inhaled nitric oxide in neonates. Lancet 1993; 341:968-969.

44. Laguenie G, Berg A, Saint-Maurice JP et al. Measurement of nitrogen dioxide formation from nitric oxide by chemiluminescence in ventilated children. Lancet 1993; 341:969.

45. Stenqvist O, Kjelltoft B, Lundin S. Evaluation of a new system for ventilatory administration of nitric oxide. Acta Anaesthesiol Scand 1993; 37:687-691.

46. Wennmalm Å, Benthin G, Petresson A-S. Dependence of the metabolism of nitric oxide (NO) in healthy human blood on the oxygenation of its red cell haemoglobin. Br J Pharmacol 1992; 106:507-508.

47. Wennmalm Å, Benthin G, Edlund A et al. Metabolism and excretion of nitric oxide in humans. An experimental and clinical study. Circ Res 1993; 73:1121-1127.

48. Iwamoto J, Krasney JA, Morin FC. Methemoglobin production by nitric oxide in fresh sheep blood. Resp Physiol 1994; 96:273-283.

49. Eriksson LE. Binding of nitric oxide to intact human erythrocytes as monitored by

electron paramagnetic resonance. Biochem Biophys Res Commun 1994; 203:176-181.

50. Kinsella JP, Abman SH. Methaemoglobin during nitric oxide therapy with high-frequency ventilation. Lancet 1993; 342: 615.

51. Rich GF, Roos CM, Anderson SM et al. Inhaled nitric oxide:dose response and the effects of blood in the isolated rat lung. J Appl Physiol 1993; 75:1278-1284.

52. Conrad KP, Joffe GM, Kruszyna H et al. Identification of increased nitric oxide biosynthesis during pregnancy in rats. FASEB J 1993; 7:566-571.

53. Iwamoto J, Morin FC. Nitric oxide inhibition varies with hemoglobin saturation. J Appl Physiol 1993; 75:2332-2336.

54. Suzuki S, Takenaka K, Kassell NF et al. Hemoglobin augmentation of interleukin-1β-induced production of nitric oxide in smooth-muscle cells. J Neurosurg 1994; 81:895-901.

55. Kosaka H, Shiga T. Detection of nitric oxide by electron paramagnetic resonance using hemoglobin. In: Feelisch M, Stamler JS. Methods in Nitric Oxide Research. John Wiley & Sons, 1996:373-381.

56. Kosaka H, Seiyama A. Physiological role of nitric oxide as an enhancer of oxygen transfer from erythrocytes to tissues. Biochem Biophys Res Commun 1996; 218:749-752.

57. Jia L, Bonaventura C, Bonaventura J et al. S-nitrosohaemoglobin: a dynamic activity of blood involved in vascular control. Nature 1996; 380:221-226.

UTILIZATION OF NITRIC OXIDE AS A PARAMAGNETIC PROBE OF THE MOLECULAR OXYGEN BINDING SITE OF METALLOENZYMES

Yann A. Henry

INTRODUCTION

The paramagnetism of many nitrosylated complexes of metalloproteins has led to numerous but limited studies by EPR spectroscopy since the early seventies. Spectral analysis can, in the best of cases, bring evidence of specific structural changes within the metal ligation sphere. However interpretations are somewhat difficult when no X-ray or other complementary spectroscopic data, such as EXAFS, resonance Raman or infrared, are available. Furthermore some of the nitrosylated metalloprotein complexes are diamagnetic (S = 0) or paramagnetic with integer spin (S = 1, 2, etc.), which are EPR-silent with conventional instrumentation. Despite its specificity EPR spectroscopy is therefore much less in general use than other spectroscopic methods, such as resonance Raman or NMR. See chapter 3 for the basic principles of EPR spectroscopy and chapters 4 and 5 for its exemplary application to hemoglobin in vitro and in red blood cells.

Nitric oxide binds in vitro to many oxygen carriers, and to oxygenases and oxidases, which contain pentacoordinated hemes (a, b, c' or d), Fe-Fe or Cu-Cu groups or mononuclear Fe or Cu protein-bound centers (overviewed in refs. 1 and 2). In most instances nitrosylation led to the enzyme inhibition, and NO was used as an EPR probe of the oxygen binding site with the implicit assumption of similarity in binding. Proteins with [Fe-S] and other clusters, which are often O_2-inactivated, have also been probed by NO.

Our aim in the present chapter is to compile an EPR spectra reference catalog or dictionary, based on in vitro experiments with

Nitric Oxide Research from Chemistry to Biology: EPR Spectroscopy of Nitrosylated Compounds, edited by Yann A. Henry, Annie Guissani and Béatrice Ducastel.

purified proteins. It represents in a way the very center of this book.

All metalloproteins listed in this chapter bind NO, but the biological relevance in vivo is in most cases null or remains to be established, as will be discussed in chapters 10 and 14. Some redundancy with chapters 7, 10 and 14 is inevitable, but we will try to take different points of view in these different chapters.

In our description of the binding of NO to hemoglobin we followed mostly a historical order. In the present chapter, in an effort to be concise, we will rapidly describe each metalloprotein potential NO binding site, as analyzed by X-ray spectroscopy (when available) and by other complementary spectroscopic methods: EXAFS, infrared, resonance Raman (RR), etc., experiments which were usually posterior to NO binding studies by EPR.

HEMOPROTEINS

Many pentacoordinate hemoproteins bind NO, in the ferric as well as in the ferrous states. Nitric oxide can also, in a few instances, displace one of the axial ligands of hexacoordinate heme iron.

MYOGLOBIN

Nitric oxide binding to myoglobin (usually from sperm-whale skeletal muscle, horse heart or horse skeletal muscle) has been studied mostly in order to constitute a simple model of nitrosyl hemoproteins, particularly HbNO. In fact EPR spectra at 77 K of MbNO obtained either with single crystals or using frozen solutions were very similar to those of HbNO, that is a mixture of the respective characters of α and β subunits.[3-6] The principal g-values of the g-tensor are g_{zz} = 2.0068, g_{yy} = 1.9850, g_{xx} = 2.0728, and two hyperfines splitting in the z-direction were measured, one due to the NO N-atom (24.5 G for ^{15}NO, 18.6 G for ^{14}NO), the other due to the N_{ϵ} of the proximal His-F8 (6.4 G). The Fe-N-O bond angle appears to be 110° similar to those found in HbNO.[4,7]

More accurate EPR studies of MbNO in frozen solutions as a function of temperature demonstrated that MbNO exists in an equilibrium between two conformations. Both are hexacoordinate conformers differing primarily in the position of the iron with respect to the ligands and to the heme plane.[8] This was confirmed by a study of MbNO single crystals at ambient temperature and at 77 K which showed that freezing induces reversible changes in the molecular orientation of the ligand without disorder of the crystal lattice (Fe-N-O angle equal to 153° at 20°C, 109° at 77 K).[9] In fact this is true also of α and β hemoglobin subunits.[8,10,11] EPR spectroscopy has also been used to probe Mb, modified Mb, metMb and ferrylmyoglobin in their NO binding properties.[12-18] Results were very similar to those obtained for hemoglobin, summarized in chapter 4.

Other physical techniques were used to probe the heme site of Mb taken as a simple prototype of hemoproteins, e.g., guanylate cyclase, such as infrared spectroscopy of NO bonding,[19] or ultrafast (picosecond) geminate recombination of NO following laser photolysis,[20-23] often used as an experimental test of advanced simulation techniques of molecular dynamics.[23,24] Laser photolysis experiments were performed in particular with a view to provide a model of *pentacoordinate* and *tetracoordinate* heme within a globin pocket without proximal and/or distal ligand, such as could be the state of the heme moiety in the turnover of guanylate cyclase.[25,26]

ANIMAL SPECIES AND PLANT OXYGEN TRANSPORTERS

The following paragraphs are, like those on Mb, a sequel of chapter 4. Animal species, mammals, fishes, insects, annelids, polychaetes, etc., have monomeric myoglobins, tetrameric hemoglobins or extracellular erythrocruorins, heme-containing oxygen carriers, with O_2 affinity, heme-heme interaction and Bohr effects (or Root effect for fishes) which can be very different from those of human adult HbA. NO binding and the resulting EPR spectra have been used to probe the heme site of these oxygen carriers.[27-32] These studies were initial attempts at structural interpretations of functional properties, which may appear anecdotic in retrospect!

The instance of leghemoglobin (LegHb) is more interesting for the understanding of the possible functions of this specific oxygen carrier. Leghemoglobins from root nodules of leguminous plants (soya-bean (*Glycine max*), cowpea (*Vigna ungriculata*), peas (*Pisum sativum, Phaseolus vulgaris*), lupine (*Lupinus luteus*), etc.) are essential for nitrogen fixation, although they do not participate directly in symbiotic N_2 fixation. The leguminous plants and specific bacteria (*Rhizobium*) are symbiotic, the bacteria being contained in nodules attached to the plant roots. N_2 fixation occurs in the nodules and is profitable to the plant. LegHb isoproteins are contained at high concentration (~ 1 mmol/l) in the cytoplasm of the plant nodule cells and outside of the bacteroids. The globin parts are coded by plant genes and heme by the bacteria; the assembly occurs at the plant cell ribosomes. The unique characteristic of LegHb is an extremely high affinity for O_2 (~ 10^7 mol^{-1} l), 10-fold that of sperm-whale Mb. Leghemoglobin's function is to facilitate O_2 diffusion at the bacteroids' surface, to provide a stable and very low free O_2 concentration and thus protect nitrogenase from destruction by O_2.

LegHb isoproteins from soya-bean and cowpeas were studied by EPR spectroscopy in the ferric (high-spin and low-spin) forms and in the ferrous nitrosylated form. The study showed that crude extracts of root nodules contained a large concentration of endogenous NO bound to ferrous LegHb (27% of total LegHb).[33] The origin of NO could be soil nitrate which the bacteria, and possibly the plant,[34] can reduce (see following chapter 7). It could also derive from L-arginine NOS enzymatic systems, which might occur in plants.[35] Thus NO binding will decrease the amount of ferrous LegHb available for facilitating O_2

diffusion to the nodule bacteria. It could equally have the interest to prevent inhibition of bacteroid nitrogenase by NO or nitrite through binding to the [4Fe-4S] and [FeMoS] clusters of the two nitrogenase components (see below).[36-40]

CYTOCHROMES *C* AND *C'*

Mitochondrial cytochrome *c* from beef heart forms a diamagnetic complex with NO in the ferric state.[41] A ferrous NO complex can be formed by alkalinization of the ferric complex, which remains stable when titrated back to neutral pH.[41] The NO molecule takes the place of Met-80 at a neutral pH or of a probable Lys residue at an alkaline pH, while His-18 remains the proximal heme iron ligand.[42] The EPR spectra obtained at pH 12 and 7, are quite similar in shape to those of HbNO and MbNO with three distinct g-values, 2.07, 2.003 and 2.01 and a well resolved SHF structure of nine lines, triplet (6.8 G) of a triplet (23.7 G for [14]NO) (Fig. 6.1).[5,42]

Another cytochrome has been probed by EPR of its nitrosylated complex, cytochrome *c'*. It is a periplasmic class II cytochrome *c*, with very peculiar heme spin states. As it is distributed only amongst bacteria and could have some function in regulating free NO concentration in denitrifying bacteria, it shall be dealt with in chapter 7.[43-45] Interestingly, it could be a model of a guanylate cyclase heme site (see chapter 10).

PEROXIDASES

Many peroxidases, cytochrome *c* peroxidase (CcP; EC 1.11.1.5) from baker's yeast,[5,46] horseradish peroxidase (HRP; EC 1.11.1.7),[5,46] isoperoxidases (P1 to P7) from turnip,[47] chloroperoxidase (CPO; EC 1.11.1.10), a P-450-like enzyme from *Caldariomyces fumago*,[48] myeloperoxidase (MPO; EC 1.11.1.7) from neutrophils and monocytes[49] and lactoperoxidase (LPO; EC 1.11.1.7) from milk,[49,50] were "spin probed" by EPR spectroscopy of their Fe$^{(II)}$-NO form, analogous to their usually transient diamagnetic Fe$^{(II)}$-O_2 form. Their EPR-silent Fe$^{(III)}$-NO complexes are reversible and

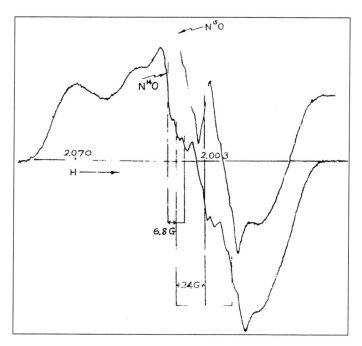

Fig. 6.1. EPR spectra of ^{14}NO- and ^{15}NO-ferrocytochrome c in frozen solution at 77 K. (Reproduced with permission from: Kon H. Electron paramagnetic resonance of nitric oxide cytochrome c. Biochem Biophys Res Commun 1969; 35:423-427.) Copyright © Academic Press, Inc, Orlando, Florida, USA.

stable but not reducible by NO in contrast to those of Mb and Hb.

EPR spectra of these compounds are the best resolved of any nitrosylated metalloprotein. Actually they were the first systems to provide a full description of the interaction of a NO unpaired electron with the iron atom and its N ligands (see chapters 3 and 4). Three distinct g-values are measured: g_x between 2.08 and 2.06, g_z at 2.004 and g_y between 1.95 and 1.97, revealing a largely rhombic symmetry $Fe^{(II)}$-NO state as in the enzyme ferric resting state. A very well resolved nine-line SHF structure is detected in the z-direction (30-25 G for ^{15}NO, 21-16 G for ^{14}NO, 6.5 for the proximal ^{14}N atom, and 6.5 G for ^{57}Fe).[5] Quite exceptional SHF structures can also be detected in the x- and y-directions (see Figure 4.4 in chapter 4).

In the case of CcP from baker's yeast, the proximal N atom is that of His-175 residue, three residues Trp-51, His-52 and Arg-48 being on the distal side of the heme b pocket.[51,52] For one of the turnip isoperoxidases, the EPR probe is sensitive to the peroxidase P1 conformational changes induced by a substrate (indolacetic

acid) or an analogue (indolbutyric acid) binding.[47] Both the symmetry of the active $Fe^{(II)}$-NO site and its SHF interaction are modified upon substrate or analogue binding to a site distinct from the $Fe^{(II)}$-NO site (Fig. 6.2).[47]

EPR results obtained on nitrosyl ferrous LPO were somewhat discordant.[49,50] While one group detected a three-line SHF structure, the other obtained a well resolved nine-line structure; the general aspect of the spectra was also different, indicating differences in symmetry. In fact the bond between the heme b iron atom and an N atom in the "fifth" coordination position could be either broken or with another ligand, possibly due to a pH drop upon freezing or to a difference in the enzyme reduction process by dithionite or ascorbate, or by electrochemistry.

Myeloperoxidase, which converts Cl⁻ into HOCl via a two-electron peroxidation step, and eosinophil peroxidase from human leukocytes gave EPR spectra in the $Fe^{(II)}$-NO state, with nine-line SHF structure and rhombic symmetry, similar to those obtained with milk LPO.[49] A recent report demonstrated an interaction of human

MPO with nitrite, characterized by a conversion from ferric high-spin to ferric low-spin as detected by UV-visible and EPR spectroscopy.[53] A complex conversion of NO_2^- to NO and peroxynitrite $ONOO^-$ by $O_2^{-\bullet}$ in activated neutrophils could also interfere with hypochlorite HClO synthesis by the formation of an even stronger oxidant, and strong nitrosylating agent, nitrosyl chloride ClNO.[54]

Chloroperoxidase isolated from the mold *Caldariomyces fumago* catalyzes the classical peroxidation and a specific halogenation reaction. It differs from the other peroxidases as it has several P-450 characters: ferrous CO complex Soret band at 443 nm, Mössbauer parameters and EPR high-spin and low-spin g-values,[55] very similar if not identical to those of P-450$_{cam}$ from *Pseudomonas putida* or P-450$_{scc}$ from bovine adrenal cortex (see below), which suggest an axial thiolate ligand to heme *b* iron. However CPO has only two half-cystine residues, neither being free sulfhydril as they form a disulfide linkage.[48] The ferrous heme site was probed by NO. The EPR spectrum is very similar to those of P-450-NO with a three-line pattern (20 G splitting) in the z-direction (g_z = 2.004) and three-line splittings in the x- and y-directions (g_x = 2.082 and g_y = 1.975).[48] This would indicate that both proteins do not have an N atom as a proximal ligand to the iron. This has been proven for P-450$_{cam}$ by the crystal structure at 1.63 Å resolution, revealing residue Cys-357 as the proximal ligand in both forms of the enzyme (see below). In the high-spin state the iron atom is pentacoordinate, the sixth ligand of the low-spin form being a hydroxide anion.[56] The iron axial ligand of CPO is not yet known.

CATALASE

Catalase (hydrogen peroxide:hydrogen peroxide oxidoreductase, CAT; EC 1.11.1.6) occurs in almost all aerobically respiring organisms, catalyzing the following reaction:

$$2H_2O_2 \rightarrow O_2 + 2H_2O$$

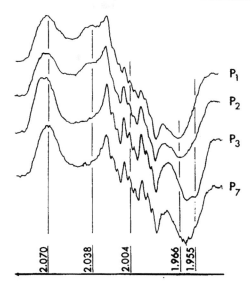

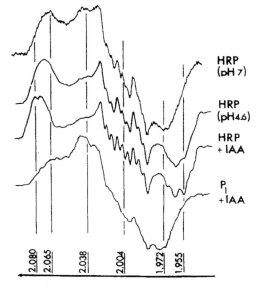

Fig. 6.2. Upper panel. EPR spectra at 77 K of nitric oxide complexes of turnip isoperoxidases P1, P2, P3 and P7. Lower panel. Effects of pH variation and indolacetic acid addition. (Reproduced with permission from: Henry Y, Mazza G. EPR studies of nitric oxide complexes of turnip and horseradish peroxidases. Biochim Biophys Acta 1974; 371:14-19.) Copyright © Elsevier Science, Sara Burgerhartstraat 25, 1055 KV Amsterdam, The Netherlands.

with no exogenous electron donor, and catalyzing also the usual peroxidation reaction.

The main sources of catalase for structural and functional studies are mammalian liver, kidney and erythrocytes.[57] Beef liver CAT contains four identical subunits (57 kDa) with ferric high-spin heme *b* which can bind exogenous ligands, F^-, NO_2^-, N_3^-, formate, etc., and stay in a high-spin state. The X-ray structure at 2.5 Å resolution indicates residue Tyr-357 as the Fe proximal ligand, while His-74 and Asp-147 are important residues on the distal side.[58]

It is not clear from the literature whether nitric oxide is able to bind to ferric heme of native CAT and whether it can reduce it to the $Fe^{(II)}$-NO form.[59,60] In fact CAT appears to be irreducible, even by dithionite. The quality of the purified CAT sample used and the existence of two conformational states could account for the variability in the ease of reduction. However the intermediate formation of Compound I with H_2O_2 followed by the oxidation of azide N_3^- or its acid form N_3H or by hydroxylamine NH_2OH, yields CAT in the EPR-detectable $Fe^{(II)}$-NO form (CAT-NO).[5,59,60] The reaction goes through an azidyl radical $N_3^•$ intermediate.[61] While the EPR spectrum of CAT-NO obtained by Yonetani et al[5] is quite specific, showing a rhombic symmetry ($g_x = 2.05$, $g_y = 1.97$, $g_z = 2.004$) and a characteristic three-line SHF interaction (21 G) due to NO (the proximal Tyr-357 ^{16}O atom having no nuclear spin), those obtained by Craven et al are quite nonspecific (see Figure 4.3 in chapter 4).[5,60]

A recent report brings a new interest in NO binding by catalase. Free NO and O_2 concentrations were measured simultaneously with independent oxygen and NO electrodes, so that the effect of NO on the catalase activity toward H_2O_2 could be simultaneously assayed.[62] NO and H_2O_2 do not seem to react appreciably at reasonably low concentrations of H_2O_2 (< 10 mmol/l). Catalase activity was reversibly inhibited by NO with an estimated $K_i = 0.18$ μmol/l,

well in the possible range (50 nmol/l to 5 μmol/l) of in vivo NO concentrations.[62] However the author also measured a direct binding of NO by catalase, with a high affinity, a finding which contradicts some previous reports using native CAT. Simultaneous reactions of NO with CAT, O_2 and possibly H_2O_2 with different order of reactions could complicate the quantitative analysis of catalase inhibition by NO.

HEME *B*-CONTAINING DIOXYGENASES

L-Tryptophan 2,3-dioxygenase (TPO, sometimes incorrectly called tryptophan pyrrolase; EC 1.13.11.11) from rat liver and from *Pseudomonas* is a tetramer containing two molecules of heme *b* per molecular weight of 120 kDa. It has a dioxygenase activity, the two atoms of the O_2 molecule being incorporated into the pyrrole ring to yield N-formylkynurenine by opening the C2-C3 double bond of the indole ring. The enzyme is inhibited by CO and NO. Dithionite-reduced TPO binds NO and gives specific EPR spectra (Fig. 6.3.).[63]

The TPO-NO complex symmetry and SHF interaction are deeply modified upon binding of the substrate L-tryptophan, and of the competitive inhibitor 5-hydroxytryptophan or of the allosteric effector α-methyltryptophan.[63] This simple example shows the possibility of the EPR-NO probing of allosteric enzymes, in a manner analogous to the probing of hemoglobin.

Indoleamine 2,3-dioxygenase (IDO, EC 1.13.11.17; often known as intestinal D-tryptophan pyrrolase) catalyzes the same reaction as TPO, that is the oxygenative cleavage of the indole ring, on various indoleamine derivatives with a wide substrate specificity, D- and L-tryptophan, D- and L-hydroxytrytophan, tryptamine and serotonin.[64] It consists of a single 42 kDa polypeptide chain with one heme *b*. It is rather widely distributed in mammals, for example in the intestine, lung, stomach, spleen and brain of rabbit, and plays an important role in the metabolism of serotonin (3-(2-aminoethyl)-indole or 5-hydroxy-tryptamine), melatonin (N-acetyl-5-methoxytryptamine), and other indoleamine

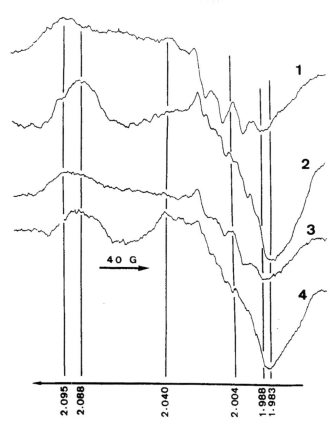

Fig. 6.3. EPR spectra of nitric oxide complex of TPO in frozen solution at 77 K. 1: TPO-NO alone; 2: TPO-NO in the presence of L-tryptophan; 3: TPO-NO in the presence of α-methyltryptophan; 4: TPO-NO in the presence of 5-hydroxytryptophan. (Reproduced with permission from: Henry Y, Ishimura Y, Peisach J. Binding of nitric oxide to reduced L-tryptophan-2,3-dioxygenase as studied by electron paramagnetic resonance. J Biol Chem 1976; 251:1578-1581.) Copyright © The American Society for Biochemistry and Molecular Biology, Inc. Bethesda, MD, USA.

40 G

2.095 2.088 2.040 2.004 1.988 1.983

derivatives.[65] The dioxygenase activity is accelerated in the presence of $O_2^{-\bullet}$.[65,66]

In the ferric and ferrous forms, IDO is a mixture of high-spin and low-spin species, the proportion of which varies with substrate binding.[67] The "fifth" ligand seems to be a His residue. The two hexacoordinate low-spin species could have another His residue or an OH⁻ as the "sixth" iron ligand.[68] The dithionite-reduced enzyme binds NO as detected by UV-visible absorption and EPR spectroscopy, indicating a rhombic symmetry (g_x = 2.08, g_y = 1.98 and g_z = 2.01) and an N sixth ligand.[64,68]

It would be worthwhile to know whether NO interferes with O_2 and $O_2^{-\bullet}$ binding by IDO and whether it is an inhibitor to a biologically significant level. In human mononuclear phagocytes and other cells types, exposure to interferon-γ causes the induction of IDO.[69,70] In the

same cells nitric oxide inhibits IDO activity, causing a functional perturbation in the linkage between arginine and tryptophan through interferon-γ induction of IDO and NOS (see chapter 14).[71]

PROSTAGLANDIN H SYNTHASE (CYCLOOXYGENASE)

Prostaglandin H (prostaglandin endoperoxide) synthase (PGHS, EC 1.14.99.1) catalyzes the conversion of arachidonic acid (5,8,11,14-eicosatetraenoic acid) to prostaglandins (PGG_2 and PGH_2), precursors of the other prostaglandins, prostacyclin and thromboxane A2. It has two different enzymatic activities: a bis oxygenase (cyclooxygenase, COX) activity by which it incorporates two molecules of O_2 into one molecule of arachidonic acid to form 15-hydroperoxy-9,11-peroxidoprosta-5,13-dienoic acid (PGG_2), and a peroxidase reaction which reduces the 15-perhydroxy group into a 15-hydroxy group to form 15-hydroxy-9,11-peroxidoprosta-5,13-dienoic acid (PGH$_2$). It is a homodimer of a glycoprotein (70 kDa) which requires one heme b per monomer to recover full enzymatic activity after purification of the apoprotein, although the protein can bind unspecifically higher amounts of heme. It is commonly purified by use of detergents from microsomes of bovine and ovine seminal glands and from sheep platelets, in which it is tightly associated to the endoplasmic reticulum membrane. PGHS is the target of nonsteroidal anti-inflammatory drugs such as indomethacin, which are reversible, competitive inhibitors at the arachidonate catalytic binding site.[72] Aspirin and related salicylic acid derivatives irreversibly inhibit PGHS by interaction at a distinct site through acetylation of residue Ser-530 which interferes with arachidonate binding.[73] The amino acid sequence has been obtained from cDNA of sheep, mouse and human PGHS.

Spectroscopic studies (optical absorption, EPR, MCD and RR) have shown that ferric resting PGHS is an equilibrium mixture of high-spin and low-spin species.[74] The nature of the heme iron axial ligands has been the subject of controversy. On the basis of the EPR spectrum of the high-spin species with g-values at 6.6 and 5.4 analogous to those of native catalase and indicating a rhombic symmetry, PGHS was proposed to have a tyrosine O atom as a proximal ligand.[75] This was substantiated by the absence of an N atom SHF structure except for that of NO in the ferrous nitrosylated PGHS EPR spectra (see below). However amino acid sequence comparisons of mouse and sheep PGHS suggested His-309 as the axial ligand of the enzyme.[73] A site-directed mutagenesis study also suggested His-309 as one of the heme axial ligands and His-207 or His-388 as possible distal heme ligands.[76] Another detailed spectroscopic study of the enzyme finally suggested that both the high-spin and the low-spin ferric species were hexacoordinated to a pair of histidine residues, the proximal being His-309, and

the distal histidine being weakly associated and exchangeable with exogenous ligands, CN$^-$, N$_3^-$, F$^-$ or imidazole.[74] The value of the midpoint potential value (-52 mV) of PGHS was found to be close to those of hemoproteins with two histidine axial ligands and different from that of catalase (-400 mV). In fact axial ligands might also change during the redox cycle mechanism.

PGHS also produces a transient tyrosyl radical, detectable by EPR, which appears during both the cyclooxygenase and the peroxidase reactions.[77-80] The tyrosyl radical assumes several EPR line-shapes and line-widths during the course of the reactions which probably arise from the same residue in various conformations of the methylene protons of the tyrosyl radical.[79] While the tyrosyl radical in one conformation (doublet EPR line-shape) probably participates in the cyclooxygenase catalysis, the other two tyrosyl conformations (singlet EPR line-shape) might lead to the enzyme self-inactivation processes.[80] This tyrosyl radical, probably Tyr-385, presents striking analogies with that found in a stable form in ribonucleotide reductase subunit R2 (see below and chapter 10).[77-80]

PGHS therefore offers at least two potential targets for reaction with NO, heme b and the tyrosyl radical. Binding of NO to ferrous heme of PGHS produces two types of NO complexes, characterized by different EPR signals: a species with rhombic symmetry at $g_x = 2.07$, $g_z = 2.01$ and $g_y = 1.97$, assigned to a hexacoordinate complex, and a stable axial species at $g = 2.12$ and $g = 2.001$, assigned to a nonspecific, i.e., pentacoordinate nitrosyl heme, similar to that of denatured HbNO or P-420-NO (see below).[75] The SHF structure consists of only three lines analogous to the catalase-NO spectrum, which does not indicate an N atom of a histidine residue as the proximal iron ligand. The issue of the actual nature of the axial ligands in PGHS in the two redox states shall only be solved by X-ray diffraction analysis. As for a possible interaction between NO and the transient tyrosyl radical of PGHS, similar to the quenching of the tyrosyl radical of ribonucleotide reductase, no report has yet appeared. The importance of a possible role of L-arginine derived NO in regulations of PGHS (COX) functions, together with that of other enzymatic pathways stemming from arachidonic acid and lipoxygenases, shall be discussed in chapter 14.

P-450 MONOOXYGENASES

The P-450 enzymes (cytochrome P-450, EC 1.14.14.1) are a widespread superfamily of heme *b*-containing monooxygenases in prokaryotes and eukaryotes; more than 250 distinct P-450 genes have been identified.[81] They play an important role in the oxidative metabolism of endogenous substrates and the oxidative detoxication of exogenous compounds.[82] Since the original finding, which defined "cytochrome P-450" as an unusual Soret band at 450 nm for the ferrous CO derivative, the identification of heme *b* proximal ligand and of the sixth ligand has kept many spectroscopists busy for 20 years until the X-ray crystallographic structure of one P-450 was described.[56,83-85] The most studied P-450s are those extracted from *Pseudomonas putida* (P-450$_{cam}$), from liver microsomes of animals (rats, rabbits) pretreated with methylcholanthrene or phenobarbital (P-450$_{LM}$), or from bovine adrenal cortex (P-450$_{scc}$).

P-450 and model compounds have been studied by many spectroscopic methods (among which EPR when possible) and in several states: ferric high- and low-spin, ferrous CO and ferrous NO. Rapidly a thiolate proximal ligand was proposed.[86-89] As P-450 can be in several spin states, high and low in both the ferric and the ferrous oxidation states, the nature of the sixth ligand *trans* to cysteinate was also debated. This ligand can be displaced in the ferric protein by CN^-, guanidine or amines, and by CO and NO in the ferrous protein. While Peisach and colleagues proposed an imidazole residue from His,[88-91] others proposed an oxygen donor atom,[92] that of a Tyr residue in particular.[93]

The crystal structures at 2.6 to 1.63 Å resolution of P-450$_{cam}$, a 45 kDa single

414 amino acid polypeptide chain, confirmed a without a doubt that the proximal ligand is Cys-357, in the most highly conserved sequence in eukaryotic P-450s, in substrate (camphor)-bound ferric, substrate-free ferric and substrate (camphor)-bound-CO ferrous forms of the enzyme. In the high-spin state, as in the ferric camphor-bound state, the iron atom is pentacoordinated. The sixth ligand of the low-spin, substrate-free form, is proposed to be a water molecule or a hydroxide anion. Seven conserved or homologous amino acids, Thr-252, Gly-248, Val-247, Leu-244, Phe-87, Tyr-96 and Phe-98, line the wide distal pocket, providing oxygen and substrate camphor binding sites.[56,83-85] The Tyr-96 residue is probably one of the two tyrosine residues which can be nitrosated.[93]

Nitric oxide binds to microsomal P-450$_{LM}$ from phenobarbital-treated rat and to P-450$_{cam}$ in both the ferric and the dithionite-reduced forms.[89,94-98] As in the case of CO binding, NO binding to the ferrous enzyme elicits an unusual Soret absorption band at 437 nm as compared to the usual absorption at 420 nm. The irreversible conversion of the unstable P-450 state to the P-420 state occurs also for the NO-bound enzyme.

EPR spectroscopy has been used to probe the Fe$^{(II)}$-NO heme site. P-450-Fe$^{(II)}$-NO EPR signal is rhombic ($g_x \sim 2.07$, $g_y \sim 1.97$, $g_z \sim 2.004$) with a definite three-line SHF structure (20 G) in the z-direction, indicating a hexacoordinate heme-NO complex, although excluding a proximal N atom as an axial ligand. The P-420-Fe$^{(II)}$-NO form has an axial symmetry ($g_x = g_y \sim 2.07$, $g_z \sim 2.01$) with a 16 G SHF triplet, similar to that of SDS-denatured HbNO, indicating a nonspecific pentacoordinate heme-NO complex (Fig. 6.4).[94,95]

Substrate binding modifies slightly the g-values in the case of P-450$_{cam}$, but large variations in symmetry occur upon binding of cholesterol and analogues to P-450$_{scc}$.[96] Although NO binding to P-450 enzymes induces their inhibition,[99,100] its extent has not been fully quantified. Thus the biological relevance of NO binding, apart from its use as a spin label, is not proven. Nitrite addition to microsomes could also lead to the inhibition of the functions, as nitrite and P-450 are enzymatically reduced by NADPH in microsomes, forming the characteristic reduced P-450-NO complex.[101] Whether this inhibition induced by nitrite occurs in vivo is not known (see chapter 14).

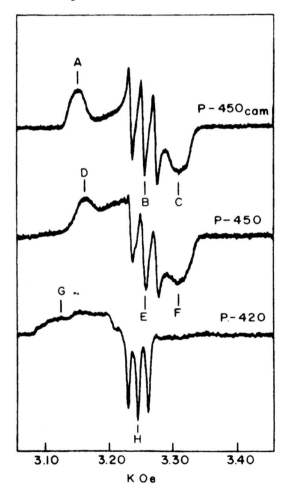

Fig. 6.4. EPR spectra at 120 K of ferrous P-450cam-NO, microsomal P-450-NO and P-420-NO. (The magnetic field is express in Oersted.) The g-values are A: 2.068; B: 2.002; C: 1.970; D: 2.062; E: 2.002; F: 1.970; G: 2.089 and H: 2.009. (Reproduced with permission from: Ebel RE, O'Keeffe DH, Peterson JA. Nitric oxide complexes of cytochrome P-450. FEBS Lett 1975; 55:198-201.) Copyright © Elsevier Science, Sara Burgerhartstraat 25, 1055 KV Amsterdam, The Netherlands.

OTHER HEMOPROTEINS

A few other hemoproteins bind NO. This binding is relevant to the enzyme function in nitrite-reductase (cytochrome cd_1),[102-105] or in hexaheme cytochrome-nitrite reductase,[106] and in NO-reductase (cytochrome bc).[107,108] This shall be dealt with specifically in chapter 7 on the role of NO as an intermediate in the denitrification process.

Another class of hemoproteins that interact with NO are the nitrophorins, nitric oxide carriers, found in salivary glands of blood-sucking insects, *Rhodnius proxilus*, a vector of Chagas' disease, or the bedbug *Cimex lectularius* (see chapter 14).[109,110]

CYTOCHROME *C* OXIDASE (CYTOCHROME *AA₃*)

Being a complex heme *a*- and copper-containing protein and thus having several potential NO binding sites, cytochrome *c* oxidase deserves a special treatment!

Cytochrome aa_3 (cyt aa_3) is a member of the heme-copper cytochrome oxidase superfamily of redox-driven proton pumps.[111-113] Other members contain other heme types: heme *b*, *d* or *o*. There are two main branches in the superfamily according to their substrate specificity: those that use cytochrome *c* as substrate (ferrocytochrome *c*-oxygen oxidoreductase, EC 1.9.3.1; cytochrome *c* oxidase, CcO) found in mitochondria (the prototype being that of bovine heart) and many bacteria (*Paracoccus denitrificans*, *Rhodobacter sphaeroides*), and those that use membrane-bound quinol (ubiquinol or menaquinol) found in bacteria (*Bacillus subtilis*, *E. coli*).[111-114] We shall restrain the discussion in this part to a brief description of the active sites of CcO and of its spin probing by NO.

CcO is part of the mitochondrial respiratory chain that catalyzes the terminal act of respiration by exchange of electrons derived from the oxydation of foodstuffs to molecular oxygen. It is called complex IV, and was called *Der Atmungsferment* by Otto Warburg and cytochrome aa_3 by David Keilin.[111] CcO is a complex integral membrane protein comprised of 2 or 3 subunits in the simplest bacterial systems and of 13 dissimilar subunits in mammals.[115] They all contain two inequivalent a hemes and two inequivalent copper atoms and catalyze the four-electron reduction of O_2 to H_2O with the reducing equivalent from cyt c:

$$4\text{Cyt } c^{2+} + O_2 + 4H^+ \rightarrow 4\text{Cyt } c^{3+} + 2H_2O$$

The electrons enter CcO, imbedded in the 40 Å thick mitochondrial inner membrane from the cytosol side, and the protons consumed are taken up from the matrix side. All four redox-active metal centers participate in the catalytic activity, with different specific functions. Cytochrome a and Cu_A (also called Cu_a) participate in the electron flow from reduced cyt c to the O_2-binding site, formed by the binuclear cluster, cytochrome a_3 and Cu_B (Cu_{a3}).[115-117]

Cytochrome a contains a hexacoordinate low-spin heme a, with two His residues (numbered His-102 and His-421 in $Rb.\ sphaeroides$ CcO) as axial ligands in both redox states and a medium to low redox potential (350 to 280 mV depending on the state of cyt a_3 and also pH-dependent). Copper A is the low-potential copper atom of CcO (285 mV). It is probably the primary electron acceptor from ferro cyt c, although that role has also been assumed previously for the cyt a moiety. Four amino acid ligands of Cu_A have been proposed on the basis of many spectroscopic, EPR, ENDOR, EXAFS and RR studies: two His and two Cys residues, constituting a highly covalent environment, with some similarity to the "blue" copper site of type 1 found in copper oxidases (see below).[115,118] More recently, it has been proposed that this Cu_A center has two equivalent copper atoms in a mixed valence [Cu(I)-Cu(II)] or rather a [Cu(1.5)-Cu(1.5)] $S = 1/2$ configuration, explaining its peculiar EPR spectrum,[119,120] like that found in nitrous oxide reductase (see chapter 7).[121] This matter is still open to debate.[118,122]

The binuclear heme a_3-Cu_B center has also been studied by a variety of spectroscopic techniques: EPR, ENDOR, EXAFS, RR, etc. It coordinates a variety of exogenous ligands: F^-, CN^-, formate and peroxide in the

oxidized state; and O_2, CO and NO in the reduced state. Cytochrome a_3 is a high-spin heme in both the oxidized and reduced forms of CcO, with one N atom of a His residue (His-284 or His-419 of $Rb.\ sphaeroides$ CcO) as an axial ligand in the position distal to Cu_B. The other axial ligand of the iron atom depends on the state of the enzyme. In the resting ferric state, the heme a_3 is strongly antiferromagnetically coupled to Cu_B yielding a net $S = 2$ paramagnetic state, which does not exhibit an EPR signal with conventional instrumentation. The bridging ligand could be a μ-oxo or a μ-hydroxyl. In the reduced state, heme a_3 is high-spin ferrous and also $S = 2$. On the open side of heme a_3 lies Cu_B with three His residues as ligands (His-333, His-334, and His-419 or His-284 for $Rb.\ sphaeroides$ CcO). While heme a_3 and Cu_B interact magnetically and behave as a unit, heme a exhibits an anticooperative interaction with heme a_3 and Cu_B, while Cu_A interacts allosterically with heme a_3.[115-117]

The spatial relationship of heme a_3 and Cu_B relative to each other and to heme a and Cu_A has been described in models, which are in disagreement over the exact assignments of His-284 and His-419 as ligands of heme a_3 and Cu_B, and as to whether the alignment is heme a-heme a_3-Cu_B or heme a-Cu_B-heme a_3, a crucial point for understanding CcO function.[117,123,124]

NITRIC OXIDE BINDING TO REDUCED CYTOCHROME c OXIDASE

Nitric oxide interacts with CcO in various ways. When hydroxylamine NH_2OH is added aerobically to oxidized CcO, heme a_3 is reduced, and a heme a_3^{2+}-NO complex is formed. The complex is more stable when the reaction is performed anaerobically. A characteristic EPR spectrum shows a rhombic symmetry ($g_x = 2.09$, $g_y = 1.98$, $g_z = 2.005$) with a nine-line SHF structure (21.1 G and 6.8 G) indicative of hexacoordinate nitrosyl heme (Fig. 6.5).[5,8,125,126] The same spectrum of the heme a_3^{2+}-NO moiety is obtained when CcO is reduced by dithionite with authentic NO or by ascorbate in the presence of nitrite, or when mitochondria are reduced by succinate in the presence of NO. This effect of NH_2OH is specific to CcO and does not occur with Hb, Mb or cyt c.

An EPR study at high resolution using a Fourier transform technique and simulation revealed the complete SHF A-tensors of ^{15}NO (A_x = 18.0 G, A_y = 17.8 G and A_z = 29.8 G), of N_ϵ of the proximal histidine in all principal directions (A_x = 6.2 G, A_y = 6.4 G and A_z = 7.3 G) and also that of N atoms of the pyrrole ring (A_x = 4.2 G, A_y = 4.2 G and A_z =

0.9 G).[127] Another EPR study coupled to ENDOR experiments and spectral simulation yielded slightly different sets of values for g-tensor and A-tensors, a difference explained by a rotation by 15-20° of the A-tensor of NO relative to the g-tensor (Fig. 6.5).[128]

NITRIC OXIDE BINDING TO OXIDIZED CYTOCHROME c OXIDASE

Nitric oxide also binds to ferric CcO as measured by the appearance of a new rhombic high-spin EPR signal (g_x = 6.16, g_y = 5.82) assigned to uncoupled

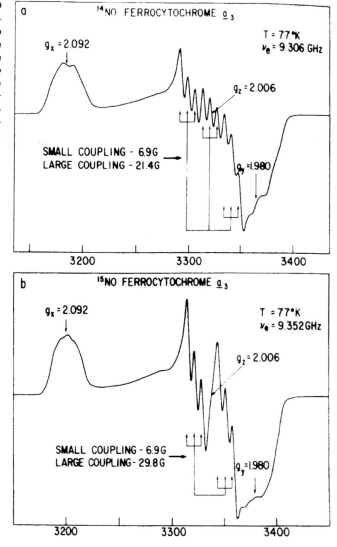

Fig. 6.5. EPR spectra of nitrosyl-CcO ligated with ^{14}NO (a), or ^{15}NO (b). (Reproduced with permission from: LoBrutto R, Wei YH, Mascarenhas R et al. Electron nuclear double resonance and electron paramagnetic resonance study on the structure of the NO-ligated heme a_3 in cytochrome c oxidase. J Biol Chem 1983; 258:7437-7448.) Copyright © The American Society for Biochemistry and Molecular Biology, Inc. Bethesda, MD, USA.

heme $a_3{}^{3+}$. The low-spin heme a and the Cu_A EPR signals remain unchanged.[129] The binding is totally reversible as the high-spin signal disappears upon NO removal, but does not occur with all CcO preparation types. The half-binding NO pressure is 65 mm Hg, i.e., ~170 μmol/l. It suggests that NO coordinates to $Cu_B{}^{2+}$ and breaks the antiferromagnetic couple by forming a heme $a_3{}^{3+}$-$Cu_B{}^{2+}$-NO complex. The $Cu_B{}^{2+}$ and NO spins ($S = 1/2$) are very strongly antiferromagnetically coupled so that no EPR signal results that can be assigned to a triplet state.

The effect of azide addition to the heme $a_3{}^{3+}$-$Cu_B{}^{2+}$-NO complex is to reduce heme a_3 to the ferrous state, followed by the binding of NO to heme $a_3{}^{2+}$, according to the proposed scheme:[129]

cytochrome $a_3{}^{3+}$ + $N_3{}^-$ + NO → cytochrome $a_3{}^{2+}$ + N_2O + N_2

The production of N_2O was effectively detected by mass spectroscopy. The reaction products were also assayed by EPR spectroscopy. The low-spin signals of cytochrome a^{3+} and that of $Cu_A{}^{2+}$ were unaffected, the heme $a_3{}^{2+}$-NO signal disappeared upon $N_3{}^-$ addition, and new EPR signals were detected near $g = 2$ ($\Delta M_S = 1$) and $g = 4.3$ ($\Delta M_S = 2$), characteristic of a triplet species with a small zero field splitting.[129] The resonance at $g = 4.3$ exhibits a four-line hyperfine pattern with a splitting of 97 G (0.020 cm^{-1} or 20 mK, see chapter 3), indicative of a mononuclear type 2 copper atom.[130] It is proposed that the triplet state signals arise from the antiferromagnetic coupling of two $S = 1/2$ sites, heme $a_3{}^{2+}$-NO and $Cu_B{}^{2+}$ arranged in a bridged $a_3{}^{2+}$-NO-$Cu_B{}^{2+}$ complex.[129] The formation of this bridged complex $a_3{}^{2+}$-NO-$Cu_B{}^{2+}$ is irreversible contrary to that of the $a_3{}^{3+}$-$Cu_B{}^{2+}$-NO complex.

These results were interpreted in the frame of the existence of three different conformations found in oxidized CcO, depending on the "resting" state of the oxidized enzyme as isolated or on the oxygenation of the reduced enzyme.[131] By photodissociation experiments coupled to EPR over a wide temperature range (15 to 77 K) the above results were extended, and their general interpretations were confirmed.[13,132,133] A model was built of the "triplet" $a_3{}^{2+}$-NO-$Cu_B{}^{2+}$ complex, giving a distance of 4.2 Å between Fe and Cu and a distance of the electron spins of heme $a_3{}^{2+}$-NO and $Cu_B{}^{2+}$.[133] This complex was proposed as a model of the cage formed by the heme $a_3{}^{2+}$-$Cu_B{}^{2+}$ pair in which O_2 is bidentately coordinated to the two metal atoms during the catalytic reaction of CcO.[133] A RR scattering study also confirmed these interpretations.[134]

REACTION CYCLES CATALYZED BY CYTOCHROME c OXIDASE

The original observation of Stevens et al[129] that N_2O evolved when a mixture of $N_3{}^-$ plus NO was added to oxidized CcO was confirmed by use of ^{15}N FT NMR and mass spectroscopy with different mixtures combining ^{14}NO, ^{15}NO, $^{14}N_3{}^-$ and $^{15}N^{14}N_2{}^-$.[135] Depending on the redox state of the enzyme when NO or a mixture of $N_3{}^-$ plus NO was added, three different catalytic cycles were hypothesized.

When NO is added to CcO fully reduced by ascorbate and p-phenylenediamine as a mediator, a two-electron reaction occurs:
$a_3{}^{2+}Cu_B{}^+$ + 2NO + $2H^+$ → $a_3{}^{3+}Cu_B{}^{2+}$ + N_2O + H_2O

In the presence of both $N_3{}^-$ and NO acting on oxidized CcO, N_2O and N_2 are formed by a one-electron reaction:
$a_3{}^{3+}$ + $N_3{}^-$ + NO → $a_3{}^{2+}$ + N_2O + N_2

The enzyme can also catalyze through its Cu_B moiety the reversible one-electron oxidation of NO to nitrite:
$Cu_B{}^{2+}$ + NO + H_2O ↔ $Cu_B{}^+$ + $NO_2{}^-$ + $2H^+$
and probably the two-electron oxidation of NO to NO_2:
$a_3{}^{3+}Cu_B{}^{2+}$ + NO + H_2O → $a_3{}^{2+}Cu_B{}^+$ + NO_2

All these reactions were observed in strictly anaerobic incubation conditions (Fig. 6.6).

The reactions are slow, requiring hours, although CcO-catalyzed. Therefore it is possible that these NO-reductase and nitrite-reductase activities are only rudimentary activities shared with other metalloproteins, laccase for instance. They however provide good models of the reduction of dioxygen to water by CcO.[134,135]

Let us recall that the experiments described above were only meant to probe the four metal centers of CcO and that the catalysis involving NO was fortuitously discovered. All three of these catalytic cycles should now be further studied in kinetic terms and compared to that for oxygen-reductase, since NO considered as a substrate has a totally new biological meaning, being an endogenous substrate with other acceptor sites along the mitochondrial respiration pathway, in particu-

lar complexes I and II (see below and chapter 10). Eliminating NO through the NO-reductase activity of CcO could be rather "advantageous" for the mitochondrial respiration on O_2 as it would avoid a blockage upstream at the electron sources on complexes I and II. The whole notion of respiration inhibition by NO should be reconsidered in kinetic terms—most of which have still to be measured quantitatively—to determine whether NO regulates the electron flows or merely competes with O_2 as an electron acceptor, or really inhibits the aerobic respiration. Experiments have

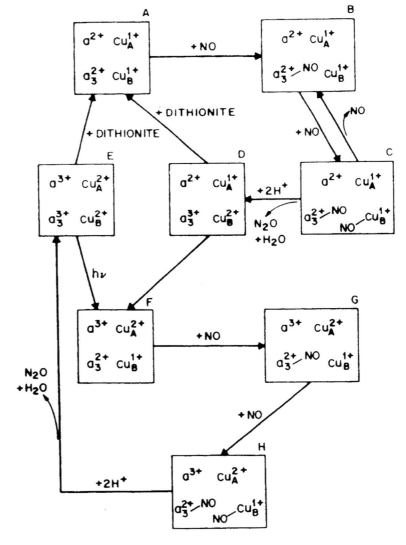

Fig. 6.6. Diagram showing the possible oxidation and ligand binding states of CcO interacting with NO and dithionite. (Reproduced with permission from: Rousseau DL, Sing S, Ching YC et al. Nitrosyl cytochrome c oxidase formation and properties of mixed valence enzyme. J Biol Chem 1988; 263:5681-5685.) Copyright © The American Society for Biochemistry and Molecular Biology, Inc. Bethesda, MD, USA.

been reported recently along this line.[136,137] This shall be further discussed in another context in chapter 14.

In fact, the belief that CcO's oxygen-reductase activity is inhibited by NO should perhaps be reversed in saying that the NO-N_2O:oxidoreductase activity is inhibited by O_2. This idea could be even more readily accepted when one considers that CcO is an enzyme older than atmospheric oxygen, i.e., it existed prior to the appearance of eubacterial oxygenic photosynthetic organisms.[114] It would be interesting to make comparisons with the timescale of the evolution of NO-synthase on the one hand and of the enzymes of the denitrification pathway on the other hand (see chapters 7 and 14).[138,139]

COPPER-CONTAINING PROTEINS

The three types of copper sites in copper-containing proteins were defined early by their specific spectroscopic EPR and visible UV absorption properties (reviewed in refs. 130 and 140). All three types were found in multicopper oxidases, while other proteins with other functions contained only one type of copper site. The first, type *1*, is characterized in its Cu^{2+} state by a strong optical absorption near 600 nm (ε = 3000 to 6000 mol^{-1} l cm^{-1}) which endows the protein with intense blue color and an anomalously small hyperfine constant of the copper atom in the EPR spectrum ($A_{//}$ = 35 to 90 G, i.e. 3.5 to 9 mK) (1 Kaiser = 1 cm^{-1}, see chapter 3). The second, type *2,* has the usual physical characteristics of simple cupric salts (ε = 100-400 mol^{-1} l cm^{-1}, $A_{//}$ = 130 to 200 G, i.e. 13 to 20 mK). The type *3* site is EPR-silent, has a strong UV absorption around 330 nm (ε = 3000 to 6000 mol^{-1} l cm^{-1}) and consists of a pair of antiferromagnetically coupled Cu^{2+} cations.[130,140] All three sites have unusually high oxidation-reduction potentials. While type *1* copper is in multicopper oxidases, the primary electron acceptor from the reducing substrate, the type *3* site interacts with molecular oxygen, and

type *2* acts as an electron "buffer". In fact type *2* and *3* coppers are now recognized to form a trinuclear center (see below). The type *1* site is also found alone in low molecular weight blue proteins, such as plastocyanin, azurin or stellacyanin, which have the sole function to transfer electrons. The binuclear type *3* site is present in the mixed-function oxidase, tyrosinase, and in the oxygen-transporting protein, hemocyanin (see below). Actually the term oxidase (oxydase in French) was coined by the discoverer of laccase and tyrosinase exactly a hundred years ago (1894-1897)—Gabriel Bertrand (1867-1962), working successively in famous French scientific institutions: Muséum d'Histoire Naturelle, Institut Pasteur and La Sorbonne.[141]

MULTICOPPER OXIDASES

Multicopper oxidases catalyze, like CcO, the four-electron reduction of dioxygen to water with concommitant one-electron oxidation of the substrate and are characterized by their deep blue color and by their content in copper atoms equal or greater than four, arranged into at least one copy of each type *1*, *2* and *3* sites; the best known are ceruloplasmin, laccase and ascorbate oxidase.[130,140,142-145]

Ceruloplasmin (CPN; ferroxidase; iron (II):oxygen oxidoreductase, EC 1.16.3.1) is a blue α_2-glycoprotein found in mammals' blood plasma at 2-3 µmol/l normal concentration. The concentration is hormonally regulated, and CPN is an acute-phase protein in the inflammatory response. It is synthesized in hepatocytes, and it binds 95% of blood plasma copper. It has multiple functions in iron and copper metabolisms, in particular that of ferroxidase, in addition to its oxidase function with a low substrate specificity: indoleamine and catechol derivatives, aminophenols, etc. It consists of a single polypeptide chain (M_r = 135 kDa) and contains 6-7 bound copper atoms.[142,145,146] There is a redundancy in the number of copies of the three different copper sites related to internal triplication of an ancestral gene,[146] which might explain CPN multiple functions.

Ascorbate oxidase (AO; L-ascorbate-oxygen oxidoreductase, EC 1.10.3.3) is found in most higher plants and is commonly purified from green zucchini (*Cucurbita pepo medullosa*) and cucumber (*Cucumis sativus*). It is highly specific of L-ascorbic acid and substrates with a lactone ring.[144] It is a homodimer (140 kDa) that contains eight copper atoms, four in each subunit. The four atoms are arranged in one blue type *1* site and a trinuclear center which serves as the O_2 binding site and reduction site. The trinuclear center consists of the two copper atoms of the antiferromagnetically coupled EPR silent type *3* site, in close proximity of the type *2* copper atom.

The X-ray crystal structure of AO has been analyzed for several states of the enzyme: oxidized, fully reduced, peroxide- and azide-bound forms and type-*2*-depleted forms.[147-149] In oxidized AO, the mononuclear type *1* copper Cu1 has two histidines (His-445 and His-512), a cysteine (Cys-507) and a methionine (Met-517) ligand, a site also found in plastocyanin and azurin, proteins containing only type *1* copper. The amazing trinuclear cluster has eight histidine ligands. One pair of copper atoms (labelled Cu2 and Cu3) has six histidine residues (His-106, His-450, His-506 for Cu2, His-62, His-104 and His-508 for Cu3) arranged as a trigonal prism, with an OH⁻ or O^{2-} bridge; it represents the putative spectroscopic type *3* site. The third copper atom, labelled Cu4, spectroscopic type *2*, is coordinated with two histidine ligands (His-60 and His-448) and a water or OH⁻ molecule, which is exchangeable by anions. The sequence His-506:Cys-507:His-508 links the type *1* and type *3* centers. The average copper-copper distance in the trinuclear site is 3.74 Å in oxidized AO, but is more variable in fully reduced AO, 5.1 Å for Cu2-Cu3, 4.4 Å for Cu2-Cu4, 4.1 Å for Cu3-Cu4. This trinuclear site provides room for O_2 binding. A complete catalytic mechanism of AO was proposed.[149]

Laccase (EC 1.12.3.1) is found in a variety of plants and fungi. The most studied laccases are purified from the lacquer tree (e.g. *Rhus vernicifera*) and from white-rot fungi of the Basidiomycetes (e.g. *Polyporus versicolor*) or of the Ascomycetes phyla (e.g. *Neurospora crassa*). Laccase is a phenol oxidase with very low specificity towards the reducing substrate: mono-, di-, polyphenols, aminophenols or diamines. Laccases from white-rot probably play an important role in lignin degradation, while those of *Rhu*s play a role in phenol polymerization to form lacquer layers.[143,145] It has a molecular mass from a minimal 64 kDa to 140 kDa, with a widely variable carbohydrate content in a single polypeptide. It contains four copper atoms arranged in one copy of each type *1*, *2* and *3*.

The three examples of multicopper oxidases have large sequence and domain structure analogies, and are probably evolved through gene duplication or triplication from a common blue copper protein of low molecular weight.[145,146,150]

Nitric oxide reacts with CPN in the reduced state,[151] the oxidized state,[152,153] and the type-*2*-copper depleted state,[154] with tree and fungal laccases,[155,156] and finally with AO.[157]

There are many similarities, with slight differences in detail between the results obtained with the three enzymes and also with those obtained with CcO. NO is able to reduce all three copper sites, but it can also oxidize reduced type *1* copper. NO forms a specific reversible complex with reduced type *2* copper. Finally triplet EPR signals are attributed to a Cu⁺-NO complex of decoupled copper atoms in the half-oxidized type *3* site. A complete anaerobic catalytic cycle yielding N_2O and NO_2^- is set up for the tree laccase, similar to that discovered for CcO.[135,156]

For the time being no such catalytic cycle with NO and nitrite has been demonstrated for CPN, an acute-phase plasma protein, which could also play a role in immune response or at least appears, through variation in plasma concentration, to be a signal of nonspecific immune response in cancer cases or in pregnancy, for instance. A kinetic study of the interaction of NO with CPN within normal concentration ranges and of the effects of NO upon the various catalytic activities of CPN would be interesting in order to ascertain any role for NO in that aspect of nonspecific inflammatory response. For instance a recent report suggested that peroxynitrite, which could derive from NO and $O_2^{-\bullet}$ in circulating blood in inflammatory conditions, causes release of copper from CPN. CPN looses its ferroxidase activity, while freely released copper could have devastating pro-oxidant effects on lipids and low-density lipoproteins, effects which are implicated in atherosclerosis.[158]

A biological role for the interaction of NO with plant or fungi multicopper oxidases, AO and laccase, has yet to be ascertained.

HEMOCYANIN AND COPPER MONOOXYGENASES

Proteins as functionally different as hemocyanin and copper monooxygenases, tyrosinase and dopamine β-monooxygenase, are here grouped together as they have in common binuclear copper sites which present structural and spectroscopic similarity to the type 3 site of multicopper oxidases.

As for study of other oxygen carriers, NO was used to probe the O_2 binding site of hemocyanin. The study of the interaction of NO with hemocyanin (Hc) can only be anecdotic as Hc is confined to a few not very evolved animals. However such studies may be interesting for a better comprehension of the evolution of oxygen carriers.[159] All the hemocyanins are found in the molluscs (gastropods, e.g. *Aplysia californica* and snail *Helix pomatia*; cephalopods, e.g. *Octopus vulgaris*) and arthropods (e.g. the horse-shoe crab *Limulus polyphemus*, lobsters and crayfish, e.g. *Panulirus interruptus*, and crabs, e.g. *Cancer magister*).[160,161] Hemocyaninss are extracellular hemolymph proteins organized into enormous polymers (3 to 24 subunits) of half a million to sev-

eral million daltons which are differently arranged into the two phyla of molluscs and arthropods. The common feature is the presence of one type 3 site for oxygen binding per subunit, which is colorless in the deoxy state and light blue (ε_{340nm} ~20 000 mol^{-1} l cm^{-1} and ε_{570nm} ~1000 mol^{-1} l cm^{-1}) in the oxyHc state formulated as $[Cu^{(II)}-O_2^{2-}-Cu^{(II)}]$. Neither deoxyHc nor oxyHc shows any EPR spectrum. Hemocyanin can be transformed to the fully oxidized (metHc) $Cu^{(II)}-Cu^{(II)}$ form and to a semi-met $Cu^{(I)}-Cu^{(II)}$ form.

The crystal structure of hexameric deoxyHc from *Panulirus interruptus* has been solved to 3.2 Å resolution.[162,163] Although the exact geometry cannot be defined at this resolution, each copper atom appears to be coordinated with two histidine residues with $Cu-N_\varepsilon$ distances of ~2.0 Å (His-194 and His-198 for one Cu, His-344 and His-348 for the other) and to a more distant histidine at ~2.6 Å (His-224 and His-384 respectively)) arranged in a trigonal antiprism structure. The Cu-Cu distance is 3.6 Å with no evidence of a bridging ligand.[162,163] Similar results were obtained with deoxyHc from another arthropod *Limulus polyphemus*.

Tyrosinase (monophenol, dihydrophenylalanine: oxygen oxidoreductase, EC 1.14.18.1) is a monooxygenase widely distributed in nature, involved in the biosynthesis of melanins and polyphenolic compounds. The main sources of purified enzyme are mushroom *Agaricus bisporus* and ascomycete *Neurospora crassa*.[164] Each tyrosinase monomeric unit has an active site structurally similar to that of Hc,[165] which can take in the enzymatic cycle the various deoxy-, oxy- and met(resting)- forms. Histidine residues are probably ligands to the copper pair.

Dopamine β-monooxygenase (EC 1.14.17.1, previously named tyrosine hydroxylase and dopamine β-hydroxylase) is functionally very different from tyrosinase as it catalyzes the β-hydroxylation of dopamine to noradrenaline. It is localized in the storage vesicles of mammalian adrenal medulla and sympathetic nerve endings.[164,166] It contains four very labile copper atoms per tetramer (290 kDa), and the copper atoms appear to be isolated mono-nuclears (EPR detectable and classified as type 2) in the oxidized form which change to a binuclear cuprous form upon reduction with ascorbate.[167,168] The copper site is therefore somewhat different from the regular type 3.

Both NO and nitrite interact with hemocyanin and tyrosinase. DeoxyHc from molluscan *Helix pomatia* and arthropodal *Cancer magister,* and tyrosinase from *Agaricus bisporus* are half oxidized and react with NO forming a dipole-dipole coupled $Cu^{(II)}-Cu^{(II)}$ ion pair in a triplet state characterized by EPR signals near g = 2 (ΔM_S = 1) and g = 4 (ΔM_S = 2), other copper atoms being decoupled in the $Cu^{(II)}$ form.[169-171] NO is presumably bound at the O_2-binding site. Computer simulation of the dipolar coupled $Cu^{(II)}-Cu^{(II)}$ pair EPR spectra allowed calculation of the $Cu^{(II)}-Cu^{(II)}$ distances to about 6 Å in all cases, which is much more than that determined by X-ray diffraction for *Panulirus* Hc. The reactions of NO with deoxyHc and of nitrite with metHc yielded N_2O, as for the reactions of CcO that we have detailed above.[172]

The physiological importance of tyrosinase and dopamine β-monooxygenase activities in mammals would warrant a completely new appraisal of their interaction with NO and of the effect of NO on their monooxygenase activities.

MONONUCLEAR IRON-CONTAINING DIOXYGENASES

Several iron-containing dioxygenases with widely different substrate specificity have in common a mononuclear Fe site. Their reactions with nitric oxide also present some similarities.

LIPOXYGENASE

Lipoxygenases are a class of non-heme non-[FeS] iron dioxygenases which catalyze the dioxygenation (hydroperoxidation) of unsaturated fatty acids with regio- and stereochemical specificity. In mammals, they play a role in the synthesis of leukotrienes

and lipoxins from arachidonic acid and are thus part of the inflammatory response. In plants, lipoxygenases play a role in germination and in defense against pathogens. Three major types of lipoxygenases have been identified, inserting dioxygen at the C-5, C-12 or C-15 positions of arachidonic acid (5,8,11,14-eicosatetraenoic acid). In human leukocytes, 5-lipoxygenase (EC 1.13.11.34) catalyzes the oxygenation of arachidonic acid into 5-hydroperoxy eicosatetraenoic acid (5-HPETE).[173,174] Soybean lipoxygenase-1 (L-1, linoleate:oxygen oxidoreductase, EC 1.13.11.12), the most commonly studied, is an arachidonic acid 15-lipoxygenase.[175-177] It serves as a prototype for all lipoxygenases. They all contain a single iron atom (per 98.5 kDa for L-1) which is believed to cycle between the ferric (high-spin) and ferrous (high-spin) forms during the catalytic cycle.

Two independent X-ray structure analyses of L-1 proposed that the iron atom is pentacoordinated to His-499, His-504, His-690, Asn-694 and Ile-839, the C-terminal residue.[176,177] The sixth coordination position is not occupied, and the site is wide open. The weak field O ligand of Asn-694 in L-1 could be substituted by a strong field N atom of a His residue in human 15-lipoxygenase.[178]

Nitric oxide binds to ferrous L-1 forming a complex with a total electronic spin of S = 3/2, in a nearly axial symmetry, as shown by EPR resonances observed at low temperature (4 to 15 K) near g = 2 and g = 4.[179-181] This species arises from an antiferromagnetic coupling between high-spin ferrous iron (S = 2) and S = 1/2 of NO. The complete theory was formulated by Salerno & Siedow.[180] The ligand binding position can be occupied by linoleic acid in competition with NO. A similar S = 3/2 species was detected in mitochondria of *N. crassa* and of higher plants upon reaction with NO; the S = 2 center from which it arises is otherwise undetectable.[182]

Lipoxygenase and prostaglandin H synthase (cyclooxygenase) are the enzymes responsible for arachidonic acid hydroperoxidation, leading to the formation of hydroxyeicosatetraenoic acid (HETE) and hydroperoxyeicosatetraenoic acid (HPETE) on the one hand, and prostaglandins, thromboxanes and 12-hydroxyheptade-catrinoic acid (12-HHT) on the other hand (see above). A recent report showed that the 12-lipoxygenase is selectively and more effectively inhibited by nitric oxide than the cyclooxygenase pathway, suggesting a regulatory mechanism rather than an on/off switch (see chapter 14).[183]

PHENOLYTIC DIOXYGENASES

Phenolytic dioxygenase enzymes which catalyze aromatic ring cleavage of catechol (pyrocatechol or 1,2-benzenediol) and protocatechuate (3,4-hydroxybenzoate) are classified as intra- or extradiol according to the site of ring opening relative to the *ortho*-hydroxyl groups of the substrate. The prototype of intradiol dioxygenase is protocatechuate 3,4-dioxygenase (EC 1.13.11.3; 3,4-PDO) purified from *Pseudomonas aeruginosa* or from *Brevibacterium fuscum*. Typical extradioldioxygenases are protocatechuate 4,5-dioxygenase (EC 1.13.11.8; 4,5-PDO) from *P. testosteroni* and catechol 2,3-dioxygenase (EC 1.13.1.2; 2,3-CDO) from *P. putida*. These enzymes have many natural products as substrates, flavonoids, alkaloids and lignins, and incorporate two atoms of molecular oxygen into dihydroxybenzenes.

The X-ray structure of *P. aeruginosa* 3,4-PDO has been solved.[184] It is an oligomer containing 12 copies each of α (22.3 kDa) and β (26.6 kDa) subunits. The ferric iron is coordinated to four β subunit ligands as an approximate trigonal bipyramid, with Tyr-147β and His-162β in the axial coordinating position and His-160β, Tyr-118β and a solvent water molecule in the equatorial plane. Such a ligation sphere is compatible with EXAFS and RR data.[185] A mechanism implying chelation of the substrate at the iron site and displacement of a protein ligand has been proposed.[185] 4,5-PDO from *P. testosteroni* has a similar αβ arrangement of unequivalent sububits (α, 17.7 kDa; β, 33.8 kDa).[186] Mössbauer spec-

tra of the ferrous (active) and EPR of the ferric (inactive) 4,5-PDO showed that both forms are high-spin.[186]

Nitric oxide has been used to probe the ferrous iron site, which is the potential oxygen binding site in all three enzymes: 4,5-PDO,[186-188] 2,3-CDO[187] and 3,4-PDO.[189] All three enzymes form $Fe^{(II)}$-NO complexes characterized by a S = 3/2 electronic spin, resulting from an antiferromagnetic coupling between high-spin iron (S = 2) and S = 1/2 of NO. EPR spectra at low temperature are very similar to those obtained with L-1($Fe^{(II)}$)-NO. Experiments with [^{17}O]-enriched water showing the EPR line broadening suggest that water is iron bound and can be displaced by substrates or inhibitors of 4,5-PDO and of 2,3-CDO.[187] NO binding has also been found to be substrate dependent and to involve two independent binding sites. By use of isotopically [^{17}O]-labeled substrates and inhibitors, the modes of ligation in intradiol dioxygenase (3,4-PDO) and extradiol dioxygenases (4,5-PDO and 2,3-CDO) and the substrates and NO to $Fe^{(III)}$ and $Fe^{(II)}$, and possibly that of O_2, were proposed.[188,189]

OTHER MONONUCLEAR IRON-CONTAINING PROTEINS

Other mononuclear iron-containing proteins have been described, some of which bind NO, such as putidamonooxin and isopenicillin N synthase.[190,191] Some model compounds have also been studied, well explaining the S = 3/2 ground state.[192]

[FeS], NI AND MO CLUSTER-CONTAINING PROTEINS

Many [FeS], Ni and Mo cluster-containing proteins, with various essential cellular functions: electron transfer, enzymatic activities, etc., play a role in widely different pathways, such as mitochondrial respiration,[111] Krebs cycle,[193] iron metabolism: IRP or IRE-binding proteins,[194-201] ferrochelatase,[202-205] etc., bacterial N_2-fixation: nitrogenase,[206-208] H_2-evolution and H_2-oxidation: hydrogenase,[209,210] DNA repair: endonuclease III,[211] and oxidative stress: SoxR protein,[212-214] etc.

Many of these cluster-containing proteins are easily inactivated by molecular oxygen through disruption of the [FeS] cluster. Interaction of the clusters with NO has been characterized in vitro on purified proteins in only a few cases. Whole cellular effects of L-arginine-derived NO have also been detected, as measured by enzymatic activities or by EPR spectroscopy (see chapters 10 and 14).

COMPONENTS OF MITOCHONDRIAL RESPIRATION

Most of the five components of the mitochondrial respiratory chain in mammal cells contain [FeS] clusters.[111] Complex I (NADH-ubiquinone oxidoreductase, EC 1.6.5.3) contains five [2Fe-2S], three [4Fe-4S] clusters and one FMN site, in a complex of 25 unlike proteins. Complex II (succinate-ubiquinone oxidoreductase, EC 1.3.5.1) has a major component, succinate dehydrogenase of the citric acid cycle, purified as a membrane-bound enzyme which can be solubilized. Complex II contains within 4 different polypeptides two [2Fe-2S] and one [4Fe-4S] cluster, a FAD site and a cytochrome b_{560}. Complex III (ubiquinol-cytochrome c oxidoreductase) contains in 9 to 10 unlike polypeptides a [2Fe-2S] Rieske cluster, two cytochromes b and one cytochrome c. Finally complex IV is cytochrome c oxidase, and complex V is ATPase.[111]

Thus the respiratory chain offers a host of 40 different potential metal targets for NO, altogether eight [2Fe-2S] clusters, four [4Fe-4S] clusters with different stoichiometry and in different proteic environments, two hemes a, three hemes b, one heme c and two copper sites (see above the description of CcO). In fact very little is known of the interaction of NO with complexes I, II and III. The interaction of NO with complex IV (CcO) is a little more documented (see above).

Succinate dehydrogenase (SDH), purified from beef heart mitochondria as an active soluble enzyme, contains as complex II two [2Fe-2S] and one [4Fe-4S] cluster and a FAD site per molecule. EPR signals

of these different [FeS] components are highly temperature dependent.[215-217] When SDH was incubated with nitrite and dithionite, new EPR signals detectable at 77 K with g-values at 2.035 and 2.010 were found (Fig. 6.7).[218]

They are similar to those arising from cysteine-iron-NO ternary complexes,[219-223] also found in vivo in carcinogen-treated rats,[224,225] and now assigned to $Fe(NO)_2(SR)_2$ complexes (see below and in chapter 10).

Another now well described potential target for NO is aconitase (citrate/isocitrate hydro-lyase, EC 4.2.1.3) of the Krebs cycle or tricarboxylic acid cycle. The enzyme is a 80 kDa protein containing a [4Fe-4S] cluster in the active form which can be converted to an inactive form with a [3Fe-4S] cluster, through the reversible exchange of a labile iron atom.[193,226-228] Although the X-ray crystal structure has been resolved and interaction with NO has been demonstrated in whole cells (see below and in chapter 10), no direct evidence of this interaction has yet been given in vitro with the purified enzyme.

The study of cytotoxic activated macrophages as a component of the immune response contributed to the discovery of the L-arginine-NO pathway. The earliest and best characterized cytostatic effect of activated macrophages was the inhibition of [FeS]-containing enzymes, particularly those involved in mitochondrial respiration. The citrate-dependent respiration (aconitase) is inhibited first ($\tau_{1/2} \sim 4$ hr), followed by the inhibition of mitochondria complex I ($\tau_{1/2} \sim 8$ hr) and complex II ($\tau_{1/2} \sim 14$ hr), while complex III with a [2Fe-2S] Rieske cluster remains unaffected (see the review by Hibbs et al in ref. 231).[229,230] EPR spectroscopic studies of these phenomena shall be described in chapter 10.[232-234]

FERREDOXINS

Ferredoxins are [Fe-S]-containing proteins assuming electron transfer functions. The [Fe-S] clusters have widely varying stoichiometry, from [Fe-S] to [6Fe-6S] or [8Fe-8S]. Only a few instances of their

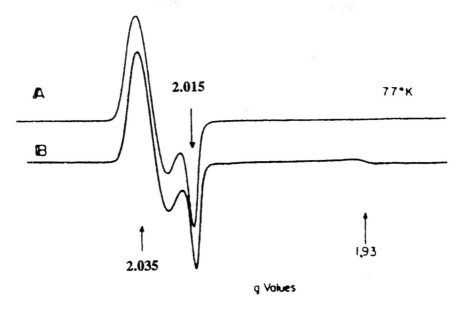

Fig. 6.7. EPR spectra at 77 K of Fe(NO)$_2$(SR)$_2$ complexes. (A): formed by the addition of nitrite to a solution of ascorbate, cysteine and ferrous sulfate; (B): formed on succinate dehydrogenase (SDH) by addition of nitrite and ascorbate. (Reproduced with permission from: Salerno JC, Ohnishi T, Lim J et al Tetranuclear and binuclear iron-sulfur clusters in succinate dehydrogenase: a method of iron quantitation by formation of paramagnetic complexes. Biochem Biophys Res Commun 1976; 73:833-840.) Copyright © Academic Press, Inc, Orlando, Florida, USA.

interaction with NO have been described. Ferredoxins from *Porphyra umbilicalis* and *Spirulina platensis*, interacting with NO produced from the reaction of nitrite with ascorbate, gave EPR signals at g = 2.040 and 2.015, assigned to Fe(NO)$_2$(SR)$_2$ complexes.[234,235] Similar signals were obtained by the bacteriostatic action of nitrite on whole vegetative cells of *Clostridium botulinum* or *Clostridium sporogenes*.[236,237]

NITROGENASE

Nitrogen fixation is part of the biological nitrogen cycle (see following chapter 7). It is the catalyzed reduction of nitrogen to ammonia which is always coupled to the obligatory synthesis of H$_2$,[206-208] requiring eight electrons:

$$N_2 + 8H^+ \rightarrow 2NH_3 + H_2$$

Nitrogenase has been well characterized in bacteria participating in nitrogen fixation systems: *Rhodopseudomonas sphaeroides*

f.sp. *denitrificans*, *Clostridium pasteurianum* and *Azotobacter vinelandii*. Nitrogenase consists of two proteins. The Fe-protein (called Cp2 in *C. pasteurianum* and Av2 in *A. vinelandii*) is a 60 kDa dimer of identical subunits bridged by a single [4Fe-4S] cluster. The Fe-protein receives electrons from electron transfer proteins, ferredoxin (Fd) and flavodoxin (Fld). It binds the nucleotides MgATP and MgADP through a *ras* p21-type nucleotide-binding site, and is implicated in the coupling of ATP hydrolysis to electron transfer. The [4Fe-4S] cluster undergoes a one-electron redox cycle between 2Fe^{2+}-2Fe^{3+} and 3Fe^{2+}-Fe^{3+}. The MoFe-protein (Cp1 in *C. pasteurianum* and Av1 in *A. vinelandii*) is an $\alpha_2\beta_2$ tetramer with a total molecular weight of ~ 240 kDa. It contains two unusual metal centers: the diamagnetic, EPR-silent P-cluster pairs and the paramagnetic (S = 3/2) FeMo-cofactor or M-center, in which the

Mo atom can be substituted by a V atom.[206] The arrangement of the 2 Mo, 30 Fe and ~34 inorganic S atoms has been determined as three-dimensional structures of the Av1 and Cp1 proteins (reviewed in refs. 206-208). Each of the two FeMo-cofactors contains 1Mo:7Fe:9S and one isocitrate molecule, arranged into two [4Fe-3S] and [Mo-3Fe-3S] partial cubanes bridged by three nonprotein S ligands. They provide the site of substrate (N_2 or C_2H_2) binding and reduction to NH_3 or C_2H_4, and of proton source for H_2 synthesis. Acetylene is an alternate substrate and is often used for assessment of nitrogenase actvity. There are two P-cluster pairs per MoFe-protein tetramers, each containing two [4Fe-4S] clusters bridged by two bidendate thiol groups, Cys-88α and Cys-95β, and a disulfide bond.

The oxidized N derivatives of the nitrogen cycle have all been tested as inhibitors of nitrogenase. N_2O is in fact reduced to NH_3 via N_2 as intermediate, while on the contrary NO_3^- is totally inactive. The results dealing with NO_2^- are not clear cut, except that it is an inhibitor. Some researchers reported that the inhibition of nitrogenase from *Rhizobium*-infected soybean bacteroids by NO_2^- is reversible and occurs through binding to the MoFe-protein.[37,238] Others reported that inhibition of nitrogenase from *C. pasteurianum* or from *Rp. sphaeroides* is irreversible through disruption of the [FeS] cluster of the Fe-protein.[38,39] In fact by a study of purified Av1 and Av2 components of nitrogenase of *A. vinelandii*, it was shown that the Fe-protein is partially and irreversibly inhibited by NO_2^-, but NO_2^- is itself a substrate reduced to NH_3 without H_2 evolution.[239] Experiments in that respect are difficult as nitrogenase is very susceptible to time-dependent inactivation in air.

Another candidate as a nitrogenase inhibitor is NO. The nitrogenase system from *C. pasteurianum*, from *Rp. sphaeroides* and from soybean bacteroids (*Glycine max* inoculated with *Rhizobium japonicum*) is inhibited by NO.[36-39] Nitric oxide binds probably to the [FeS] cluster of the Fe-pro-

tein as the EPR signals of the [FeS] cluster disappear (g = 2.02 and 1.94) and new EPR components appear at g = 2.035 and 2.019, since assigned to a $Fe(NO)_2(SR)_2$ complex.[39] A more recent study of the purified Av1 and Av2 components demonstrated a very complex mechanism for the loss of actitivty of Av2, highly dependent on NO concentration and time; the inactivation of the Av1 component is even more complex.[40] The only clear-cut point is that NO reacts with the [FeS] clusters of both components.

Hydrogenase

Hydrogenases are unusual non-heme iron-containing proteins which may contain various type of clusters: the usual [4Fe-4S] clusters or Ni centers with [3Fe-4S] clusters or [NiFeSe] centers, in which S atoms can be substituted by endogenous natural ^{79}Se and by ^{77}Se (S = 1/2) atoms.[209] Hydrogenase is a key enzyme of the anaerobic sulfate reducing chemotrophic and phototrophic bacteria, like: *Desulfovibrio vulgaris* Hildenborough,[240] *D. desulfuricans* Norway,[241] *D. baculatus*[209,242] and *D. gigas*.[210,243] It catalyzes the reversible oxidation of H_2 and is inhibited by O_2, CO, C_2H_2 and NO.[240,244] A given species may have only one or several different hydrogenases, which are found in different locations: periplasm, membrane or cytoplasm.

The active sites of several [NiFeSe] hydrogenases have been studied by EPR,[209,242,243] and the X-ray structure of the *D. gigas* [NiFeSe] periplasmic hydrogenase has been solved to 2.8 Å resolution.[210] The latter enzyme is certainly the best studied hydrogenase; it is a heterodimer of 60 and 28 kDa subunits. The 60 kDa subunit contains the Ni atom bound to four Cys (S/Se) residues and another adjacent metal center tentatively assigned to Fe.[210] The 28 kDa subunit contains one [3Fe-4S] and two [4Fe-4S] clusters. The structure suggests plausible electron and proton transfer pathways.[210]

Very little is known of hydrogenase inhibition by NO.[240,244] The various hydrogenases of the *Desulfovibrio* anaerobic genus are inhibited by very low concentrations of NO (full inhibition for 0.02 to 0.4 µmol/l NO). Inhibition is instantaneous and fully reversible by sparging with an inert gas.[240] Hydrogenase of free-living aerobic, N_2-fixating bacterium *A. vinelandii* is membrane-bound and also contains nickel and [FeS] clusters. Inhibition by NO of this hydrogenase exhibits two components.[244] The H_2-oxidizing activity is reversibly inhibited under turnover conditions (i.e. in the presence of H_2 and of an electron acceptor, methylene blue). In this case the effect of NO is not on the Ni site believed to be interacting with H_2, neither on the final electron acceptor binding site. The second inhibitory effect, under non-turnover conditions, requires higher NO concentrations, is irreversible and results in a time-dependent loss of enzymatic activity. It may well result from the disruption of an intermediate [FeS] cluster.[244]

METALLOTHIONEIN

We have introduced metallothionein (MT) in this chapter, although it does not contain the usual [FeS] cluster considered as a prosthetic group, because it forms a $Fe(NO)_2(SR)_2$ complex similar to those of [FeS]-containing proteins.

MT is found in many animals and all mammals in variable concentrations depending on the cell metabolism. For instance many tumor cells contain high concentrations of MT. Mammalian MT is a small protein of only 61 or 62 amino acids (6-7 kDa), 20 of which are evolutionarily conserved cysteine residues, and can bind with high affinity 7 (and up to 12) divalent Group II metal ions per polypeptide. The physiological function is probably to contribute to metal homeostasis particularly that of Cu and Zn. It also binds toxic and nonessential metals, such as Cd, Hg, Ag or Au, or drug cis-dichloro-diammine-$Pt^{(II)}$. It may form metal complexes with many different stoichiometries: for instance Cd_7-MT, Cd_5Zn_2-MT, Cd_nZn_{7-n}-MT, Zn_7-MT, etc.

The crystal structure of Cd_5Zn_2-MT at 2.0 Å resolution agrees with NMR 3D structure to define two domains, α (residues 1-29) and β (residues 33-61) with a linker (30-32), with clusters of stoichiometry ($[M_4(CysS)_{11}]^{3-}$ and $[M_3(CysS)_9]^{3-}$) with adamantane-like geometry. Each metal is coordinated by cysteine residues in tetrahedral tetrathiolate coordination, and thiolates act both as ligands and as bidendate bridges; the average M-S-M internal angle is 103-104°, and the average M-M distance is 3.9 Å.[245-247]

MT also binds iron(II) forming a Fe_7-MT complex, which has been studied by Mössbauer spectroscopy of $^{57}Fe(II)_7$-MT and magnetic susceptibility.[248] All individual $Fe^{(II)}$ ions in Fe_7-MT are in the high-spin (S = 2) state, but they are segregated into a diamagnetic part (4 atoms over 7) and a paramagnetic part (3 atoms). The diamagnetic part has been attributed to a four-metal cluster with a strong antiferromagnetic exchange coupling resulting in a ground state with a total spin S = 0, consistent with the $[M_4(CysS)_{11}]^{3-}$ cluster.[248] The other component consists of the other three $Fe^{(II)}$ ions in the high-spin state with only a very weak, and very unusual (with these thiol ligands), exchange coupling between adjacent $Fe^{(II)}$ ions corresponding probably to the $[M_3(CysS)_9]^{3-}$ cluster, with a pseudo-planar geometry, in the β domain.[248]

The possibility that MT could be involved in the cellular response to NO has been tested in vitro by following the interaction of pure Fe_7-MT complex with NO).[249] An EPR spectrum at 77 K with g-values at 2.039 and 2.013 was obtained, characteristic of a $Fe(NO)_2(SR)_2$ complex (in which incidently the iron atom is formally $Fe^{(I)}$ rather than $Fe^{(III)}$) (Fig. 6.8).[249]

Fe^{2+} ions, in the presence of NO can displace Zn^{2+} from the Zn_7-MT complex, which Fe^{2+} ions cannot do alone. The EPR signal of the $Fe(NO)_2(SR)_2$ complex in mitochondrial aconitase cannot be transfered to metallothionein in the Zn_7-MT form.[249]

Thus two interesting features can been drawn from these experiments. Firstly MT may be a physiologically important thiol source to form $Fe(NO)_2(SR)_2$ complexes in tumor cells, together with other [FeS]-

containing proteins. Secondly, the EPR signal assigned to $Fe(NO)_2(SR)_2$ complexes appears to be rather unspecific, and many other crossed experiments using enzymatic activity tests, molecular weight determination, measure of iron contents, etc., have to be performed before correct assignments can be suggested.

Recently one instance for a new biological role has been attributed to the chelation of Fe and NO by metallothionein; it could protect cells from the cytotoxic and DNA-damaging effects of nitric oxide.[250] Overexpression of MT in NIH 3T3 cells reduces the sensitivity of these cells to NO, as checked by the formation of EPR-detected $Fe(NO)_2(SR)_2$ complexes within the cells. This shall be further discussed in chapter 10.

BINUCLEAR IRON-CONTAINING PROTEINS

Hemerythrin is the first protein for which a binuclear iron oxygen binding site

was discovered. It presents both striking analogies and differences with active sites of proteins with completely different catalytic activities, uteroferrin, purple acid phosphatase, protein A of methane monooxygenase, the nucleation center of ferritin and ribonucleotide reductase subunit R2 (see below and chapter 10). Comparisons of sequence and of iron coordination in these proteins have been performed.[251,252]

HEMERYTHRIN

Hemerythrin (Hr) is a non-heme, non-[FeS] binuclear iron oxygen carrier found exclusively in small pink erythrocytes of the coelomic fluid of some marine vertebrates, Sipunculoidea, Polychaeta, etc., such as *Sipunculus nudus*, *Phascolposis gouldii*, and *Themiste* species. While Hr is oligomeric (Mr = 108 kDa), usually octameric, but also trimeric, an analogous monomeric protein, myohemerythrine, can store oxygen. These proteins have been widely studied by all kinds of spectroscopic methods as

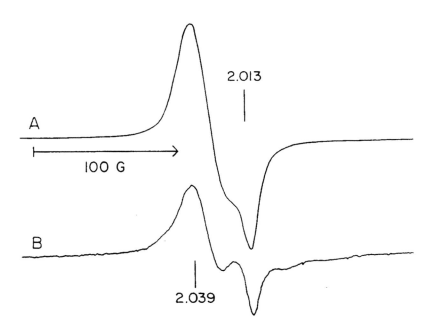

Fig. 6.8. EPR spectra of $Fe(NO)_2(SR)_2$ complex upon addition of Fe^{2+} and NO to apoMT (A) and to Zn-MT (B) at 15 K (A) and 77 K (B). (Reproduced with permission from: Kennedy MC, Gan T, Antholine WE et al Metallothionein reacts with Fe^{2+} and NO to form products with g = 2.039 ESR signal. Biochem Biophys Res Commun 1993; 196:632-635.) Copyright © Academic Press, Inc, Orlando, Florida, USA.

they possess one dimeric iron atom unit per subunit as the O_2 binding site.

Each O_2 binding site consists of two iron atoms bound to amino acid residues, and bridged together by a μ-oxo bridge (μ-OH⁻ in deoxyHr, μ-O^{2-} in oxyHr) and in some conformations by common carboxyl (Glu, Asp) ligands. For instance Hr from *Themiste dyscritum* has one iron atom (Fe1) bound by His-101, His-73 and His-77, and the other (Fe2) by His-25 and His-54, while the carboxyl residues of Asp-106 and Glu-58 bridge the two atoms, together with an O atom.[253] The iron atoms are in the Fe(III) oxidation state in oxygenated and met (oxidized) forms and in the Fe(II) state in the deoxy form:

$$[Fe^{II}(\mu\text{-}OH^-)Fe^{II}] + O_2 \leftrightarrow [Fe^{III}(\mu\text{-}O^{2-})Fe^{III}O_2H^-]$$

Two so-called "semi-met" forms {Fe(II)-Fe(III)} also exist. The binuclear iron sites in oxyHr and metHr have ground states with S = 0, due to strong antiferromagnetic coupling between the iron atoms. The coupling is weaker in deoxyHr (S = 0) and semi-metHr (S = 1/2). The Fe2 site can also bind anions: N_3^-, SCN^- and CN^-.

This biological material, along with multicopper oxidases and CcO, has certainly been for twenty years one of the most favored metalloproteins for biospectroscopists as it can occur in many different conformational and electronical states. The degree of complexity can be multiplied ad infinitum, and all spectra have to be interpreted in terms of distances, angles, etc.; paradise for emeritus spectroscopists, hell for nonspecialists! (see ref. 254).

Nitric oxide binds reversibly to deoxyHr.[255-257] NO is directly coordinated to one of the iron atoms, probably at the vacant coordination site of Fe2. The EPR signal, observed only below 30 K, indicates an axial symmetry ($g_{//}$ = 2.77 and $g_\perp$ = 1.84). The strongly antiferromagnetically coupled iron pair is decoupled, and the EPR signal is explained by the magnetic interaction of a Fe(II) (S = 2) center and a {FeNO}⁷ (S = 3/2) center at Fe2 observed by Mössbauer spectroscopy (7 total elec-

trons: 6 metal *d* electrons plus 1 NO π* electron). See chapter 2 for this nomenclature of the electron configuration.[258,259] Semi-metHr also binds NO to form an EPR-silent adduct, the iron pair being antiferromagnetically coupled Fe(III) (S = 5/2) and {FeNO}⁷ (S = 3/2) giving an integer spin ground state, likely S^{eff} = 1.[255-257]

Nitrite, or its acid form HNO_2, oxidizes deoxyHr to a semi-metHr-NO adduct which dissociates slowly to form a semi-metHr-NO_2^- adduct, instead of the expected fully oxidized metHr form:[255-257]

$$[Fe^{II}(\mu\text{-}OH^-)Fe^{II}] + HONO \rightarrow [Fe^{III}(\mu\text{-}O^{2-})\{FeNO\}^7]$$

$$[Fe^{III}(\mu\text{-}O^{2-})\{FeNO\}^7] + HONO \rightarrow [Fe^{II}(\mu\text{-}OH^-)Fe^{III}NO_2^-] + NO$$

NO does not bind to metHr. There is no catalytic cycle implying NO_2^- and NO, as is the case for CcO and laccase. All these reactions provide a clear model for the reversible O_2 binding which is not accompanied by autoxidation.

RIBONUCLEOTIDE REDUCTASE SUBUNIT R2

Ribonucleotide reductase (RNR, EC 1.17.4.1) is a key enzyme in DNA biosynthesis as it is rate limiting in the enzymatic conversion of ribonucleotides to deoxyribonucleotides, and it provides, through allosteric regulations, a balance of each of the four deoxyribonucleotides (reviewed in refs. 260-264). Ribonucleotide reductases are ubiquitous. Three classes of RNR are known which all use a protein radical in the catalytic cycle.[263,264] We shall deal only with class I RNRs in the following paragraphs. Class II RNRs are adenosyl cobalamin-dependent reductases found in many prokaryotes, for instance *Lactobacillus leichmannii*. Class III RNRs are anaerobic ribonucleotide reductases containing two EPR-detected centers, a [FeS] center and a free glycine radical. The prototype of this class is purified from anaerobically grown *E. coli*.[264-266] Class I RNRs are iron dimer-containing and tyrosyl radical-containing enzymes, using ribonucleotide diphosphates as substrates. They are found in mammal cells, in cells host to DNA-virus (for instance T4 phage, herpes virus, RNR being encoded by viral DNA) and in *E. coli* grown aerobically, the latter being the best characterized. Class I RNRs have an α2β2 protein structure. They are dimers of two inequivalent homodimers, named R1 and R2 for the *E. coli* enzyme.[260-263]

R1 protein (protomer 85.7 kDa) binds substrates and allosteric effectors and provides the reducing equivalents for substrate reduction through two thiol pairs (Cys-754 and Cys-759, Cys-225 and Cys-462) undergoing RSH/RSSR reactions with thioredoxin or glutathione-glutathione reductase systems and one thiol residue which might form a thiyl radical (Cys-439), which in turn could interact with Tyr-122 of the R2 subunit by means of long-range electron transfer.[263] The R1 protein of *E. coli* has been crystallized and analyzed by X-ray diffraction.[267,268] Two redox-active cysteine residues (Cys-225 and Cys-462) can readily form a disulfide bridge; residue Cys-439 is close-by but not enough to form any disulfide bridge, and the two other conserved cysteine residues (754 and 759) are located at the carboxyl end in a flexible arm.[268]

The R2 subunit (protomer 43.4 kDa) in the ferric active form contains a (μ-oxo)-bridged diferric center and a stable tyrosyl radical at position Tyr-122 close to the binuclear center. R2 protein can exist in several states: the active form (two ferric atoms and a radical Tyr-122), the reduced form $R2_{red}$ (two ferrous iron atoms and a normal Tyr-122 residue), the oxidized form $R2_{met}$ (two ferric atoms and a normal Tyr-122 residue), the apo form and finally the semi-met EPR-detectable forms.[269]

The X-ray crystallographic analysis of *E. coli* $R2_{met}$ dimer has determined the very unusual coordination of the two equivalent [Fe1-Fe2] dimer centers in each subunit, where the iron atoms are 3.3 Å apart and have inequivalent coordination: respectively the two O atoms of the carboxylate group of Asp-84, one N atom of His-118, one O atom of car-

boxylate of bridging Glu-115 residue, the (μ-oxo) bridge and one water molecule for the Fe1 atom, and one O atom of Glu-238 and of Glu-204, one N atom of His-241, the other O atom of carboxylate of bridging Glu-115 residue, the (μ-oxo) bridge and another water molecule for the Fe2 atom, respectively.[252,270] The crystal structure of apo-RNR has also been determined, explaining well the reversible iron exchanges.[271] The coordination of active R2 and $R2_{met}$ iron atoms is probably the same.

The stable phenoxy free radical of Tyr-122 residue, which is formed during the reaction of dioxygen with the diiron(II) center, is located 5.3 Å away from the closest (Fe1) iron atom and is buried 10 Å from the protein surface in a hydrophobic pocket.[252] The pocket is lined with aromatic residues,[252] some of which could also form free radicals.[272-274] In the *E. coli* enzyme the Y122 tyrosyl radical plays an essential role in RNR catalysis. It is probably also the case of the Y177 residue in the mouse enzyme, in view of the 0.5% residual activity found with the Y177F-R2 mutant as compared to wild-type R2.[273] It also plays a role in the inhibition of R2 subunit by hydroxyurea,[275] and in the incorporation of $Fe^{(II)}$ and O_2 to R2 apoprotein to form the diiron center and the μ-oxo bridge.[276] The Tyr-122 free radical is easily detectable by EPR spectroscopy, the ^1H hyperfine structure being fully resolved around 20 K and appearing as a doublet at 77 K and above.[277-279] Recently a high-frequency (139.5 GHz) EPR spectroscopy study was performed which resolved the three principal g-values and A-tensors of both the β-methylene ^1H and 3,5-phenol ring ^1H atoms.[280]

RNR contains therefore many potential targets for reaction with nitric oxide: five catalytically active conserved cysteine groups (over 11) per R1 monomer, the iron pair and the tyrosyl radical of each R2 monomer. In fact evidence has been brought up that RNR is reversibly inhibited in vitro and in vivo by NO,[281-285] and most if not all these centers could be effective targets.[286,287]

Reaction of NO with the diiron(II) center of $R2_{red}$ has been used to probe the putative dioxygen binding site and the formation of a possible diiron peroxide intermediate in the catalytic cycle of R2.[287] A mononuclear complex with the electronic structure $\{FeNO\}^7$ (S = 3/2) is formed, characterized by new absorbption bands at 450 nm and 620 nm and EPR signals detected at 10 K at g = 4.06, 3.98 and 2.02, which could represent 10-20% of the iron. A minority $[Fe(NO)_2(SR)_2]$ signal was detected at g = 2.03 together with a large signal at g ~ 1.97 attributed to the so-called matrix bound-NO in solution.

The $\{FeNO\}^7$ complex is described as high-spin $Fe^{(III)}$ antiferromagnetically coupled to NO^-.[62,192] As there is no exchange interaction with a nearby $Fe^{(II)}$ atom and since the enzyme does not contain adventitious iron, the $\{FeNO\}^7$ complex formation could indicate some destruction of the original diiron center during incubation with NO. Mössbauer spectroscopy detected this mononuclear $\{FeNO\}^7$ complex and also another dimeric species $\{FeNO\}_2$. This complex must have an integer electronic spin and could be a diiron center containing two exchange-coupled S = 3/2 $\{FeNO\}^7$ sites. $R2_{met}$ form is also detected with time at the expense of the $\{FeNO\}_2$ dimer. Correlatively N_2O gas is detected in the reaction chamber headspace by gas chromatography. These reactions provide a good model for the formation of the (μ-oxo)-diiron(III) center from $R2_{red}$ through a diiron(III) peroxide intermediate.[287] These authors did not try to establish any correlation between their results and RNR inhibition by NO, which could occur through the Tyr radical quenching.[282-286] See also chapter 10, where the interaction of RNR and L-arginine derived NO in whole mammal cells is described.

FERRITIN AND BACTERIOFERRITIN

Describing ferritin in the present place might seem awkward, but ferritin does contain binuclear μ-oxo bridged iron centers as Hr and RNR, and does interact with NO in vitro to form EPR-detectable nitrosylated complexes.

Ferritin (FTN) is responsible for iron storage in many tissues. It is found in serum, in intestinal mucosa where iron absorption occurs, in bone marrow for hematopoietic cell synthesis, in embryo red cells for short-term storage, in liver, spleen and kidney for long-term storage, and in most cells providing iron necessary for intracellular synthesis.[288-291] FTN, a large, spherical, hollow protein (450 kDa) forms a shell of 12 nm diameter and 1 nm thickness around a mineral mass of ferric iron hydrates made of up to 4500 Fe atoms. ApoFTN is an eicosatetramer of two types of subunits with similar mass, L (20 kDa) and H (21 kDa), associated by noncovalent interactions in variable proportions depending on the cellular source; for instance there is more L type in liver and spleen, and more H type in heart. Recombinant rat and human H and L subunits are now available and have been crystallized. *E. coli* contains two types of ferritins, heme-containing bacterioferritin (BFT) and conventional ferritin, both with structures analogous to those of mammal cells' FTN. There is a high degree of conservation of amino acid sequence of FTN from prokaryotes, plants and animals.

Iron absorption occurs in the $Fe^{(II)}$ form which enters the cavity through selective 3-fold and 4-fold symmetry channels, and is followed by several steps. Iron oxidation by molecular oxygen is catalyzed by a site of the H subunits active as a ferroxidase center. Ferric iron takes several forms: protein-bound mononuclear EPR-detectable $Fe^{(III)}$, binuclear (μ-oxo)-bridged $Fe^{(III)}$-$Fe^{(III)}$ centers, mixed valence EPR-detectable $Fe^{(II)}$-$Fe^{(III)}$ dimers, nucleation clusters and finally the ferric hydrate core.[292] Iron can be released in the $Fe^{(II)}$ form after iron reduction by many agents.[291]

The structure of the ferroxidase and $Fe^{(III)}$-$Fe^{(III)}$ binding site in human H FTN has been solved by X-ray crystallography associated to site-directed mutagenesis.[293-295] The site resembles that of the RNR R2 subunit in its iron ligands. The amino acids' ligands are Glu-27 and His-65 for FeA, Glu-61 and Glu-107 for FeB,

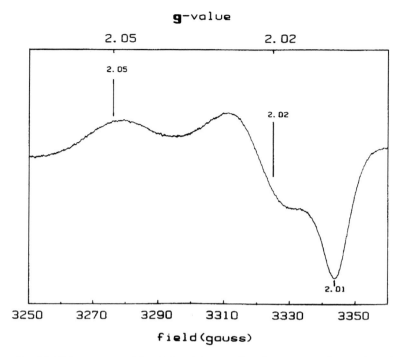

Fig. 6.9. EPR spectrum of the nitric oxide complex of BFR at 40 K. (Reproduced with permission from: LeBrun NE, Cheesman RM, Thomson AJ et al. An EPR investigation of non-haem iron sites in Escherichia coli *bacterioferritin and their interaction with phosphate. A study using nitric oxide as a spin probe. FEBS Lett 1993; 323:261-266.) Copyright © Elsevier Science, Sara Burgerhartstraat 25, 1055 KV Amsterdam, The Netherlands.*

with bridging Glu-62 and the (μ-oxo)-bridge between FeA and FeB. The dimer site also has a water molecule and enough room for O_2 binding, and finally contains neighboring H-bonded Tyr-137, Tyr-34 and Gln-141 residues. The corresponding region in the horse L subunit which has no ferroxidase activity, does not offer any channel or any possible ligation site for iron.[296] A Fe[(III)]-tyrosinate complex, probably with Tyr-34 residue, has also been detected by UV-visible and RR spectroscopy in the course of biomineralization of H subunit, which does not occur on L FTN.[296,297] A third possible iron binding site close to the binuclear iron site has been recently discovered, at least in *E. coli* conventional FTN, lying on the inner surface of the protein shell, with four Glu residues as ligands (Glu-61, Glu-140, Glu-143 and Glu-144).[298] Of this third iron site,

only Glu-61 and Glu-140 are conserved in mammalian FTN. This site could be the monomeric Fe[(III)] observed by Mössbauer spectroscopy.[298] The analogy with the RNR R2 binuclear site is made even more striking, as a tyrosyl radical, perhaps from Tyr-34 residue, is formed during the oxidative deposition of iron in human apoferritin.[299] The complete X-ray analysis of both H and L subunits has not yet been completely solved.

The interaction of FTN with NO is the object of much research work, yielding sometimes contradictory results as to whether NO directly mediates iron release from FTN.[300,301] Independently of this important point that shall be discussed in chapter 10, NO has been used as an EPR probe of FTN organic iron sites.[234,235,302-304]

Bacterioferritin (BFR) from *E. coli* is an eicosatetramer of identical subunits

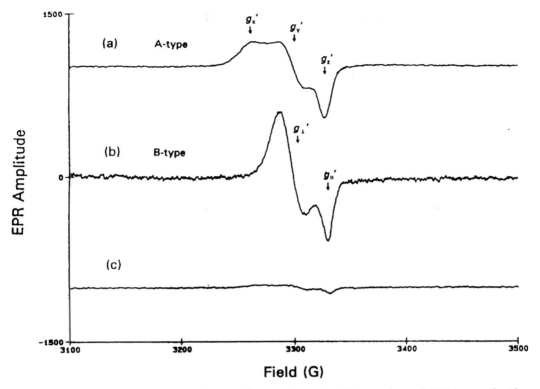

Fig. 6.10. EPR spectra at 77 K of A- and B-type horse spleen FTN-NO complexes. (a) FTN treated with p-(chloromercuri)benzoate (PMB) containing 48 Fe²⁺/molecule flushed with NO under anaerobic conditions, $g'_x = 2.055$, $g'_y = 2.033$, $g'_z = 2.015$; (b) FTN treated with diethyl pyrocarbonate (DEP) containing 120 Fe²⁺/molecule flushed with NO under anaerobic conditions, $g_\perp = 2.033$ and $g_{//} = 2.014$; (c) FTN treated with both PMB and DEP containing 48 Fe²⁺/molecule flushed with NO under anaerobic conditions. (Reproduced with permission from: Lee M, Arosio P, Cozzi A et al Identification of the EPR-active iron-nitrosyl complexes in mammalian ferritins. Biochemistry 1994; 33:3679-3687.) Copyright © The American Chemical Society, Columbus, OH, USA.

(18.5 kDa), exhibits also a ferroxidase activity and contains also up to 12 *b*-type methionine-ligated low-spin (S = 1/2) hemes. EPR spectroscopic studies showed that other signals are detectable at 10 K, corresponding to slightly distorted rhombic mononuclear high-spin (S = 5/2) iron centers (g = 5.04, 4.27 and 3.48). Reaction of ascorbate-reduced BFR with NO forms two EPR-detectable complexes distinguishable by different symmetry, rhombic and axial. Both complexes are S = 3/2 species derived from antiferromagnetic coupling between NO and non-heme iron and can be assigned as [Fe⁽ᴵᴵᴵ⁾-NO⁻] or Fe⁽ᴵᴵ⁾-NO.[302] NO does not seem to bind to the ferric form of non-heme iron. In anaerobiosis BFR reacts also with NO to form a rhombic S = 1/2 complex, with g-values at 2.05, 2.02 and 2.01, likely NO[Fe⁽ᴵᴵ⁾-Fe⁽ᴵᴵ⁾], as in Hc from *Phascolopsis gouldii* (Fig. 6.9).[256]

Thus NO can be considered as a good probe of the first intermediate sites of iron deposition in bacterioferritin in the following sequence: binding of Fe⁽ᴵᴵ⁾ probably at the dimeric ferroxidase center, oxidation of the Fe⁽ᴵᴵ⁾ dimer, production of mononuclear Fe⁽ᴵᴵᴵ⁾ and iron nucleation.[302,303]

Similar results were obtained with mammalian FTN.[234,235,304] The detailed study by Lee et al indicates three different EPR-detectable species resulting from the interaction of NO with horse spleen FTN, recombinant human H chains and site-directed mutants.[304]

Table 6.1. Potential NO targets

	References
A priori all metalloproteins	1, 2, 306-310
Hemoglobin, hemoproteins (*Inhibition*)	3-13, 306, 310
Fe-S cluster proteins (*Inhibition*)	229-244
Ribonucleotide reductase (R2) (*Inhibition*)	281-287
Guanylate cyclase (*Activation*)	311-315
NO-synthase (*Inhibition*)	316-326
P-450 cytochrome	89, 94-100
Indolamine dioxygenase (*Inhibition*)	64-71
Prostaglandin H synthase (*Activation*) (cyclooxygenase)	75
Lipoxygenase (*Inhibition*)	179-183
Cytochrome *c* oxidase (*Inhibition*)	125-137
Iron metabolism proteins: -Ferritin, ferroxidases -Transferrin, transferrin receptor -"iron responsive factor" or "iron responsive element-binding protein" -Metallothionein -Ferrochelatase	300-304 194-201 249, 250 202-205

A-type complex (g = 2.055, 2.033 and 2.05) is assigned to a rhombic S = 1/2 species, with His-128 and possibly His-118 residues as ligands. B-type complex ($g_\perp$ = 2.033 and $g_{//}$ = 2.014) is assigned to an axial S = 1/2 species with Cys-130 an iron ligand. The third C-type complex ($g_\perp$ = 4 and $g_{//}$ = 2) with S = 3/2 could be nonspecific (Fig. 6.10).[304]

The existence of B-type complex upon incubation of FTN with NO is another instance of the ubiquity and nonspecificity of the [Fe(NO)$_2$(SR)$_2$] EPR signal, already described in [FeS] cluster-containing proteins, metallothionein and small molecular weight iron-thiol-NO ternary complexes.

None of these NO interacting sites seem to be related to the ferroxidase site of H subunit of FTN described by Lawson et al,[293-295] or to the third iron site of *E. coli* FTN.[298] On the contrary, residues His-128 and Cys-130 in the vicinity of the 3-fold channels leading to the interior of FTN are important in the formation of iron-nitrosyl complexes. It could be a way of storing together NO and iron in the form of Fe(NO)$_2$(SR)$_2$ complexes.[304]

CONCLUSION

Nitric oxide is in principle able to bind to all transition metals through their *d* electron orbitals (see chapter 2). To our knowledge, no example of NO binding to Mo and Ni-containing proteins has yet been described, although such bindings are in principle possible. However nitrogenase and hydrogenase, which contain respectively Mo and Ni in addition to [FeS] clusters, can be inhibited. It is not yet decided through which metal center this inhibition occurs. The only instance of binding of NO to manganese is that of Mn-protoporphyrin IX-substituted CcP.[5] The EPR signal of high-spin (d^5; $S = 5/2$) $Mn^{(II)}$-CcP disappears upon NO addition by the formation of EPR-silent $Mn^{(II)}$-CcP-NO (either $S = 0$, or S integer). $Mn^{(III)}$-protoporphyrin IX-substituted peroxidases, Hb or Mb did not react with NO. Similar results were obtained with $Co^{(II)}$-protoporphyrin IX-substituted hemoglobin. There is no instance, even in inorganic chemistry, of NO binding to Zn^{2+}, which has an alkaline earth metal character like Ca^{2+} or Mg^{2+}, rather than that of a transition metal.

A simple conclusion can be drawn from this chapter: ***all metalloproteins are potential targets for nitric oxide (Table 6.1).*** The EPR spectra catalog compiled since 1968 by many biospectroscopists[1-11,305-310] may allow in the best of cases (or considered best by biospectroscopists!) detection of some of these nitrosyl metalloprotein complexes in vitro and in vivo, in cell cultures, in animal models and in humans in normal physiological or in physiopathological conditions. That is the subject of the following chapters 10 to 12. Chapter 14 shall be a sequel of these, being a discussion of which metalloproteins might be interesting and biologically relevant targets of nitric oxide.

REFERENCES

1. Henry Y, Ducrocq C, Drapier J-C, et al. Nitric oxide, a biological effector. Electron paramagnetic resonance detection of nitrosyl-iron-protein complexes in whole cells. Eur Biophys J 1991; 20:1-15.
2. Richter-Addo GB, Legzdins P. Metal nitrosyls. Oxford University Press, Oxford, UK, 1992.
3. Kon H. Paramagnetic resonance study of nitric oxide hemoglobin. J Biol Chem 1968; 243:4350-4357.
4. Dickinson LC, Chien JCW. An electron paramagnetic resonance study of nitrosylmyoglobin. J Am Chem Soc 1971; 93:5036-5040.
5. Yonetani T, Yamamoto H, Erman JE et al. Electron properties of hemoproteins. V. Optical and electron paramagnetic resonance characteristics of nitric oxide derivatives of metalloporphyrin-apo-hemoprotein complexes. J Biol Chem 1972; 247:2447-2455.
6. Henry Y, Banerjee R. Electron paramagnetic studies of nitric oxide haemoglobin derivatives: isolated subunits and nitric oxide hybrids. J Mol Biol 1973; 73:469-482.
7. Chien JCW. Electron paramagnetic resonance study of the stereochemistry of nitrosylhemoglobin. J Chem Phys 1969; 51:4220-4227.
8. Morse RH, Chan SI. Electron paramagnetic resonance studies of nitrosyl ferrous heme complexes. Determination of an equilibrium between two conformations. J Biol Chem 1980; 255:7876-7882.
9. Hori H, Ikeda-Saito M, Yonetani T. Single crystal EPR of myoglobin nitroxide. Freezing-induced reversible changes in the molecular orientation of the ligand. J Biol Chem 1981; 256:7849-7855.
10. Shiga T, Hwang R-J, Tyuma I. Electron paramagnetic resonance studies of nitric oxide hemoglobin derivatives. I. Human hemoglobin subunits. Biochemistry 1969; 8:378-383.
11. Nagai K, Hori H, Yoshida S et al. The effect of quaternary structure on the state of the α and β subunits within nitrosyl haemoglobin. Low temperature photodissociation and the ESR spectra. Biochim Biophys Acta 1978; 532:17-28.
12. Sharma VS, Isaacson RA, John ME et al. Reaction of nitric oxide with heme proteins: studies on metmyoglobin, opossum methemoglobin, and microperoxidase. Biochemistry 1983; 22:3897-3902.
13. LoBrutto R, Wei YH, Yoshida S et al. Electron paramagnetic resonance- (EPR-) resolved kinetics of cryogenic nitric oxide recombination to cytochrome *c* oxidase and myoglobin. Biophys J 1984; 45:473-479.
14. Bazylinski DA, Hollocher TC. Metmyoglobin and methemoglobin as efficient traps for nitrosyl hydride (nitroxyl) in neutral aqueous solution. J Am Chem Soc 1985; 107:7982-7986.
15. Bazylinski DA, Goretski J, Hollocher TC. On the reaction of trioxodinitrate (II) with

hemoglobin and myoglobin. J Am Chem Soc 1985; 107:7986-7989.

16. Neto LM, Nascimento OR, Tabak M et al. The mechanism of reaction of nitrosyl with met- and oxymyoglobin: an ESR study. Biochim Biophys Acta 1988; 956:189-196.

17. De Sanctis G, Falcioni G, Polizio F et al. Mini-myoglobin: native-like folding of the NO-derivative. Biochim Biophys Acta 1994; 1204:28-32.

18. Gorbunov NV, Osipov AN, Day BW et al. Reduction of ferrylmyoglobin and ferryl-hemoglobin by nitric oxide: a protective mechanism against ferryl hemoprotein-induced oxidations. Biochemistry 1995; 34:6689-6699.

19. Zhao XJ, Sampath V, Caughey WS. Infra-red characterization of nitric oxide bonding to bovine heart cytochrome *c* oxidase and myoglobin. Biochem Biophys Res Commun 1994; 204:537-543.

20. Jongeward KA, Marsters JC, Mitchell MJ et al. Picosecond geminate recombination of nitrosylmyoglobins. Biochem Biophys Res Commun 1986; 140:962-966.

21. Petrich JW, Lambry J-C, Balasubramanian S et al. Ultrafast measurements of geminate recombination of NO with site-specific mutants of human myoglobin. J Mol Biol 1994; 238:437-444.

22. Walda KN, Liu XY, Sharma VS et al. Geminate recombination of diatomic ligands CO, O_2, NO with myoglobin. Biochemistry 1994; 33:2198-2209.

23. Carlson ML, Regan R, Elber R et al. Nitric oxide recombination to double mutants of myoglobin: role of ligand diffusion in a fluctuating heme pocket. Biochemistry 1994; 33:10597-10606.

24. Schaad O, Zhou HX, Szabo A et al. Simulation of the kinetics of ligand binding to a protein by molecular dynamics: geminate rebinding of nitric oxide to myoglobin. Proc Natl Acad Sci USA 1993; 90:9547-9551.

25. Das TK, Mazumdar S, Mitra S. Micelle-induced release of haem-NO from nitric oxide complex of myoglobin. J Chem Soc Chem Commun 1993; 18:1447-1448.

26. Duprat AF, Traylor TG, Wu G-Z et al. Myoglobin-NO at low pH: free four-co-ordinated heme in the protein pocket. Biochemistry 1995; 35:2634-2644.

27. Antholine WE, Mauk AG, Swartz HM et al. Electron spin resonance spectra of feline NO-hemoglobins. FEBS Lett 1973; 36:199-202.

28. Trittelvitz E, Sick H, Gersonde K et al. Reduced Bohr effect in NO-ligated *Chironomus* haemoglobin. Eur J Biochem 1973; 35:122-125.

29. Brunori M, Falcioni G, Rotilio G. Kinetic properties and electron paramagnetic resonance spectra of the nitric oxide derivative of hemoglobin componenets of trout (*Salmo irideus*). Proc Natl Acad Sci USA 1974; 71:2470-2472.

30. Scholler DM, Wang MR, Hoffman BM. Resonance Raman and EPR of nitrosyl human hemoglobin and chains, carp hemoglobin, and model compounds. Implications for the nitrosyl heme coordination state. J Biol Chem 1979; 254:4072-4078.

31. Caracelli I, Meirelles NC, Tabak M et al. An ESR study of nitrosyl-*Aplysia brasiliana* myoglobin and nitrosyl annelidae *Glossoscolex paulistus* erythrocruorin. Biochim Biophys Acta 1988; 955:315-320.

32. Tsuneshige A, Imai K, Hori H et al. Spectroscopic, electron paramagnetic resonance and oxygen binding studies on the hemoglobin from the marine polychaete *Perinereis aibuhitensis* (Grübe): comparative physiology of hemoglobin. J Biochem 1989; 106:406-417.

33. Maskall CS, Gibson JF, Dart PJ. Electron paramagnetic resonance studies of leghaemoglobins from soya-bean and cowpea root nodules. Identification of nitrosyl-leghaemoglobin in crude leghaemoglobin preparations. Biochem J 1977; 167:435-445.

34. Garcia-Plazaola JI, Arrese-Igor C, Langara L et al. Denitrification in intact lucerne plants. J Plant Physiol 1995; 146:563-565.

35. Sen S, Cheema IR. Nitric oxide synthase and calmodulin immunoreactivity in plant embryonic tissue. Biochem Arch 1995; 11:221-227.

36. Lockshin A, Burris RH. Inhibitors of nitrogen fixation in extracts from *Clostridium pasteurianum*. Biochim Biophys Acta 1965; 111:1-10.

37. Trinchant JC, Rigaud J. Nitrite inhibition of nitrogenase from soybean bacteroids. Arch Microbiol 1980; 124:49-54.

38. Meyer J. Comparison of carbon monoxide, nitric oxide, and nitrite as inhibitors of the nitrogenase from *Clostridium pasteurianum*. Arch Biochem Biophys 1981; 210:246-256.

39. Michalski WP, Nicholas DJD. Inhibition of nitrogenase by nitrite and nitric oxide in *Rhodopseudomonas sphaeroides* f. sp. *denitrificans*. Arch Microbiol 1987; 147:304-308.

40. Hyman MR, Seefeldt LC, Morgan TV et al. Kinetic and spectroscopic analysis of the inactivating effects of nitric oxide on the individual components of *Azotobacter vinelandii* nitrogenase. Biochemistry 1992; 31:2947-2955.

41. Ehrenberg A, Szczepkowski TW. Properties and structure of the compounds formed between cytochrome *c* and nitric oxide. Acta Chem Scand 1960; 14:1684-1692.

42. Kon H. Electron paramagnetic resonance of nitric oxide cytochrome *c*. Biochem Biophys Res Commun 1969; 35:423-427.

43. Yoshimura T, Suzuki S, Nakahara A et al. Spectral properties of nitric oxide complexes of cytochrome *c'* from *Alcaligenes* sp. NCIB 11015. Biochemistry 1986; 25:2436-2442.

44. Yoshimura T, Iwasaki H, Shidara S et al. Nitric oxide complex of cytochrome *c'* in cells of denitrifying bacteria. J Biochem 1988; 103:1016-1019.

45. Yoshimura T, Shidara S, Ozaki T et al. Five coordinated nitrosylhemoprotein in whole cells of denitrifying bacterium, *Achromobacter xylosoxidans* NCIB 11015. Arch Microbiol 1993; 160:498-500.

46. Yonetani T, Yamamoto H. Optical and electron paramagnetic resonance properties of the nitric oxide compounds of cytochrome *c* peroxidase and horseradish peroxidase. In: King TE, Mason HS, Morrison M, eds. Oxidases and Related Redox Systems, Vol I, University Park Press, Baltimore, 1973; 279-298.

47. Henry Y, Mazza G. EPR studies of nitric oxide complexes of turnip and horse-radish peroxidases. Biochim Biophys Acta 1974; 371:14-19.

48. Chiang R, Makino R, Spomer WE et al. Chloroperoxidase: P-450 type absorption in the absence of sulfhydryl groups. Biochemistry 1975; 14: 4166-4171.

49. Bolscher BGJM, Wever R. The nitrosyl compounds of ferrous animal haloperoxidases. Biochim Biophys Acta 1984; 791:75-81.

50. Sievers G, Peterson J, Gadsby PMA et al. The nitrosyl compound of ferrous lactoperoxidase. Biochim Biophys Acta 1984; 785:7-13.

51. Poulos TL, Freer ST, Alden RA et al. The crystal structure of cytochrome *c* peroxidase. J Biol Chem 1980; 255:575-580.

52. Finzel BC, Poulos TL, Kraut J. Crystal structure of yeast cytochrome *c* peroxidase refined at 1.7 Å resolution. J Biol Chem 1984; 259:13027-13036.

53. Cooper CE, Odell E. Interaction of human myeloperoxidase with nitrite. FEBS Lett 1992; 314:58-60.

54. Koppenol WH. Thermodynamic consideration on the formation of reactive species from hypochlorite, superoxide and hydrogen monoxide. Could nitrosyl chloride be produced by neutrophils and macrophages? FEBS Lett 1994; 347:5-8.

55. Champion PM, Münck E, DeBrunner PG et al. Mössbauer investigations of chloroperoxidase and its halide complexes. Biochemistry 1973; 12:426-435.

56. Poulos TL, Finzel BC, Howard AJ. High-resolution crystal structure of cytochrome P450cam. J Mol Biol 1987; 195:687-700.

57. Deisseroth A, Dounce AL. Catalase: physical and chemical properties, mechanism of catalysis, and physiological role. Physiol Rev 1970; 50:319-375.

58. Reid TJ, Murhy MRN, Sicignano A et al. Structure and heme environment of beef liver catalase at 2.5 Å resolution. Proc Natl Acad Sci USA 1981; 78:4767-4771.

59. Nicholls P. The reactions of azide with catalase and their significance. Biochem J 1964; 90:331-343.

60. Craven PA, DeRubertis FR, Pratt DW. Electron spin resonance study of the role of NO-catalase in the activation of guanylate cyclase by NaN_3 and NH_2OH. Modulation of enzyme responses by heme proteins and their nitrosyl derivatives. J Biol Chem 1979; 254:8213-8222.

61. Kalyanaraman B, Janzen EG, Mason RP. Spin trapping of the azidyl radical in azide/catalase/H_2O_2 and various azide/peroxidase/H_2O_2 peroxidizing systems. J Biol Chem 1985; 260:4003-4006.

62. Brown GC. Reversible binding and inhibition of catalase by nitric oxide. Eur J Biochem 1995; 232:188-191.

63. Henry Y, Ishimura Y, Peisach J. Binding of nitric oxide to reduced L-tryptophan-2,3-dioxygenase as studied by electron paramagnetic resonance. J Biol Chem 1976; 251:1578-1581.

64. Shimizu T, Nomiyama S, Hirata F et al. Indoleamine 2,3-dioxygenase. Purification and some properties. J Biol Chem 1978; 253:4700-4706.

65. Hirata F, Hayaishi O. Studies on indoleamine 2,3-dioxygenase. I. superoxide anion as substrate. J Biol Chem 1975; 250:5960-5966.

66. Taniguchi T, Hirata F, Hayaishi O. Intracellular utilization of superoxide anion by indoleamine 2,3-dioxygenase of rabbit enterocytes. J Biol Chem 1977; 252:2774-2776.

67. Sono M, Tanigushi T, Watanabe et al. Indoleamine 2,3-dioxygenase. Equilibrium studies of the tryptophan binding to the ferric, ferrous, and CO-bound enzymes. J Biol Chem 1980; 255:1339-1345.

68. Sono M, Dawson JH. Extensive studies of the heme coodination structure of indoleamine 2,3-dioxygenase and of tryptophan binding with magnetic and natural circular dichroism and electron paramagnetic resonance. Biochim Biophys Acta 1984; 789:170-187.

69. Werner ER, Bitterlich G, Fuchs D et al. Human macrophages degrade tryptophan upon induction by interferon-gamma. Life Sci 1987; 41:273-280.

70. Werner ER, Werner-Felmayer G, Fuchs D et al. Paralllel induction of tetrahydrobiopterin biosynthesis and indoleamine 2,3-dioxygenase activity in human cells and cell lines by interferon-γ. Biochem J 1989; 262:861-866.

71. Thomas SR, Mohr D, Stocker R. Nitric oxide inhibits indoleamine 2,3-dioxygenase activity in interferon-γ primed mononuclear phagocytes. J Biol Chem 1994; 269:14457-14464.

72. Humes JL, Winter CA, Sadowski SJ et al. Multiple sites on prostaglandin cyclooxygenase are determinants in the action of nonsteroidal antiinflammatory agents. Proc Natl Acad Sci USA 1981; 78:2053-2056.

73. DeWitt DL, El-Harith EA, Kraemer SA et al. The aspirin and heme-binding sites of ovine and murine prostaglandin endoperoxide synthases. J Biol Chem 1990; 265:5192-5198.

74. Tsai AL, Kulmacz RJ, Wang JS et al. Heme coordination of prostaglandin H synthase. J Biol Chem 1993; 268:8554-8563.

75. Karthein R, Nastainczyk W, Ruf HH. EPR study of ferric prostaglandin H synthase and its ferrous NO derivative. Eur J Biochem 1987; 166:173-180.

76. Shimokawa T, Smith WL. Essential histidines of prostaglandin endoperoxide synthase. His-309 is involved in heme binding. J Biol Chem 1991; 266:6168-6173.

77. Karthein R, Dietz R, Nastainczyk W et al. Higher oxidation states of prostaglandin H synthase. EPR study of a transient tyrosyl radical in the enzyme during the peroxidase reaction. Eur J Biochem 1988; 171:313-320.

78. Shimokawa T, Kulmacz RJ, DeWitt DL et al. Tyrosine 385 of prostaglandin endoperoxide synthase is required for cyclooxygenase catalysis. J Biol Chem 1990; 265:20073-20076.

79. Lassman G, Odenwaller R, Curtis JF et al. Electron spin resonance investigation of tyrosyl radicals of prostaglandin H synthase. Relation to enzyme catalysis. J Biol Chem 1991; 266:20045-20055.

80. Tsai AL, Palmer G, Kulmacz. Prostaglandin H synthase. Kinetics of tyrosyl radical formation and of cyclooxygenase catalysis. J Biol Chem 1992; 267;17753-17759.

81. Degtyarenko KN, Archakov AI. Molecular evolution of P450 superfamily and P450-containing monooxygenase systems. FEBS Lett 1993; 332:1-8.

82. Mansuy D, Battioni P, Battioni JP. Chemical model systems for drug-metabolizing cytochrome-P-450-dependent monooxygenases. Eur J Biochem 1989; 184:267-285.

83. Poulos TL, Finzel BC, Gunsalus IC et al. The 2.6-Å crystal structure of *Pseudomonas putida* cytochrome P-450. J Biol Chem 1985; 260:16122-16130.

84. Poulos TL, Finzel BC, Howard AJ. Crystal structure of substrate-free *Pseudomonas putida* cytochrome P-450. Biochemistry 1986; 25:5314-5322.

85. Raag R, Poulos TL. Crystal structure of the carbon monoxide-substrate-cytochrome P-450$_{CAM}$ ternary complex. Biochemistry 1989; 28:7586-7592.

86. Hill HAO, Röder A, Williams RJP. The chemical nature and reactivity of cytochrome P-450. In: Hemmerich P, Jorgensen CK, Neilands JB et al, eds. Structure and Bonding, Vol 8, 1970; 123-151.

87. Peisach J, Blumberg WE. Electron paramagnetic resonance study of the high- and low-spin forms of cytochrome P-450 in liver and in liver microsomes from a methylcholanthrene-treated rabbit. Proc Natl Acad Sci USA 1970; 67:172-179.

88. Peisach J, Mims WB. Linear electric field-induced shifts in electron paramagnetic resonance: a new method for study of the ligands of cytochrome P-450. Proc Natl Acad Sci USA 1973; 70:2979-2982.

89. Stern JO, Peisach J. A model compound for nitrosyl cytochrome P-450; further evidence for mercaptide sulfur ligation to heme. FEBS Lett 1976; 62:364-368.

90. Peisach J, Stern JO, Blumberg WE. Optical and magnetic probes of the structure of cytochrome P-450's. Drug Metab Dispos 1973; 1:45-61.

91. Chevion M, Peisach J, Blumberg WE. Imidazole, the ligand *trans* to mercaptide in ferric cytochrome P-450. An EPR study of proteins and model compounds. J Biol Chem 1977; 252:3637-3645.

92. Dawson JH, Andersson LA, Sono M. Spectroscopic investigations of ferric cytochrome P-450-CAM ligand complexes. Identification of the ligand *trans* to cysteinate in the native enzyme. J Biol Chem 1982; 257:3606-3617.

93. Jänig GR, Dettmer R, Usanov SA et al. Identification of the ligand trans to thiolate in cytochrome P-450 LM2 by chemical modification. FEBS Lett 1983; 159:58-62.

94. Ebel RE, O'Keeffe DH, Peterson JA. Nitric oxide complexes of cytochrome P-450. FEBS Lett 1975; 55:198-201.

95. O'Keeffe DH, Ebel RE, Peterson JA. Studies of the oxygen binding site of cytochrome P-450. Nitric oxide as a spin-label probe. J Biol Chem 1978; 253:3509-3516.

96. Tsubaki M, Hiwatashi A, Ichikawa Y et al. Electron paramagnetic resonance study of ferrous cytochrome P-450$_{scc}$-nitric oxide complexes: effects of cholesterol and its analogues. Biochemistry 1987; 26:4527-4534.

97. Hori H, Masuya F, Tsubaki M et al. Electronic and stereochemical characterizations of intermediates in the photolysis of ferric cytochrome P450$_{scc}$ nitrosyl complexes. Effects of cholesterol and its analogues on ligand binding structures. J Biol Chem 1992; 267:18377-18381.

98. Masuya F, Tsubaki M, Makino R et al. EPR studies on the photoproducts of ferric cytochrome P450$_{cam}$ (CYP101) nitrosyl complexes: effects of camphor and its analogues on ligand-bound structures. J Biochem 1994; 116:1146-1152.

99. Wink DA, Osawa Y, Darbyshire JF et al. Inhibiton of cytochromes P450 by nitric oxide and a nitric oxide-releasing agent. Arch Biochem Biophys 1993; 300:115-123.

100. Stadler J, Trockfeld J, Schmalix WA et al. Inhibition of cytochromes P4501A by nitric oxide. Proc Natl Acad Sci USA 1994; 91:3559-3563.

101. Kahl R, Wulff U, Netter KJ. Effect of nitrite on microsomal cytochrome P-450. Xenobiotica 1978; 8:359-364.

102. LeGall J, Payne WJ, Morgan TV et al. On the purification of nitrite reductase from *Thiobacillus denitrificans* and its reaction with nitrite under reducing conditions. Biochem Biophys Res Commun 1979; 87:355-362.

103. Bessières P, Henry Y. Etude de la réduction du nitrite par le NADH, catalysée par la nitrite réductase de *Pseudomonas aeruginosa*. C R Acad Sc Paris 1980; 290:1309-1312.

104. Bessières P, Henry Y. Stoichiometry of nitrite reduction catalyzed by *Pseudomonas aeruginosa* nitrite-reductase. Biochimie 1984; 66:313-318.

105. Liu M-C, Huynh B-H, Payne WJ et al. Optical, EPR and Mössbauer spectroscopic

studies on the NO derivatives of cytochrome cd$_1$ from *Thiobacillus denitrificans*. Eur J Biochem 1987; 169:253-258.

106. Liu M-C, Liu M-Y, Peck HD et al. Comparative EPR studies on the nitrite reductases from *Escherichia coli* and *Wolinella succinogenes*. FEBS Lett 1987; 218:227-230.

107. Dermastia M, Turk T, Hollocher TC. Nitric oxide reductase. Purification from *Paracoccus denitrificans* with use of a single column and some characteristics. J Biol Chem 1991; 266:10899-10905.

108. Kastrau DHW, Heiss B, Kroneck PMH et al. Nitric oxide reductase from *Pseudomonas stutzeri*, a novel cytochrome *bc* complex. Phospholipid requirement, electron paramagnetic resonance and redox properties. Eur J Biochem 1994; 222:293-303.

109. Ribeiro JMC, Hazzard JMH, Nussenveig RH et al. Reversible binding of nitric oxide by a salivary heme protein from a bloodsucking insect. Science 1993; 260:539-541.

110. Valenzuela JG, Walker FA, Ribeiro JMC. A salivary nitrophorin (nitric-oxide-carrying hemoprotein) in the bedbug *Cimex lectularius*. J Exp Biol 1995; 198:1519-1526.

111. Hatefi Y. The mitochondrial electron transport and oxidative phosphorylation system. Annu Rev Biochem 1985; 54:1015-1069.

112. Anraku Y. Bacterial electron transport chains. Annu Rev Biochem 1988; 57:101-132.

113. Calhoun MW, Thomas JW, Gennis RB. The cytochrome oxidase superfamily of redoxdriven proton pumps. Trends in Biol Sci 1994; 19:325-330.

114. Castresana J, Lübben M, Saraste M et al. Evolution of cytochrome oxidase, an enzyme older than atmospheric oxygen. EMBO J 1994; 13:2516-2525.

115. Chan SI, Li PM. Cytochrome *c* oxidase: understanding nature's design of a proton pump. Biochemistry 1990; 29:1-12.

116. Larsen RW, Pan L-P, Musser SM et al. Could Cu$_B$ be the site of redox linkage in cytochrome *c* oxidase? Proc Natl Acad Sci USA 1992; 89:723-727.

117. Shapleigh JP, Hosler JP, Tecklenburg MMJ et al. Definition of the catalytic site of cytochrome *c* oxidase: specific ligands of heme *a* and the heme a_3-Cu$_B$ center. Proc Natl Acad Sci USA 1992; 89:4786-4790.

118. Li PM, Malmström BG, Chan SI. The nature of Cu$_A$ in cytochrome *c* oxidase. FEBS Lett 1989; 248:210-211.

119. Lappalainen P, Aasa R, Malmström BG et al. Soluble Cu$_A$-binding from the *Paracoccus* cytochrome *c* oxidase. J Biol Chem 1993; 268:26416-26421.

120. Lappalainen P, Saraste M. The binuclear Cu$_A$ centre of cytochrome oxidase. Biochim Biophys Acta 1994; 1187:222-225.

121. Kroneck PMH, Antholine WA, Riester J et al. The cupric site in nitrous oxide contains a mixed-valence [Cu(II),Cu(I)] binuclear center: a multifrequency electron paramagnetic resonance investigation. FEBS Lett 1988; 242:70-74.

122. Kroneck PMH, Antholine WA, Riester J et al. The nature of the cupric site in nitrous oxide reductase and of Cu$_A$ in cytochrome *c* oxidase. FEBS Lett 1989; 248:212-213.

123. Ohnishi T, Harmon HJ, Waring AJ. Electron-paramagnetic-resonance studies on the spatial relationship of redox components in cytochrome oxidase. Biochem Soc Transac 1985; 13:607-611.

124. Hosler JP, Kim Y, Shapleigh J et al. Vibrational characteristic of mutant and wild-type carbon monoxide cytochrome *c* oxidase: evidence for a linear arrangement of heme *a*, a_3, and Cu$_B$. J Am Chem Soc 1994; 116:5515-5516.

125. Blokzijl-Homan MFJ, Van Gelder BF. Biochemical and biophysical studies on cytochrome aa_3. III. The EPR spectrum of NO-ferrocytochrome a_3. Biochim Biophys Acta 1971; 234:493-498.

126. Stevens TH, Bocian DF, Chan SI. EPR studies of ^{15}NO-ferrocytochrome a_3 in cytochrome *c* oxydase. FEBS Lett 1979; 97:314-316.

127. Twilfer H, Gersonde K, Christahl M. Resolution enhancement of EPR spectra using the Fourier transform technique. Analysis of nitrosyl cytochrome *c* oxidase in frozen solution. J Magn Res 1981; 44:470-478.

128. LoBrutto R, Wei YH, Mascarenhas R et al. Electron nuclear double resonance and electron paramagnetic resonance study on the structure of the NO-ligated heme a_3

in cytochrome *c* oxidase. J Biol Chem 1983; 258:7437-7448.

129. Stevens TH, Brudvig GW, Bocian FP et al. Structure of cytochrome a₃-Cua₃ couple in cytochrome *c* oxidase as revealed by nitric oxide binding studies. Proc Natl Acad Sci USA 1979; 76:3320-3324.

130. Fee JA. Copper proteins systems containing the "blue" copper center. Structure and Bonding 1975; 23:1-60.

131. Brudvig GW, Stevens TH, Morse RH et al. Conformations of oxidized cytochrome *c* oxidase. Biochemistry 1981; 20:3912-3921.

132. Boelens R, Rademaker H, Pel R et al. EPR studies of the photodissociation reactions of cytochrome *c* oxidase-nitric oxide complexes. Biochim Biophys Acta 1982; 679:84-94.

133. Boelens R, Rademaker H, Wever R et al. The cytochrome *c* oxidase-azide-nitric oxide complex as a model for oxygen-binding site. Biochim Biophys Acta 1984; 765:196-209.

134. Rousseau DL, Sing S, Ching YC et al. Nitrosyl cytochrome *c* oxidase. Formation and properties of mixed valence enzyme. J Biol Chem 1988; 263:5681-5685.

135. Brudvig GW, Stevens TH, Chan SI. Reactions of nitric oxide with cytochrome *c* oxidase. Biochemistry 1980; 19:5275-5285.

136. Cleeter MWJ, Cooper JM, Darley-Usmar VM et al. Reversible inhibition of cytochrome *c* oxidase, the terminal enzyme of the mitochondrial respiratory chain, by nitric oxide. Implications for neurodegenerative diseases. FEBS Lett 1994; 345:50-54.

137. Brown GC, Cooper CE. Nanomolar concentrations of nitric oxide reversibly inhibit synaptosomal respiration by competing with oxygen at cytochrome oxidase. FEBS Lett 1994; 356:295-298.

138. Saraste M, Castresana J. Cytochrome oxidase evolved by tinkering with denitrification enzymes. FEBS Lett 1994; 341:1-4.

139. Van der Oost J, de Boer APN, de Gier JW et al. The heme-copper oxidase family consists of three distinct types of terminal oxidases and is related to nitric oxide reductase. FEMS Microbiol Lett 1994; 121:1-10.

140. Malkin R, Malmström BG. The state and function of copper in biological systems. Adv Enzymol 1970; 33:177-244.

141. Lehn J-M, Malmström BG, Selin E et al. Metal analysis of the laccase of Gabriel Bertrand (1897). Trends in Biol Sci 1986; 11:228-230.

142. Ryden L. Ceruloplasmin. In: Lontie R, ed. Copper Proteins and Copper Enzymes, Vol III. CRC Press, Boca Raton, Florida, USA, 1984:37-100.

143. Reinhammar B. Laccase. In: Lontie R, ed. Copper Proteins and Copper Enzymes, Vol III. CRC Press, Boca Raton, Florida, USA, 1984:1-35.

144. Mondovi B, Avigliano L. Ascorbate oxidase. In: Lontie R, ed. Copper Proteins and Copper Enzymes, Vol III. CRC Press, Boca Raton, Florida, USA, 1984:101-118.

145. Messerschmidt A, Huber R. The blue oxidases, ascorbate oxidase, laccase and ceruloplasmin. Modelling and structural relationships. Eur J Biochem 1990; 187:341-352.

146. Takahashi N, Ortel TL, Putnam FW. Single-chain structure of human ceruloplasmin: the complete amino acid sequence of the whole molecule. Proc Natl Acad Sci USA 1984; 81:390-394.

147. Messerschmidt A, Ladenstein R, Huber R et al. Refined crystal structure of ascorbate oxidase at 1.9 Å resolution. J Mol Biol 1992; 224:179-205.

148. Messerschmidt A, Steigemann W, Huber R et al. X-ray crystallographic characterisation of type-2-depleted ascorbate oxidase from zucchini. Eur J Biochem 1992; 209:597-602.

149. Messerschmidt A, Luecke H, Huber R. X-ray structures and mechanistic implications of functional derivatives of ascorbate oxidase from zucchini: reduced, peroxide and azide forms. J Mol Biol 1993; 230:997-1014.

150. Ryden L. Evolution of blue copper proteins. In: King T, Mason HS, Morrison M, eds. Oxidases and Related Redox Systems. Alan R. Liss, Inc, New York. 1988; 349-366.

151. Van Leeuwen FXR, Wever R, Van Gelder BF. EPR study of nitric oxide-treated ceruloplasmin. Biochim Biophys Acta 1973; 315:200-203.

152. Wever R, Van Leeuwen FXR, Van Gelder BF. The reaction of nitric oxide with ceruloplasmin. Biochim Biophys Acta 1973; 302:236-239.

153. Van Leeuwen FXR, Van Gelder BF. A spectroscopic study of nitric-oxide-treated ceruloplasmin. Eur J Biochem 1978; 87:305-312.

154. Musci G, Di Marco S, Bonaccorsi di Patti M et al. Interaction of nitric oxide with ceruloplasmin lacking an EPR-detectable type 2 copper. Biochemistry 1991; 30:9866-9872.

155. Rotilio G, Morpurgo L, Graziani MT et al. The reaction of nitric oxide with *Rhus vernicifera* laccase. FEBS Lett 1975; 54:163-166.

156. Martin CT, Morse RH, Kanne RM et al. Reactions of nitric oxide with tree and fungal laccase. Biochemistry 1981; 20:5147-5155.

157. Van Leeuwen FXR, Wever R, Van Gelder BF et al. The interaction of nitric oxide with ascorbate oxidase. Biochim Biophys Acta 1975; 403:285-291.

158. Swain JA, Darley-Usmar V, Gutteridge JMC. Peroxynitrite releases copper from ceruloplasmin: implications for atherosclerosis. FEBS Lett 1994; 342:49-52.

159. Volbeda A, Hol WGJ. Pseudo 2-fold symmetry in the copper-binding domain of arthropodan haemocyanins. Possible implications for the evolution of oxygen transport proteins. J Mol Biol 1989; 206:531-546.

160. Van Holde KE, Miller KI. Haemocyanins. Quarterly Rev Biophys 1982; 15:1-129.

161. Ellerton HD, Ellerton NF, Robinson HA. Hemocyanin- a current perspective. Prog Biophys Molec Biol 1983; 41:143-248.

162. Gaykema WPJ, Hol WGJ, Vereijken JM et al. 3.2 Å structure of the oxygen-carrying protein *Palunirus interruptus* haemocyanin. Nature 1984; 309:23-29.

163. Volbeda A, Hol WGJ. Crystal structure of hexameric haemocyanin from *Panulirus interruptus* refine at 3.2 Å resolution. J Mol Biol 1989; 209:249-279.

164. Lerch K. Copper monooxygenases: tyrosinase and dopamine β-monooxygenase. In: Sigel H, ed. Metal Ions in Biological Systems, Vol 13, Copper proteins. Marcel Dekker, New York, 1981; 143-186.

165. Solomon EI. Binuclear copper active site: hemocyanin, tyrosinase, and type 3 copper oxidases. In: Spiro TG, ed. Copper Proteins.

166. Villafranca JJ. Dopamine β-hydroxylase. In: Spiro TG, ed. Copper Proteins. Wiley Interscience, New York, 1981: 263-289.

167. Blackburn NJ, Mason HS, Knowles PF. Dopamine β-hydroxylase: evidence for binuclear copper sites. Biochem Biophys Res Commun 1980; 95:1275-1281.

168. Hasnain SS, Diakun GP, Knowles FP et al. Direct structural information for the copper site of dopamine β-monooxygenase obtained by using extended X-ray-absorption fine structure. Biochem J 1984; 221:545-548.

169. Schoot-Uiterkamp AJM. Monomer and magnetic dipole-coupled Cu^{2+} EPR signals in nitrosylhemocyanin. FEBS Lett 1972; 20:93-96.

170. Schoot-Uiterkamp AJM, Mason HS. Magnetic dipole-dipole coupled Cu(II) pairs in nitric-oxide-treated tyrosinase: a structural relationship between the active sites of tyrosinase and hemocyanin. Proc Natl Acad Sci USA 1973; 70:993-996.

171. Schoot-Uiterkamp AJM, Van Der Deen H, Berendsen HCJ et al. Computer simulation of the EPR spectra of mononuclear and dipolar coupled Cu(II) ions in nitric oxide- and nitrite-treated hemocyanins and tyrosinase. Biochim Biophys Acta 1974; 372:407-425.

172. Verplaetse J, Van Tornout P, Defreyn G et al. The reaction of nitrogen monoxide and of nitrite with deoxyhaemocyanin and methaemocyanin of *Helix pomatia*. Eur J Biochem 1979; 95:327-331.

173. Percival MD. Human 5-lipoxygenase contains an essential iron. J Biol Chem 1991; 266:10158-10061.

174. Chasteen ND, Grady JK, Skorey KI et al. Characterization of the non-heme iron center of human 5-lipoxygenase by electron paramagnetic resonance, fluorescence, and ultraviolet-visible spectroscopy: redox cycling between ferrous and ferric states. Biochemistry 1993; 32:9763-9771.

175. Gaffney BJ, Mavrophilipos DV, Doctor KS. Access of ligands to the ferric center in lipoxygenase-1. Biophys J 1993; 64:773-783.

176. Boyington JC, Gaffney BJ, Amzel LM. The

three-dimensional structure of an arachidonic acid 15-lipoxygenase. Science 1993; 260:1482-1486.

177. Minor W, Steczko J, Bolin JT et al. Crystallographic determination of the active site iron and its ligands in soybean lipoxygenase L-1. Biochemistry 1993; 32:6320-6323.

178. Zhang Y, Gan Q-F, Pavel EG et al. EPR definition of the non-heme ferric active sites of mammalian 15-lipoxygenase: major difference relative to human 5-lipoxygenases and plant lipoxygenases and their ligand field origin. J Am Chem Soc 1995; 117:7422-7427.

179. Galpin JR, Veldink GA, Vliegenthart JFG et al. The interaction of nitric oxide with soybean lipoxygenase-1. Biochim Biophys Acta 1978; 536:356-362.

180. Salerno JC, Siedow JN. The nature of the nitric oxide complexes of lipoxygenase. Biochim Biophys Acta 1979; 579:246-251.

181. Nelson MJ. The nitric oxide complex of ferrous soybean lipoxygenase-1. Substrate, pH and ethanol effects on the active site iron. J Biol Chem 1987; 262:12137-12142.

182. Rich PR, Salerno JC, Leigh JS et al. A spin 3/2 ferrous-nitric oxide derivative of an iron-containing moiety associated with *Neurospora crassa* and higher plant mitochondria. FEBS Lett 1978; 93:323-326.

183. Nakatsuka M, Osawa Y. Selective inhibition of the 12-lipoxygenase pathway of arachidonic acid metabolism by L-arginine or sodium nitroprusside in intact human platelets. Biochem Biophys Res Commun 1994; 200:1630-1634.

184. Ohlendorf DH, Lipscomb JD, Weber PC. Structure and assembly of protocatechuate 3,4-dioxygenase. Nature 1988; 336:403405.

185. True AE, Orville AM, Pearce LL et al. An EXAFS study of the interaction of substrate with the ferric active site of protocatechuate 3,4-dioxygenase. Biochemistry 1990; 29:10847-10854.

186. Arciero DM, Lipscomb JD, Huynh BH et al. EPR and Mössbauer studies of protocatechuate 4,5-dioxygenase. Characterization of a new Fe^{2+} environment. J Biol Chem 1983; 258:14981-14991.

187. Arciero DM, Orville AM, Lipscomb JD. [17]O water and nitric oxide binding by protocatechuate 4,5-dioxygenase and catechol 2,3-dioxygenase. J Biol Chem 1985; 260:14035-14044.

188. Arciero DM, Lipscomb JD. Binding of [17]O-labeled substrate and inhibitors to protocatechuate 4,5-dioxygenase-nitrosyl complex. Evidence for direct substrate binding to the active Fe^{2+} of extradiol dioxygenase. J Biol Chem 1986; 261:2170-2178.

189. Orville AM, Lipscomb JD. Simultaneous binding of nitric oxide and isotopically labeled substrates or inhibitors by reduced protocatechuate 3,4-dioxygenase. J Biol Chem 1993; 268:8596-8607.

190. Bill E, Bernhardt FH, Trautwein AX et al. Mössbauer investigation of the cofactor iron of putidamonooxin. Eur J Biochem 1985; 147:177-182.

191. Chen VJ, Orville AM, Harpel MR et al. Spectroscopic studies of isopenicillin N synthase. A mononuclear nonheme Fe^{2+} oxidase with metal coordination sites for small molecules and substrate. J Biol Chem 1989; 264:21677-21681.

192. Brown CA, Pavlosky MA, Westre TE et al. Spectroscopic and theoretical description of the electronic structure of $S = 3/2$ iron-nitrosyl complexes and their relation to O_2 activation by non-heme iron enzyme active sites. J Am Chem Soc 1995; 117:715-732.

193. Beinert H. Recent developments in the field of iron-sulfur proteins. FASEB J 1990; 4:2483-2491.

194. Drapier J-C, Hirling H, Wietzerbin J et al. Biosynthesis of nitric oxide activates iron regulatory factor in macrophages. EMBO J 1993; 12:3643-3649.

195. Melefors Ö, Hentze MW. Iron regulatory factor—the conductor of cellular iron regulation. Blood Reviews 1993; 7:251-258.

196. Weiss G, Goossen B, Doppler W et al. Translational regulation via iron-responsive elements by the nitric oxide/NO-synthase pathway. EMBO J 1993; 12:3651-3657.

197. Klausner RD, Rouault TA, Harford JB. Regulating tha fate of mRNA: the control of cellular iron metabolism. Cell 1993; 72:19-28.

198. Philpott CC, Klausner RD, Rouault TA. The bifunctional iron-responsive element binding protein/cytosolic aconitase: the role

of active-site residues in ligand binding and regulation. Proc Natl Acad Sci USA 1994; 91:7321-7325.

199. Hirling H, Henderson BR, Kühn LC. Mutational analysis of the [4Fe-4S]-cluster converting iron regulatory factor from its RNA-binding form to cytoplasmic aconitase. EMBO J 1994; 13:453-461.

200. Ward RJ, Kühn LC, Kaldy P et al. Control of cellular iron homeostasis by iron-responsive elements *in vivo*. Eur J Biochem 1994; 220:927-931.

201. Weiss G, Werner-Felmayer G, Werner ER et al. Iron regulates nitric oxide synthase activity by controlling nuclear transcription. J Exp Med 1994; 180:969-976.

202. Dailey HA, Finnegan MG, Johnson MK. Human ferrochelatase is an iron-sulfur protein. Biochemistry 1994; 33:403-407.

203. Ferreira GC. Mammalian ferrochelatase. Overexpression in *Escherichia coli* as a soluble protein, purification and characterization. J Biol Chem 1994; 269:4396-4400.

204. Kohno H, Okuda M, Furukawa T et al. Site-directed mutagenesis of human ferrochelatase: identification of histidine-263 as a binding site for metal ions. Biochim Biophys Acta 1994; 1209:95-100.

205. Ferreira GC. Ferrochelatase binds the iron-responsive element present in the erythroid 5-aminolevulinate synthase mRNA. Biochem Biophys Res Commun 1995; 214:875-878.

206. Smith BE, Eady RR. Metalloclusters of the nitrogenases. Eur J Biochem 1992; 205:1-15.

207. Kim J, Rees DC. Nitrogenase and biological nitrogen fixation. Biochemistry 1994; 33:389-397.

208. Howard JB, Rees DC. Nitrogenase: a nucleotide-depedent molecular switch. Annu Rev Biochem 1994; 63:235-264.

209. He SH, Teixeira M, LeGall J et al. EPR studies with [77]Se-enriched (NiFeSe) hydrogenase of *Desulfovibrio baculatus*. Evidence for a selenium ligand to the active site nickel. J Biol Chem 1989; 264:2678-2682.

210. Volbeda A, Charon M-H, Piras C et al. Crystal structure of the nickel-iron hydrogenase from *Desulfovibrio gigas*. Nature 1995; 373:580-587.

211. Cunningham RP, Asahara H, Bank JF et al. Endonuclease III is an iron-sulfur protein. Biochemistry 1989; 28:4450-4455.

212. Nunoshiba T, deRojas-Walker T, Wishnok JS et al. Activation by nitric oxide of an oxidative-stress response that defends *Escherichia coli* against activated macrophages. Proc Natl Acad Sci USA 1993; 90:9993-9997.

213. Hidalgo E, Demple B. An iron-sulfur center essential for transcriptional activation by the redox sensing SoxR protein. EMBO J 1994; 13:138-146.

214. Hidalgo E, Bollinger JM, Bradley TM et al. Binuclear [2Fe-2S] clusters in the *Escherichia coli* SoxR protein and role of the metal centers in transcription. J Biol Chem 1995; 270:20908-20914.

215. Beinert H, Sands RH. Studies on succinic acid and DPNH dehydrogenase preparations by paramagnetic resonance (EPR) spectroscopy. Biochem Biophys Res Commun 1960; 3:41-46.

216. King TE, Howard RL, Mason HS. An electron spin resonance study of soluble succinic dehydrogenase. Biochem Biophys Res Commun 1961; 5:329-333.

217. Ohnishi T, Salerno JC, Winter DB et al. Thermodynamic and EPR characteristics of two ferredoxin-type iron-sulfur centers in the succinate-ubiquinone reductase segment of the respiratory chain. J Biol Chem 1976; 251:2094-2104.

218. Salerno JC, Ohnishi T, Lim J et al. Tetranuclear and binuclear iron-sulfur clusters in succinate dehydrogenase: a method of iron quantitation by formation of paramagnetic complexes. Biochem Biophys Res Commun 1976; 73:833-840.

219. McDonald CC, Phillips WD, Mower HF. An electron spin resonance of some complexes of iron, nitric oxide, and anionic ligands. J Am Chem Soc 1965; 87: 3319-3326.

220. Vanin AF. Identification of divalent iron complexes with cysteine in biological systems by the EPR method. Biokhimia 1967; 32:277-282 (English translation 228-232).

221. Woolum JC, Tiezzi E, Commoner B. Electron spin resonance of iron-nitric oxide complexes with amino acids, pep-

tides and proteins. Biochim Biophys Acta 1968; 160:311-320.

222. Vanin AF, Vakhnina LV, Chetverikov AG. Nature of the EPR signals of a new type found in cancer tissues. Biofizika 1970; 15:1044-1051 (English translation 1082-1089).

223. Butler AR, Glidewell C, Li M-H. Nitrosyl complexes of iron-sulfur clusters. Adv Inorg Chem 1988; 32:335-393.

224. Woolum JC, Commoner B. Isolation and identification of a paramagnetic complex from the livers of carcinogen-treated rats. Biochim Biophys Acta 1970; 201:131-140.

225. Nagata C, Ioki Y, Kodama M et al. Free radical induced in rat liver by a chemical carcinogen, *N*-methyl-*N'*-nitro-*N*-nitrosoguanidine. Ann N Y Acad Sci 1973; 222:1031-1047.

226. Beinert H, Kennedy MC. Engineering of protein bound iron-sulfur clusters. A tool for the study of protein and cluster chemistry and mechanism of iron-sulfur enzymes. Eur J Biochem 1989; 186:5-15.

227. Robbins AH, Stout CD. The structure of aconitase. Proteins 1989; 5:289-312.

228. Robbins AH, Stout CD. Structure of activated aconitase: formation of the [4Fe-4S] cluster in the crystal. Proc Natl Acad Sci USA 1989; 86:3639-3643.

229. Drapier J-C, Hibbs JB. Murine cytotoxic activated macrophages inhibit aconitase in tumor cells. Inhibition involves the iron-sulfur prosthetic group and is reversible. J Clin Invest 1986; 78:790-797.

230. Drapier J-C, Hibbs JB. Differentiation of murine macrophages to express nonspecific cytotoxicity for tumor cells results in L-arginine-dependent inhibition of mitochondrial iron-sulfur enzymes in the macrophage effector cells. J Immunol 1988; 140:2829-2838.

231. Hibbs JB, Taintor RR, Vavrin Z et al. Synthesis of nitric oxide from a terminal guanidino nitrogen atom of L-arginine: a molecular mechanism regulating cellular proliferation that targets intracellular iron. In: Moncada S and Higgs EA, eds. Nitric Oxide from L-arginine: a Bioregulatory System. Amsterdam: Elsevier Science Publishers BV, 1990: 189-223.

232. Pellat C, Henry Y, Drapier J-C. IFN-γ-activated macrophages: detection by electron paramagnetic resonance of complexes between L-arginine-derived nitric oxide and non-heme iron proteins. Biochem Biophys Res Commun 1990; 166:119-125.

233. Lancaster JR, Hibbs JB. EPR demonstration of iron-nitrosyl complex formation by cytotoxic activated macrophages. Proc Natl Acad Sci USA 1990; 87:1223-1227.

234. Drapier J-C, Pellat C, Henry Y. Generation of EPR-detectable nitrosyl-iron complexes in tumor target cells cocultured with activated macrophages. J Biol Chem 1991; 266:10162-10167.

235. Drapier J-C, Pellat C, Henry Y. Characterization of the nitrosyl-iron complexes generated in tumour cells after co-culture with activated macrophages. In: Moncada S, Marletta MA, Hibbs JB et al, eds. The Biology of Nitric Oxide. London, UK: Portland Press, 1992: 72-76.

236. Reddy D, Lancaster JR, Cornforth DP. Nitrite inhibition of *Clostridium botulinum*: electron spin resonance detection of iron-nitric oxide complexes. Science 1983; 221:769-770.

237. Payne MJ, Woods LFJ, Gibbs P et al. Electron paramagnetic resonance spectroscopic investigation of the inhibition of the phosphoroclastic system of *Clostridium sporogenes* by nitrite. J Gen Microbiol 1990; 136:2067-2076.

238. Trinchant JC, Rigaud J. Nitrite and nitric oxide as inhibitors of nitrogenase from soybean bacteroids. Appl Environ Microbiol 1982; 44:1386-1388.

239. Vaughn SA, Burgess BK. Nitrite, a new substrate for nitrogenase. Biochemistry 1989; 28:419-424.

240. Berlier Y, Fauque GD, LeGall J et al. Inhibition studies of three classes of *Desulfovibrio* hydrogenase: application to the further characterization of the multiple hydrogenases found in *Desulfovibrio vulgaris* Hildenborough. Biochem Biophys Res Commun 1987; 146:147-153.

241. Bianco P, Haladjian J, Bruschi M et al. Reactivity of [Fe] and [Ni-Fe-Se] hydrogenases with their oxido-reduction partner: the tetraheme cytochrome c_3. Biochem Biophys

Res Commun 1992; 189:633-639.

242. Wang C-P, Franco R, Moura JJG et al. The nickel site in active *Desulfovibrio baculatus* [NiFeSe] hydrogenase is diamagnetic. Multifield saturation magnetization measurement of the spin state of Ni(II). J Biol Chem 1992; 267:7378-7380.

243. Guigliarelli B, More C, Fournel A et al. Structural organization of the Ni and (4Fe-4S) centers in the active form of *Desulfovibrio gigas* hydrogenase. Analysis of the magnetic interactions by electron paramagnetic resonance spectroscopy. Biochemistry 1995; 34:4781-4790.

244. Hyman MR, Arp DJ. Kinetic analysis of the interaction of nitric oxide with the membrane-associated, nickel and iron-sulfur-containing hydrogenase from *Azotobacter vinelandii*. Biochim Biophys Acta 1991; 1076:165-172.

245. Robbins AH, McRee DE, Williamson M et al. Refined crystal structure of Cd,Zn metallothionein at 2.0 Å resolution. J Mol Biol 1991; 221:1269-1293.

246. Messerle BA, Schäffer A, Vasak M et al. Three-dimensional structure of human [^{113}Cd$_7$] metallothionein-2 in solution determined by n.m.r. spectroscopy. J Mol Biol 1990; 214:765-779.

247. Zhu Z, DeRose EF, Mullen GP et al. Sequential proton resonance assignments and metal cluster topology of lobster metallothionein-1. Biochemistry 1994; 33:8858-8865.

248. Ding XQ, Butzlaff C, Bill E et al. Mössbauer and magnetic susceptibility studies on iron(II) metallothionein from rabbit liver. Evidence for the existence of an unusual type of [M_3(CysS)$_9$]$^{3-}$ cluster. Eur J Biochem 1994; 220:827-837.

249. Kennedy MC, Gan T, Antholine WE et al. Metallothionein reacts with Fe^{2+} and NO to form products with g = 2.039 ESR signal. Biochem Biophys Res Commun 1993; 196:632-635.

250. Schwarz MA, Lazo JS, Yalowich JC et al. Metallothionein protects against the cytotoxic and DNA-damaging effects of nitric oxide. Proc Natl Acad Sci USA 1995; 92:4452-4456.

251. Sanders-Loehr J. Binuclear iron proteins. In: Loehr TM, ed. Iron Carriers and Iron Proteins. New York, VHC publishers, 1989:375-466.

252. Nordlund P, Eklund H. Structure and function of the *Escherichia coli* ribonucleotide reductase protein R2. J Mol Biol 1993; 232:123-164.

253. Stenkamp RE, Siecker LC, Jensen LH et al. Structure of the binuclear iron complex in metazidohaemerythrin from *Themiste dyscritum* at 2.2 Å resolution. Nature 1981; 291:263-264.

254. Pulver S, Froland WA, Fox BG et al. Spectroscopic studies of the coupled binuclear non-heme iron active site in the fully reduced hydroxylase component of methane monooxygenase: comparison to deoxy and deoxy-azide hemerythrin. J Am Chem Soc 1993; 115:12409-12422.

255. Nocek JM, Kurtz DM, Pickering RA et al. Oxidation of deoxyhemerythrin to semimethemoglobin by nitrite. J Biol Chem 1984; 259:12334-12338.

256. Nocek JM, Kurtz DM, Sage JT et al. Nitric oxide adducts of the binuclear iron center in deoxyhemerythrin from *Phascolopsis gouldii*. Analogue of a putative intermediate in the oxygenation reaction. J Am Chem Soc 1985; 107:3382-3384.

257. Nocek JM, Kurtz DM, Sage JT et al. Nitric oxide adducts of the binuclear iron site of hemerythrin: spectroscopy and reactivity. Biochemistry 1988; 27:1014-1024.

258. Enemark JH, Feltham RD. Stereochemical control of valence and its application to the reduction of coordinated NO and N$_2$. Proc Natl Acad Sci USA 1972; 69:3534-3536.

259. Enemark JH, Feltham RD. Principles of structure, bonding, and reactivity for metal nitrosyl complexes. Coord Chem Reviews 1974; 13:339-406.

260. Stubbe J. Ribonucleotide reductases. Adv Enzymol Related Areas Mol Biol 1990; 63:349-419.

261. Stubbe J. Ribonucleotide reductases: amazing and confusing. J Biol Chem 1990; 265; 5329-5332.

262. Fontecave M, Nordlund P, Eklund H et al. The redox centers of ribonucleotide reductase of *Escherichia coli*. Adv Enzymol Re-

lated Areas Mol Biol 1992; 65:147-183.

263. Reichard P. From RNA to DNA, why so many ribonucleotide reductases? Science 1993; 260:1773-1777.

264. Reichard P. The anaerobic ribonucleotide reductase from *Escherichia coli*. J Biol Chem 1993; 268:8383-8386.

265. Sun X, Harder J, Krook M et al. A possible glycine radical in anaerobic ribonucleotide reductase from *Escherichia coli*: nucleotide sequence of the cloned nrdD gene. Proc Natl Acad Sci USA 1993; 90;577-581.

266. Mulliez E, Fontecave M, Gaillard J et al. An iron-sulfur center and a free radical in the active anaerobic ribonucleotide reductase of *Escherichia coli*. J Biol Chem 1993; 268:2296-2299.

267. Uhlin U, Uhlin T, Eklund H. Crystallization and crystallographic investigations of ribonucleotide reductase protein R1 from *Escherichia coli*. FEBS Lett 1993; 336:148-152.

268. Uhlin U, Eklund H. Structure of ribonucleotide protein R1. Nature 1994; 370:533-539.

269. Davidov R, Kuprin S, Gräslund A, et al. Electron paramagnetic resonance study of the mixed-valent diiron center in *Escherichia coli* ribonucleotide reductase produced by reduction of radical-free protein R2 at 77 K. J Am Chem Soc 1994; 116:11120-11128.

270. Nordlund P, Sjöberg BM, Eklund H. Three-dimensional structure of the free radical protein of ribonucleotide reductase. Nature 1990; 345:593-598.

271. Åberg A, Nordlund P, Eklund H. Unusual clustering of carboxyl side chains in the core of iron-free ribonucleotide reductase. Nature 1993; 361:276-278.

272. Sahlin M, Lassmann G, Pötsch S et al. Tryptophan radicals formed by iron/oxygen reaction with *Escherichia coli* ribonucleotide reductase protein R2 mutant Y122F. J Biol Chem 1994; 269:11699-11702.

273. Henriksen MA, Cooperman BS, Salem JS et al. The stable tyrosyl radical in mouse ribonucleotide reductase is not essential for enzymatic activity. J Am Chem Soc 1994; 116:9773-9774.

274. Ormö M, Regnström K, Wang Z et al. Residues important for radical stability in

ribonucleotide reductase from *Escherichia coli*. J Biol Chem 1995; 270:6570-6576.

275. Nyholm S, Thelander L, Gräslund A. Reduction and loss of the iron center in the reaction of the small subunit of mouse ribonucleotide reductase with hydroxyurea. Biochemistry 1993; 32:11569-11574.

276. Ling J, Sahlin M, Sjöberg BM et al. Dioxygen is the source of the μ-oxo bridge in iron ribonucleotide reductase. J Biol Chem 1994; 269:5595-5601.

277. Sahlin M, Petersson L, Gräslund A et al. Magnetic interaction between the tyrosyl free radical and the antiferromagnetically coupled iron center in ribonucleotide reductase. Biochemistry 1987; 26:5541-5548.

278. Mann GJ, Gräslund A, Ochiai EI et al. Purification and characterization of recombinant mouse and herpes simplex virus ribonucleotide reductase R2 subunit. Biochemistry 1991; 30:1939-1947.

279. Galli C, Atta M, Andersson KK et al. Variations of the diferric exchange coupling in the R2 subunit of ribonucleotide reductase from four species as determined by saturation-recovery EPR spectroscopy. J Am Chem Soc 1995; 117:740-746.

280. Gerfen GJ, Bellew BF, Sun U et al. High-frequency (139.5 GHz) EPR spectroscopy of the tyrosyl radical in *Escherichia coli* ribonucleotide reductase. J Am Chem Soc 1993; 115:6420-6421.

281. Lepoivre M, Chenais B, Yapo A et al. Alterations of ribonucleotide reductase activity following induction of the nitrite-generating pathway in adenocarcinoma cells. J Biol Chem 1990; 265:14143-14149.

282. Lepoivre M, Fieschi F, Coves J et al. Inactivation of ribonucleotide reductase by nitric oxide. Biochem Biophys Res Commun 1991; 179:442-448.

283. Kwon NS, Stuehr DJ, Nathan CF. Inhibition of tumor cell ribonucleotide reductase by macrophage-derived nitric oxide. J Exp Med 1991; 174:761-767.

284. Lepoivre M, Flaman J-M, Henry Y. Early loss of the tyrosyl radical in ribonucleotide reductase of adenocarcinoma cells producing nitric oxide. J Biol Chem 1992; 267:22994-23000.

285. Lepoivre M, Flaman J-M, Bobé P et al.

Quenching of the tyrosyl free radical of ribonucleotide reductase by nitric oxide. Relationship to cytostasis induced in tumor cells by cytotoxic macrophages. J Biol Chem 1994; 269:21891-21897.

286. Roy B, Lepoivre M, Henry Y et al. Inhibition of ribonucleotide reductase by nitric oxide derived from thionitrites: reversible modifications of both subunits. Biochemistry 1995; 34:5411-5418.

287. Haskin CJ, Ravi N, Lynch JB et al. Reaction of NO with the reduced R2 protein of ribonucleotide reductase from Escherichia coli. Biochemistry 1995; 34:11090-11098.

288. Theil EC. Ferritin: structure, gene regulation, and cellular function in animals, plants and microorganisms. Annu Rev Biochem 1987; 56:289-315.

289. Theil EC. The ferritin family of iron storage proteins. Adv Enzymol 1989; 63:421-449.

290. Andrews SC, Arosio P, Bottke W et al. Structure, function, and evolution of ferritins. J Inorg Biochem 1992; 47:161-174.

291. Reif DW. Ferritin as a source of iron for oxidative damage. Free Rad Biol Med 1992; 12:417-427.

292. Sun S, Chasteen ND. Rapid kinetics of the EPR-active species formed during initial iron uptake in horse spleen apoferritin. Biochemistry 1994;33:15095-15102.

293. Lawson DM, Treffry A, Artymiuk PJ et al. Identification of the ferroxidase centre in ferritin. FEBS Lett 1989; 254:207-210.

294. Lawson DM, Artymiuk PJ, Yewdall SJ et al. Solving the structure of human H ferritin by genetically engineering intermolecular crystal contacts. Nature 1991; 349:541-544.

295. Treffry A, Hirzmann J, Yewdall SJ et al. Mechanism of catalysis of Fe(II) oxidation by ferritin H chains. FEBS Lett 1992; 302:108-112.

296. Bauminger ER, Harrison PM, Hechel D et al. Iron (II) oxidation and early intermediates of iron-core formation in recombinant human H-chain ferritin. Biochem J 1993; 296:709-719.

297. Waldo GS, Ling J, Sanders-Loehr J et al. Formation of an Fe(III)-tyrosinate complex during biomineralization of H-subunit ferritin. Science 1993; 259:796-798.

298. Hempstead PD, Hudson AJ, Artymiuk PJ et al. Direct observation of the iron binding sites in a ferritin. FEBS Lett 1994; 350:258-262.

299. Chen-Barrett Y, Harrison PM, Treffry A et al. Tyrosyl radical formation during the oxidative deposition of iron in human apoferritin. Biochemistry 1995; 34:7847-7853.

300. Reif DW, Simmons RD. Nitric oxide mediates iron release from ferritin. Arch Biochem Biophys 1990; 283:537-541.

301. Laulhère JP, Fontecave M. Nitric oxide does not promote iron release from ferritin. Biometals 1996; 9:10-14.

302. LeBrun NE, Cheesman RM, Thomson AJ et al. An EPR investigation of non-haem iron sites in Escherichia coli bacterioferritin and their interaction with phosphate. A study using nitric oxide as a spin probe. FEBS Lett 1993; 323:261-266.

303. LeBrun NE, Wilson MT, Andrews SC et al. Kinetic and structural characterization of an intermediate in the biomineralization of bacterioferritin. FEBS Lett 1993; 333:197-202.

304. Lee M, Arosio P, Cozzi A et al. Identification of the EPR-active iron-nitrosyl complexes in mammalian ferritins. Biochemistry 1994; 33:3679-3687.

305. Kon H, Kataoka N. Electron paramagnetic resonance of nitric oxide-protoheme with some nitrogenous base. Model systems of nitric oxide hemoproteins. Biochemistry 1969; 8:4757-4762.

306. Henry Y, Lepoivre M, Drapier J-C et al. EPR characterization of molecular targets for NO in mammalian cells and organelles. FASEB J 1993; 7:1124-1134.

307. Wilcox DE, Smith RP. Detection and quantification of nitric oxide using electron magnetic resonance spectroscopy. Methods: a companion to Methods in Enzymology 1995; 7:59-70.

308. Henry YA, Singel DJ. Metal-nitrosyl interactions in nitric oxide biology probed by electron paramagnetic resonance spectroscopy. In: Feelisch M, Stamler J, eds. Methods in Nitric Oxide Research. John Wiley and Sons, 1996: 357-372.

309. Singel DJ, Lancaster JR. Electron paramag-

netic resonance spectroscopy and nitric oxide biology. in: Feelisch M, Stamler J, eds. Methods in Nitric Oxide Research. John Wiley and Sons, 1996:341-356.

310. Kosaka H, Shiga T. Detection of nitric oxide by electron paramagnetic resonance using hemoglobin. In: Feelisch M, Stamler JS, eds. Methods in nitric oxide research. John Wiley & Sons. 1996:373-381.

311. Arnold WP, Mittal CK, Katsuki S et al. Nitric oxide activates guanylate cyclase and increases guanosine 3':5'-cyclic monophosphate levels in various tissue preparations. Proc Natl Acad Sci USA 1977; 74:3203-3207.

312. Murad F, Mittal CK, Arnold WP et al. Guanylate cyclase: activation by azide, nitro compounds, nitric oxide, and hydroxyl radical and inhibition by hemoglobin and myoglobin. Adv Cyclic Nucleotide Res 1978; 9:145-157.

313. Stone JR, Marletta MA. Soluble guanylate cyclase from bovine lung: activation with nitric oxide and carbon monoxide and spectral characterization of the ferrous and ferric states. Biochemistry 1994; 33:5636-5640.

314. Stone JR, Sands RH, Dunham WR et al. Electron paramagnetic resonance spectral evidence for the formation of a penta-coordinated nitrosyl-heme complex on soluble guanylate cyclase. Biochem Biophys Res Commun 1995; 207:572-577.

315. Stone JR, Marletta MA. Spectral and kinetic studies on the activation of soluble guanylate cyclase by nitric oxide. Biochemistry 1996; 35:1093-1099.

316. Rogers NE, Ignarro LJ. Constitutive nitric oxide synthase from cerebellum is reversibly inhibited by nitric oxide formed from L-arginine. Biochem Biophys Res Commun 1992; 189:242-249.

317. Rengasamy A, Johns RA. Regulation of nitric oxide synthase by nitric oxide. Mol Pharmacol 1993; 44:124-128.

318. Rengasamy A, Johns RA. Inhibition of nitric oxide synthase by a superoxide generating system. J Pharmacol Exp Ther 1993; 267:1024-1027.

319. Assreuy J, Cunha FQ, Liew FY et al. Feedback inhibition of nitric oxide synthase activity by nitric oxide. Br J Pharmacol 1993; 108:833-837.

320. Buga H, Griscavage JM, Rogers NE et al. Negative feedback regulation of endothelial cell function by nitric oxide. Cir Res 1993; 73:808-812.

321. Griscavage JM, Rogers NE, Sherman MP et al. Inducible nitric oxide synthase from a rat alveolar macrophage cell line is inhibited by nitric oxide. J Immunol 1993; 151:6329-6337.

322. Griscavage JM, Fukuto JM, Komori Y et al. Nitric oxide inhibits neuronal nitric oxide synthase by interacting with the heme prosthetic group. Role of tetrahydrobiopterin in modulating the inhibitory action of nitric oxide. J Biol Chem 1994; 269:21644-21649.

323. Wang J, Rousseau DL, Abu-Soud HM et al. Heme coordination of NO in NO synthase. Proc Natl Acad Sci USA 1994; 91:10512-10516.

324. Abu-Soud HM, Wang JL, Rousseau DL et al. Neuronal nitric oxide synthase self-inactivates by forming a ferrous-nitrosyl complex during aerobic catalysis. J Biol Chem 1995; 270:22997-23006.

325. Hurshman AR, Marletta MA. Nitric oxide complexes of inducible nitric oxide synthase: spectral characterization and effect on catalytic activity. Biochemistry 1995; 34:5627-5634.

326. Ravichandran LV, Fohns RA, Rengasamy A. Direct and reversible inhibition of endothelial nitric oxide synthase by nitric oxide. Am J Physiol 1995; 268:H2216-H2223.

CHAPTER 7

NITRIC OXIDE, AN INTERMEDIATE IN THE DENITRIFICATION PROCESS AND OTHER BACTERIAL PATHWAYS, AS DETECTED BY EPR SPECTROSCOPY

Yann A. Henry

INTRODUCTION: A HISTORICAL BACKGROUND

The importance of the denitrification pathway within the nitrogen cycle was recognized very early (1882) by two of Louis Pasteur's (1822-1895) younger colleagues, Léonard-Ulysse Gayon (1842-1920) and Auguste-Gabriel Dupetit (1861-1886). Professor William Jackson Payne has recently provided us with a very fine and illuminating historical analysis (in French) of the invention of the concepts of the three main processes forming the nitrogen cycle that he recommends calling "Reiset cycle", after Jules Reiset (died 1896):[1] firstly bacterial denitrification of nitrate and nitrite into nitrogen gas, secondly bacterial nitrification of ammonia to nitrite and nitrate and thirdly nitrogen fixation by free-living bacteria and bacterial-macrosymbiontic plants into ammonia and amines. These concepts resulted mostly from the work of French scientists over a thirty year period, between 1856 and 1886.[1] A very complete description of the fundamental microbiologic, metabolic and genetic aspects of denitrification was previously offered in a book by the same author,[2] pointing in the last chapter to the promising areas of research that have been undertaken in the last fifteen years. Two chapters of the book were focused on the two reactions involving nitric oxide, the first "that defines denitrification":

Nitric Oxide Research from Chemistry to Biology: EPR Spectroscopy of Nitrosylated Compounds, edited by Yann A. Henry, Annie Guissani and Béatrice Ducastel. © 1997 R.G. Landes Company.

$2NO_2^- \rightarrow 2NO$

and the "second defining reaction":

$2NO \rightarrow N_2O$

More recently, a review of the first step of denitrification described the nitrite-reducing enzymes in denitrifying, assimilatory and dissimilatory pathways of bacteria.[3] Two other reviews dealt with the unique role of nitric oxide in bacterial denitrification and other pathways such as nitrosation or the effect of NO on nitrogenase.[4,5]

We shall focus here on the use of EPR spectroscopy in the elucidation of basic questions which led to a long controversy over the role of nitric oxide as an obligate free or bound intermediate in denitrifiers. We shall describe its interaction with nitrite reductases (NiR) and with nitric oxide reductases (NoR). As we shall see, EPR spectroscopy has only played a minor historical role in this controversy as compared to other powerful analytical methods, such as manometry, gas chromatography, use of isotopic tracers in mass spectrometry and emission spectrometry,[2] and more recently genetics.[4] Probably the only advantage of EPR spectroscopy lies in its high specificity: a well characterized EPR spectrum of a nitrosyl-metalloprotein is in general unambiguous. It first proves that NO is effectively bound to a metal atom; secondly the spectrum is often characteristic of a specific metalloprotein or at least of a class of prosthetic groups. We shall however give a few specific instances of the contrary. The absence of any EPR spectrum is however a completely neutral finding, as many NO complexes are either diamagnetic or EPR silent for various reasons (see chapters 2 and 3).

General Ecology of the Nitrogen Cycle: Denitrification Versus Nitrification

The nitrogen cycle is one of the most important bases of global ecology (Fig. 7.1). It consists of several chains of oxidative or reductive processes over all the various nitrogen oxidation numbers. The processes are very finely regulated, depending mostly on the degree of anaerobiosis of a medium and on the presence of nitrate and nitrite acting as gene transcription activators.[4-7] They are performed by microorganisms which can often be closely located in soils, marshes, oceans, etc.[8,9] In denitrification, the reduction of nitrate and nitrite is performed in a respiratory process by facultative anaerobes such as *Pseudomonas, Paracoccus* or *Alcaligenes* and leads to nitrogen gas.[4-6,10-12] In other bacteria such as *Clostridium* or *Wolinella*, nitrate and nitrite reduction proceeds to ammonia with hydroxylamine as an intermediate (dissimilative reduction). Nitrogen can be fixed to ammonia by other anaerobes such as Pseudomonads or *Clostridium pastorianum*. and free-living or Leguminosae-associated *Rhizobium*.[13-16] Finally an oxidative pathway oxidizes ammonia and hydroxylamine to nitrite (nitrification) with nitric oxide and nitrous oxide as intermediates, such as performed by *Nitrosomonas europea* or *Alcaligenes faecalis*.[17,18] Some heterotrophic bacteria can combine nitrification and denitrification;[19] while *Alcaligenes faecalis* for instance can denitrify under anaerobic conditions and nitrify in aerobic ones;[17] *Thiosphaera pantotropha* can couple nitrification to aerobic denitrification[18] (Fig. 7.1).

Respiration on nitrate and nitrite, like that on oxygen, can be coupled to a continuous supply of free energy for the bacterial growth and maintenance under the form of chemicals such as ATP, phosphoenolpyruvate, organic phosphates, acetylcoenzyme A, etc. A free enthalpy diagram (Fig. 7.2) shows the cascade in the reduction pathway to the energetic deep well, represented by the nitrogen gas.[11]

All steps are catalyzed by metalloenzymes. Several of these enzymes (nitrate reductase, NaR; nitric oxide reductase, NoR) are coupled to energy transduction and proton translocation in membranes under the form of ATP or other phosphate's synthesis, while others (nitrite reductase, NiR; nitrous oxide reductase, N_2OR or NoS—a most unfortunate abbreviation as it is also used for L-arginine-dependent nitric oxide synthase in the form NOS) are located in the periplasmic space.

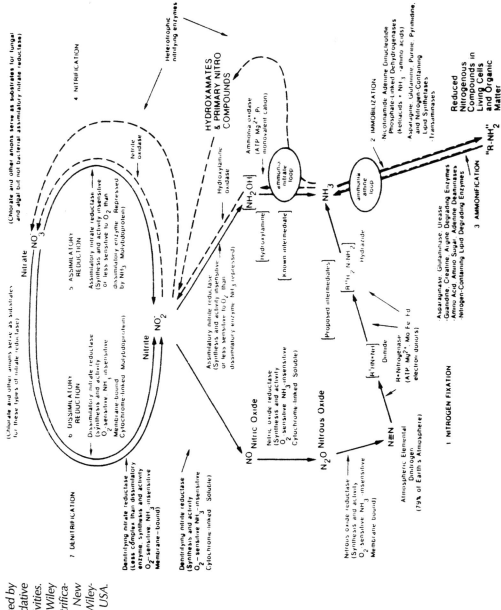

Fig. 7.1. The nitrogen cycle as depicted by Payne. Dashed curves indicate oxidative activities; solid curves, reductive activities. (Reproduced by permission of John Wiley & Sons, Ltd, from: Payne WJ. Denitrification. Wiley-Interscience Publication, New York,' USA, 1981.) Copyright © Wiley-Interscience Publication, New York, USA.

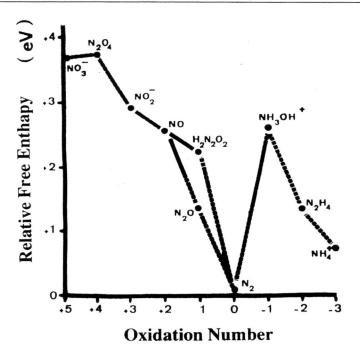

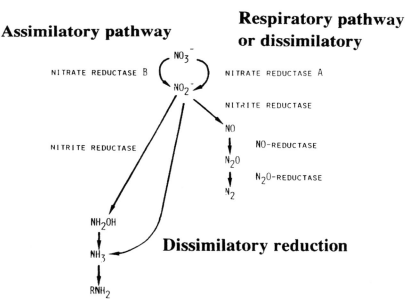

Fig. 7.2. Upper panel: Diagram of the relative free enthalpy of formation of various nitrogen derivatives as a function of the oxidation number. The free enthalpy relative to that of the dinitrogen molecule, is derived from the redox potential of the various couples at pH 7 multiplied by the number of electrons involved in the stoichiometric reactions per nitrogen atom. The redox potential values were taken from the Handbook of Chemistry and Physics, CRC Press, 1973. Lower panel: Comparison of various processes of reduction of nitrate and nitrite. (Reproduced with permission from: Henry Y, Bessières P. Denitrification and nitrite reduction: Pseudomonas aeruginosa nitrite reductase. Biochimie 1984; 66:259-289.) Copyright © Editions Masson, Paris, France.

Incidently this bacterial process, all important in the general ecology of soils, rivers, marshes, oceans, etc, is also relevant to animals and to human beings as well. In fact, according to Luckey[20,21] a human being is made of 10^{13} cells and a "healthy" human being is host to 10^{14} bacterial cells, a large proportion of which are anaerobes.

CONTROVERSY OVER NITRIC OXIDE AS AN OBLIGATORY, FREE OR BOUND INTERMEDIATE IN DENITRIFICATION

Since nitrite reduction leads to nitrous oxide and nitrogen, the question was raised early whether nitric oxide was a stable intermediate. The question became more acute when nitrite reductase was recognized (see below) to be a cytochrome cd_1, a homodimer containing both heme c and a pentacoordinated heme d_1. People questioned whether nitric oxide as an intermediate product would not inhibit its own synthesis and that of subsequent nitrous oxide, through simple binding to heme c and/or heme d_1. A controversy raged for over ten years from the late 70s to the 80s, as to whether NO is an *obligate* (or obligatory) and *free* or a *bound* intermediate in bacterial denitrification.

Various schemes were proposed for the reduction of nitrite into nitric oxide and nitrous oxide, by Payne and coworkers,[22,23] Hollocher and coworkers,[24-26] Averill, Tiedje et al,[27,28] as summarized by Henry and Bessières (Fig. 7.3).[11]

The controversy finally seemed to be settled in the late 1980s by the recognition that NO was definitely an intermediate, through two main lines of evidence: denitrifying bacteria clearly possess a nitric oxide reductase activity,[29-33] and NO reduction to N_2O is associated with proton translocation, while that of nitrite is not.

That NO is a *free diffusible* intermediate between nitrite and N_2O during denitrification by many bacteria: *P. aeruginosa, P. stutzeri, Pa. denitrificans,* and *Achromobacter cycloclastes*, was definitely proven by its trapping by extracellular hemoglobin.[30,34] With an extracellular concentration of NO measured at 2 nmol/l,

its intracellular concentration could be estimated at a little less than 1 μmol/l. A steady-state solution concentration lower than 0.4 μmol/l could be extrapolated from kinetic and isotopic data.[35,36] This quantification of NO produced was confirmed by sparging denitrifier suspensions to remove gaseous products of nitrate reduction.[37] A full discussion of this problem was proposed recently by Zumft.[4]

Even though NO is an intermediate[38-40] and a concurrent NoR activity exists,[3-6] the ability of NiR to reduce nitrite directly to N_2O remains however an open possibility. Its actual mechanism is still debated.[41-44]

EARLY PROOF BY EPR SPECTROSCOPY OF NO PRODUCTION BY DENITRIFIERS

EPR spectroscopy was used very early (1971) to test the possibility that NO was an intermediate product of denitrification, soon after the first EPR spectra of nitrosyl hemoglobin were recorded by Kon and by Shiga (1968-1969) (see chapter 4). First to be tested the binding of NO to cytochrome c_3 purified from *Desulfovibrio vulgaris* (Fig. 7.4.).[45]

These spectra were however poorly resolved and showed only an unspecific three-line hyperfine structure indicative of pentacoordinated nitrosyl heme, either weakly bound or not bound to the apoprotein. The same year Cox et al[46] were the first to show that *Pseudomonas perfectomarinus*, a nitrate-grown denitrifying bacteria, contained both a nitrite reductase and a NO-reductase (a c-type cytochrome), both able to bind NO as shown by their EPR spectra (Fig. 7.5).[46]

These EPR spectra were also poorly resolved and did not reveal anything about the enzymes involved, except that they were hemoproteins.

RESPIRATORY AND ASSIMILATORY NITRITE REDUCTASES

Four different types of enzymes with a function of nitrite reductase have been discovered, falling into two classes, those involved in denitrification (cytochrome cd_1

Denitrification schemes

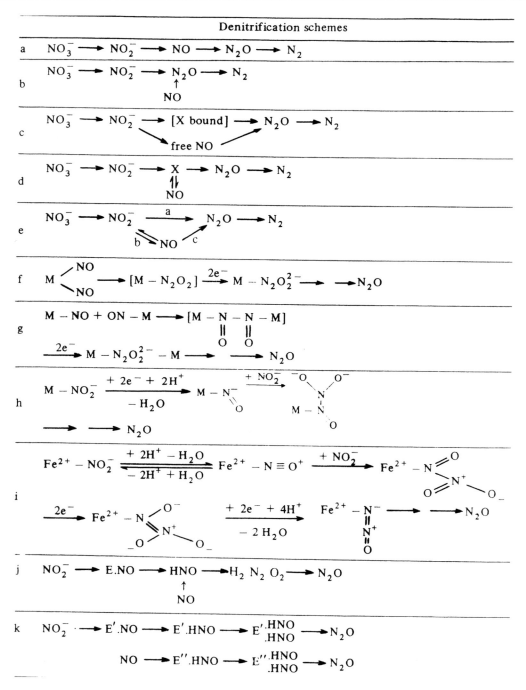

Fig. 7.3. Summary of the various denitrification schemes proposed for different denitrifying bacteria. M represents a metal center, E, E', E'' enzymes. a (ref. 22, 23, 27), b (ref. 24), c (ref. 10 , 27), d (ref. 22, 27), e (ref. 25), f-i (ref. 28), j,k (ref. 26). (Reproduced with permission from: Henry Y, Bessières P. Denitrification and nitrite reduction: Pseudomonas aeruginosa nitrite reductase. Biochimie 1984; 66:259-289.) Copyright © Editions Masson, Paris, France.

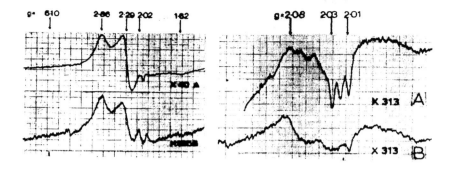

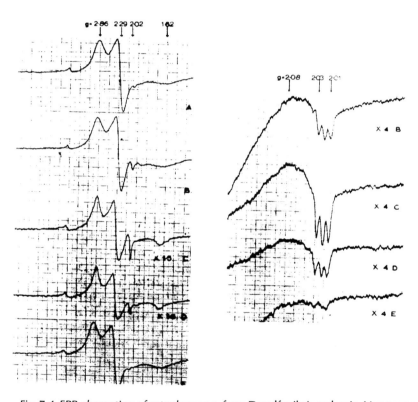

Fig. 7.4. EPR absorption of cytochrome c_3 from Desulfovibrio vulgaris. *Upper panel:* A: Initial ferric spectrum of cytochrome c_3 (1.92 mM), pH 7.6, reacted with NO for 1hr at 25°C; B: after reduction with sodium dithionite under helium and partial reoxidation. Lower panel: A: Initial ferric spectrum obtained in NH_2OH; B: reduced with excess dithionite under helium; C: further reduced; D: still further reduced; E: almost reoxidized. Spectra on right are identical with corresponding traces on left side but at 10-fold expanded field setting and where indicated at higher gain. (Reproduced with permission from: DerVartanian DV, LeGall J. Electron paramagnetic studies on the reaction of exogenous ligands with cytochrome c_3 from Desulfovibrio vulgaris. Biochim Biophys Acta 1971; 243:53-65.) Copyright © Elsevier Science, Sara Burgerhartstraat 25, 1055 KV Amsterdam, The Netherlands.

Fig. 7.5. EPR absorption of nitrite reductase and NO-reductase from Pseudomonas perfectomarinus. Right panel: A: Nitrite reductase; B: + NADH + FMN + FAD + nitrite. Left panel: A: Nitric oxide reductase; B: + NO + NADH for 5 min at 25°C; C: as B, 25 min at 25°C. (Reproduced with permission from: Cox CD, Payne WJ, DerVartanian DV. Electron paramagnetic resonance studies on the nature of hemoproteins in nitrite and nitric oxide reduction. Biochim Biophys Acta 1971; 253:290-294.) Copyright © Elsevier Science, Sara Burgerhartstraat 25, 1055 KV Amsterdam, The Netherlands.

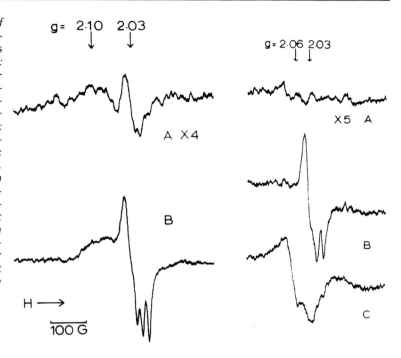

and copper-containing NiR) and those (assimilatory and dissimilatory) which reduce nitrite directly to ammonia (siroheme-containing nitrite reductase, hexaheme and tetraheme nitrite reductases).[3] The first, long called cytochrome oxidase as it catalyzes electron transfer from a cytochrome c (c_{551}) and oxygen reduction to water, has been studied in detail mostly as a soluble simple model of mitochondrial cytochrome oxidase (cytochrome aa_3). The second, a blue copper-containing NiR, has also many structural and functional analogies with blue copper oxidases such as ascorbate oxidase, laccase or ceruloplasmin.[47] These two kinds of NiR seem to be mutually exclusive and to have the same frequency of occurrence in denitrifying bacteria. The two other types, siroheme and hexaheme ammonia-producing NiR seem to be structurally and functionally unique types of enzymes.

CYTOCHROME CD_1

Nitrite reductase-cytochrome cd_1 (EC 1.9.3.2) was the first detected form of NiR in denitrifiers. It is a homodimer with an individual molecular mass of 63 kDa, each

subunit containing one heme c and one heme d_1. It has been purified from many species, e.g., *Pseudomonas aeruginosa*, *P. stutzeri*, *Pa. denitrificans*, *Thiobacillus denitrificans*,[43] *Alcaligenes eutrophus*,[49] and that from *Thiosphaera pantotropha* has been recently crystallized and found suitable for X-ray analysis.[50-52] Both heme types are low-spin (S = 1/2) in the ferric form, the axial ligands of heme c being methionine and histidine and those of heme d_1 (oxidized) being probably two histidines. In the ferrous state heme c remains low-spin (S = 0) and heme d_1 is high-spin (S = 2) penta-coordinated. Four groups performed nearly simultaneous independent EPR studies of NO binding by NiR from various sources. The work of LeGall et al[53] showed that NiR from *Thiobacillus denitrificans* produced NO which was partially bound to heme d and to heme c (Fig. 7.6). It was historically the first definitive evidence for the mandatory liganding of NO with pure NiR during nitrite reduction and for a continuous release of NO from the active site.

These findings were confirmed and extended simultaneously by three independent groups (Fig. 7.7-7.10).[39,54,55] The British group showed that multiple species were formed upon reduction of nitrite by ascorbate catalyzed by the *P. aeruginosa* enzyme, with a strong dependence upon the pH value and the nature of the buffer used (Figs. 7.7-7.8). Spectral analysis and quantification of the EPR signals indicated a nitrogen atom (possibly a histidine) as the ligand *trans* to NO in heme d_1 and accounted for NO bound to the heme d_1 in both subunits.[54]

Similar results were obtained by Bessières and Henry with the same enzyme

using NADH as a nitrite reducer, plus phenazine methosulfate (PMS) as an electron mediator (Fig. 7.9) and by Muhoberac and Wharton studying the binding of pure NO to the ascorbate-reduced enzyme (Fig. 7.10).[39,55]

While these EPR studies characterized NO as *heme-bound*, Wharton and Weintraub in the same year[38] identified both NO and N_2O by gas chromatography-mass spectrometry as *free gaseous* products of nitrite reduction by a mixture of NADH, FMN and NADH-quinone reductase catalyzed by NiR from *Pseudomonas*. Furthermore

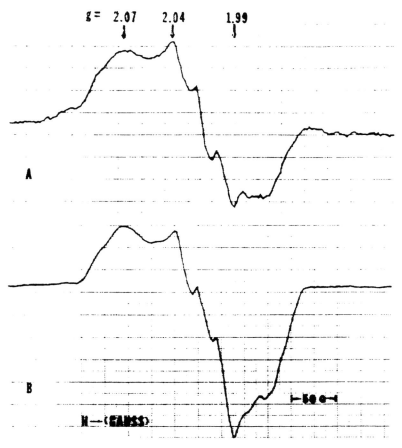

Fig. 7.6. EPR spectra at 26 K of nitrite reductase (cytochrome cd_1) from Thiobacillus denitrificans. A: Nitrite reductase + nitrite + ascorbate + phenazine methosulfate for 1 min; B: Nitrite reductase + nitrite + sulfide. (Reproduced with permission from: LeGall J, Payne WJ, Morgan TV et al. On the purification of nitrite reductase from Thiobacillus denitrificans and its reaction with nitrite under reducing conditions. Biochem Biophys Res Commun 1979; 87:355-362.) Copyright © Academic Press, Inc, Orlando, Florida, USA.

they demonstrated that the enzyme catalyzed also the reduction of NO into N_2O. Further evidence for an intermediate role of NO in the reduction of nitrite to nitrous oxide was provided in stoichiometry experiments in which the reducer was either the NADH + PMS system,[39] or cytochrome c_{551}, the natural electron donor to cytochrome cd_1.[40] In gas chromatography experiments both gases were detected as products (Fig. 7.11).

Finally in work on NiR from *Thiobacillus denitrificans* Liu et al, using both EPR and Mössbauer spectroscopy,[48] demonstrated that the binding of NO to heme *c* is very pH dependent and concentration-dependent, no heme *c* being NO-bound at pH 7.6 and half of them bound at pH 5.8.

Heme d_1 could be the substrate-product binding site while heme *c* could play a regulatory role through a concentration-dependent inhibition by NO binding.

COPPER-CONTAINING ENZYME

A copper-containing NiR (EC 1.7.99.3) was discovered as early as 1972 (see refs. 3-6 for reviews) in several denitrifiers: *Achromobacter cycloclastes*,[47,56-58] *Bacillus halodenitrificans*,[59] *Alcaligenes xylosoxidans* formerly called *Alcaligenes* sp. NCIB 11015 and *Pseudomonas denitrificans*,[60] *Alcaligenes faecalis*,[61] *Bacillus stearothermophilus*,[12] etc. It was recently discovered in a denitrifying fungus, *Fuzarium oxysporum*.[62] NiR from *Achromobacter cycloclastes* and *Bacillus halodenitrificans* contains

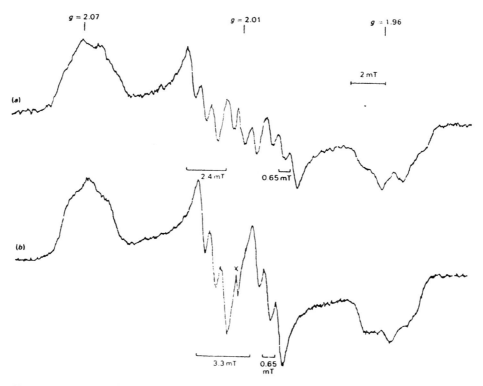

Fig. 7.7. EPR spectra (30 K) of ¹⁴NO and ¹⁵NO-bound Pseudomonas nitrite reductase at pH 5.8 (0.1 M Tris/HCl buffer). a: 23 µM protein + 1.8 mM nitrite + 20 mM ascorbate. b: 25 µM protein + 1.8 mM ¹⁵N nitrite + 20 mM ascorbate. (Reproduced with permission from: Johnson MK, Thomson AJ, Walsh TA et al. Electron paramagnetic studies on Pseudomonas nitrosyl nitrite reductase. Evidence for multiple species in the electron paramagnetic resonance spectra of nitrosyl haemoproteins. Biochem J 1980; 189:285-294.) Copyright © The Biochemical Society and Portland Press, London, UK.

two kinds of copper atoms, type *1* and type *2*, as defined by their EPR properties, in a variable ratio depending on the enzyme source and means of purification.[59,63,64] It has been sequenced[47] and analyzed by X-ray diffraction;[58] it is a trimer of identical subunits, each containing one "blue" copper atom of type *1* (His-95 and His-145, Cys-136 and Met-150 as ligands) and one type *2* copper site (His-100, His-135, His-306 and one solvent molecule as copper ligands). Cu-*1* and Cu-*2* are approximately 12.5 Å apart. Cu-*2* is bound not within a single subunit but by residues from two subunits of the trimer. The Cu-*2* site is at the bottom of a 12 Å deep solvent channel

and is the site to which nitrite binds.[58,60,64] Sequence and structural comparisons revealed domain similarities to cupredoxins such as plastocyanin and multi-copper oxidases, ascorbate oxidase, laccase and ceruloplasmin.[47,58]

The copper-containing NiR forms a nitrosyl bound intermediate like cytochrome *cd₁*.[57] The reactions of NO with the blue copper-containing NiR from *Achromobacter cycloclastes* and *Alcaligenes sp.* NCIB 11015 were studied by EPR spectroscopy.[65] The copper EPR signals of the oxidized NiR obtained at 77 K disappeared in the presence of NO, together with a decrease of the visible absorbance, and were

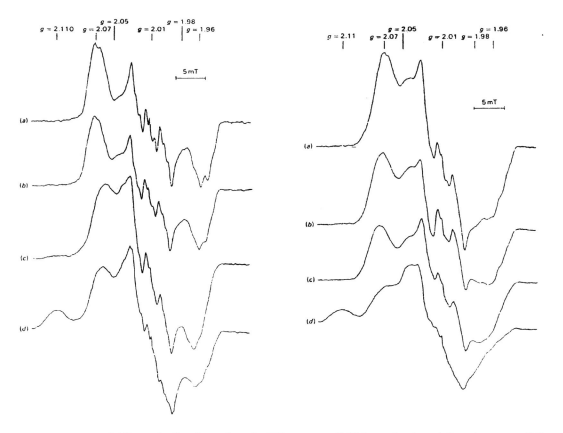

Fig. 7.8. Effect of different buffers (0.1 M) on the EPR spectra of NO-bound reduced Pseudomonas nitrite reductase at pH 5.8 (right panel) and pH 8.0 (left panel). a: Tris/HCl; b :MOPS; c: phosphate; d: HEPES. (Reproduced with permission from: Johnson MK, Thomson AJ, Walsh TA et al. Electron paramagnetic studies on Pseudomonas nitrosyl nitrite reductase. Evidence for multiple species in the electron paramagnetic resonance spectra of nitrosyl haemoproteins. Biochem J 1980; 189:285-294.) Copyright © The Biochemical Society and Portland Press, London, UK.

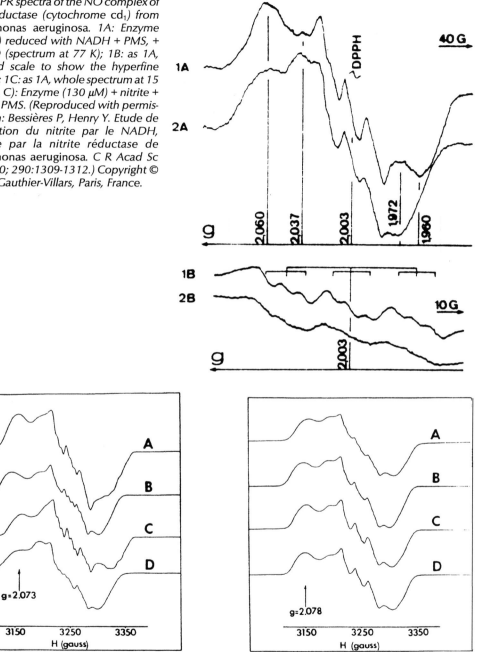

Fig. 7.9. EPR spectra of the NO complex of nitrite reductase (cytochrome cd_1) from Pseudomonas aeruginosa. 1A: Enzyme (110 µM) reduced with NADH + PMS, + pure NO (spectrum at 77 K); 1B: as 1A, expanded scale to show the hyperfine structure; 1C: as 1A, whole spectrum at 15 K. 2(A, B, C): Enzyme (130 µM) + nitrite + NADH + PMS. (Reproduced with permission from: Bessières P, Henry Y. Etude de la réduction du nitrite par le NADH, catalysée par la nitrite réductase de Pseudomonas aeruginosa. C R Acad Sc Paris 1980; 290:1309-1312.) Copyright © Editions Gauthier-Villars, Paris, France.

Fig. 7.10. Left panel: EPR spectra of the NO complex of Pseudomonas cytochrome oxidase. A: cytochrome oxidase (0.62 mM) + ascorbic acid + NO (pH 7.0); B: horse heart cytochrome c (6.3 mM) + NO; C: cytochrome oxidase (0.62 mM) + ascorbic acid + contaminating protein-bound nitrite; D :heme d_1-depleted cytochrome oxidase (0.43 mM) + NO (pH 9.3). Right panel: EPR spectra of NO complexes of heme d_1 in the presence of various nitrogenous bases. A: heme d_1 + imidazole + NO; B: heme d_1 + pyridine + NO; C: heme d_1 + propanol + NO; D :heme d_1 + butylamine + NO. (Reproduced with permission from: Muhoberac BB, Wharton DC. EPR study of heme.NO complexes of ascorbic acid-reduced Pseudomonas cytochrome oxidase and corresponding model complexes. J Biol Chem 1980; 255:8437-8442.) Copyright © The American Society for Biochemistry and Molecular Biology, Inc, Bethesda, MD, USA.

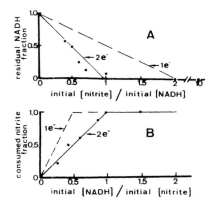

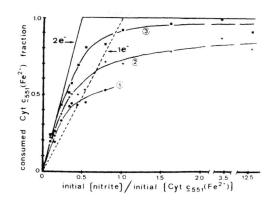

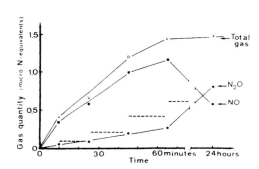

Fig. 7.11. *Production of NO and N₂O in the reduction of nitrite catalzed by* Pseudomonas aeruginosa *nitrite-reductase. Upper panel, left: Stoichiometry of the reduction of nitrite by NADH, in the presence of PMS, catalyzed by* Pseudomonas aeruginosa *nitrite-reductase. Upper panel, right: Stoichiometry of the reduction of nitrite by reduced cytochrome c₅₅₁ catalyzed by* Pseudomonas aeruginosa *nitrite-reductase. 1: 15 sec, 2: 10 min, 3: 2 hr reaction. Lower panel: Measurement by gas chromatography of the production of NO and N₂O in the reduction of nitrite by NADH, in the presence of PMS, catalyzed by* Pseudomonas aeruginosa *nitrite-reductase. (Reproduced with permission from: Bessières P, Henry Y. Etude de la réduction du nitrite par le NADH, catalysée par la nitrite réductase de* Pseudomonas aeruginosa. *C R Acad Sc Paris 1980; 290:1309-1312, Copyright © Editions Gauthier-Villars, Paris, France, and from: Bessières P, Henry Y. Denitrification and nitrite reduction:* Pseudomonas aeruginosa *nitrite reductase. Biochimie 1984; 66:313-318. Copyright © Editions Masson, Paris, France.)*

fully restored by its removal suggesting the formation of a diamagnetic cuprous nitrosyl $[Cu^+-NO^+]$ complex in equilibrium with a possible EPR silent (at least at 77 K) spin-coupled $[Cu^{2+}-NO^\bullet]$ complex. In that respect NiR, which has only type *1* (blue) and *2* (non-blue) copper sites, is largely different from copper-oxidases like laccase or ceruloplasmin which bind NO through their type *3* site, forming EPR-detectable complexes (See chapter 6).

The intermediate role of NO in the catalytic reduction of nitrite by the NADH + PMS system catalyzed by copper-containing NiR from *Achromobacter cycloclastes* was rather clear cut by gas chromatography experiments,[66] contrary to the case of the cytochrome *cd₁* + NoR system. After a rapid production of NO, its rate levels off while that of N_2O remains slow and constant. Addition of NO to nitrite increases the N_2O yield. Conversely trapping NO in the form of nitrosyl hemoglobin or removal of NO by sparging resulted in undetectable N_2O production. These results are quite compatible with the EPR results and consistent with the formation of a $[Cu^+-NO^+]$ complex which decomposes to NO. N_2O production requires the binding of a second nitrite anion or a second NO molecule or of another derived form, according to the following schemes (Fig. 7.12).[66]

Several lines of evidence showed that type *2* (non-blue) copper centers of NiR are the site of nitrite reduction.[64] By selective depletion and reconstitution of the type *2* Cu sites a series of enzyme preparations with varying type *1* to type *2* ratios and a constant content of type *1* (3.0 copper atoms per trimer) was prepared. The NiR activity was found to be proportional to the type *2* content.[64] An EPR and ENDOR study of NiR from *Alcaligenes xylosoxidans* indicates also that the type *2* site binds nitrite.[60] Finally another recent study by X-ray analysis and site-directed mutagenesis confirmed this role for the type *2* site, while the type *1* site was assigned the role of electron transfer site from pseudoazurin, a member of the cupredoxin

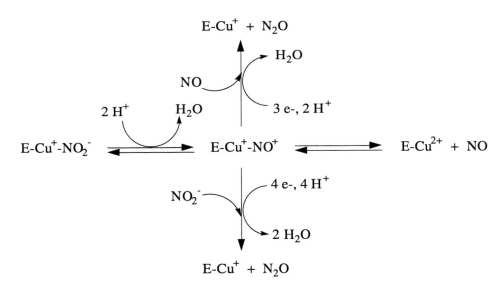

Fig. 7.12. Two alternative schemes of production of NO and N_2O by reduction of nitrite catalyzed by nitrite reductase from Achromobacter cycloclastes. (Reproduced with permission from: Jackson MA, Tiedje JM, Averill BA. Evidence for an NO-rebound mechanism for production of N_2O from nitrite by the copper-containing nitrite reductase from Achromobacter cycloclastes. FEBS Lett 1991; 291:41-44, Copyright © Elsevier Science, Sara Burgerhartstraat 25, 1055 KV Amsterdam, The Netherlands.)

family, acting as an electron donor.[61] An interesting model of the active site of copper NiR has been recently proposed.[67]

HEXAHEME CYTOCHROME-NITRITE REDUCTASE

Assimilatory and dissimilatory nitrite reductases catalyze the six-electron reduction of nitrite to ammonia and are often called assimilative or dissimilative ammonia-forming NiR. One type (EC 1.7.7.1) found in spinach and *E. coli* for instance, contains a single siroheme associated with one [4Fe-4S] center and one molecule of bound FAD. It is assimilative as nitrite is reduced to ammonia for nutritional purpose. The other type is a respiratory (dissimilative) ammonia-producing NiR functioning in anaerobic conditions. It contains several, usually six, c-type hemes.[3] This type of NiR was first purified from *Achromobacter fischeri*,[68] from the strictly anaerobic sulfate-reducing *Desulfovibrio desulfuricans*,[69-71] and subsequently from *E. coli* and *Wolinella succinogenes*.[72,73] The membrane-bound enzymes from *Desulfovibrio desulfuricans*, *E. coli* and *Wolinella succinogenes* contain six c-type heme groups per molecule with a minimal $M_r = 66$ kDa,[70] and have been thoroughly studied by Mössbauer and EPR spectroscopy.[69,72,73] In the ferric form, one of the hemes is a distorted rhombic high-spin (g-values at 9.7, 3.7), the five others being low-spin with different g-values at 2.9, 2.3 and 1.5 on the one hand, and g = 3.2 and 2.14 on the other. Distinct crystal field and hyperfine parameters of all six hemes could be estimated from the Mössbauer and EPR spectra.[71]

Addition of nitrite to the fully reduced enzyme reoxidized all five low-spin hemes, while the high-spin heme formed a complex with NO. This suggests that the high-spin heme could be the substrate binding site and that NO could be an intermediate present in an enzyme-bound form during the six-electron reduction of nitrite to ammonia (Fig. 7.13).[72,73]

The hexaheme NiR also forms a ferrous heme-NO complex when reacted with hydroxylamine.[72] It can also reduce NO to either ammonia or N_2O.[73] Here again nitric oxide plays a key regulatory role as a free or bound intermediate in nitrite reduction.

Other ammonia-forming NiRs containing two or four c-type hemes have been isolated from other species: *A. fisheri*,[68] *Sulfurospirillum deleyianum* and *Wolinella succinogenes*,[74] for instance.

ASSIMILATORY NITRITE REDUCTASE

Called ammonia:ferredoxin oxidoreductase (EC 1.7.7.1), it contains one siroheme (common to sulfite reductase) and one [4Fe-4S] cluster per 61 kDa and can use NADH, reduced viologen dyes or ferredoxin as electron donors. The siroheme is in the high-spin ferric state (g = 6.78, 5.17 and 1.98) in the enzyme as isolated and, when reduced, it interacts with NO as shown by EPR spectroscopy (Fig. 7.14).[75]

The heme-NO spectrum with a nearly axial symmetry (g = 2.066 and 2.010) with a distinct ^{14}N three-line hyperfine structure (A = 2.1 mT) at g = 2.01, is detected upon addition of nitrite with reductant (reduced methylviologen or dithionite) or hydroxylamine (with or without added reductant) or a reductant plus NO itself to the enzyme. The axial symmetry is unusual for a ferroheme-NO complex but can be explained by the unusual structure of siroheme itself.[76]

NITRIC OXIDE REDUCTASES

CYTOCHROME *BC* COMPLEX

The final step in solving the controversy about the obligate character of NO-independently of the nature of the nitrite reductase, cytochrome cd_1 or copper-containing enzyme, was the discovery and purification of a membrane-bound nitric oxide reductase (NoR) (EC 1.7.99.7). NoR has been characterized from *P. stutzeri*,[29,77,78] *Pa. denitrificans*[31,33,77] and *Achromobacter cycloclastes*.[79] NoR is a cytochrome *bc* complex with two subunits (17 kDa and 38 kDa) with high-spin ferric cytochrome *b* and low-spin ferric cytochrome *c* heme components

(Fig. 7.15). The mol ratio of cytochrome c/b varied from 1 to 2 depending on the bacterial species (*A. cycloclastes*, *Pa. denitrificans* or *P. stutzeri*). It is expressed under anaerobic conditions, and its mutational loss is a conditional lethal event.[32] NoR is a very efficient scavenger and prevents the build-up of toxic concentrations of NO. It is also reversibly inhibited by NO and has a very low K_m (< 1 μmol/l).

The EPR spectrum of the nitrosylated reduced enzyme (Fig. 7.15) is again typical of a pentacoordinated nitrosyl heme, but the absence of a structure except for the three-line hyperfine coupling of 1.7 mT does not allow one to infer the existence of the heme proximal ligand or whether NO is bound to heme *b* or heme *c*.

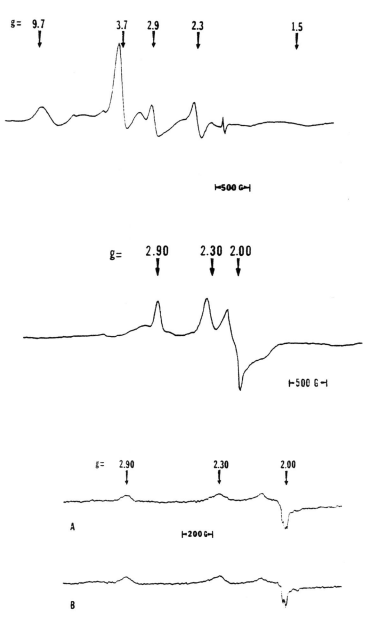

Fig. 7.13. EPR spectra (12 K) of nitrite reductase from Wolinella succinogenes and its NO complex. Upper panel: EPR spectrum of nitrite reductase (50 μM), pH 7.6, showing the high-spin and low-spin ferric hemes. Middle panel: EPR spectrum of nitrite reductase (50 μM), pH 7.6, reduced by with a slight excess of sodium dithionite and then reacted with sodium nitrite (10 mM). Lower panel: EPR spectrum of nitrite reductase (50 μM), pH 7.6, reduced by with a slight excess of sodium dithionite and then reacted with 10 mM $^{14}NO_2^-$ (trace A) or $^{15}NO_2^-$ (trace B). The field scale is expanded. (Reproduced with permission from: Liu M-C, Liu M-Y, Peck HD et al. EPR studies on the nitrite reductases from Escherichia coli and Wolinella succinogenes. FEBS Lett 1987; 218:227-230.) Copyright © Elsevier Science, Sara Burgerhartstraat 25, 1055 KV Amsterdam, The Netherlands.

CYTOCHROME P450NOR/NITRIC OXIDE REDUCTASE

A new feature in the denitrification process followed the discovery that the fungus *Fusarium oxysporum*, a eukaryotic cell, contained both a microsomal and a soluble cytochrome P-450—then a novelty for eukaryotes—and that the latter was induced by nitrate or nitrite.[80] Furthermore the fungus exibits a distinct denitrifying ability which results in an anaerobic production of nitrous oxide from nitrate or nitrite.[81] The NiR activity was strongly inhibited by CO, CN- and O_2, suggesting the involvment of the P-450 in the denitrifying reduction. In fact *Fusarium oxysporum* P-450 exhibits a specific and potent NoR activity to form N_2O with NADH but not NADPH as electron donor. The study of the reaction mechanism suggested that the

decomposition of the ferrous P-450-NO complex might be rate-limiting and demonstrated the first instance where the electron transport occurs directly from NADH without any intermediate,[82] according to the following alternate schemes (Fig.7.16) involving one or two electron transfer from NADH.[82,83]

To our knowledge no EPR spectrum of the P-450 (FeII)-NO has yet been reported. A preliminary X-ray structure has recently been published.[84] The same fungus contains a blue-colored type *1* and *2* copper-containing protein which exhibits a NiR activity to form mainly NO.[62] Thus NO seems to be an obligatory free intermediate in *Fusarium oxysporum*, as in most denitrifying bacteria.

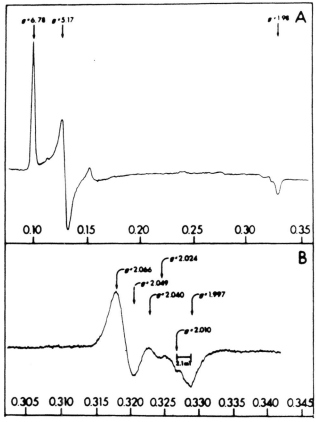

Fig. 7.14. EPR spectra (20 K, 9.19 GHz) of spinach nitrite reductase (75 mM, pH 7.7) and of its NO complex. A: oxidized enzyme; B: NO complex. (Reproduced with permission from: Lancaster JR, Vega JM, Kamin H et al. Identification of the iron-sulfur center of spinach ferredoxin-nitrite reductase as a tetranuclear center, and preliminary EPR studies of mechanism. J Biol Chem 1979; 254:1268-1272.) Copyright © The American Society for Biochemistry and Molecular Biology, Inc. Bethesda, MD, USA.

OTHER CYTOCHROMES INTERACTING WITH NITRIC OXIDE

Two other cytochromes, with no obvious cellular function, are able to interact with NO, cytochrome c' and an unusual green hemoprotein from *Bacillus halodenitrificans*.

CYTOCHROME c'

Cytochrome c' is a periplasmic class II cytochrome c, with low redox potentials, widely distributed amongst bacteria. Its prosthetic group has been very thoroughly studied by sophisticated spectroscopic methods because of its unique spin state in the ferric form, not being the usual high-spin (S = 5/2) or low-spin (S = 1/2), but a quantum-mechanically admixed S = 5/2,3/2 state. The ferrous form is also unusual for a cytochrome c, being high-spin pentacoordinated, with a proximal His residue as an iron-ligand in strong interaction (hydrogen bound) with the protein moiety.

Fig. 7.15. EPR spectra of nitric oxide reductase from Pseudomonas stutzeri, *a cytochrome bc complex. Upper panel: Oxidized complex, showing high-spin heme b and low-spin heme c (9 K); Lower panel: NO derivative of the ascorbate/ PMS reduced complex (T < 20 K). (Reproduced with permission from: Kastrau DHW, Heiss B, Kroneck PMH et al. Nitric oxide reductase from* Pseudomonas stutzeri, *a novel cytochrome bc complex. Phospholipid requirement, electron paramagnetic resonance and redox properties. Eur J Biochem 1994; 222:293-303.) Copyright © European Journal of Biochemistry, Zürich, Switzerland.*

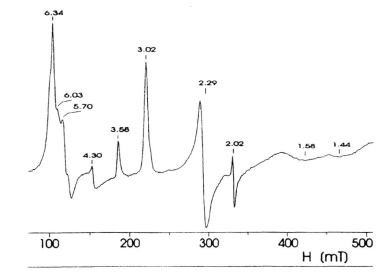

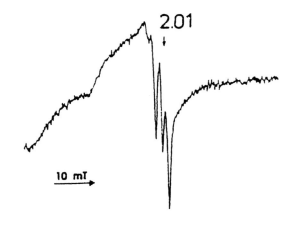

Cytochrome c' is isolated from *Alcaligenes sp.* NCIB 11015 now called *Achromobacter xylosoxidans*, bacteria utilizing a copper-containing NiR,[85-87] from phototrophic bacteria: *Rhodopseudomonas capsulata*,[88] *Rhodobacter capsulatus*,[88] etc. Reduced cytochrome c' forms complexes with NO, and NO is able to reduce by itself the ferric form. As shown by specific EPR spectra (Fig. 7.17) indicative of a mixture of pentacoordinated and hexacoordinated nitrosyl heme in varying proportion depending on the bacterial species, the heme iron to histidine (N_ϵ) bond is very weak and cleaved upon the coordination of NO.[85,86,88]

Similar EPR spectra assigned to NO-cytochrome c' complex, were detected in whole cells of bacteria cultured under denitrifying conditions in the presence of nitrate, but not under non-denitrifying conditions.[87,90] (Fig. 7.18).

This finding allows one to suggest a function for cytochrome c', that of a NO-scavenger in the cellular periplasmic space in order to keep free cellular NO concentration as low as possible between its source NiR and its metabolizing enzyme NoR.[87,90] This has been substantiated by the fact that cytochrome c' is found largely in the ferrous form in whole cells.[89]

BACILLUS HALODENITRIFICANS GREEN HEMOPROTEIN

Recently a novel and unusual green hemoprotein capable of reversible binding

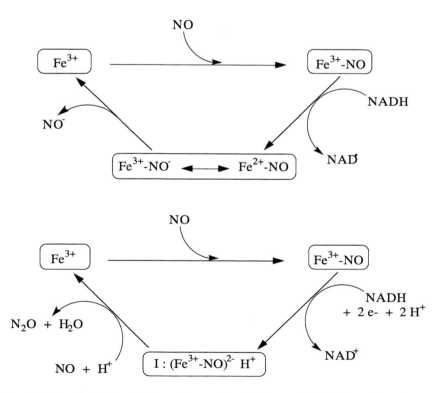

Fig. 7.16. Possible NO reduction schemes by cytochrome P450-nitric oxide reductase from Fusarium oxysporum. (Reproduced with permission from: Shiro Y, Fujii M, Iizuka T et al. Spectroscopic and kinetic studies on reaction of cytochrome P450nor with nitric oxide. Implication for this nitric oxide reduction mechanism. J Biol Chem 1995; 270:1617-1623.) Copyright © The American Society for Biochemistry and Molecular Biology, Inc. Bethesda, MD, USA.

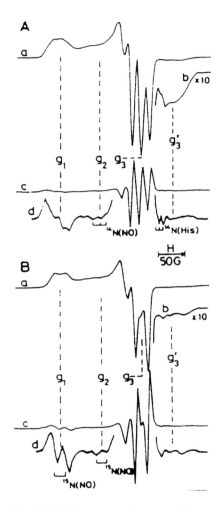

Fig. 7.17. EPR spectra of NO-cytochrome c' from Alcaligenes sp. NBIC 11015 at 77 K and at pH 7.2. A: [14]NO complex, B: [15]NO complex, c: second derivatives of EPR spectra. (Reproduced with permission from: Yoshimura T, Suzuki S, Nakahara A et al. Spectral properties of nitric oxide complexes of cytochrome c' from Alcaligenes sp. NCIB 11015. Biochemistry 1986; 25:2436-2442.) Copyright © The American Chemical Society, Columbus, OH, USA.

of NO was purified from the gram-positive *Bacillus halodenitrificans*.[91] Having a molecular mass of 64 kDa, it contains 6.2 mol *b*-type protoheme per mol protein. It presents many spectral characterizations similar to those of soluble guanylate cyclase and cytochrome c', including the reversible binding of NO. The reduced but not the oxidized form binds CN^-, CO and NO. NO failed to reduce the protein. The EPR spectra of the oxidized cytochrome are characteristic of high-spin heme, with g-values at 6.83, 5.14 and 1.98. The EPR spectra of the ferrous nitrosylated cytochrome are like those of cytochrome c' characteristic of a pentacoordinated nitrosyl heme, with g-values at 2.11 and 2.01, which appeared also in whole cells grown on nitrate or added with nitrite and dithionite.[91] (Fig. 7.19).

An essential function of detoxification of NO produced in denitrifiers could be assigned to these NO-binding cytochromes, cytochrome c',[90] or the green NO-binding hemoprotein of *Bacillus halodenitrificans*;[91] this hypothesis remains however to be substantiated.

NITROUS OXIDE REDUCTASES

Nitrous oxide reductases (N_2OR) are only mentioned here as part of the respiratory denitrification process, but they do not seem a priori to interact with NO.

N_2OR the periplasmic terminal oxidoreductase of N_2O respiration catalyzing its two electron reduction to N_2 has been purified from anaerobic denitrifiers *P. perfectomarinus*,[92] now designated *P. stutzeri*,[94,95] and *P. aeruginosa*,[96,97] and from aerobic denitrifier *Thiosphaera pantotropha*,[98] etc. N_2OR from *P. perfectomarinus* contains eight copper atoms per 120-140 kDa and is a dimer of identical subunits (4 Cu per subunit).[92,93] Several spectroscopically distinct 'pink', 'purple', 'blue', etc., forms were identified. EPR spectra suggested the presence of an unusual type 1 copper center with a specific seven-line hyperfine pattern, with no type 2 copper. A multifrequency EPR investigation of N_2OR from *P. stutzeri* performed at 2.4, 3.4, 9.3 and 35 GHz demonstrated the presence of a mixed-valence [Cu(II),Cu(I)] binuclear center, better described as a [Cu(1.5)-Cu(1.5)] S = 1/2 species, with the unpaired electron delocalized between two equivalent Cu nuclei.[94,99]

Several beautifully convergent spectroscopic investigations, utilizing EPR,[94,99,100] resonance Raman,[101] pulsed EPR,[102] EXAFS[97] and MCD,[95,100] indicated [Cu(II)-S_2(Cys)$_2$N(His)] sites in the resting (oxidized) enzyme, wholly similar to the so-called Cu$_A$-type site of mitochondrial cytochrome oxidase (COX) and compatible with primary sequence alignment. A further spectroscopic study of the oxidized state, a semi-reduced form and the fully-reduced state of the enzyme allowed one to propose a model comprising two distinct copper centers, a center A similar to the Cu$_A$ site of COX, assigned to an electron-transfer function, and a center Z, assigned to a catalytic substrate binding site.[100] Both centers are binuclear copper complexes. The binuclear character of these sites was clearly confirmed by a more detailed multifrequency EPR investigation of N$_2$OR and COX.[103]

The structure of N$_2$OR, the nature of its prosthetic groups, the strong analogy of the enzyme with mitochondrial cytochrome aa_3 which interacts with NO and produces N$_2$O,[104-106] and the finding of Frunzke and Zumft[107] that NO inhibits N$_2$O respiration in a denitrifying bacterium, would warrant careful functional and spectroscopic (mostly EPR and Mössbauer) studies of a possible interaction of NO with N$_2$OR.

A second type of N$_2$OR has been purified from *Wolinella succinogenes* which contains one low-spin heme c and four Cu atoms per subunit of Mr=88 kDa.[108]

A NOTE ON ENZYME EVOLUTION

The reader is asked to excuse the following "a parte", only explained by the naïve curiosity of the author concerning the comparison of the many enzymes rapidly described in this chapter.

The first point reflects on why each of the successive enzymatic activities: NiR,

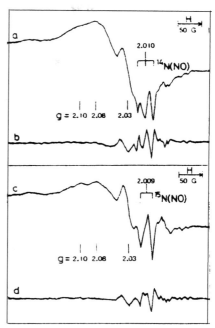

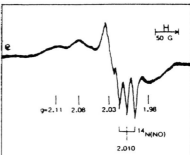

Fig. 7.18. EPR spectra at 77 K of acetone-dried cells of Alcaligenes sp. NBIC 11015 cultured with (a, b) K^{14}NO$_3$ and (c, d) K^{15}NO$_3$: (a, c) first derivative; (b, d) second derivative spectra. (e) EPR spectrum of extract obtained from acetone-dried cells of Alcaligenes sp. NBIC 11015 in phosphate buffer at pH 7.2. (Reproduced with permission from the authors: Yoshimura T, Iwasaki H, Shidara S et al. Nitric oxide complex of cytochrome c' in cells of denitrifying bacteria. J Biochem 1988; 103:1016-1019.) Copyright © the authors.

NoR and N_2OR of denitrification and other bacterial processes has received, like oxidases having O_2 as substrate, two or more structural "recipes" to functional or catalytic problems in terms of active sites or prosthetic groups, always based on the same "ingredients": hexacoordinated heme *c*, pentacoordinated heme *b* or *d* or siroheme, type *1* copper, type *2* copper, binuclear copper center, [Fe-S] clusters, etc. How is the observed substrate specificity achieved, a nearly absolute specificity, thus implying cross-inhibition?

The recent availability of an abundance of protein sequences has allowed many comparisons between heme and copper-containing terminal oxidases utilizing NO_2^-, NO, N_2O and O_2 as substrates, which have revealed fascinating analogies implying a pathway for enzyme evolution. It provides an explanation for the origin of aerobic respiration developing from the anaerobic, denitrifying respiratory system.[109,110]

Fig. 7.19. Effect of NO on the EPR spectrum (10 K) of the green hemoprotein from Bacillus halodenitrificans. A: protein as prepared; B: dithionite + nitrite added. B': enlargement of B in the g=2 region; C: the sample from B left overnight in the EPR tube. C': enlargement of C in the g=2 region; D: the sample from C gently shaken under air. (Reproduced with permission from: Denariaz G, Ketchum PA, Payne WJ et al. An unusual hemoprotein capable of reversible binding of nitric oxide from the gram-positive Bacillus halodenitrificans. Arch Microbiol 1994; 162:316-322.) Copyright © Springer Verlag, New York, NY, USA.

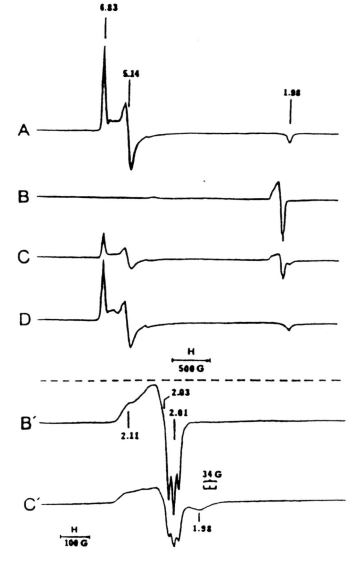

POSSIBLE ROLES OF NITRIC OXIDE IN NITRIFICATION AND NITROGEN-FIXATION

As already mentioned, nitrification by bacteria such as *Nitrosomonas europea* or *Thiosphaera pantotropha* and denitrification, regarded usually as a separate phenomenon carried out by different bacteria in segregated media, can in fact be combined.[17-19] While NO and N_2O are mostly produced by denitrifiers in anaerobic conditions, the same chemicals are produced by nitrifiers in aerobic soils. The exact mechanism of production of NO by the enzymatic sequence (ammonia-monooxygenase, hydroxylamine oxidase, multi-heme-*c* containing enzymes yielding nitrite as product) is however not known.

A similar "ambidextrous" character is observed in *Pseudomonas* species which can combine denitrification and N_2-fixation.[16] This conclusion has to be reconciled with the fact that in other species, nitrite and NO inhibit nitrogen fixation.[111] In fact, inhibition of nitrogenase, a two protein complex containing $[Fe_4S_4]$ clusters and Mo, either by nitrite or by NO, has been documented in many species: soybean bacteroid (*Rhizobium*),[15] *Clostridium pastorianum*,[112] *Rhodopseudomonas sphaeroides* f. sp. *denitrificans*,[113] *Azotobacter vinelandii*,[114] etc.

BACTERIAL NITRIC OXIDE SYNTHASE

Recently, strong evidence was presented for the first time that bacteria of the genus *Nocardia* contained a nitric oxide synthase similar to mammalian neutrophil NOS, with the same substrates and cofactors requirements. This interesting finding could be related to the fact that these species can survive for a long period in macrophages.[115]

DISCUSSION AND CONCLUSION

We have discussed at length in this chapter the obligate intermediate character of both NO and N_2O in the bacterial supported nitrogen cycle. EPR spectroscopy has brought factual evidence for this character, essential because of its absolute non-ambiguity. These findings have direct consequences and applications over two important aspects of global and human ecology, among others already mentioned in chapter 1 and in the introductory part of the present chapter, namely the use of microorganisms and immobilized enzymes for depollution and the role of bacterial catalyzed nitrosation in cancerogenesis.

DEGRADATION OF NITRATE, NITRITE AND NITROGEN OXIDES BY MICROORGANISMS

The use of microorganisms in their capacity to convert pollutant such as nitrate, nitrite or nitrogen oxides (NO_x) receives more and more attention, of which we shall cite only two instances. The first is the use of enzymes, nitrate, nitrite and N_2O reductases immobilized together with electron-carrying dyes in a polymer matrix.[116] The second is the use of denitrifying bacteria for the removal of nitrogen oxides from combustion gases.[117] The main problem is to avoid leaks in the process, for instance to avoid forming the pollutant N_2O, while getting rid of nitrite, a pollutant of a different niche. In fact organisms able to combine nitrification and denitrification processes are, as we have seen above, of great interest for waste water treatment. On the one hand they are able to remove nitrate and NO_x pollution, on the other hand they can deal with nitrous oxide, another serious pollutant, implicated in most current environmental problems: acid rain, greenhouse effect, ozone depletion, etc.[19]

ENZYME CATALYZING NITROSAMINE FORMATION

A detailed understanding of the denitrification process, particularly in microaerophilic conditions is of great importance in understanding why and how microorganisms such as *P. aeruginosa*, *Neisseria mucosae*, *E. coli*, etc., catalyze nitrosamine formation leading to potential carcinogenesis in humans at specific infected sites such as the bladder and stomach.[118-120]

ADDITIONAL RECENT BIBLIOGRAPHY

A few interesting articles have appeared since this chapter was completed. A cytochrome cd_1-nitrite reductase has been purified and characterized from the marine denitrifier *Pseudomonas nautica*.[121] The structure determination of cytochrome cd_1-nitrite reductase by Fülöp et al[50] has been completed.[122] The crystal structure analysis provides a structural basis for the dual function of the enzyme, the reduction of oxygen to water and the synthesis of NO.[122]

Cytochrome cd_1-nitrite reductase from *P. aeruginosa* has been demonstrated to be one enzyme responsible for the catalysis of nitrosation of secondary amines by bacteria, leading to the synthesis of carcinogenic *N*-nitroso compounds.[123] This reaction may play an important role in the etiology of human cancer.[123]

The P-450nor from the fungus *Fusarium oxysporum* has been further characterized, in particular its NO complex.[124]

REFERENCES

1. Payne WJ. Pasteur, Gayon, Dupetit et le cycle de Reiset. Annales de l'Institut Pasteur/Actualités. 1990; 1:31-44.
2. Payne WJ. Denitrification, Wiley-Interscience Publication, New York, 1981.
3. Brittain T, Blackmore R, Greenwood C et al. Bacterial nitrite-reducing enzymes. Eur J Biochem 1992; 209:793-802.
4. Zumft WG. The biological role of nitric oxide in bacteria. Arch Microbiol 1993; 160:253-264.
5. Ye RW, Averill BA, Tiedje JM. Denitrification: production and consumption of nitric oxide. Appl Environm Microbiol 1994; 60:1053-1058.
6. Ferguson SJ. Denitrification: a question of the control and organisation of electron and ion transport. Trends in Biol Sci 1987; 12:354-357.
7. Arai H, Igarashi Y, Kodama T. Nitrite activates the transcription of the *Pseudomonas aeruginosa* nitrite reductase and cytochrome *c-551* operon under anaerobic conditions. FEBS Lett 1991; 288:227-228.
8. Brock TD. Biology of microorganisms.

Prentice-Hall, Englewood Cliffs, NJ, USA, 1979.
9. Vincent WF, Downes MT, Vincent CL. Nitrous oxide cycling in Lake Vade, Antarctica. Nature 1981; 292:618-620.
10. Zumft WG, Cardenas J. The inorganic biochemistry of nitrogen bioenergetic processes. Naturwissenschaften 1979; 66:81-88.
11. Henry Y, Bessières P. Denitrification and nitrite reduction: *Pseudomonas aeruginosa* nitrite reductase. Biochimie 1984; 66:259-289.
12. Ho TP, Jones AM, Hollocher TC. Denitrification enzymes of *Bacillus stearothermophilus*. FEMS Microbiol Lett 1993; 114:135-138.
13. Quispel A. The biology of nitrogen fixation. North Holland Pub, Amsterdam, 1974.
14. Burns RC, Hardy RWF. Nitrogen Fixation in Bacteria and Higher Plants, Molecular Biology, Biochemistry and Biophysics, Vol. 21, Springer-Verlag, Berlin, 1975.
15. Trinchant JC, Rigaud J. Nitrite inhibition of nitrogenase from soybean bacteroids. Arch Microbiol 1980; 124:49-54.
16. Chan Y-K, Barraquio WL, Knowles R. N_2-fixing pseudomonads and related soil bacteria. FEMS Microbiol Rev 1994; 13:95-118.
17. Anderson IC, Poth M, Homstead J et al. A comparison of NO and N_2O production by the autotrophic nitrifier *Nitrosomonas europaea* and the heterotrphic nitrifier *Alcalignes faecalis*. Appl Environm Microbiol 1993; 59:3525-3533.
18. Wehrfritz J-M, Reilly A, Spiro S et al. Purification od hydroxylamine oxidase from *Thiosphaera pantotropa*. Identification of electron acceptors that couple heterotrophic nitrification to aerobic denitrification. FEBS Lett 1993; 335:246-250.
19. Kuenen JG, Robertson LA. Combined nitrification-denitrification processes. FEMS Microbiol Rev 1994;15:109-117.
20. Luckey TD. Introduction to intestinal microecology. Am J Clin Nutr 1972; 25:1292-1294.
21. Hong R, Trudell JR, O'Neil JR et al. Metabolism of nitrous oxide by human and rat intestinal contents. Anesthesiol 1993; 52:16-19.
22. Payne WJ. Reduction of nitrogenous oxides by microorganisms. Bacteriol Rev 1973; 37:409-452.

23. Rowe JJ, Sherr BF, Payne WJ et al. A unique nitric oxide-binding complex formed by denitrifying *Pseudomonas aeruginosa*. Biochem Biophys Res Commun 1977; 77:253-258.

24. St John RT, Hollocher TC. Nitrogen 15 tracer studies on the pathway of denitrifcation in *Pseudomonas aeruginosa*. J Biol Chem 1977; 252:212-218.

25. Garber EAE, Hollocher TC. ^{15}N tracer studies on the role of NO in denitrification. J Biol Chem 1981; 256:5459-5465.

26. Garber EAE, Hollocher TC. $^{15}N,^{18}O$ tracer studies on the activation of nitrite by denitrifying bacteria. J Biol Chem 1982; 257:8091-8097.

27. Firestone MK Firestone RB, Tiedje JM. Nitric oxide as an intermediate in denitrification: evidence from nitrogen-13 isotope exchange. Biochem Biophys Res Commun 1979; 91:10-16.

28. Averill BA, Tiedje JM. The chemical mechanism of microbial denitrification. FEBS Lett 1982; 138:8-12.

29. Heiss B, Frunzke K, Zumft WG. Formation of the N-N bond from nitric oxide by a membrane-bound cytochrome *bc* complex of nitrate-respiring (denitrifying) *Pseudomonas stutzeri*. J Bacteriol 1989; 171:3288-3297.

30. Carr GJ, Page MD, Ferguson SJ. The energy-conserving nitric-oxide-reductase system in *Paracoccus denitrificans*. Distinction from the nitrite reductase that catalyses synthesis of nitric oxide and evidence from trapping experiments for nitric oxide as a free intermediate during denitrification. Eur J Biochem 1989; 179:683-692.

31. Carr GJ, Ferguson SJ. The nitric-oxide reductase of *Paracoccus denitrificans*. Biochem J 1990; 269:423-429.

32. Braun C, Zumft WG. Marker exchange of the structural genes for nitric oxide reductase blocks the denitrifcation pathway of *Pseudomonas stutzeri* at nitric oxide. J Biol Chem 1991; 266:22785-22788.

33. Dermastia M, Turk T, Hollocher TC. Nitric oxide reductase. Purification from *Paracoccus denitrificans* with use of a single column and some characteristics. J Biol Chem 1991; 266:10899-10905.

34. Goretski J, Hollocher TC. Trapping of nitric oxide produced during denitrification by extracellular hemoglobin. J Biol Chem 1988; 263:2316-2323.

35. Zafiriou OC, Hanley QS, Snyder G. Nitric oxide and nitrous oxide production and cycling during dissimilatory nitrite reduction by *Pseudomonas perfectomarina*. J Biol Chem 1989; 264:5694-5699.

36. Goretski J, Hollocher TC. The kinetic and isotopic competence of nitric oxide as an intermediate in denitrification. J Biol Chem 1990; 265:889-895.

37. Kalkowski I, Conrad R. Metabolism of nitric oxide in denitrifying *Pseudomonas aeruginosa* and nitrate-respiring *Bacillus cereus*. FEMS Microbiol Lett 1991; 82:107-112.

38. Wharton DC, Weintraub ST. Identification of nitric oxide and nitrous oxide as products of nitrite reductase by *Pseudomonas* cytochrome oxidase (nitrite reductase). Biochem Biophys Res Commun 1980; 97:236-242.

39. Bessières P, Henry Y. Etude de la réduction du nitrite par le NADH, catalysée par la nitrite réductase de *Pseudomonas aeruginosa*. C R Acad Sc Paris 1980; 290:1309-1312.

40. Bessières P, Henry Y. Stoichiometry of nitrite reduction catalyzed by *Pseudomonas aeruginosa* nitrite-reductase. Biochimie 1984; 66:313-318.

41. Shearer G, Kohl DH. Nitrogen isotopic fractionation and ^{18}O exchange in relation to the mechanism of denitrification of nitrite by *Pseudomonas stutzeri*. J Biol Chem 1988; 263:13231-13245.

42. Weeg-Aerssens E, Tiedje JM, Averill BA. Evidence from isotope labeling studies for a sequential mechanism for dissimilatory nitrite reduction. J Am Chem Soc 1988; 110:6851-6856.

43. Goretski J, Hollocher TC. Catalysis of nitrosyl transfer by denitrifying bacteria is facilitated by nitric oxide. Biochem Biophys Res Commun 1991; 175:901-905.

44. Ye RW, Toro-Suarez I, Tiedje JM et al. $H_2^{18}O$ isotope exchange studies on the mechanism of reduction of nitric oxide and nitrite to nitrous oxide by denitrifying bacteria. Evidence for an electrophilic nitrosyl during reduction of nitric oxide. J Biol Chem 1991; 266:12848-12851.

45. DerVartanian DV, LeGall J. Electron paramagnetic studies on the reaction of exogenous ligands with cytochrome c₃ from *Desulfovibrio vulgaris*. Biochim Biophys Acta 1971; 243:53-65.

46. Cox CD, Payne WJ, DerVartanian DV. Electron paramagnetic resonance studies on the nature of hemoproteins in nitrite and nitric oxide reduction. Biochim Biophys Acta 1971; 253:290-294.

47. Fenderson FF, Kumar S, Adman ET et al. Amino acid sequence of nitrite reductase: a copper protein from *Achromobacter cycloclastes*. Biochemistry 1991; 30:7180-7185.

48. Liu M-C, Huynh B-H, Payne WJ et al. Optical, EPR and Mössbauer spectroscopic studies on the NO derivatives of cytochrome cd₁ from *Thiobacillus denitrificans*. Eur J Biochem 1987; 169:253-258.

49. Sann R, Kostka S, Friedrich B. A cytochrome cd₁-type nitrite reductase mediates the first step of denitrification in *Alcaligenes eutrophus*. Arch Microbiol 1994; 161:453-459.

50. Fülöp V, Moir JWB, Ferguson SJ et al. Crystallization and preliminary crystallographic study of cytochrome cd₁ nitrite reductase from *Thiosphaera pantotropa*. J Mol Biol 1993; 232:1211-1212.

51. Moir JWB, Baratta D, Richardson DJ et al. The purification of a cd₁-type nitrite reductase from, and the absence of a copper-type nitrite reductase from, the aerobic denitrifier *Thiosphaera pantotropha*; the role of pseudoazurin as an electron donor. Eur J Biochem 1993; 212: 377-385.

52. Silvestrini MC, Falcinelli S, Ciabatti I et al. *Pseudomonas aeruginosa* nitrite reductase (or cytochrome oxidase): an overview. Biochimie 1994; 76:641-654.

53. LeGall J, Payne WJ, Morgan TV et al. On the purification of nitrite reductase from *Thiobacillus denitrificans* and its reaction with nitrite under reducing conditions. Biochem Biophys Res Commun 1979; 87:355-362.

54. Johnson MK, Thomson AJ, Walsh TA et al. Electron paramagnetic studies on *Pseudomonas* nitrosyl nitrite reductase. Evidence for multiple species in the electron paramagnetic resonance spectra of nitrosyl haemoproteins. Biochem J 1980; 189:285-294.

55. Muhoberac BB, Wharton DC. EPR study of heme-NO complexes of ascorbic acid-reduced *Pseudomonas* cytochrome oxidase and corresponding model complexes. J Biol Chem 1980; 255:8437-8442.

56. Iwasaki H, Matsubara T. A nitrite reductase from *Achromobacter cycloclastes*. J Biochem 1972; 71:645-652.

57. Hulse CL, Averill BA, Tiedje JM. Evidence for a copper-nitrosyl intermediate in denitrification by the copper-containing nitrite reductase of *Achromobacter cycloclastes*. J Am Chem Soc 1989; 111:2322-2323.

58. Godden JW, Turley S, Teller DC et al. The 2,3 Angstrom X-ray structure of nitrite reductase from *Achromobacter cycloclastes*. Science 1991; 253:438-442.

59. Denariaz G, Payne WJ, LeGall J. The denitrifying nitrite reductase of *Bacillus halodenitrificans*. Biochim Biophys Acta 1991; 1056:225-232.

60. Howes BD, Abraham ZHL, Lowe DJ et al. EPR and electron nuclear double resonance (ENDOR) studies show nitrite binding to the type 2 copper centers of the dissimilatory nitrite reductase of *Alcaligenes xylosoxidans* (*NCIMB 11015*). Biochemistry 1994; 33:3171-3177.

61. Kukimoto M, Nishiyama M, Murphy MEP et al. X-ray structure and site-directed mutagenesis of a nitrite reductase from *Alcaligenes faecalis* S-6: roles of two copper atoms in nitrite reduction. Biochemistry 1994; 33:5246-5252.

62. Kobayashi M, Shoun H. The copper-containing dissimilatory nitrite reductase involved in the denitrifying system of the fungus *Fusarium oxysporum*. J Biol Chem 1995; 270:4146-4151.

63. Iwasaki H, Noji S, Shidara S. *Achromobacter cycloclastes* nitrite reductase. The function of copper, amino acid composition and ESR spectra. J Biochem 1975; 78:355-361.

64. Libby E, Averill BA. Evidence that the type 2 copper centers are the site of nitrite reduction by *Achromobacter cycloclastes* nitrite reductase. Biochem Biophys Res Commun 1992; 187:1529-1535.

65. Suzuki S, Yoshimura T, Kohzuma T et al. Spectroscopic evidence for a copper-nitrosyl intermediate in nitrite reduction

by blue copper-containing nitrite reductase. Biochem Biophys Res Commun 1989; 164:1366-1372.

66. Jackson MA, Tiedje JM, Averill BA. Evidence for an NO-rebound mechanism for production of N_2O from nitrite by the copper-containing nitrite reductase from *Achromobacter cycloclastes*. FEBS Lett 1991; 291:41-44.

67. Halfen JA, Tolman WB. Synthetic model of the substrate adduct to the reduced active site of copper nitrite reductase. J Am Chem Soc 1994; 116:5475-5476.

68. Prakash OM, Sadana JC. Purification, characterization and properties of nitrite reductase of *Achromobacter fischeri*. Arch Biochem Biophys 1972; 148:614-632.

69. Liu MC, DerVartanian DV, Peck HD. On the nature of the oxidation-reduction properties of nitrite reductase from *Desulfovibrio desulfuricans*. Biochem Biophys Res Commun 1980; 96:278-285.

70. Liu M-C, Peck HD. The isolation of a hexaheme cytochrome from *Desulfovibrio desulfuricans* and its identification as a new type of nitrite reductase. J Biol Chem 1981; 256:13159-13164.

71. Costa C, Moura JJG, Moura I et al. Hexaheme nitrite reductase from *Desulfovibrio desulfuricans*. Mössbauer and EPR characterization of the heme groups. J Biol Chem 1990; 265:14382-14387.

72. Liu M-C, Liu M-Y, Peck HD et al. Comparative EPR studies on the nitrite reductases from *Escherichia coli* and *Wolinella succinogenes*. FEBS Lett 1987; 218:227-230.

73. Costa C, Macedo A, Moura I et al. Regulation of the hexaheme nitrite/nitric oxide reductase of *Desulfovibrio desulfuricans*, *Wolinella succinogenes* and *Escherichia coli*. A mass spectrometric study. FEBS Lett 1990; 276:67-70.

74. Schumacher W, Hole U, Kroneck PMH. Ammonia-forming cytochrome *c* nitrite reductase from *Sulfurospirillum deleyianum* is a tetraheme protein: new aspects of the molecular composition and spectroscopic properties. Biochem Biophys Res Commun 1994; 205:911-916.

75. Lancaster JR, Vega JM, Kamin H et al. Identification of the iron-sulfur center of

spinach ferredoxin-nitrite reductase as a tetranuclear center, and preliminary EPR studies of mechanism. J Biol Chem 1979; 254:1268-1272.

76. Murphy MJ, Siegel LM, Tove SR et al. Siroheme: a new prosthetic group participating in six-electron reduction reactions catalyzed by both sulfite and nitrite reductases. Proc Natl Acad Sci USA 1974; 71:612-616.

77. Kastrau DHW, Heiss B, Kroneck PMH et al. Nitric oxide reductase from *Pseudomonas stutzeri*, a novel cytochrome *bc* complex. Phospholipid requirement, electron paramagnetic resonance and redox properties. Eur J Biochem 1994; 222:293-303.

78. Zumft WG, Braun C, Cuypers H. Nitric oxide reductase from *Pseudomonas stutzeri*. Primary structure and gene organisation of a novel bacterial cytochrome *bc* complex. Eur J Biochem 1994; 219:481-490.

79. Jones AM, Hollocher TC. Nitric oxide reductase of *Achromobacter cycloclastes*. Biochim Biophys Acta 1993; 1144:359-366.

80. Shoun H, Suyama W, Yasui T. Soluble, nitrate/nitrite-inducible cytochrome P-450 of the fungus, *Fusarium oxysporum*. FEBS Lett 1989; 244:11-14.

81. Shoun H, Tanimoto T. Denitrification by the fungus *Fusarium oxysporum* and involvement of cytochrome P-450 in the respiratory nitrite reduction. J Biol Chem 1991; 266:11078-11082.

82. Nakahara K, Tanimoto T, Hatano K et al. Cytochrome P-450 55A1(P-450dNIR) acts as nitric oxide reductase employing NADH as the direct electron donor. J Biol Chem 1993; 268:8350-8355.

83. Shiro Y, Fujii M, Iizuka T et al. Spectroscopic and kinetic studies on reaction of cytochrome P450nor with nitric oxide. Implication for tis nitric oxide reduction mechanism. J Biol Chem 1995; 270:1617-1623.

84. Nakahara K, Shoun H, Adachi S et al. Crystallization and preliminary X-ray diffraction studies of nitric oxide reductase cytochrome P450nor from *Fusarium oxysporum*. J Mol Biol 1994; 239:158-159.

85. Yoshimura T, Suzuki S, Nakahara A et al. Spectral properties of nitric oxide complexes

of cytochrome *c'* from *Alcaligenes* sp. NCIB 11015. Biochemistry 1986; 25:2436-2442.

86. Iwasaki H, Yoshimura T, Suzuki S et al. Spectral properties of *Achromobacter xylosoxidans* cytochrome *c'* and their NO complexes. Biochim Biophys Acta 1991; 1058:79-82.

87. Yoshimura T, Shidara S, Ozaki T et al. Five coordinated nitrosylhemoprotein in whole cells of denitrifying bacterium, *Achromobacter xylosoxidans* NCIB 11015. Arch Microbiol 1993; 160:498-500.

88. Yoshimura T, Suzuki S, Iwasaki H et al. Spectral properties of nitric oxide complex of cytochrome *c'* from *Rhodopseudomonas capsulata* B100. Biochem Biophys Res Commun 1987; 145:868-875.

89. Monkara F, Bingham SJ, Kadir FHA et al. Spectroscopic studies of *Rhodobacter capsulatus* cytochrome *c'* in the isolated state and intact cells. Biochim Biophys Acta 1992; 1100:184-188.

90. Yoshimura T, Iwasaki H, Shidara S et al. Nitric oxide complex of cytochrome *c'* in cells of denitrifying bacteria. J Biochem 1988; 103:1016-1019.

91. Denariaz G, Ketchum PA, Payne WJ et al. An unusual hemoprotein capable of reversible binding of nitric oxide from the gram-positive *Bacillus halodenitrificans*. Arch Microbiol 1994; 162:316-322.

92. Zumft WG, Matsubara T. A novel kind of multi-copper protein as terminal oxidoreductase of nitrous oxide respiration in *Pseudomonas perfectomarinus*. FEBS Lett 1982; 148:107-112.

93. Coyle CL, Zumft WG, Kroneck PMH et al. Nitrous oxide reductase from denitrifying *Pseudomonas perfectomarina*. Purification and properties of a novel multicopper enzyme. Eur J Biochem 1985; 153:459-467.

94. Riester J, Zumft WG, Kroneck PMH. Nitrous oxide reductase from *Pseudomonas stutzeri*. Redox properties and spectroscopic characterization of different forms of the multicopper enzyme. Eur J Biochem 1989; 178:751-762.

95. Scott RA, Zumft WG, Coyle CL et al. *Pseudomonas stutzeri* N₂O reductase contains CuA-type sites. Proc Natl Acad Sci USA 1989; 86:4082-4086.

96. SooHoo CK, Hollocher TC. Purification and characterization of nitrous oxide reductase from *Pseudomonas aeruginosa* strain P2. J Biol Chem 1991; 266:2203-2209.

97. SooHoo CK, Hollocher TC, Kolodziej AF et al. Extended X-ray absorption fine structure and electron paramagnetic resonance of nitrous oxide reductase from *Pseudomonas aeruginosa* strain P2. J Biol Chem 1991; 266:2210-2218.

98. Berks BC, Baratta D, Richardson DJ et al. Purification and characterization of a nitrous oxide reductase from *Thiosphaera pantotropha*. Implications for the mechanism of aerobic nitrous oxide reduction. Eur J Biochem 1993; 212:467-476.

99. Kroneck PMH, Antholine WA, Riester J et al. The cupric site in nitrous oxide contains a mixed-valence [Cu(II),Cu(I)] binuclear center: a multifrequency electron paramagnetic resonance investigation. FEBS Lett 1988; 242:70-74.

100. Farrar JA, Thomson AJ, Cheesman MR et al. A model of the copper centres of nitrous oxide reductase (*Pseudomonas stutzeri*). Evidence from optical, EPR and MCD spectroscopy. FEBS Lett 1991; 294:11-15.

101. Dooley DM, Moog RS, Zumft WG. Characterization of the copper sites in *Pseudomonas perfectomarina* nitrous oxide reductase by resonance Raman spectroscopy. J Am Chem Soc 1987; 109:6730-6735.

102. Jin H, Thomann H, Coyle CL et al. Copper coordination in nitrous oxide reductase from *Pseudomonas stutzeri*. J Am Chem Soc 1989; 111:4262-4269.

103. Antholine WE, Kastrau DHW, Steffens GCM et al. A comparative EPR investigation of the multicopper nitrous-oxide reductase and cytochrome *c* oxidase. Eur J Biochem 1992; 209:875-881.

104. Brudvig GW, Stevens TH, Chan SI. Reactions of nitric oxide with cytochrome *c* oxidase. Biochemistry 1980; 19:5275-5285.

105. LoBrutto R, Wei YH, Mascarenhas R et al. Electron nuclear double resonance and elctron paramagnetic resonance study on the structure of the NO-ligated heme *a₃* in cytochrome *c* oxidase. J Biol Chem 1983; 258:7437-7448.

106. Rousseau DL, Singh S, Ching Y et al.

Nitrosyl cytochrome *c* oxidase. Formation and properties of mixed valence enzyme. J Biol Chem 1988; 263:5681-5685.

107. Frunzke K, Zumft WG. Inhibition of nitrous-oxide respiration by nitric oxide in the denitrifying bacterium *Pseudomonas perfectomarina*. Biochim Biophys Acta 1986; 852:119-125.

108. Zhang C, Hollocher TC, Kolodziej AF et al. Electron paramagnetic observations on the cytochrome *c*-containing nitrous oxide reductase from *Wolinella succinogenes*. J Biol Chem 1991; 266:2199-2202.

109. Saraste M, Castresana J. Cytochrome oxidase evolved by tinkering with denitrification enzymes. FEBS Lett 1994; 341:1-4.

110. van der Oost J, de Boer APN, de Gier JW et al. The heme-copper oxidase family consists of three distinct types of terminal oxidases and is related to nitric oxide reductase. FEMS Microbiol Lett 1994; 121:1-10.

111. Lockshin A, Burris RH. Inhibitors of nitrogen fixation in extracts from *Clostridium pasteurianum*. Biochim Biophys Acta 1965; 111:1-10.

112. Meyer J. Comparison of carbon monoxide, nitric oxide, and nitrite as inhibitors of the nitrogenase from *Clostridium pasteurianum*. Arch Biochem Biophys 1981; 210:246-256.

113. Michalski WP, Nicholas DJD. Inhibition of nitrogenase by nitrite and nitric oxide in *Rhodopseudomonas sphaeroides* f. *sp. denitrificans*. Arch Microbiol 1987; 147:304-308.

114. Hyman MR, Seefeldt LC, Morgan TV et al. Kinetic and spectroscopic analysis of the inactivating effects of nitric oxide on the individual components of *Azotobacter vinelandii* nitrogenase. Biochemistry 1992; 31:2947-2955.

115. Chen Y, Rosazza JPN. A bacterial nitric oxide synthase from a *Nocardia* species. Biochem Biophys Res Commun 1994; 203:1251-1258.

116. Mellor RB, Ronnenberg J, Campbell WH et al. Reduction of nitrate and nitrite in water by immobilized enzymes. Nature 1992; 355:717-719.

117. Apel WA, Turick CE. The use of denitrifying bacteria for the removal of nitrogen oxides from combustion gases. Fuel 1993; 72:1715-1718.

118. Bartsch H, Ohshima H, Pignatelli B. Inhibitors of endogenous nitrosation mechanisms and implications in human cancer prevention. Mutation Res 1988; 202:307-324.

119. Calmels S, Dalla Venezia N, Bartsch H. Isolation of an enzyme catalysing nitrosamine formation in *Pseudomonas aeruginosa* and *Neisseria mucosae*. Biochem Biophys Res Commun 1990; 171:655-660.

120. Dalla Venezia N, Calmels S, Bartsch H. Production of polyclonal and monoclonal antibodies for specific detection of nitrosation-proficient denitrifying bacteria in biological fluids. Biochem Biophys Res Commun 1991; 176:262-268.

121. Besson S, Carneiro C, Moura JJG et al. A cytochrome cd_1-type nitrite reductase isolated from the marine denitrifier *Pseudomonas nautica* 617: purification and characterization. Anaerobe 1995; 1:219-226.

122. Fülöp V, Moir JWB, Ferguson SJ et al. The anatomy of a bifunctional enzyme: structural basis for reduction of oxygen to water and synthesis of nitric oxide by cytochrome cd_1. Cell 1995; 81:369-377.

123. Calmels S, Ohshima H, Henry Y et al. Characterization of bacterial cytochrome cd_1-nitrite reductase as one enzyme responsible for catalysis of nitrosation of secondary amines. Carcinogenesis 1996; 17:533-536.

124. Shiro Y, Fujii M, Isogai S et al. Iron-ligand structure and iron redox property of nitric oxide reductase cytochrome P450nor from *Fusarium oxysporum*: relevance to its NO reduction activity. Biochemistry 1995; 34:9052-9058.

NITRIC OXIDE BIOSYNTHESIS IN MAMMALS

Sandrine Vadon-Le Goff and Jean-Pierre Tenu

The aim of this chapter is not to give a detailed view of everything that is known about NO biosynthesis, but to give the reader some clues to understand the next chapters (for more detailed information see the recent reviews of Kerwin et al,[1] Knowles and Moncada,[2] Marletta,[3] Stuehr and Griffith,[4] and Nathan.[5]

Since the early eighties, evidence has accumulated demonstrating that NO is produced by mammalian cells,[6,7] and plays a key role in blood pressure regulation, neurotransmission and cellular defense mechanisms. After the first work of Furchgott,[8] showing that a substance, called endothelium-derived relaxing factor (EDRF), was released by stimulated endothelial cells and was responsible for the relaxation of smooth muscles, several previously separate biological fields of research (toxicology, cardiovascular pharmacology, immunology and neurobiology) converged in 1987-1988 to the same point: nitric oxide. NO was identified as the EDRF, regulating smooth muscle relaxation, platelet aggregation and blood pressure.[9-11] NO is liberated by macrophages in response to inflammation and acts as a defense mechanism against tumor cells or infections.[12,13] NO is also produced in the central and peripheral nervous systems, and acts as a second messenger molecule.[14] L-arginine is the precursor of NO in all these processes. The enzymes which catalyze L-arginine transformation to nitric oxide and L-citrulline began to be purified in 1987 and were cloned in the early nineties.

In mammalian cells, three isoforms of NADPH-dependent flavoproteins, called nitric oxide synthases (NOS I, II and III) appear to be responsible for the biosynthesis of NO by oxidation of the guanidino group of L-arginine.[13,15] L-arginine, molecular oxygen and NADPH are cosubstrates; NO, L-citrulline, NADP and H_2O are coproducts; stoichiometry for O_2 and NADPH consumption and H_2O synthesis is 4:3:4. Five cofactors are required: flavins FAD and FMN, heme

Nitric Oxide Research from Chemistry to Biology: EPR Spectroscopy of Nitrosylated Compounds, edited by Yann A. Henry, Annie Guissani and Béatrice Ducastel.
© 1997 R.G. Landes Company.

(Fe-protoporphyrin IX), tetrahydrobiopterin (BH$_4$) and calmodulin (CaM).[16,17]

CONSTITUTIVE AND INDUCIBLE NITRIC OXIDE SYNTHASES

Nitric oxide synthases can be classified into two main groups (Table 8.1):[18] constitutive (NOS I and NOS III) activated by the Ca^{2+}/CaM complex and inducible (NOS II). NOS I and NOS III were first characterized respectively in neurons and in vascular endothelial cells and therefore are referred to as neuronal and endothelial NOS, whereas NOS II was first isolated from murine macrophages stimulated with interferon-γ (IFNγ) and lipopolysaccharide (LPS) (reviewed in ref. 4).

A fundamental difference between isoforms concerns the NOS activity in situ. The constitutive enzymes producing short-lasting (a few seconds to minutes) and small quantities of NO (pmol/min/mg of cellular protein), without any lag, have a regulatory role: NO synthesized by NOS I behaves as a neuromediator,[14] whereas NO produced by NOS III acts as an EDRF,[11] regulates platelet aggregation,[19] and prevents leukocyte adhesion.[20] When the Ca^{2+}-independent CaM-containing enzyme, for example the murine macrophage enzyme, is induced by immunomodulators such as cytokines, or bacterial-derived molecules such as LPS, large quantities of NO (nmol/min/mg of cellular protein) are produced over a long period (hours to days) after a delay of several hours.[12,13] Nitric oxide produced in this way has a physiopathological role as an antibacterial,[21] antiparasitic,[22] antitumoral[23] and antiviral agent.[24] It is the main effector molecule of the antiproliferative effect exerted by rodent activated macrophages.[12] NO is also involved in apoptosis,[25,26] immunosuppression,[27] allograft rejection[28,29] and production of carcinogenic nitrosamines.[30]

Nitric oxide synthases are encoded by three different genes corresponding to NOS I (constitutive neuronal isoform, Mr 161 kDa), NOS III (constitutive endothelial isoform, Mr 133 kDa) and NOS II (inducible isoform, Mr 131 kDa). Human and bovine NOS III display 93% homology, and there is 80% homology between NOS II from murine macrophages and human hepatocytes.[2,31] Human NOS I, II and III genes are located on chromosomes 12,[32] 17[33] and 7,[34] respectively. At the C-terminal end (Fig. 8.1) are found consensus sequences for NADPH, FAD and FMN binding.[2,35,36] This C-terminal domain has a large homology with mammalian P-450 reductase.[35] Moving towards the N-terminal end is found the CaM binding site between the reductase and oxygenase domains. This N-terminal part of the polypeptide chain also contains threonine and serine phosphorylation sites, and a heme binding site including a cysteine residue as in cytochrome P-450,[37,38] probably Cys-194 in murine NOS II, Cys-415 and Cys-184, for NOS I and NOS III, respectively.[38,39] Binding sites for BH$_4$ and L-arginine are not yet well defined. The N-terminal end of NOS III bears a myristoylation site which can explain its interaction with membranes.[40,41]

The NOS enzymes seem to behave as self-sufficient P-450 cytochromes since they have their reductase and heme domains as a part of the same polypeptide.[17,37]

As to kinetic parameters, K$_m$ values for arginine are in the micromolar range for all three NOS; V$_{max}$ is about 1300 nmol/min/mg protein for NOS II, 730 nmol/min/mg protein for NOS I and one order of magnitude lower for NOS III, so that k$_{cat}$ values are in the range of 1-2 s^{-1} for NOS I and NOS II. Therefore, differences in NO production levels in tissues are more related to various abundances and to the "pulsed" versus continuous synthesis than to large differences between kinetic parameters.[5,42] Commercial polyclonal and monoclonal antibodies raised against all three isoforms are now available.

ENZYMATIC MECHANISM OF NITRIC OXIDE SYNTHASES

The reactional mechanism (Fig. 8.2) involves two steps.[2-4] It has been shown, using $^{18}O_2$, that the oxygen atom incorporated

Table 8.1. Similarities and differences between the three main classes of nitric oxide synthases

Endothelial const. (type III)	Neuronal const. (type I)	Inducible (type II)
7q 35-36 (human) 133 kDa (1203 aa)	12q24.2 (human) 161 kDa (1433 aa)	17cen-q12 (human) 131 kDa (1144 aa)
Cytosolic/particulate NADPH dependent P-450 type hemoprotein Inhibited by L-arginine analogues	Cytosolic/particulate NADPH dependent P-450 type hemoprotein Inhibited by L-arginine analogues	Cytosolic NADPH dependent P-450 type hemoprotein Inhibited by L-arginine analogues
Ca^{2+}/calmodulin dependent	Ca^{2+}/calmodulin dependent Ca^{2+}-independent, Regulated by transcription	Calmodulin is a subunit
Picomols NO released /min/mg protein (basal level) Short-lasting release	Picomols NO released /min/mg protein (basal level) Short-lasting release (hours lag)	Nanomols NO released /min/mg protein (high level) Long-lasting release
Unaffected by glucocorticoids	Unaffected by glucocorticoids	Induction inhibited by glucocorticoids
Responding to acetylcholine, bradykinin, glutamate, etc.,	Responding to acetylcholine, bradykinin, glutamate, etc.,	Induction by endotoxins (LPS, BCG) or cytokines (IFNγ, TNFα, IL-1)
Accounts for the EDRF effects in: Vascular endothelial cells, platelets Found in platelets, vascular endothelial cells	Found in: cerebellum, cerebellar cortex, central nervous system, brain endothelial cells, peripheral nervous system, adrenal gland (cortex and medulla), astrocytes, NANC neurons, skeletal muscle	found in: liver, lung, aorta, peritoneal macrophages, PMNs, T lymphocytes, hepatocytes,osteoclasts, Kupffer cells, kidney epithelial cells, retina epithelial cells, fibroblasts, keratinocytes vascular endothelial cells, vascular smooth muscle cells, chondrocytes, mesangial cells, brain endothelial cells, microglia cells, astrocytes, Langerhans β-islets, tumor cells, HL-60 leukemia, neuroblastoma, EMT-6, TA3 adenocarcinoma cells

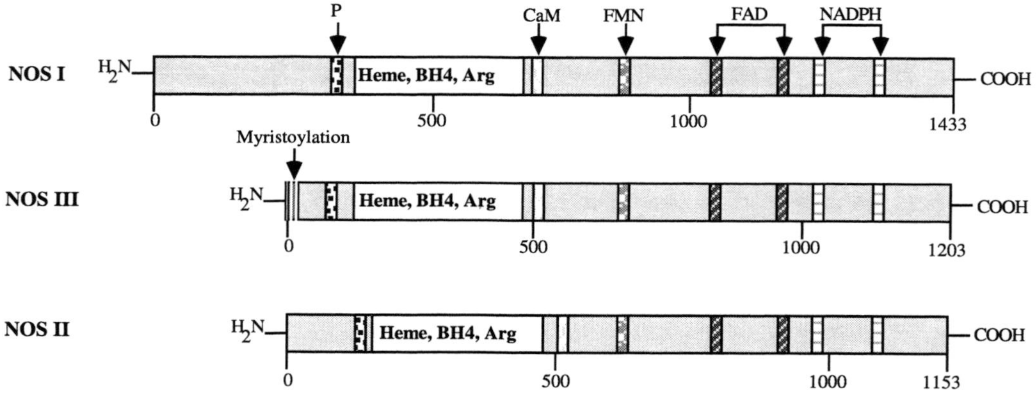

Fig. 8.1. Primary structure of NO synthases. (P = Phosphorylation site).

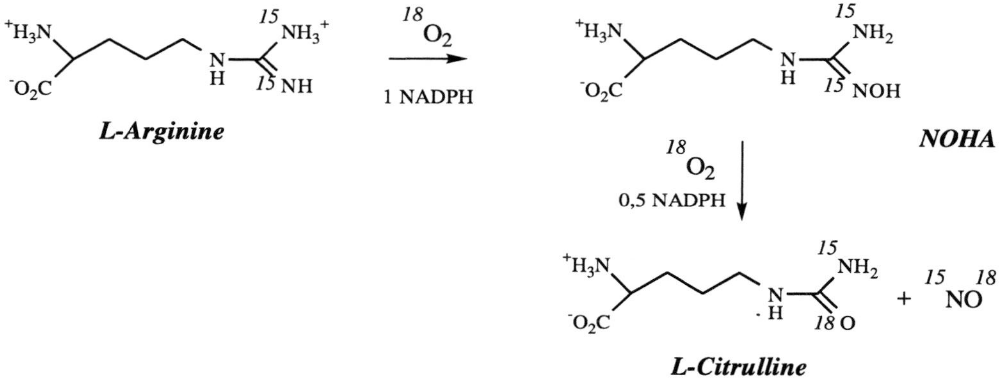

Fig. 8.2. Metabolization of L-arginine catalyzed by NO synthases.

into NO and L-citrulline originates from molecular oxygen and not from water.[43,44]

The first step of the reaction appears as a classical P-450 dependent N-hydroxylation which leads to the formation of N^ω-hydroxy-L-arginine (NOHA).[16] This intermediate is barely detectable in constitutive NOS-catalyzed reactions but can be liberated in the reactional medium in fairly large amounts by NOS II from several cell types such as EMT6 murine adenocarcinoma.[45] In this step, 1 mol of NADPH and 1 mol of O_2 are consumed, and 1 mol of H_2O is produced. The mechanism probably involves the high-valence iron-oxo species $Fe^V=O$, as in classical P-450 reactions.[46,47]

The second step involves the oxidative cleavage of the C=NOH bond of the intermediate NOHA and leads to NO and L-citrulline. It is a three-electron oxidation of NOHA, and it requires a second mol of oxygen, but only 0.5 mol of NADPH,[16] which is different from classical monoxygenation reactions. The mechanism is still controversial, but probably involves the iron-peroxo intermediate NOS-Fe^{III}-O-O$^\bullet$, which is generated during the P-450-redox cycle. Two mechanisms have been proposed, which both involve the fragmentation of the same tetrahedral intermediate (Fig. 8.3). In the first mechanism, the iron-peroxo complex abstracts a hydrogen atom of the C=NOH group of NOHA to gener-

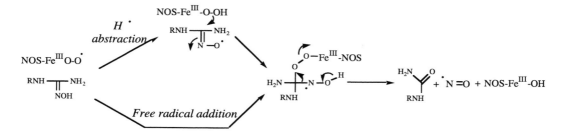

Fig. 8.3 . Mechanisms proposed for the oxidation of N^ω-hydroxy-L-arginine by NO synthases.

ate an iminoxyl radical and a hydroperoxo species NOS-Fe^{III}-O-OH, which could carry out a nucleophilic attack on the guanidino carbon of the radical and generate a tetrahedral intermediate that rearranges to yield products.[48] The second mechanism involves the direct addition of the NOS-Fe^{III}-O-O[•] intermediate to the guanidino carbon of NOHA, which directly leads to the same tetrahedral intermediate.[47]

ROLE OF N^ω-HYDROXY-L-ARGININE

N^ω-hydroxy-L-arginine can appear as an endogenous NO donor, as it can generate citrulline and nitrite not only with NOS, but also with other hemoproteins such as hepatic P-450 3A,[49] peroxidases[50] and a lipoxygenase.[51] The existence of these last pathways at a physiological level remains to be established, but recent results show that NOHA can be liberated from its parent cell.[45,52,53] Moreover, NOHA can induce the death of tumoral cells devoid of NOS, which could be explained by the metabolization of NOHA into NO by other enzymatic pathways (see chapter 9).[54]

It has also recently appeared that N^ω-hydroxy-L-arginine is the best arginase inhibitor reported so far.[55-57] Thus, interesting cross regulations between the arginase and NOS pathways might occur (see chapter 14).

NITRIC OXIDE SYNTHASES' REGULATIONS

The regulation of NOS activity can be envisioned at different levels: the enzyme is regulated by its cofactors (flavins, BH_4, Ca^{2+}/CaM and heme). Its activity can be modulated at a post-transcriptional stage (phosphorylations) but can also be directly regulated at the products or substrate level. Finally, the enzyme may be regulated at a transcriptional level.

REGULATION BY COFACTORS

Active forms of NOS are dimeric. Stuehr et al have shown that two monomers containing CaM, FAD and FMN have to interact with two heme and tetrahydrobiopterin molecules to build the active dimer.[58,59] Thus BH_4 is involved in the generation and the stabilization of active NOS, whereas its redox role remains controversial.[60]

Calmodulin seems to regulate electron flow from NADPH to heme by acting as an on/off switch between the reductase and oxygenase domains.[61,62] An essential distinction between constitutive and inducible NOS lies in the following: NOS I and NOS III bind CaM in a reversible Ca^{2+}-dependent manner and hence are susceptible to activation by agonists which increase the intracellular Ca^{2+} concentration; in contrast NOS II binds CaM so tightly at the concentration of Ca^{2+} found in resting cells that its activity is not susceptible to control by transient variations in Ca^{2+} concentration.[5] In NOS I, a Ca^{2+} concentration increase results in the immediate binding of CaM and triggering of the electron flow in the presence or absence of arginine (Fig. 8.4).

Electron flow (NADPH consumption) is even higher in the absence of arginine, and in this case NOS shows a NADPH

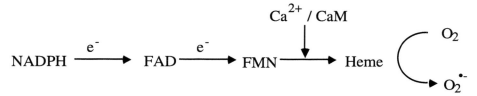

Fig. 8.4. Electron transfer in NO synthase, in the absence of L-arginine.

oxidase activity ($O_2^{\bullet-}$ production), which is controlled by Ca^{2+} concentration.[63]

In contrast, CaM can be considered as a permanently bound NOS II subunit.[64] In the absence of arginine, a very low electron flow is observed (low NADPH consumption) resulting in a very low production of $O_2^{\bullet-}$. Electron flow is dramatically increased upon arginine binding. This NOS II behavior is reminiscent of P-450 properties.[59,65]

The calcium EC_{50} of NOS I is about 200-500 nmol/l. NOS I is inactive below 80 nmol/l, the free Ca^{2+} concentration found in synaptosomes. Agonists such as excitatory amino acids (glutamate and glycine) can bind to receptor-associated channels of the N-methyl-D-aspartate type, resulting in a transient free Ca^{2+} increase and thus in a very fast increase of NOS I activity. The same scheme applies for NOS III, Ca^{2+} influx being triggered by vasodilator compounds (bradykinin, acetylcholine, etc.).[4] NOS III can also be activated by shear stress forces generated in the blood stream.[66]

All NOS isoforms are also inhibited by flavoprotein inhibitors such as diphenyleniodonium,[67] but only constitutive NOS isoforms are inhibited by calmodulin antagonists such as trifluoroperazine.

POST-TRANSCRIPTIONAL MODIFICATIONS OF THE ENZYME

Phosphorylation of constitutive NOS on serine or threonine residues was proposed as a mean of negative regulation of NOS activity.[68,69]

INHIBITORS OF NITRIC OXIDE SYNTHASES

Active site inhibitors should be the most suitable tools to achieve selective inhibition of NOS. The first N-substituted arginine found to inhibit NOS activity was N^{ω}-monomethyl-L-arginine (NMMA).[12] Since then, numerous arginine analogues have been synthesized with the hope of finding isoform-specific inhibitors (Table 8.2).

NMMA is a good inhibitor of all isoforms, whereas N^{ω}-nitro-L-arginine is relatively selective for the constitutive isoforms.[76] Recently, thiocitrulline and S-alkyl-isothiocitrulline, new analogues of arginine containing a sulfur atom, have been synthesized and evaluated. They are the most potent arginine analogues to inhibit both classes of NOS. S-methylisothiocitrulline is particularly interesting as it appears to be selective for NOS I in vitro. However, it seems to be much less potent in vivo.[77] Some of these molecules such as NMMA behave as irreversible mechanism-based inhibitors.[78] Thiocitrulline and N^{ω}-nitroarginine compete for the arginine active site, and both inhibit electron transfer from the flavins to the heme. They are probably acting as heme iron ligands.[72,79] When considering the activity of NOS inhibitors in whole cells, one must also keep in mind that there exists another regulatory mechanism, involving problems of transport and penetration in cells. In particular, NMMA is a good inhibitor of arginine uptake, via the Y^+ transporter.[80]

Recently, simple non-amino acid molecules have been found to be very potent inhibitors of NOS. Aminoguanidine is a potent and selective inhibitor of NOS II (K_i = 16 µmol/l whereas K_i = 830 µmol/l for NOS I),[81] and is also selective for NOS II in vivo.[82] Interestingly, very simple isothioureas and *bis*isothioureas appear to be the best NOS inhibitors known to date.

For example, *S*-ethylisothiourea inhibits human NOS I, II and III with K_i values of 29, 19 and 39 nmol/l respectively.[83] The study of Garvey et al[83] shows that, depending on the substituent, selectivity towards NOS II can be achieved. Little is known about the mechanism of action of these isothioureas, but they may bind at the guanidine portion of the arginine site. They may also play a role by influencing the enzyme dimerization. Unfortunately, they are not very potent at inhibiting NO production in human cells, possibly because of a poor cellular penetration.[83]

Table 8.2. Effects of arginine analogues on NOS activity

Arginine analogues: $^-OOC{-}CH{-}(CH_2)_3{-}R$ (^+H_3N)	R :	NOS I	NOS II
N^ω-methylarginine (NMMA)	$-N(H){-}C(=NH){-}NH{-}CH_3$	Ki = 0.7 μM (70)	Ki = 2.5 μM (70)
N^ω-nitroarginine (NO2Arg)	$-N(H){-}C(=NH){-}NH{-}NO_2$	Ki = 0.2 μM (70)	Ki = 8.7 μM (70)
N^ω-aminoarginine (AminoArg)	$-N(H){-}C(=NH){-}NH{-}NH_2$	Ki = 1.2 μM (70)	Ki = 1.7 μM (70)
N^δ-iminoethylornithine (NIO)	$-N(H){-}C(=NH){-}CH_3$	IC50 = 3.9 μM (71)	IC50 = 2.2 μM (71)
N^ε-iminoethyllysine (NIL)	$-CH_2{-}N(H){-}C(=NH){-}CH_3$	IC50 = 92 μM (71)	IC50 = 3.3 μM (71)
N^ω-allylarginine	$-N(H){-}C(=NH){-}NH{-}CH_2{-}CH{=}CH_2$	Ki = 0.85 μM (71)	Ki = 8.5 μM (71)
N^ω-methoxyarginine	$-N(H){-}C(=NH_2){-}N{=}O{-}CH_3$	Ki = 6 μM (71)	Ki = 20 μM (71)
Thiocitrulline	$-N(H){-}C(=NH_2){=}S$	Ki = 60 nM (72)	Ki = 3.6 μM (72)
S-methyl*iso*thiocitrulline	$-N(H){-}C(=NH){-}S{-}CH_3$	Ki = 5 nM (73)	Ki = 840 nM (73)
N^ω-hydroxycitrulline	$-N(H){-}C(=O){-}NH{-}OH$	— (74)	— (74)
N^ω-hydroxyindospicine	$-CH_2{-}C(=NH_2){-}N{-}OH$	— (74)	— (74)
Indospicine	$-CH_2{-}C(=NH_2){=}NH$	— (75)	— (75)

M for mol/l; –: compounds with IC$_{50}$ > 1 mmol/l

7-Nitroindazole (7-NI) and its analogues have also been described as very good inhibitors of NOS.[84] 7-NI competes both with arginine and BH_4,[85] which suggests that the binding sites of arginine and BH_4 may be very close to each other.[84] 7-NI inhibits NOS I activity in vivo,[86] but it may also have non-NOS-related activities.[87]

The reaction product itself, NO, was reported to be a reversible inhibitor of constitutive[88,89] and inducible NOS.[90] So, feedback inhibition by NO may be an endogenous mechanism of NOS regulation. In vitro, BH_4 seems to attenuate NOS inhibition by NO. This phenomenon could be another role for BH_4 in NO biosynthesis (see chapter 14).[60,89]

REGULATIONS AT THE TRANSCRIPTIONAL LEVEL

Distinction between the constitutive and inducible character of NO synthases tends to become obsolete as induction of all three isoforms has been observed to some extent.

Although being considered as constitutive, NOS III (endothelial) contains within its promoter consensus sequences for response to estrogens and, depending on the species, NOS III expression can be increased by estradiol.[91,92]

There is no efficient regulation of NOS II after its biosynthesis, except by substrate or cofactor avaibility; thus its regulation at the transcriptional level is crucial. In mice and rodents, most inducer signals (overviewed in ref. 93) are cytokines: interferon-γ (IFNγ), tumor necrosis factor-α (TNFα), interleukin-1 (IL-1), interleukin-2 (IL-2) and bacterial-derived molecules: lipopolysaccharide (LPS), muramyl dipeptide (MDP), phosphatidylinositolmannosides[94] and toxic shock syndrome toxin.[95] Induction is inhibited by other cytokines such as: interleukin 4 (IL-4),[96] interleukin-10 (IL-10),[97] interleukin-8 (IL-8) in neutrophils,[98] transforming growth factor-β (TGFβ);[99] and also by glucocorticoids,[100] prostaglandin E_2[101] and trans retinoic acid.[102] Several growth factors can prevent NOS II induction, e.g., epidermal growth factor in keratinocytes,[103] platelet-derived growth factor in renal mesangial cells[104] and fibroblast growth factor in retinal pigmented epithelial cells.[105] Another regulation is possible at the level of cofactor availability. It is known, for example that IFNγ induces GTP-cyclohydrolase I, a key enzyme in BH_4 synthesis.[4]

Messenger RNA for NOS II can be detected by Northern blot using specific probes. NOS II mRNA, in the RAW 264.7 cell line, is barely detectable two hours after exposure to inducer signals IFNγ and LPS; its level increases rapidly for up to 8 hours and more slowly thereafter. NOS II mRNA, the half-life of which is about 3 hours, requires protein synthesis ability,[106] as well as protein tyrosine kinase activity,[107] to be synthesized. A similar time course of NOS II mRNA induction by IFNγ and LPS was observed in the murine adenocarcinoma cell line EMT-6.[108] NOS II could be induced in EMT-6 cells by IFNγ synergistically with LPS, TNFα or IL-1. In murine inflammatory macrophages, MDP was also effective as a second signal. All these activating agents were unable to induce NOS when used alone, except at very high concentrations and after a long period of stimulation. It has been shown that a specific sequence of stimuli is required to induce NOS II. As a general rule, two distinct types of signals, A and B, are necessary to achieve NOS II induction. The A-type stimulus is IFNγ, and it must act first. Run-on experiments have shown that IFNγ increases the rate of transcription. Its effect seems irreversible; as a result it can be present only transiently. The B-type signal (LPS, IL-1, TNFα, MDP, etc.) must act concomitantly or after the A-type signal. Their action seems to be reversible; thus they have to be present permanently to maintain continuous stimulation of transcription. LPS and TNFα increase mRNA stability.[109,110] As pointed out by Lorsbach et al, the absolute requirement for the B-type stimulus to trigger or maintain NOS II mRNA transcription is probably important in preventing inappropriate synthesis of NO; the order in which cells encounter A-type and B-type stimuli can deeply influence the extent to which NO synthesis and antiproliferative activity are induced in these cells.[106]

Inhibition of NOS II induction by TGFβ and IL-4[96] in murine macrophages and endothelial cells can be explained by an increased rate of mRNA degradation, whereas IL-10[97] and trans retinoic acid[102] could prevent synthesis of TNFα, an inducing cytokine.

Murine NOS II promoter has been cloned and sequenced.[111,112] The mRNA initiation site is preceded by a TATA box and at least 22 oligonucleotide elements homologous to consensus sequences for the binding of transcription factors. These include ten copies of IFNγ response element (γ-IRE), three copies of γ-activated site, two copies each of nuclear factor-κB (NFκB), IFNα-stimulated response, activating protein-1 and TNF-response element; and finally one X box. Plasmids in which the downstream one-third of the promoters (containing the NFκB binding site close to the TATA box), linked to a reporter gene encoding chloramphenicol transacetylase or luciferase, were transfected into the RAW 264.7 macrophage-like cell line and conferred inducibility by LPS. The actual importance of NFκB translocation for expressing the NOS II gene has also been demonstrated by using antioxidants such as dithiocarbamate derivatives and an electrophoretic mobility shift assay.[113] In epithelial cells, $O_2^{\bullet-}$ was able to promote NFκB translocation and to induce NOS II.[114] It has been shown that there are physical associations and functional antagonism between the p65 subunit of NFκB and the glucocorticoid receptor.[115] The other clearly identified transcription factor is the interferon-response factor-1 (IRF-1), since macrophages from mice with a disruption of the IRF-1 gene produced little or no NO, and synthesized barely detectable amounts of NOS II mRNA.[116]

Human NOS II promoter has also been sequenced.[33] A sequence of about 400 base-pairs, upstream of the human NOS II transcriptional initiation site, is 66% identical to the mouse NOS II promoter. Analysis of the putative human promoter revealed the presence of three γ-IRE, one NFκB site conserved within the murine promoter, a palindromic TNF-response element and one NF-IL-6 response element. The human and murine promoters also contain an A-activator binding site (AABS),

Table 8.3. Constitutive NOS in humans

Cells	References
Cerebellum	32
Cavernosal smooth muscle	119
Endothelium	34
Umbilical vein endothelial cells	4
Skeletal muscle	120
Placenta	121
Lung epithelial cells	122

Table 8.4. Inducible NOS in humans

In cells	References
Hepatocytes	31,118
Vascular smooth muscles	123
Neutrophils	124
Chondrocytes	125, 126
Astrocytes	127
Mesangial cells	128
Pancreatic β-cells	129
Lung epithelial cell	114, 122, 130
Retinal pigmented epithelial cells	131
Osteoblasts	132
Keratinocytes	133
Monocytes/macrophages	134-138

In pathological situations	References
Sepsis	139
Interleukin-2 immunotherapy	140
Synovial fluid in rheumatic diseases (arthritis, osteoarthritis)	141
HIV-1 infected monocytes	142
Tumor cell line	143
Glioblastoma cell line	144
Expired (exhaled) air	145

reported to mediate liver-specific gene expression. Interestingly, the human NOS II promoter contains a shear-stress responsive element (GAGACC) which was also found to exist in human NOS III promoter but not in murine NOS promoter.[117]

As far as NOS II induction is concerned, contrasted situations have been reported depending on species, strains, tissues and cells: i.e., induction by apparently a single stimulus (IL-1 for rabbit chondrocytes, human and rodent pancreatic β-islet, rat smooth musle cells, etc.) or in contrast, strong synergy between stimuli (IFNγ, IL-1, TNFα and LPS in human hepatocytes;[118] IFNγ, IL-1, TNFα and IL-6 in a human adenocarcinoma cell line). We may assume that differences in susceptibility to stimuli could be explained, at least in part, by the history of the cells (e.g. uncontrolled contact with bacterial-derived molecules in nonpathogen-free animals), or by a particular ability to synthesize an inducing factor (e.g. TNFα) to achieve a paracrine loop in response to a first stimulus.

Table 8.5. NO in pathophysiology

NO inhalation	References
Persistent pulmonary hypertension of the newborn	147
Chronic pulmonary hypertension	149
Adult respiratory distress syndrome	148

Use of NO donors	References
Hypertension	153
Male impotence	151
Adult heart disease	154

Use of inhibitors	References
Septic shocks	155

CELLULAR LOCALIZATION OF NITRIC OXIDE SYNTHASES

Nitric oxide synthases have been found in numerous cells, tissues and species.[4,18,93] Table 8.3 and Table 8.4 summarize some data for humans.

There are still conflicting opinions about NOS II expression in human monocytes/macrophages.[134] Despite great efforts by many investigators, biosynthesis of NO could barely be observed under the conditions leading to induction of murine NOS II. Recently, it has been shown that stimulation of human macrophages through CD69,[135] or by IL-4,[136] and IGE or anti-CD 23 antibodies,[137] resulted in nitrite production after several days of culture. Other authors detected an increased level of NOS II mRNA in human monocytes/macrophages upon stimulation with IFNγ/LPS, by using the reverse transcription-polymerase chain reaction but could not detect significant levels of nitrite release.[138]

As pointed out by Nussler and Billiar:[146] "What evolutionary pressures preserved NOS II expression in hepatocytes and reduced NOS II expression in macrophages in humans are not clear. Perhaps the benefit derived from the antimalarial or other antiparasitic actions of NO in the liver may have preserved NO expression in hepatocytes whereas a low threshold for macrophage NOS II expression may have had adverse consequences (e.g host cell injury and carcinogen formation)". In addition, these authors published data showing that, compared with macrophages, hepatocytes are equipped to efficiently resist metabolic inhibition by endogenous NO.[146]

CONCLUSION

There are situations in mammals in which NO is either underproduced or overproduced. The first case deals with diseases such as persistent hypertension of the newborn,[147] adult respiratory distress syndrome[148,149] or male impotence (Table 8.5).[119,150,151]

The palliatives to underproduction of NO are chemically diverse: inhaled NO and NO donors. On the opposite side, overproduction of NO was detected in septic shock[152] or diabetes mellitus.[129] Significant increases in circulating nitrate concentrations were observed in septic humans.[139,152] In addition, it was recently shown that the circulating concentration of NOHA increased up to 5-fold in the serum of rats treated with bacterial LPS. Thus the serum level of NOHA may represent a useful diagnostic parameter for the monitoring of patients with septic shock.[52] IL-2 therapy in cancer patients results in sustained hypotension. Evidence for increased NO formation in humans after IL-2 therapy has been reported.[140] NMMA, a mechanism-based NOS inhibitor, was used to successfully reverse hypotension in septic patients.[155]

All these aspects will be developed in chapters 11 and 12.

REFERENCES

1. Kerwin JF, Lancaster JR, Feldman PL. Nitric oxide: A new paradigm for second messengers. J Med Chem 1995; 38:4343-4362.
2. Knowles RG, Moncada S. Nitric oxide synthases in mammals. Biochem J 1994; 298:249-258.
3. Marletta MA. Approaches toward selective inhibition of nitric oxide synthase. J Med Chem 1994; 37:1899-1907.
4. Stuehr DJ, Griffith OW. Mammalian nitric oxide synthases. Adv Enzymol Relat Areas Mol Biol 1992; 65:287-346.
5. Nathan C. Nitric oxide as a secretory product of mammalian cells. FASEB J 1992; 6:3051-3064.
6. Green LC, Ruiz de Luzuriaja K, Wagner DA et al. Nitrate biosynthesis in man. Proc Natl Acad Sci USA 1981; 78:7764-7768.
7. Green LC, Wagner DA, Glowiski J et al. Analysis of nitrate, nitrite and (15 N) nitrate in biological fluids. Anal Biochem 1982; 126:131-138.
8. Furchgott RF, Zawadzki JV. The obligatory role of endothelial cells in the relaxation of arterial smooth muscle by acetylcholine. Nature 1980; 288:373-376.
9. Furchgott RF. Studies on relaxation of rabbit aorta by sodium nitrite: the basis for the proposal that the acid-activatable inhibitory factor from bovine retractor penis is inorganic nitrite and the endothelium-derived relaxing factor is nitric oxide. In: Vanhoutte PM, ed. Vasodilatation: Vascular Smooth Muscle, Peptides, Autonomic Nerves, and Endothelium. Raven Press, New York. 1988;401-414.
10. Ignarro LJ, Buga GM, Wood KS et al. Endothelium-derived relaxing factor produced and released from artery and vein is nitric oxide. Proc Natl Acad Sci USA 1987; 84:9265-9269.
11. Palmer RMJ, Ferridge AG, Moncada S. Nitric oxide release accounts for the biological activity of endothelium-derived relaxing factor. Nature 1987; 327:524-526.
12. Hibbs JB, Taintor RR, Vavrin Z. Macrophage cytotoxicity: Role for L-arginine deiminase and imino nitrogen oxidation to nitrite. Science 1987; 235:473-476.
13. Marletta MA, Yoon PS, Iyengar R et al. Macrophage oxidation of L-arginine to nitrite and nitrate: Nitric oxide is an intermediate. Biochemistry 1988; 27:8706-8711.
14. Garthwaite J, Charles SL, Chess-Williams R. Endothelium-derived relaxing factor release on activation of NMDA receptors suggests role as intercellular messenger in the brain. Nature 1988; 336:385-388.
15. Palmer RMJ, Ashton DS, Moncada S. Vascular endothelial cells synthesize nitric oxide from L-arginine. Nature 1988; 333:664-666.
16. Stuehr DJ, Kwon NS, Nathan CF et al. N^{ω}-Hydroxy-L-arginine is an intermediate in the biosynthesis of nitric oxide from L-arginine. J Biol Chem 1991; 266:6259-6263.
17. White KA, Marletta MA. Nitric oxide is a cytochrome P-450 type hemoprotein. Biochemistry 1992; 31:6627-6631.
18. Förstermann U, Closs EI, Pollock JS et al. Nitric oxide synthase isozymes - Characterization, purification, molecular cloning, and functions. Hypertension 1994; 23:1121-1131.
19. Radomski MW, Palmer RMJ, Moncada S. An L-arginine-nitric oxide pathway present in human platelets regulates aggregation. Proc Natl Acad Sci USA 1990; 87:5193-5197.
20. Kubes P, Suzuki M, Granger DN. Nitric oxide: an endogenous modulator of leuko-

cyte adhesion. Proc Natl Acad Sci USA 1991; 88:4651-4655.

21. Bookvar KS, Granger DL, Poston RM et al. Nitric oxide produced during murine listeriasis is protective. Infection Immunity 1994; 62:1089-1100.

22. Nussler A, Drapier J-C, Renia L et al. L-arginine-dependent destruction of intra hepatic malaria parasites in response to tumor necrosis factor and/or interleukin-6 stimulation. Eur J Immunol 1991; 21:227-230.

23. Farias-Eisner R, Sherman MP, Aeberhard E et al. Nitric oxide is an important mediator for tumoricidal activity in vivo. Proc Natl Acad Sci USA 1994; 91:9407-9411.

24. Karupiah G, Xie QW, Buller RML et al. Inhibition of viral replication by interferon γ induced nitric oxide synthase. Science 1993; 261:1445-1448.

25. Albina JE, Cui S, Mateo RB et al. Nitric oxide - mediated apoptosis in murine peritoneal macrophages. J Immunol 1994; 150:5080-5085.

26. Sarih M, Souvannavong V, Adam A. Nitric oxide synthase induces macrophage death by apoptosis. Biochem Biophys Res Commun 1993; 191:503-508.

27. Candolfi E, Hunter CA, Remington JS. Mitogen and antigen specific proliferation of T cells in murine toxoplasmosis is inhibited by reactive nitrogen species. Infection Immunity. 1994; 62:1995-2001.

28. Langrehr JM, Murase N, Markus PM et al. Nitric oxide production in host vs graft and graft vs host reaction in the rat. J Clin Invest 1992; 90:679-683.

29. Benvenuti C, Bories PN, Loisance D. Increased serum nitrite concentration in cardiac transplant patients. Transplantation 1996; 61:745-749.

30. Ohshima H, Tsuda M, Adachi H et al. L-arginine dependent formation of N-nitrosamines by the cytosol of macrophages activated with lipopolysaccharide and interferon-γ. Carcinogenesis. 1991; 12:1217-1220.

31. Geller DA, Lowenstein CJ, Shapiro RA et al. Molecular cloning and expression of inducible nitric oxide synthase from human hepatocytes. Proc Natl Acad Sci USA 1993; 90:3491-3495.

32. Kishimoto J, Spurr N, Liao M et al. Localization of brain nitric oxide synthase (NOS) to human chromosome 12. Genomics 1992; 14:802-804.

33. Chartrain NA, Geller DA, Koty PP et al. Molecular cloning, structure, and chromosomal localization of the human inducible nitric oxide synthase gene. J Biol Chem 1994; 269:6765-6772.

34. Marsden PA, Heng HHQ, Scherer SW et al. Structure and chromosomal localization of the human constitutive endothelial nitric oxide synthase gene. J Biol Chem 1993; 268:17478-17488.

35. Bredt DS, Hwang PM, Glatt CE et al. Cloned and expressed nitric oxide synthase structurally resembles cytochrome P-450 reductase. Nature 1991; 351:714-718.

36. Xie Q, Cho HJ, Calaycay J et al. Cloning and characterization of inductible nitric oxide synthase from mouse macrophages. Science 1992; 256:225-228.

37. Marletta MA. Nitric oxide synthase structure and mechanism. J Biol Chem 1993; 268:12231-12234.

38. McMillan K, Masters BSS. Prokaryotic expression of the heme- and flavin-binding domains of rat neuronal nitric oxide synthase as distinct polypeptides: Identification of the heme-binding proximal thiolate ligand as cysteine-415. Biochemistry 1995; 34:3686-3693.

39. Renaud J-P, Boucher J-L, Vadon S et al. Particular ability of liver P450s 3A to catalyze the oxidation of N^ω-hydroxyarginine to citrulline and nitrogen oxides and occurrence in NO synthases of a sequence very similar to the heme-binding sequence in P450s. Biochem Biophys Res Commun 1993; 192:53-60.

40. Lamas S, Marsden PA, Li GK et al. Endothelial nitric oxide synthase: Molecular cloning and characterization of a distinct constitutive enzyme isoform. Proc Natl Acad Sci USA 1992; 89:6348-6352.

41. Sessa WC, Babel CM, Lynch KR et al. Mutation of N-myristoylation sites converts endothelial cell nitric oxide synthase from a membrane to a cytosolic protein. Circ Res 1993; 72:921-924.

42. Feldman PL, Griffith OW, Stuehr DJ. The

surprising life of nitric oxide. Chem Eng News 1993; 71:26-38.

43. Kwon NS, Nathan CF, Gilker C et al. L-citrulline production from L-arginine by macrophage nitric oxide synthase: The ureido oxygen derives from dioxygen. J Biol Chem 1990; 265:13442-13445.

44. Leone AM, Palmer RMJ, Knowles RG et al. Constitutive and inductible nitric oxide synthases incorporate molecular oxygen into both nitric oxide and citrulline. J Biol Chem 1991; 266:23790-23795.

45. Chenais B, Yapo A, Lepoivre M et al. High-performance liquid chromatographic analysis of the unusual pathway of oxidation of L-arginine to citrulline and nitric oxide in mammalian cells. J Chromatogr 1991; 539:433-441.

46. Clement B, Jung F. N-hydroxylation of the antiprotozoal drug pentamidine catalyzed by rabbit liver cytochrome P-450 2C3 or human liver microsomes, microsomal retroreduction, and further oxidative transformation of the formed amidoximes - Possible relationship to the biological oxidation of arginine to N^G-hydroxyarginine, citrulline, and nitric oxide. Drug Metab Dispos 1994; 22:486-497.

47. Mansuy D, Boucher J-L, Clement B. On the mechanism of nitric oxide formation upon oxidative cleavage of C=NOH bonds by NO synthases and cytochromes P450. Biochimie 1995; 77:661-667.

48. Korth HG, Sustmann R, Thater C et al. On the mechanism of the nitric oxide synthase-catalyzed conversion of N^ω-hydroxy-L-arginine to citrulline and nitric oxide. J Biol Chem 1994; 269:17776-17779.

49. Boucher J-L, Genet A, Vadon S et al. Cytochrome P450 catalyzes the oxidation of N^ω-hydroxy-L-arginine by NADPH and O_2 to nitric oxide and citrulline. Biochem Biophys Res Commun 1992; 187: 880-886.

50. Boucher J-L, Genet A, Vadon S et al. Formation of nitrogen oxides and citrulline upon oxidation of N^ω-hydroxy-L-arginine by hemeproteins. Biochem Biophys Res Commun 1992; 184: 1158-1164.

51. Boucher J-L, Chopard C, Vadon S et al. Nitric oxide formation by oxidation of Nω-hydroxy-L-arginine by linoleic acid hydro-peroxide catalyzed by soybean lipoxygenase L1. Endothelium 1993; 1:s 17.

52. Hecker M, Schott C, Bucher B et al. Increase in serum N^G-hydroxy-L-arginine in rats treated with bacterial lipopolysaccharide. Eur J Pharmacol 1995; 275:R1-R3.

53. Schott CA, Bogen CM, Vetrovsky P et al. Exogenous N^G-hydroxy-L-arginine causes nitrite production in vascular smooth muscle cells in the absence of nitric oxide synthase activity. FEBS Lett 1994; 341:203-207.

54. Chenais B, Yapo A, Lepoivre M et al. N^ω-hydroxy-L-arginine, a reactional intermediate in nitric oxide biosynthesis, induces cytostasis in human and murine tumor cells. Biochem Biophys Res Commun 1993; 196:1558-1565.

55. Boucher J-L, Custot J, Vadon S et al. N^ω-hydroxy-L-arginine, an intermediate in the L-arginine to nitric oxide pathway, is a strong inhibitor of liver and macrophage arginase. Biochem Biophys Res Commun 1994; 203:1614-1621.

56. Custot J, Boucher J-L, Vadon S et al. N^ω-hydroxylamino-α-aminoacids as a new class of very strong inhibitors of arginases. J Biol Inorg Chem 1996; 1:73-82.

57. Daghigh F, Fukuto JM, Ash DE. Inhibition of rat liver arginase by an intermediate in NO biosynthesis, N^G-hydroxy-L-arginine: Implications for the regulation of nitric oxide biosynthesis by arginase. Biochem Biophys Res Commun 1994; 202:174-180.

58. Baek KJ, Thiel BA, Lucas S et al. Macrophage nitric oxide synthase subunits. Purification, characterization, and role of prosthetic groups and substrate in regulating their association into a dimeric enzyme. J Biol Chem 1993; 268:21120-21129.

59. Ghosh DK, Stuehr DJ. Macrophage NO synthase: Characterization of isolated oxygenase and reductase domains reveals a head-to-head subunit interaction. Biochemistry 1995; 34:801-807.

60. Mayer B, Werner ER. In search of a function for tetrahydrobiopterin in the biosynthesis of nitric oxide. Naunyn-Schmiedeberg's Arch Pharmacol 1995; 351:453-463.

61. Abu-Soud HM, Stuehr DJ. Nitric oxide synthase reveals a role for calmodulin in

controlling electron transfer. Proc Natl Acad Sci USA 1993; 90: 10769-10772.

62. Abu-Soud HM, Yoho LL, Stuehr DJ. Calmodulin controls neuronal nitric-oxide synthase by a dual mechanism. Activation of intra- and interdomain electron transfer. J Biol Chem 1994; 269:32047-32050.

63. Pou S, Pou WS, Bredt DS et al. Generation of superoxide by purified brain nitric oxide synthase. J Biol Chem 1992; 267:24173-24176.

64. Cho HJ, Xie QW, Calaycay J et al. Calmodulin is a subunit of nitric oxide synthase from macrophages. J Exp Med 1992; 176:599-604.

65. Matsuoka A, Stuehr DJ, Olson JS et al. L-arginine and calmodulin regulation of the heme iron reactivity in neuronal nitric oxide synthase. J Biol Chem 1994; 269:20335-20339.

66. Kuchan MJ, Frangas JA. Shear stress regulates endothelin-1 release via protein kinase C and cGMP in cultured endothelial cells. Am J Pysiol 1993; 264:H150-H156.

67. Stuehr DJ, Fasehun OA, Kwon NS et al. Inhibition of macrophage and endothelial cell nitric oxide synthase by diphenyleneiodonium and its analogs. FASEB J 1991; 5:98-103.

68. Bredt DS, Ferris CD, Snyder SH. Nitric oxide synthase regulatory sites. Phosphorylation by cyclic AMP-dependent protein kinase, protein kinase C, and calcium/calmodulin protein kinase; identification of flavin and calmodulin binding sites. J Biol Chem 1992; 267:10976-10981.

69. Venema RC, Sayegh HS, Arnal JF et al. Role of the enzyme calmodulin-binding domain in membrane association and phospholipid inhibition of endothelial nitric oxide synthase. J Biol Chem 1995; 270:14705-14711.

70. Komori Y, Wallace GC, Fukuto JM. Inhibition of purified nitric oxide synthase from rat cerebellum and macrophage by L-arginine analogs. Arch Biochem Biophys 1994; 315:213-218.

71. Moore WM, Webber RK, Jerome GM et al. L-N^6-(1-iminoethyl)lysine: A selective inhibitor of inducible nitric oxide synthase. J Med Chem 1994; 37:3886-3888.

72. Frey C, Narayanan K, McMillan K et al. L-Thiocitrulline. A stereospecific, heme-bind-

ing inhibitor of nitric-oxide synthases. J Biol Chem 1994; 269: 26083-26091.

73. Narayanan K, Spack L, McMillan K et al. S-alkyl-L-thiocitrullines. Potent stereoselective inhibitors of nitric oxide synthase with strong pressor activity in vivo. J Biol Chem 1995; 270:11103-11110.

74. Vadon S, Custot J, Boucher J-L et al. Synthesis and effects on arginase and NO synthase of two novel analogs of N^ω-hydroxy-arginine, N^ω-hydroxy-indospicine and p-hydroxyamidino-phenylalanine. J Chem Soc Perkin Trans 1 1996; in press.

75. Feldman PL, Chi S, Sennequier N et al. Synthesis of the L-arginine congener L-indospicine and evaluation of its interaction with nitric oxide synthase. Bioorg Medicinal Chem Lett 1996; 6:111-114.

76. Furfine ES, Harmon MF, Paith JE et al. Selective inhibition of constitutive nitric oxide synthase by L-N^G-nitroarginine. Biochemistry 1993; 32:8512-8517.

77. Furfine ES, Harmon MF, Paith JE et al. Potent and selective inhibition of human nitric oxide synthases. Selective inhibition of neuronal nitric oxide synthase by S-methyl-L-thiocitrulline and S-ethyl-L-thiocitrulline. J Biol Chem 1994; 269:26677-26683.

78. Olken NM, Osawa Y, Marletta MA. Characterization of the inactivation of nitric oxide synthase by N^G-methyl-L-arginine: Evidence for heme loss. Biochemistry 1994; 33:14784-14791.

79. Abu-Soud HM, Feldman PL, Clark P et al. Electron transfer in the nitric-oxide synthases. Characterization of L-arginine analogs that block heme iron reduction. J Biol Chem 1994; 269:32318-32326.

80. Baydoun AR, Mann GE. Selective Targeting of nitric oxide synthase inhibitors to system Y(+) in activated macrophages. Biochem Biophys Res Commun 1994; 200:726-731.

81. Wolff DJ, Lubeskie A. Aminoguanidine is an isoform-selective, mechanism-based inactivator of nitric oxide synthase. Arch Biochem Biophys 1995; 316:290-301.

82. Corbett JA, Mikhael A, Shimizu J et al. Nitric oxide production in islets from nonobese diabetic mice: Aminoguanidine-

sensitive and -resistant stages in the immunological diabetic process. Proc Natl Acad Sci USA 1993; 90:8992-8995.

83. Garvey EP, Oplinger JA, Tanoury GJ et al. Potent and selective inhibition of human nitric oxide synthases. Inhibition by non-amino acid isothioureas. J Biol Chem 1994; 269:26669-26676.

84. Mayer B, Klatt P, Werner ER et al. Molecular mechanisms of inhibition of porcine brain nitric oxide synthase by the antinociceptive drug 7-nitro-indazole. Neuropharmacol 1994; 33:1253-1259.

85. Wolff DJ, Gribin BJ. The inhibition of the constitutive and inducible nitric oxide synthase isoforms by indazole agents. Arch Biochem Biophys 1994; 311:300-306.

86. Moore PK, Babbedge RC, Wallace P et al. 7-Nitro indazole, an inhibitor of nitric oxide synthase, exhibits anti-nociceptive activity in the mouse without increasing blood pressure. Br J Pharmacol 1993; 108:296-297.

87. Medhurst AD, Greenlees C, Parsons AA et al. Nitric oxide synthase inhibitors 7- and 6-nitroindazole relax smooth muscle in vitro. Eur J Pharmacol 1994; 256:R5-R6.

88. Abu-Soud HM, Wang J, Rousseau DL et al. Neuronal nitric oxide synthase self-inactivates by forming a ferrous complex during aerobic catalysis. J Biol Chem 1995; 270:22997-23006.

89. Griscavage JM, Fukuto JM, Komori Y et al. Nitric oxide inhibits neuronal nitric oxide synthase by interacting with the heme prosthetic group. Role of tetrahydrobiopterin in modulating the inhibitory action of nitric oxide. J Biol Chem 1994; 269:21644-21649.

90. Hobbs AJ, Fukuto JM, Ignarro LJ. Formation of free nitric oxide from L-arginine by nitric oxide synthase: Direct enhancement of generation by superoxide dismutase. Proc Natl Acad Sci USA 1994; 91:10992-10996.

91. Weiner CP, Lizasoain I, Baylis SA et al. Induction of calcium-dependent nitric oxide synthases by sex hormones. Proc Natl Acad Sci USA 1994; 91:5212-5216.

92. Nadaud S, Bonnardeaux A, Lathrop M et al. Gene structure, polymorphism and mapping of the human endothelial nitric oxide synthase gene. Biochem Biophys Res Commun 1994; 198:1027-133.

93. Kröncke KD, Fehsel K, Kolb-Bachofen V. Inducible nitric oxide synthase and its product nitric oxide, a small molecule with complex biological activities. Biol Chem Hoppe-Seyler 1995; 376:327-343.

94. Tenu J-P, Sekkai D, Yapo A et al. Phosphatidylinositolmannoside-based liposomes induce NO synthase in primed mouse peritoneal macrophages. Biochem Biophys Res Commun 1995; 208:295-301.

95. Zembowicz A, Vane JR. Induction of nitric oxide synthase activity by toxic shock syndrome toxin-1 in a macrophage-monocyte cell line. Proc Natl Acad Sci USA 1992; 89:2051-2055.

96. Bogdan C, Vodovotz Y, Paik J et al. Mechanism of suppression of nitric oxide synthase expression by interleukin-4 in primary mouse macrophages. J Leuk Biol 1994; 55:227-233.

97. Oswald IP, Wynn TA, Sher A et al. Interleukin-10 inhibits macrophage microbicidal activity by blocking the endogenous production of TNF-α required as a costimulatory factor for IFN-γ induced activation. Proc Natl Acad Sci USA 1992; 89:8676-8680.

98. McCall TB, Palmer RMJ, Moncada S. Interleukin-8 inhibits the induction of nitric oxide synthase in rat peritoneal neutrophiles. Biochem Biophys Res Commun 1992; 186:680-685.

99. Gilbert RS, Herschman HR. Transforming growth factor β differentially modulates the inducible nitric oxide synthase gene in distinct cell types. Biochem Biophys Res Commun 1993; 195:380-384.

100. Geller DA, Nussler AK, DiSilvio M et al. Cytokines, endotoxin and glucocorticoids regulate the expression of inducible nitric oxide synthase in hepatocytes. Proc Natl Acad Sci USA 1993; 90:552-526.

101. Marotta P, Sautelin L, DiRosa M. Modulation of the induction of nitric oxide synthase by eicosanoids in the murine macrophage cell line J 774. Br J Pharmacol 1992; 107:640-647.

102. Metha K, McQween T, Tucket S et al. Inhibition by all trans retinoic acis of tumor

necrosis factor and nitric oxide production by peritoneal macrophages. J Leuk Biol 1994; 55:336-342.

103. Heck DE, Laskino DL, Garner CL et al. Epidermal growth factor suppresses nitric oxide and hydrogen peroxide production in keratinocytes. Potential role for nitric oxide in the regulation of wound healing. J Biol Chem 1992; 267:21277-21280.

104. Pfeilschifter J. Platelet-derived growth factor inhibits cytokine induction of nitric oxide synthase in rat mesangial cells. Eur J Pharmacol 1991; 208:339-340.

105. Goureau O, Lepoivre M, Becquet F et al. Differential regulation of inducible nitric oxide synthase by fibroblast growth factor and transforming growth factor β in bovine retinal pigmented epithelial cells: Inverse correlation with cellular proliferation. Proc Natl Acad Sci USA 1993; 90:4276-4280.

106. Lorsbach RB, Murphy WJ, Lowenstein CJ et al. Expression of the nitric oxide synthase gene in mouse macrophages activated for tumor cell killing. Molecular basis for the synergy between interferon-γ and lipopolysaccharide. J Biol Chem 1993; 268:1908-1913.

107. Dong Z, Qi X, Xie K et al. Protein tyrosine kinase inhibitors decrease induction of nitric oxide synthase activity in lipopolysaccharide-responsive and lipopolysaccharide-non responsive murine macrophages. J Immunol 1993; 12:2717-2714.

108. Chenais B, Tenu J-P. Involvement of nitric oxide synthase in antiproliferative activity of macrophages: Induction of the enzyme requires two different kinds of signal acting synergistically. Int J Immunopharmacol 1994; 16:401-406.

109. Nathan C, Xie QW. Regulation of biosynthesis of nitric oxide. J Biol Chem 1993; 269:13725-13727.

110. Nathan C, Xie QW. Nitric oxide synthases: Roles, tolls and controls. Cell 1994; 78:915-918.

111. Xie QW, Whisnant R, Nathan C. Promoter of the mouse gene encoding calcium-independent nitric oxide synthase confers inducibility by interferon-γ and bacterial lipopolysaccharide. J Exp Med 1993; 177:1779-1784.

112. Lowenstein CJ, Alley EW, Ravel P et al. Macrophage nitric oxide synthase gene: Two upstream regions mediate induction by interferon γ and lipopolysaccharide. Proc Natl Acad Sci USA 1993; 90:9730-9734.

113. Mülsch A, Schray-Utz B, Mordvintcev PI et al. Diethyldithiocarbamate inhibits induction of macrophage NO synthase. FEBS Lett 1993; 321:215-218.

114. Adcock IM, Brown CR, Kwon O et al. Oxidative stress induces NFκB DNA binding and inducible NOS mRNA in human epithelial cells. Biochem Biophys Res Commun 1994; 199:1518-1524.

115. Ray A, Prefontaine KE. Physical association and functional antagonism between the p65 subunit of transcription factor NFκB and the glucocorticoid receptor. Proc Natl Acad Sci USA 1994; 91:752-756.

116. Kamijo R, Harada H, Matsuyama T et al. Requirement for transcription factor IRF-I in NO synthase induction in macrophages. Science 1994; 263:1612-1615.

117. Nunokawa Y, Ishida N, Tanaka S. Promoter analysis of human inducible nitric oxide synthase gene associated with cardiovascular homeostasis. Biochem Biophys Res Commun 1994; 200:802-807.

118. Nussler AK, Di Silvio M, Billiar TR et al. Stimulation of the nitric oxide synthase pathway in human hepatocytes by cytokines and endotoxin. J Exp Med 1992; 176:261-264.

119. Rajer J, Aronson WJ, Bush PA et al. Nitric oxide as a mediator of the corpus cavernosum in response to nonadrenergic non cholinergic transmission. N Engl J Med 1992; 326:90-94.

120. Nakane M, Schmidt HHHW, Pollock JS et al. Cloned human brain nitric oxide synthase is highly expressed in skeletal muscle. FEBS Lett 1993; 316:175-180.

121. Garvey EP, Tuttle JV, Covington K et al. Purification and characterization of the constitutive nitric oxide synthase from human placenta. Arch Biochem Biophys 1994; 311:235-241.

122. Asano K, Chee C, Gaston B et al. Constitutive and inducible nitric oxide synthase gene expression, regulation and activity in human lung epithelial cells.

Proc Natl Acad Sci USA 1994; 91:10089-10093.

123. Scott-Burden T, Schini VB, Elizondo E et al. Platelet-derived growth factor suppresses and fibroblast growth factor enhances cytokine-induced production of nitric oxide by cultured smooth muscle cells. Circ Res 1992; 71:261-264.

124. Malavista SE, Montgomery RR, Van Baricom G. Evidence for reactive nitrogen intermediate in killing of staphylococci in human neutrophil cytoblast. J Clin Invest 1992; 90:631-636.

125. Palmer RMJ, Hickery MS, Charles IG et al. Induction of nitric oxide synthase in human chondrocytes. Biochem Biophys Res Commun 1993; 193:398-405.

126. Charles IG, Palmer RMJ, Hickery MS et al. Cloning, characterization, and expression of a cDNA encoding an inducible nitric oxide synthase from the human chondrocyte. Proc Natl Acad Sci USA 1993; 90:11419-11423.

127. Lee SC, Dickson DW, Liu W et al. Induction of nitric oxide synthase in human astrocytes by interleukin-1 β and interferon γ. J Neuroimmunol 1993; 46:19-24.

128. Nicolson AG, Haites NE, McKay NG et al. Induction of nitric oxide synthase in human mesangial cells. Biochem Biophys Res Commun 1993; 193:1269-1274.

129. Corbett JA, Sweetland MA, Wang JL et al. Nitric oxide mediates cytokine induced inhibition of insulin secretion by human islets of Langerhans. Proc Natl Acad Sci USA 1993; 90:1731-1735.

130. Tracey WR, Xue C, Klinghofer V et al. Immunochemical detection of inducible NO synthase in human lung. Am J Physiol 1994; 266:L722-L727.

131. Goureau O, Hicks D, Courtois Y. Human retinal pigmented epithelial cells produce nitric oxide in response to cytokines. Biochem Biophys Res Commun 1994; 198:120-126.

132. Ralston SH, Todd D, Helfrich M et al. Human osteoblast-like cells produce nitric oxide and express inducible nitric oxide synthase. Endocrinology 1994; 135:330-336.

133. Kolb-Baschofen V, Fehsel K, Michel G et al. Epidermal keratinocyte expression of inducible nitric oxide synthase in skin lesions of psoriasis vulgaris. Lancet 1994; 344:139.

134. Denis M. Human monocytes/macrophages: NO or not NO? J Leuk Biol 1994; 55:682-684.

135. De Maria R, Cifone MG, Trotta R et al. Triggering of human monocytes activation through CD 69, a member of the NKC family of signal transducing receptors. J Exp Med 1994; 180:1999-2004.

136. Kolb J-P, Paul-Eugène N, Damais C et al. Interleukin-4 stimulates cGMP production by IFN-γ-activated human monocytes. Involvement of the nitric oxide synthase pathway. J Biol Chem 1994; 269:9811-9816.

137. Dugas B, Paul-Eugène N, Kolb J-P et al. Ligation of CD 23 activates soluble guanylcyclase in human monocytes via an L-arginine-dependent mechanism. J Leuk Biol 1995; 57:160-167.

138. Reilling N, Ulmer AJ, Duchrow ME et al. Nitric oxide synthase: mRNA expression of different isoforms in human monocytes/macrophages. Eur J Immunol 1994; 24:1941-1944.

139. Thiermerman C. The role of L-arginine-nitric oxide pathway in circulatory shock. Adv Pharmacol 1994; 28:45-79.

140. Hibbs JB, Westenfelder C, Taintor R et al. Evidence for cytokine-inducible nitric oxide synthesis from L-arginine in patients receiving interleukin-2 therapy. J Clin Invest 1992; 89:867-877.

141. Stefanovic-Racic M, Stadler J, Evans CH. Nitric oxide and arthritis. Arthritis Rheum 1993; 36:1036-1044.

142. Bukrinsky MI, Nottet HSLM, Schmidtmayerova H et al. Regulation of nitric oxide synthase activity in human immunodeficiency virus type 1 (HIV-1)-infected monocytes: Implications for HIV-associated neurological disease. J Exp Med 1995; 181:735-745.

143. Sherman PA, Laubach VE, Reep BR et al. Purification and cDNA sequence of an inducible nitric oxide synthase from a human tumor cell line. Biochemistry 1993; 32:11600-11605.

144. Hokari A, Zeniya M, Esumi H. Cloning and functional expression of human inducible

nitric oxide synthase (NOS) cDNA from a glioblastoma cell line A-172. J Biochem 1994; 116:575-581.

145. Gaston B, Reilly J, Drazen JM et al. Endogenous nitrogen oxides and bronchodilator S-nitrosothiols in human airways. Proc Natl Acad Sci USA 1993; 90:10957-10961.

146. Nussler AK, Billiar TR. Inflammation, immunoregulation and inducible NOS. J Leuk Biol 1993; 54:171-178.

147. Lonquist PA, Winberg P, Lundell B et al. Inhaled nitric oxide in neonates and children with pulmonary hypertension. Acta Paediatrica 1994; 83:1132-1136.

148. Rossaint R, Falke KJ, Lopez F et al. Inhaled nitric oxide in adult respiratory distress syndrome. N Engl J Med 1993; 328:399-405.

149. Wessel DL, Adatia I, Thompson JE et al. Delivery and monitoring of inhaled nitric oxide in patients with pulmonary hypertension. Crit Care Med 1994; 22:930-938.

150. Burnett AL, Lowenstein CJ, Bredt DS et al. Nitric oxide: A physiologic mediator of penile erection. Science 1992; 257:401-403.

151. Heaton JPW, Morales A, Adams MA et al. Recovery of erectile functionality by oral administration of apomorphine. Urology 1995; 45:200-206.

152. Petros A, Bennett D, Vallance P. Effect of nitric oxide synthase inhibitors on hypotension in patients with septic shock. Lancet 1991; 338:1557-1558.

153. Lefer AM, Lefer DJ. Therapeutic role of nitric oxide donnors in the treatment of cardiovascular disease. Drug Future 1994; 19:665-672.

154. Ohtsuka M, Honlo T, Esumi K. A new vasodilatator with nitric oxide-donating ability. Cardiovasc Drugs Rev 1994; 12:2-15.

155. Petros A, Lamb G, Leone A et al. Effect of nitric oxide synthase inhibitors in humans with septic shock. Cardiovasc Res 1994; 28:34-39.

CHAPTER 9

THE USE OF EPR SPECTROSCOPY FOR THE IDENTIFICATION OF THE NATURE OF ENDOTHELIUM-DERIVED RELAXING FACTOR

Yann A. Henry

THE NATURE OF ENDOTHELIUM-DERIVED RELAXING FACTOR: A CONTROVERSY

The discovery by Furchgott and Zawadzki in 1980[1] of an obligatory entity derived from endothelial cells necessary for the relaxation of underlying arterial smooth muscle by acetylcholine was followed by the hypothesis made simultaneously by Furchgott and by Ignarro in 1986 that the endothelium-derived relaxing factor (EDRF) was in fact NO.[2-5] This was immediately confirmed by Ignarro and Moncada's groups,[6-9] explaining the prime functional importance of the stimulation of guanylate cyclase by NO, earlier discovered by Murad et al and Ignarro et al.[10-12] Vanhoutte questioned whether it was the end of the quest for EDRF.[13] Read this thrilling part of scientific history recently related by Furchgott![14]

It soon followed that EDRF/NO was derived from L-arginine,[15,16] the same origin as that of NO synthesized in activated macrophages and tumor cells—a simultaneous discovery of several independent laboratories.[17-25] This was anticipated slightly over the discovery that EDRF/NO was also a neuronal messenger.[26]

The quest for EDRF had not quite ended, as immediately a debate was opened, based on functional properties, as to whether NO was *the* EDRF or merely an EDRF among others.[13,27] While many experiments confirmed NO's function as an EDRF,[28-30] several other EDRF candidates were proposed: NO-containing or NO-releasing compounds

Nitric Oxide Research from Chemistry to Biology: EPR Spectroscopy of Nitrosylated Compounds, edited by Yann A. Henry, Annie Guissani and Béatrice Ducastel.
© 1997 R.G. Landes Company.

and intermediary compounds implicated in NO biosynthesis.

Dinitrosyl-Fe^{2+}-cysteine complex (DNIC),[31-33] S-nitrosocysteine, S-nitroso-glutathione and other S-nitrosothiols,[34-37] S-nitrosohemoglobin,[38] nitroxyl NO^-,[39] hydroxylamine,[40] alkylhydroxylamines RNHOH and oximes RNOH,[41] have been proposed to account for the vasorelaxant properties of EDRF.

N^ω-hydroxy-L-arginine, the stable intermediate in the hydroxylation of L-arginine,[42] is released as a stable product in culture supernatant of some mammalian cells (EMT6 and TA3 M2 murine adenocarcinomas) after induction of NOS II (iNOS).[43] It is also transported through cell membranes by the L-arginine Y^+ transporter and is metabolized into NO by NOS itself, by P-450 and other enzymes (hemoproteins and lipoxygenase L1).[44-46] N^ω-hydroxy-L-arginine was proposed and was found to have EDRF-like properties,[47-51] as well as cytostatic and cytotoxic properties.[52]

Since then, still other candidates were proposed as intercellular "messengers" having EDRF-like effects: tetrahydrobiopterin[53] and a smooth muscle- or macrophage-derived relaxing factor (MDRF) identified as a N^ω-hydroxy-L-arginine-nitric oxide adduct.[48-50,54]

The matter has been argued by many authors,[9,31-33,55-61] with pros and cons which are quite outside of our scope in the present book and of our own expertise. We shall first report the open debate on the distance over which NO can diffuse in relationship to the experimental half-life time of EDRF. We shall then outline the role played by EPR spectroscopy in these argumentations about any identification between NO and EDRF.

DIFFUSION OF ENDOGENOUS NITRIC OXIDE

An important point in question is the half-life of EDRF in the bioassay cascade, ranging from 3-5 seconds for many authors,[6,7,28,30] to 30 seconds by others,[35] in agreement with the half-life of authentic

NO and S-nitrosocysteine or S-nitrosocysteamine in the same bioassay. This argument led Feelisch et al to reject DNIC, hydroxylamine and nitroxyl, which had much longer half-lives, as possible EDRF.[30] Another factor in the argumentation lies in the dose dependence (potency) of a given compound to produce the same biological effect than EDRF.

The capacity of NO to act as EDRF relies on its ability to diffuse from the endothelial cell layer to one or several smooth muscle cells before it reaches its protracted molecular target, the heme site of guanylate cyclase. If NO is in itself stable as it does not react with itself, its overall stability depends on many factors, in particular the local concentrations of multiple reactants: O_2, $O_2^{-\bullet}$, cysteine, glutathione, protein thiols, metalloproteins, etc., which often can only be roughly estimated (see chapter 2 for an overview of the reactivity of NO). This has been discussed by many authors with quite variable conclusions.[62-68] Independently of the possibility of the production of cytotoxic $NO_2^\bullet$ and $ONOO^-$, NO at low rates of production or low steady-state concentrations (≤ 1 µmol/l) can diffuse over long distances (~ 100-200 µm) within its bioassay-determined half-life (full range 0.5-50 s), which means a large enough volume of tissue to perform its effect as the EDRF, or for neural signaling.[65,69,70] Although the existence of "NO carriers" is possible, it is not necessary to explain NO/EDRF properties.[65] Cytostatic and cytotoxic effects of NO require similar levels of NO (≥ 1 µmol/l) produced, over much longer time periods (hours to days) after an induction lag (hours), by NOS II in generator cells (smooth muscle cells, hepatocytes or macrophages, for instance), and require also long-range diffusion to reach target cells.[70,71]

NITRIC OXIDE AS AN EDRF DETECTED BY EPR SPECTROSCOPY

Identification of NO to EDRF implies the use of suitable analytical methods such as the spectrophotometrical assay of NO by the oxyhemoglobin reaction or the chemiluminescence reaction with ozone, in order to make kinetic comparisons with the bioassays measuring EDRF effects. EPR spectroscopy in frozen solutions at 77 K appeared to be another possible method to quantitate nitric oxide trapped by deoxyhemoglobin in perfusion liquid or in plasma. Two independent sets of EPR experiments using NO binding to deoxyhemoglobin led to opposite conclusions.[55,56]

Bovine hemoglobin covalently bound to agarose and reduced by dithionite was conditioned into small cylindrical columns adapted for direct EPR measurements at 77 K.[56] Human plasma samples were reduced by dithionite in order to reduce nitrite back to NO and poured over the deoxy-Hb agarose column (Φ_{ext} = 4 mm). Typical EPR spectra of HbNO in the T-state (low affinity) with three-line superhyperfine structure were recorded at liquid nitrogen temperature (77 K) (Fig. 9.1). The method required a 10 ml plasma sample volume, and the spectrum amplitude was linear in the 1-100 nmol range with a 1 nmol threshold. In healthy subjects the venous plasma nitrite level varied in the 0-0.6 µmol/l range. After 5 min of forearm or leg ischemia, plasma nitrite level as detected by the EPR signal of HbNO, increased significantly (Fig. 9.1).[56]

A similar NO trap consisting of agarose-bound hemoglobin, deoxygenated under argon, was found to be effective to trap authentic free NO at a 10 nmol/l concentration (20 ml sample volume) with a stated threshold of 1 nmol/l, much lower than that stated by Wennmalm et al.[55] In contrast, the effluent of canine femoral arteries (EDRF donor tissue) stimulated with acetylcholine, and containing superoxide dismutase and ibuprofen (inhibitor of prostacyclin synthesis), which effectively relaxed the bioassay tissue (U46619 precontracted endothelium-rubbed canine coronary arteries), did not give any detectable HbNO EPR signal when collected over deoxyHb-agarose columns (Fig. 9.2).[55] Control experiments with authentic NO

showed that its corresponding HbNO signal was not attenuated after passage through the donor tissue.

Thus the authors' conclusion was that the EDRF released from canine femoral artery could not be identified as free NO.[55] The authors discussed the possibility that EDRF might be a labile precursor of NO, in equilibrium with small amounts (EPR undetectable) of NO or which could be converted into free NO by the target tissue (vascular smooth muscle or platelets). As mentioned above, such a precursor could be an *S*-nitrosothiol, such as *S*-nitrosoglutathione, glutathione concentrations in cells and intracellular media being very high (1-10 mmol/l).[37] *S*-nitrosohemoglobin carrying NO on residue Cys-β93, could also well be one of these NO donors.[38]

FE(SR)$_2$(NO)$_2$ TERNARY IRON COMPLEXES AS PLAUSIBLE EDRF OR NITRIC OXIDE TRANSPORTERS

We have recalled at the end of chapters 1 and 2 the pioneer results of Vanin et al and of Commoner et al demonstrating by use of EPR spectroscopy the presence of nitrosylated complexes in mammalian cells.[72-76] The EPR signals with two g-values, $g_{//}$ and $g_{\perp}$, at 2.04 and 2.015, respectively, are characteristic of the Fe^I(SR)$_2$(NO)$_2$ complexes with axial symmetry (see chapter 3, Fig. 3.4 and chapter 6).

Ever since the identification of EDRF with NO itself has been proposed, Vanin made the hypothesis that EDRF could be a dinitrosyl-iron complex (DNIC) with low molecular weight thiol ligands.[31] This hypothesis has in fact been substantiated. Quite relevant to the definition of EDRF, it was shown that the dinitrosyl-iron-L-cysteine complex was a potent activator of guanylate cyclase, and it relaxed noradrenaline-precontracted segments of endothelium-denuded rabbit femoral arteries. It was suggested that NO is bound to protein thiols in vascular tissue and may be released from cells by low molecular weight thiols through ligand exchange.[32,33] Furthermore endothelial cells following stimulation

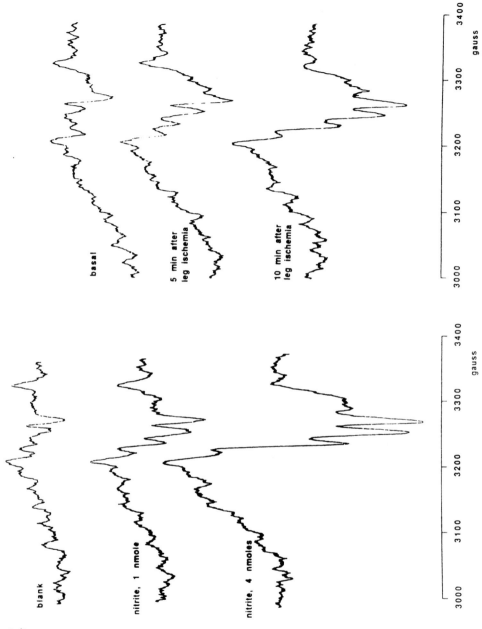

Fig. 9.1. Spin trapping of EDRF/ NO on Hb-agarose columns. Left panel: EPR spectra at 77 K of 10-ml samples of saline (blank, with 1 nmol and 4 nmols nitrite respectively) passed over the Hb-agarose columns. Right panel: EPR spectra (77 K) of 10-ml samples of human plasma passed over the Hb-agarose columns. The upper tracing represents the basal level of nitrite in plasma, and the two lower tracings are from samples collected from the same individual 5 and 10 minutes after the end of a 5-min period of leg ischemia. (Reproduced with permission from: Wennmalm Å, Lanne B, Petersson A-S. Detection of endothelium-derived factor in human plasma in the basal state and following ischemia using electron paramagnetic resonance spectrometry. Anal Biochem 1990: 187:359-363.) Copyright©Academic Press, Inc., Orlando, Florida, USA.

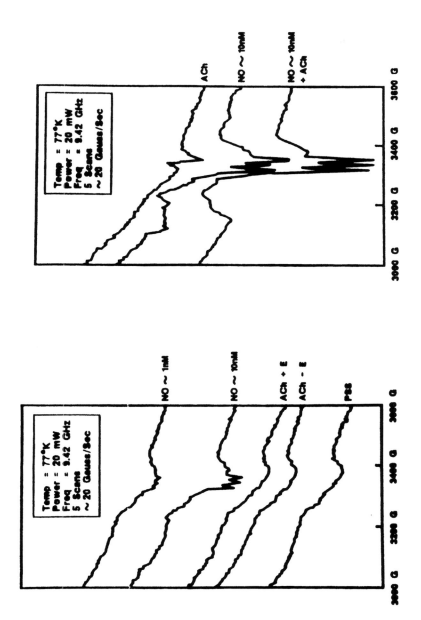

Fig. 9.2. Spin trapping of EDRF/NO on Hb-agarose columns. Left panel: EPR spectra of Hb-agarose exposed to approximately 1 and 10 nmol/l NO and effluents from rubbed (-E) and nonrubbed (+E) (containing EDRF) canine femoral arteries stimulated with acetylcholine (ACh) and physiological saline solution (PSS). Right panel: EPR spectra of Hb-agarose exposed to effluent from intact canine femoral arteries stimulated with acetylcholine (ACh) (containing EDRF), approximately 10 nmol/l NO and approximately 10 nmol/l NO infused through the donor tissue in the presence of ACh (NO + ACh). Passage through the donor tissue did not attenuate the EPR signal due to HbNO. (Reproduced with permission from: Greenberg SS, Wilcox DE, Rubanyi GM. Endothelium-derived relaxing factor released from canine femoral artery by acetylcholine cannot be identified as free nitric oxide by electron paramagnetic resonance spectroscopy. Circ Res 1990; 67:1446-1452.) Copyright © Academic Press, Inc., Orlando, Florida, USA.

of constitutive NOS III by bradykinin or calcium ionophore A23187 form such a ternary complex as detected by EPR spectroscopy of entire frozen cells (Fig. 9.3).[59,77] Such complexes are also found to be released in the culture medium, as detected by EPR spectroscopy.

Activated macrophages could also release such low molecular weight DNIC after activation of NOS II.[78] Formation of these ternary DNIC complexes could be cytoprotective against the toxicity of "free" NO and "free" ferrous ions. Finally it was recently shown that NO could be transferred to low molecular weight DNIC with the usual axial symmetry, from another ternary complex of bovine serum albumin (BSA) having a rhombic symmetry, detectable by EPR with three g-values: 2.046, 2.03 and 2.012,[79] in accordance with the g-values found earlier for the BSA-Fe-NO complex.[80] Upon acidification NO could be reversibly transferred to form S-nitrosothiols.[79] According to Vanin et al the source of iron in such complexes is the so-called labile iron pool and not [Fe-S]-cluster-containing proteins.[81] The influence of the stability and of the oxidation of DNIC on their EDRF properties have recently been discussed.[61]

These results are somewhat contradictory with others, where no EPR signals could be found in culture medium of activated macrophages alone or cocultivated with tumor cells (Drapier J-C & Henry Y, unpublished results; Lepoivre M & Henry Y, unpublished results). Finally no EPR signal was found in L1210 cells cultured in BCG-activated macrophage conditioned medium.[80,82,83] Such identification of EDRF with a ternary dinitrosyl-iron-thiol complex has also to be harmonized with other convincing results demonstrating that EDRF is the NO free radical.[28] Finally, as seen in chapter 6, many different thiol-containing proteins give the same poorly specific EPR signals with $g_{//}$ at 2.04 and $g_{\perp}$ at 2.015, characteristic of any $Fe(SR)_2(NO)_2$ complex.

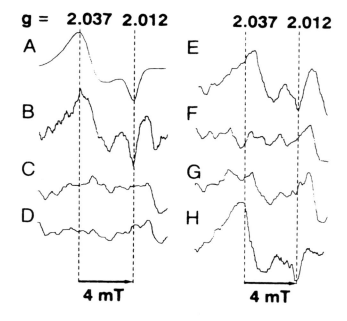

Fig. 9.3. EPR spectra of endothelial cells and cell media. A) dinitrosyl iron complex of BSA (10 μmol/l); B) porcine aortic endothelial cells stimulated by A23187 (0.1 μmol/l); C) unstimulated cells; D) cells stimulated by A23187 in the presence of N^{G}nitro-L-arginine (1 mmol/l); E) endothelial cells stimulated by bradykinin (1 μmol/l); F) medium from E); G) cells stimulated by bradykinin in the presence of N-acetyl-L-cysteine (10 mmol/l) and albumin (0.3 mmol/l); H) medium from G). The g-values of the dinitrosyl iron complex are indicated. (Reproduced with permission from: Mülsch A, Mordvintcev P, Vanin AF et al. Formation and release of dinitrosyl iron complexes by endothelial cells. Biochem Biophys Res Commun 1993; 196:1303-1308.) Copyright © Academic Press, Inc., Orlando, Florida, USA.

Attempts to reconcile these apparently contradictory results should certainly begin with precise assessments of total and "free" iron, firstly in culture mediums, secondly in cells, to better define the so-called labile iron pool, and of the presence of serum transferrin and metallothionein. Other measurements to be cautiously made are the concentration of low molecular weight thiols and their redox states in media in the course of cell cultures. The relevance of these measurements stem also from the fact that free cations: Cu^{2+}, Fe^{2+} or Fe^{3+} catalyze the *S*-nitrosothiols' decomposition.[84,85]

CONCLUSION

The various assessments of EDRF nature as paramagnetic NO or paramagnetic $Fe(SR)_2(NO)_2$ ternary compound by EPR spectroscopy are not all clear cut. The main analytical defect of the method performed on frozen solutions is the lack of time resolution. This can be overcome by some other NO spin-trapping methods (see an overview in chapter 13).[86-88]

We have mentioned at the beginning of this chapter the vast possibilities of sophisticated electrochemical porphyrin microsensors of NO, as developed by Malinski et al,[69,70,89,90] by Devynck et al,[91-93] Cespuglio et al[94,95] and others, methods well adapted to make direct correlations between biological assays and NO quantitation. A recent direct measurement of nitric oxide in superficial hand veins of healthy volunteers has provided further evidence that EDRF is NO.[96] An electrochemical porphyrinic microsensor was inserted into a hand vein, and the vessel was stimulated with acetylcholine or bradykinin. No basal release of NO was detected. Upon infusion of bradykinin or acetylcholine, NO levels as high as 250 nmol/l were measured,[96] which is slightly lower but in the same concentration range as measured in vitro upon animal prepared arteries or above cell culture layers.[69,70,89]

REFERENCES

1. Furchgott RF, Zawadzki JV. The obligatory role of endothelial cells in the relaxation of arterial smooth muscle by acetylcholine. Nature 1980; 288:373-376.

2. Furchgott RF. Studies on relaxation of rabbit aorta by sodium nitrite: the basis for the proposal that the acid-activatable inhibitory factor from bovine retractor penis is inorganic nitrite and the endothelial-derived relaxing factor is nitric oxide. In: Vanhoutte PM, ed. Vasodilatation: Vascular Smooth Muscle, Peptides, Autonomic Nerves, and Endothelium. New York, Raven Press, 1988:401-414.

3. Khan MT, Furchgott RF. Additional evidence that endothelium-derived relaxing factor is nitric oxide. In: Rand MJ, Raper C, eds. Pharmacology. Amsterdam, Elsevier, 1987:341-344.

4. Ignarro LJ, Wood KS, Byrns RE. Pharmacological and biochemical properties of endothelium-derived relaxing factor (EDRF): evidence that EDRF is closely related to nitric oxide (NO) radical. Circulation 1986; 74:II-287 (Abstr).

5. Ignarro LJ, Byrns RE, Wood KS. Biochemical and pharmacological properties of endothelium-derived relaxing factor and its similarity to nitric oxide radical. In: Vanhoutte PM, ed. Vasodilatation: Vascular Smooth Muscle, Peptides, Autonomic Nerves, and Endothelium. New York, Raven Press, 1988:427-435.

6. Ignarro LJ, Buga GM, Wood KS et al. Endothelium-derived relaxing factor produced and released from artery and vein is nitric oxide. Proc Natl Acad Sci USA 1987; 84:9265-9269.

7. Palmer RMJ, Ferrige AG, Moncada S. Nitric oxide release accounts for the biological activity of endothelium-derived relaxing factor. Nature 1987; 327:524-526.

8. Radomski MW, Palmer RMJ, Moncada S. The role of nitric oxide and cGMP in platelets adhesion to vascular endothelium. Biochem Biophys Res Commun 1987; 148:1482-1489.

9. Moncada S, Radomski MW, Palmer RMJ. Endothelium-derived relaxing factor. Identification as nitric oxide and role in the control of vascular tone and platelet function. Biochem Pharmacol 1988; 37:2495-2501.

10. Arnold WP, Mittal CK, Katsuki S et al. Nitric oxide activates guanylate cyclase and increases guanosine3':5'-cyclic monophosphate levels in various tissue preparations. Proc Natl Acad Sci USA 1977; 74:3203-3207.

11. Murad F, Mittal CK, Arnold WP et al. Guanylate cyclase: activation by azide, nitro compounds, nitric oxide, and hydroxyl radical and inhibition by hemoglobin and myoglobin. Adv Cyclic Nucleotide Res 1978; 9:145-158.

12. Gruetter CA, Barry BK, McNamara DB et al. Relaxation of bovine coronary artery and activation of coronary arterial guanylate cyclase by nitric oxide, nitroprusside and a carcinogenic nitrosoamine. J Cyclic Nucleotide Res 1979; 5:211-224.

13. Vanhoutte PM. The end of the quest? Nature 1987; 327:459-460.

14. Furchgott RF. A research trail over half a century. Annu Rev Pharmacol Toxicol 1995; 35:1-27.

15. Palmer RMJ, Rees DD, Ashton DS et al. L-arginine is the physiological precursor for the formation of nitric oxide in endothelium dependent relaxation. Biochem Biophys Res Commun 1988; 153:1251-1256.

16. Schmidt HHHW, Nau H, Wittfoht W et al. Arginine is a physiological precursor of endothelium-derived nitric oxide. Eur J Pharmacol 1988; 154:213-216.

17. Hibbs JB, Taintor RR, Vavrin Z. Macrophage cytotoxicity: role for L-arginine deiminase and imino nitrogen oxidation to nitrite. Science 1987; 235:473-476.

18. Hibbs JB, Vavrin Z, Taintor RR. L-arginine is required for expression of the activated macrophage effector mechanism causing selective metabolic inhibition in target cells. J Immunol 1987; 138:550-565.

19. Stuehr DJ, Marletta MA. Mammalian nitrate biosynthesis: mouse macrophages produce nitrite and nitrate in response to *Escherichia coli* lipopolysaccharide. Proc Natl Acad Sci USA 1985; 82:7738-7742.

20. Stuehr DJ, Marletta MA. Induction of nitrite/nitrate synthesis in murine macrophages by BCG infection, lymphokines, or interferon-γ. J Immunol 1987; 139:518-525.

21. Iyengar R, Stuehr DJ, Marletta MA. Macrophage synthesis of nitrite, nitrate, and N-nitrosamines: precursors and role of the respiratoty burst. Proc Natl Acad Sci USA 1987; 84:6369-6373.

22. Marletta MA. Mammalian synthesis of nitrite, nitrate, nitric oxide, and N-nitrosating agents. Chem Res Toxicol 1988; 1:249-257.

23. Marletta MA, Yoon PS, Iyengar R et al. Macrophage oxidation of L-arginine to nitrite and nitrate: nitric oxide is an intermediate. Biochemistry 1988; 27:8706-8711.

24. Stuehr DJ, Nathan CF. Nitric oxide. A macrophage product responsible for cytostasis and respiratory inhibition in tumor target cells. J Exp Med 1989; 169:1543-1555.

25. Stuehr DJ, Gross SS, Sakuma I et al. Activated murine macrophages secrete a metabolite of arginine with the bioactivity of endothelium-derived relaxing factor and the chemical reactivity of nitric oxide. J Exp Med 1989; 169:1011-1020.

26. Garthwaite J, Charles SL, Chess-Williams R. Endothelium-derived relaxing factor release on activation of NMDA receptors suggests role as intercellular messenger in the brain. Nature 1988; 336:385-388.

27. Furchgott RF, Vanhoutte PM. Endothelium-derived relaxing and contracting factors. FASEB J 1989; 3:2007-2018.

28. Kelm M, Feelisch M, Spahr R et al. Quantitative and kinetic characterisation of nitric oxide and EDRF released from cultured endothelial cells. Biochem Biophys Res Commun 1988; 154:236-244.

29. Kelm M, Schrader J. Control of coronary vascular tone by nitric oxide. Circ Res 1990; 66:1561-1575.

30. Feelisch M, te Poel M, Zamora R et al. Understanding the controversy over the identity of EDRF. Nature 1994; 368:62-65.

31. Vanin AF. Endothelium-derived relaxing factor is a nitrosyl iron complex with thiol ligands. FEBS Lett 1991; 289:1-3.

32. Mülsch A, Mordvintcev P, Vanin AF et al. The potent vasodilating and guanylyl cyclase activating dinitrosyl-iron(II) complex is stored in a protein-bound form in vascular tissue and is released by thiols. FEBS Lett 1991; 294:252-256.

33. Vedernikov YP, Mordvintcev PI, Malenkova IV et al. Similarity between the vasorelaxing activity of dinitrosyl iron cysteine complexes and endothelium-derived relaxing factor. Eur J Pharmacol 1992; 211:313-317.

34. Rubanyi GM, Johns A, Harrison D et al. Evidence that endothelium-derived relaxing factor may be identical with an *S*-nitrosothiol and not with free nitric oxide. Circulation 1988; 80 (Suppl II):Abstr II-281.

35. Myers PR, Minor RL, Guerra R et al. Vasorelaxant properties of the endothelium-derived relaxing factor more closely resemble *S*-nitrosocysteine than nitric oxide. Nature 1990; 345:161-163.

36. Kowaluk E, Fung H-L. Spontaneous liberation of nitric oxide cannot account for *in vitro* vascular relaxation by *S*-nitrosothiols. J Pharmacol Exp Ther 1990; 255:1256-1264.

37. Hogg N, Singh RJ, Kalyanaraman B. The role of glutathione in the transport and catabolism of nitric oxide. FEBS Lett 1996; 382:223-228.

38. Jia L, Bonaventura C, Bonaventura J et al. *S*-nitrosohaemoglobin: a dynamic activity of blood involved in vascular control. Nature 1996; 380:221-226.

39. Fukuto JM, Chiang K, Hszieh R et al. The pharmacological activity of nitroxyl: a potent vasodilator with activity similar to nitric oxide and/or endothelium-derived relaxing factor. J Pharmacol Exp Ther 1992; 263:546-551.

40. DeMaster EG, Raij L, Archer SL et al. Hydroxylamine is a vasorelaxant and a possible intermediate in the oxidative conversion of L-arginine to nitric oxide. Biochem Biophys Res Commun 1989; 163:527-533.

41. Thomas G, Ramwell PW. Vascular relaxation mediated by hydroxylamines and oximes: their conversion to nitrites and mechanism of endothelium dependent vascular relaxation. Biochem Biophys Res Commun 1989; 164:889-893.

42. Stuehr DJ, Kwon NS, Nathan CF et al. N^{ω}-hydroxy-L-arginine is an intermediate in the biosynthesis of nitric oxide from L-arginine. J Biol Chem 1991; 266:6259-6263.

43. Chenais B, Yapo A, Lepoivre M et al. High-performance liquid chromatographic analysis of the unusual pathway of oxidation of L-arginine to citrulline and nitric oxide in mammalian cells. J Chromatogr 1991; 539:433-441.

44. Boucher J-L, Genet A, Vadon S et al. Cytochrome P450 catalyzes the oxidation of N^{ω}-hydroxy-L-arginine by NADPH and O_2 to nitric oxide and citrulline. Biochem Biophys Res Commun 1992; 187:880-886.

45. Boucher J-L, Chopard C, Vadon S et al. Nitric oxide formation by oxidation of N^{ω}-hydroxy-L-arginine by linoleic acid hydroperoxide catalyzed by soybean lipoxygenase L1. Endothelium 1993; 1:S17.

46. Renaud J-P, Boucher J-L, Vadon S et al. Particular hability of liver P450s3A to catalyze the oxidation of N^{ω}-hydroxy-L-arginine to citrulline and nitrogen oxides and occurence in NO synthases of a sequence very similar to the heme-binding sequence in P450s. Biochem Biophys Res Commun 1993; 192:53-60.

47. Zembowicz A, Hecker M, Macarthur H et al. Nitric oxide and another potent vasodilator are formed from N^G-hydroxy-L-arginine by cultured endothelial cells. Proc Natl Acad Sci USA 1991; 88:11172-11176.

48. Zembowicz A, Swierkosz TA, Southan GJ et al. Mechanisms of the endothelium-dependent relaxation induced by N^G-hydroxy-L-arginine. Cardiovasc Pharmacol 1992; 20:S57-S59.

49. Zembowicz A, Swierkosz TA, Southan GJ et al. Potentiation of the vasorelaxant activity of nitric oxide by hydroxyguanidine: implications for the nature of endothelium-derived relaxing factor. Br J Pharmacol 1992; 107:1001-1007.

50. Zembowicz A, Chlopicki S, Radziszewski W et al. N^G-hydroxy-L-arginine and hydroxyguanidine potentiate the biological activity of endothelium-derived relaxing factor released from the rabbit aorta. Biochem Biophys Res Commun 1992; 189:711-716.

51. Schott CA, Bogen CM, Vetrovsky P et al. Exogenous N^G-hydroxy-L-arginine causes nitrite production in vascular smooth muscle cells in the absence of nitric oxide synthase activity. FEBS Lett 1994; 341:203-207.

52. Chenais B, Yapo A, Lepoivre M et al. N^ω-hydroxy-L-arginine, a reactional intermediate in nitric oxide biosynthesis, induces cytostasis in human and murine tumor cells. Biochem Biophys Res Commun 1993; 196:1558-1565.

53. Schaffner A, Blau N, Schneemann M et al. Tetrahydrobiopterin as another EDRF in man. Biochem Biophys Res Commun 1994; 205:516-523.

54. Hecker M, Boese M, Schini-Kerth VB et al. Characterization of the stable L-arginine-derived relaxing factor released from cytokine-stimulated vascular smooth muscle cells as an N^G-hydroxy-L-arginine-nitric oxide adduct. Proc Natl Acad Sci USA 1995; 92:4671-4675.

55. Greenberg SS, Wilcox DE, Rubanyi GM. Endothelium-derived relaxing factor released from canine femoral artery by acethycholine cannot be identified as free nitric oxide by electron paramagnetic resonance spectroscopy. Circ Res 1990; 67:1446-1452.

56. Wennmalm Å, Lanne B, Petersson A-S. Detection of endothelium-derived factor in human plasma in the basal state and following ischemia using electron paramagnetic resonance spectrometry. Anal Biochem 1990; 187:359-363.

57. Ignarro LJ. Biosynthesis and metabolism of endothelium-derived nitric oxide. Annu Rev Pharmacol Toxicol 1990; 30:535-560.

58. Rosenblum W. Endothelium-derived relaxing factor in brain blood vessels is not nitric oxide. Stroke 1992; 23:1527-1532.

59. Busse R, Mülsch A, Fleming I et al. Mechanisms of nitric oxide release from the vascular endothelium. Circulation 1993; 87[suppl V]:V-18-V-25.

60. Zamora Pino R, Feelisch M. Bioassay discrimination between nitric oxide (NO•) and nitroxyl (NO-) using L-cysteine. Biochem Biophys Res Commun 1994; 201:54-62.

61. Vanin AF. On the stability of the dinitrosyl-iron-cysteine complex, a candidate for the endothelium-derived relaxation factor. Biochemistry (Moscow) 1995; 60:225-230.

62. Saran M, Michel C, Bors W. Reaction of NO with O_2^-. Implications for the action of endothelium-derived relaxing factor (EDRF). Free Rad Res Comms 1990; 10:221-226.

63. Lancaster JR. Simulation of the diffusion and reaction of endogenously produced nitric oxide. Proc Natl Acad Sci USA 1994; 91:8137-8141.

64. Saran M, Bors W. Signalling by $O_2^{-•}$ and NO•: how far can either radical, or any specific reaction product, transmit a message under in vivo conditions? Chem-Biol Interact 1994; 90:35-45.

65. Wood J, Garthwaite J. Models of the diffusional spread of nitric oxide: implications for neural nitric oxide signalling and its pharmacological properties. Neuropharmacol 1994; 33:1235-1244.

66. Vanderkooi JM, Wright WW, Erecinska M. Nitric oxide diffusion coefficients in solutions, proteins and membranes determined by phosphorescence. Biochim Biophys Acta 1994; 1207:249-254.

67. Garthwaite J, Boulton CL. Nitric oxide signaling in the central nervous system. Annu Rev Physiol 1995; 57:683-706.

68. Squadrito GL, Pryor WA. The formation of peroxynitrite in vivo from nitric oxide and superoxide. Chem-Biol Interact 1995; 96:203-206.

69. Malinski T, Taha Z, Grunfeld S et al. Diffusion of nitric oxide in the aorta wall monitored in situ by porphyrinic microsensors. Biochem Biophys Res Commun 1993; 193:1076-1082.

70. Malinski T, Kapturczak M, Dayharsh et al. Nitric oxide synthase activity in genetic hypertension. Biochem Biophys Res Commun 1993; 194:654-658.

71. Laurent M, Lepoivre M, Tenu J-P. Kinetic modelling of the nitric oxide gradient generated in vitro by adherent cells expressing inducible nitric oxide synthase. Biochem J 1996; 314:109-113.

72. Vanin AF. Identification of divalent iron complexes with cysteine in biological systems by the EPR method. Biokhimiya 1967; 32:277-282 (English translation 228-232).

73. Vanin AF, Vakhnina LV, Chetverikov AG.

Nature of the EPR signals of a new type found in cancer tissues. Biofizika 1970; 15:1044-1051 (English translation 1082-1089).

74. Vithayathil AJ, Ternberg JL, Commoner B. Changes in electron spin resonance signals of rat liver during chemical carcinogenesis. Nature 1965; 207:1246-1249.

75. Woolum JC, Tiezzi E, Commoner B. Electron spin resonance of iron-nitric oxide complexes with amino acids, peptides and proteins. Biochim Biophys Acta 1968; 160:311-320.

76. Woolum JC, Commoner B. Isolation and identification of a paramagnetic complex from the livers of carcinogen-treated rats. Biochim Biophys Acta 1970; 201:131-140.

77. Mülsch A, Mordvintcev P, Vanin AF et al. Formation and release of dinitrosyl iron complexes by endothelial cells. Biochem Biophys Res Commun 1993; 196:1303-1308.

78. Vanin AF, Mordvintcev PI, Hauschildt S et al. The relationship between L-arginine-dependent nitric oxide synthesis, nitrite release and dinitrosyl-iron complex formation by activated macrophages. Biochim Biophys Acta 1993; 1177:37-42.

79. Vanin AF, Malenkova IV, Mordvintcev PI et al. Dinitrosyl iron complexes with thiol-containing ligands and their reversible conversion intro nitrosothiols. Biokhimya 1993; 58:1094-1103 (English Translation 773-779).

80. Drapier J-C, Pellat C, Henry Y. Generation of EPR-detectable nitrosyl-iron complexes in tumor target cells cocultured with activated macrophages. J Biol Chem 1991; 266:10162-10167.

81. Vanin AF, Men'shikov GB, Moroz IA et al. The source of non-heme iron that binds nitric oxide in cultivated macrophages. Biochim Biophys Acta 1992; 1135:275-279.

82. Pellat C, Henry Y, Drapier J-C. Detection of nitrosyl-iron complexes in tumor target cells after coculture with activated macrophages. In: Melzer MS, Mantovani A, eds. Cellular and Cytokine Networks in Tissue Immunity. Wiley-Liss, 1991: 229-234.

83. Drapier J-C, Pellat C, Henry Y. Characterization of the nitrosyl-iron complexes generated in tumour cells after co-culture with activated macrophages. In: Moncada S, Marletta MA, Hibbs JB et al, eds. The Biology of Nitric Oxide. London, UK: Portland Press, 1992: 72-76.

84. McAninly J, Williams DLH, Askew SC et al. Metal ion catalysis in nitrosothiols (RSNO) decomposition. J Chem Soc Chem Commun 1993:1758-1759.

85. Askew SC, Barnett DJ, McAninly J et al. Catalysis by Cu^{2+} of nitric oxide release from S-nitrosothiols (RSNO). J Chem Soc Perkin Trans 2 1995:741-745.

86. Misra HP, Sata T, Kubota E et al. ESR spectroscopic studies of endothelial-dependent relaxation factor in guinea pig pulmonary artery. J Vascul Med Biol 1989; 1:189 (Abstr).

87. Forray C, Arroyo CM, El-Fakahany E et al. L-arginine related spin adducts generated during muscarinic receptor-mediated activation of guanylate cyclase. Arch Int Pharm Ther 1990; 305:Abs 42, 245.

88. Arroyo CM, Forray C, El-Fakahany E et al. Receptor-mediated generation of an EDRF-like intermediate in a neuronal cell line detected by spin trapping techniques. Biochem Biophys Res Commun 1990; 170:1177-1183.

89. Kanai AJ, Strauss HC, Truskey GA et al. Shear stress induces ATP-dependent transient nitric oxide release from vascular endothelial cells, measured directly with a porphyrinic microsensor. Circ Res 1995; 77:284-293.

90. Malinski T, Taha Z. Nitric oxide release from a single cell measured in situ by a porphyrinic-based microsensor. Nature 1992; 358:676-678.

91. Bedioui F, Trévin S, Devynck J. The use of gold electrodes in the electrochemical detection of nitric oxide in aqueous solution. J Electroanal Chem 1994; 377:295-298.

92. Lantoine F, Trévin S, Bedioui F et al. Selective and sensitive electrochemical measurement of nitric oxide in aqueous solution: discussion and new results. J Electroanal Chem 1995; 392:85-89.

93. Lantoine F, Brunet A, Bedioui F et al. Direct measurement of nitric oxide production in platelets: relationship with cytosolic Ca^{2+}

concentration. Biochem Biophys Res Commun 1995; 215:842-848.

94. Cespuglio R, Burlet S, Marinesco S et al. NO voltammetric detection in the rat brain. Variations of the signal throughout the sleep-waking cycle. C R Acad Sci Paris 1996; 319:191-200.

95. Buguet A, Burlet S, Auzelle F et al. Dual intervention of NO in experimental African trypanosomiasis. C R Acad Sci Paris 1996; 319:201-207.

96. Vallance P, Patton S, Bhagat K et al. Direct measurement of nitric oxide in human beings. Lancet 1995; 346:153-154.

ENZYMATIC TARGETS OF NITRIC OXIDE AS DETECTED BY EPR SPECTROSCOPY WITHIN MAMMAL CELLS

Yann A. Henry, Béatrice Ducastel and Annie Guissani

INTRODUCTION

We concluded in chapter 6 that *all metalloproteins are potential targets of NO*, as proven by EPR spectroscopy, which is too unspecific to be a really interesting statement. We now turn to some more biologically relevant aspects of NO binding. Its production from L-arginine catalyzed by the inducible NO-synthase (iNOS, NOS II) in murine macrophages and their tumoral target cells was the simplest case to characterize at the molecular level by EPR spectroscopy. [FeS]-containing proteins and ribonucleotide reductase responsible for basic vital cellular functions such as mitochondrial respiration and DNA replication were the earliest enzymatic targets characterized in whole cultured mammal cells.

Historically the discovery of the interaction of NO with [FeS]-containing proteins was preceded by several years the demonstration of the activation of soluble guanylate cyclase (sGC) by NO. Guanylate cyclase was the first discovered target and remains so far, with the iron-regulatory protein (see below), one of the only two proven positive "receptors" for NO. It is certainly the most important for its roles in intercellular regulations, as its activation leads to an increase of guanosine cyclic 3',5'-monophosphate (cGMP) levels in target cells. cGMP acts as a second messenger regulating various protein kinases and ion channels. Its intracellular concentration is regulated by the various GC isozymes and by cGMP-degrading enzymes, cGMP-phosphodiesterases. The two

Nitric Oxide Research from Chemistry to Biology: EPR Spectroscopy of Nitrosylated Compounds, edited by Yann A. Henry, Annie Guissani and Béatrice Ducastel.
© 1997 R.G. Landes Company.

main biological effects of increased cGMP levels are the relaxation of vascular smooth muscle and the inhibition of platelet aggregation (reviewed in refs. 1-3).

The attempts at EPR characterization of these cellular phenomena related to *NO interaction with specific metalloproteins, within cultured mammal cells*, followed rapidly. Two French groups led respectively by Dr. Jean-Claude Drapier (INSERM U 365, Institut Curie, Paris) and by Dr. Michel Lepoivre (URA 1116 CNRS, Institut de Biochimie, Orsay) were particularly active and successful in this field; it is through our association with them for the EPR spectroscopy experiments that we are able to write the present chapter. Their results and those of many other groups are summarized and discussed here. Except for the case of guanylate cyclase which follows, the structural and spectral characteristics of NO binding to these metalloproteins in a pure state in test tubes have been described in chapter 6.

SOLUBLE GUANYLATE CYCLASE

Guanylate cyclases (guanylyl cyclase and guanosine 5'-triphosphate (GTP) pyrophosphate-lyase (cyclizing); EC 4.6.1.2) are a family of enzymes that catalyze the cyclization of GTP to cGMP, subdivided into two very different classes: particulate GC which possesses receptors for different peptides, natriuretic peptides and guanylins (reviewed in ref. 4), and soluble GC which contains heme and recognizes NO (reviewed in refs. 1-3).

ACTIVATION OF SOLUBLE GUANYLATE CYCLASE BY NITRIC OXIDE

Discovered in 1975, the activation of sGC by nitric oxide, nitroso derivatives, nitrite, and organic nitrates has received wide attention for twenty years,[5-13] but its importance has only been fully understood with the later discovery of NO biosynthesis by the constitutive NO-synthases (nNOS or NOS I, eNOS or NOS III). This was a highly rewarding "loan and investment" (NO activates sGC, 1975) and "cash repayment" (EDRF is L-arginine-derived NO, 1987) process.

Activation of sGC by carbon monoxide and hydroxyl radicals was also suggested for the enzyme in vitro and in cellular systems.[7,8,14-18] The activation of sGC by CO increases the level of cGMP by a small extent (1.4- to 4-fold) while that by NO could attain 30-fold with an increase of V_{max} reaching 100- to 400-fold on the purest enzyme preparations,[19-21] so the physiological implications of these activations could be completely different, a true activation of sGC for NO and perhaps a cross-regulation with heme oxygenase in the case of CO (see chapter 14).[17,18,22,23]

NITRIC OXIDE BINDING TO SOLUBLE GUANYLATE CYCLASE

sGC is a heterodimer consisting of one α subunit of a mass ranging from 73 to 88 kDa and one β subunit with a mass ranging from 70-76 kDa.[20] Two α and two β subunits (α1, α2, β1 and β2) have been identified in rat and bovine tissues.[24] It contains heme *b*.[25] The heme *b* stoichiometry, originally found to be one per heterodimer,[25,26] has been proposed recently to be two per heterodimer in the native enzyme purified by the latest methods.[20] sGC is certainly a most labile hemoprotein, but the discovery of its activation by protoporphyrin IX, heme *b* and nitrosyl heme rapidly followed that of its activation by NO itself.[1,2,27-32]

It was rapidly suggested, in particular by EPR studies, that activation of sGC required an exchange of the paramagnetic heme(Fe^{II})-NO entity rather than NO itself. The donor enzymes could be nitrosylated hemoglobin or catalase.[31,33-35]

It has only been in the last few years that modifications by Stone and Marletta of earlier purification procedures have allowed the preparation of the enzyme in a fully native form and several spectroscopic characterizations of the heme site.[19-21,25,26,28,32] From the UV-visible spectra, the heme *b* site of sGC seems to be quite unique and wholly different from that of hemoglobin or most of known hemoproteins. In the ferrous state of sGC, the heme is a high-spin pentacoordinate with probably a histidine residue as the axial ligand. Quite exceptionally the ferrous heme does not bind O_2 and is not autoxidizable. It readily binds NO to form a pentacoordinate complex or CO to form a hexacoordinate complex. Binding of NO seems therefore to break the imidazole-iron bond *trans* of NO.[26] This finding would make consistent previous models of sGC activation by NO.[36-38] The existence of a pentacoordinate heme *b*-NO complex was fully confirmed by resonance Raman (RR) spectroscopy,[39] and by EPR spectroscopy of

the ferrous-[^{14}N/^{15}N]nitrosyl complexes of sGC. Their EPR spectra have essentially three-line SHF structure, similar to those of nitrosyl-cytochrome *c'* (see chapter 7), HbNO in the T-state (see chapter 4), P-420-NO or PGHS-NO (see chapter 6) (Fig. 10.1).[19] There could be a small hexacoordinate heme-NO contribution as indicated by the small dip around g = 1.98.[19]

As indicated by the UV-visible spectra, the nitrosyl heme complex of sGC is formed during the in vitro activation process of sGC in the presence of GTP under

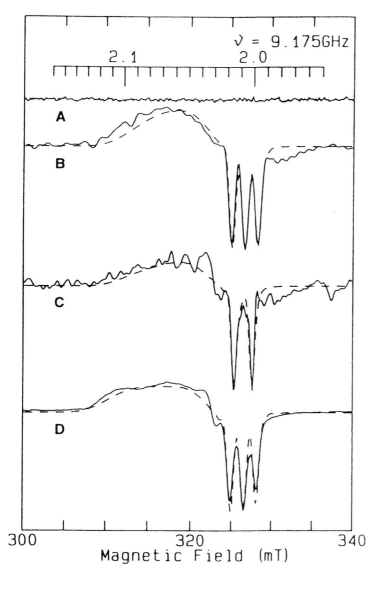

Fig. 10.1. EPR spectra of sGC at 25 K (—) original spectrum, (- - - -) computer simulation. A: sGC (3 µmol/l) under nitrogen; B: sGC under 2% [^{14}N]NO in nitrogen; C: sGC under 2% [^{15}N]NO in nitrogen; D: pentacoordinate nitrosyl heme model compound in SDS micelles. (Reproduced with permission from: Stone JR, Sands RH, Dunham WR et al. Electron paramagnetic resonance spectral evidence for the formation of a pentacoordinated nitrosyl-heme complex on soluble guanylate cyclase. Biochem Biophys Res Commun 1995; 207:572-577.) Copyright © Academic Press, Inc, Orlando, Florida, USA.

low concentration of NO (~ 8 μmol/l NO in solution in the presence of 0.4 μmol/l sGC).[19] In the ferric form, following oxidation by ferricyanide, sGC heme *b* is a high-spin pentacoordinate as indicated by UV-visible and EPR spectroscopy, similarly to cytochrome *c'*.[26,40] Also like cytochrome *c'*, it binds cyanide with a very low affinity.[26,40] Ferric sGC binds azide weakly to form a most unusual high-spin complex and does not bind fluoride.[40] Other spectroscopic (UV-visible and MCD) studies of heme-reconstituted sGC led to similar results, with differences in some details.[41] One histidine residue which could be assigned to heme *b* binding is His-105 in the bovine β1 subunit; a mutation at this site yields an enzyme able to catalyze the conversion of GTP to cGMP, but which does not respond to NO activation.[24]

A still more refined preparation procedure of sGC led Stone and Marletta to propose a native heme stoichiometry of two per heterodimer, one per subunit.[20] Spectrally only one type of heme *b* is observed, suggesting a similar environment in homologous subunits and a conserved heme binding site. Known sequence alignments suggest that the histidine residue that coordinates the heme iron is α290/β220 or α407/β346 in the bovine α1β1 isoform of sGC rather than His-105β1.[20] Studies of CO binding and CO dissociation confirmed a weak (3- to 4-fold) (or null) activation process of purified sGC by CO, by a still unknown mechanism.[21,41,42]

In a recent article, Stone and Marletta have reported the rates of binding of NO to reduced heme of sGC measured by stopped-flow spectrophotometry.[43] NO first binds reversibly to heme to form a hexacoordinate complex, which converts by two different slower reversible processes into a pentacoordinate nitrosylated complex. An upper limit for the equilibrium dissociation constant of NO for the activation of the enzyme was estimated to be 0.25 μmol/l, four orders of magnitude lower than the mean dissociation of NO from hemoglobin.[43]

The tests of the previously proposed model of imidazole-based activation of sGC

by NO or CO seem to be in general agreement with most of these recent results, but still remain controversial on some points.[21,36-38,41,42]

The UV-visible, MCD, RR and EPR spectroscopic studies of the purified enzyme are very recent, and no attempt has yet been reported of any EPR detection of nitrosylated guanylate cyclase within cultured cells.

If the activation of sGC by NO itself is universally accepted and certainly of prime importance, a recent finding that NO synthesis in hepatocytes provokes reversible loss and irreversible degradation of enzyme-bound heme has to be substantiated.[44] Its implication concerning the heme binding to sGC, or its dissociation from sGC, has to be ascertained and extended to other cell types. A better understanding of the enzyme activation would also be required to explain such facts as the reduced response to NO of sGC in the superior temporal cortex of patients with Alzheimer's disease.[45]

[FeS] CLUSTER-CONTAINING ENZYME

We have mentioned in chapter 6 the various cellular functions performed by the [FeS]-containing proteins in bacteria, plants and mammal cells. In many instances these functions have been shown to be inhibited by NO through its binding to the [FeS] cluster demonstrated in vitro.

ACTIVATION OF MACROPHAGES

The effects of NOS II (iNOS) induction following the activation of murine macrophages by *E. coli* lipopolysaccharide (LPS), by *Bacillus Calmette-Guérin* (BCG) or by cytokines, e.g., γ-interferon (IFNγ) and tumor necrosis factor-α (TNFα), are multiple, following a time sequence over several hours: an early reversible impairment of DNA replication through the reversible inhibition of ribonucleotide reductase ($\tau_{1/2}$ ~ 4-7 hr depending on cell density), reversible inhibition of mitochondrial aconitase ($\tau_{1/2}$ ~ 4-5 hr), inhibition of mitochondrial respiration through the inhibition of complex I ($\tau_{1/2} \sim 8$ hr) and complex II ($\tau_{1/2} \sim 14$ hr), massive loss of intracellular iron ($\tau_{1/2} \sim 14$ hr after a lag of 4 hr), etc. They occur both in "generator" macrophages and in target cells (such as guinea pig L10 hepatoma and mouse L1210 leukemia cells) cocultivated with activated macrophages.[46-62]

EPR spectroscopy analysis of whole mouse peritoneal macrophages or cultured continuous cell line RAW 264.7, activated by LPS/IFNγ and frozen "live" in liquid nitrogen (77 K), provided some more evidence for the reversible formation of nitrosylated complexes of [FeS] cluster-containing enzymes: mitochondrial aconitase (EC 4.2.1.3) in the Krebs cycle, complex I (NADH:ubiquinone oxidoreductase; EC 1.6.5.3) and complex II (succinate: ubiquinone oxidoreductase; EC 1.3.5.1) of the mitochondrial respiratory chain, which were previously shown to be inhibited by L-arginine-derived NO.[54,63-66] These nitrosyl-iron-thiolate complexes are characterized by $g_\perp = 2.041$ and $g_{//} = 2.015$ axial EPR signal, often mentioned as the $g_{av} = 2.03$ or 2.04 signal (see chapter 6) (Figs. 10.2 and 10.3).

This coordination can explain their reversible inhibition discovered earlier in LPS, BCG or IFNγ-activated macrophages. A demonstrative counter-experiment was performed with the P388D1 cell line that cannot be induced by IFNγ or LPS to produce NO. The same EPR signal was detected in digitonin-permeabilized macrophages not stimulated by IFNγ, in the presence of NO resulting from a mixture of nitrite and ascorbate.[63,64,66] The NO coordination may impair the substrate/product binding to aconitase or affect the active [4Fe-4S] to inactive [3Fe-4S] equilibrium of the enzyme.[67,68] The observed EPR properties (g-values, linewidth, microwave-power saturation properties) allowed one to discriminate the signal obtained with [FeS]-containing proteins such as [2Fe-2S]-containing *Spirulina platensis* and *Porphyra umbilicalis* ferredoxins, from those of NO complexes of ferritin and transferrin (see below).[69,70]

PARACRINE SYSTEMS OF MACROPHAGES AND TARGET CELLS IN COCULTURE

The intercellular communication between the NO-generator cell (activated murine macrophage) and a target tumoral cell (murine tumor cell L1210) was finely demonstrated in coculture experiments followed by EPR analysis of both types of cells separated after coculture.[69,71] The demonstration was made easy by the fact that L1210 cells have no NOS II pathway, contrary to other tumoral cells such as EMT6 or TA3 cells in which cytokines activate the NOS II pathway (see below).[59,61] NO synthesized by NOS II in activated macrophages is detected in nitrosylated iron-thiolate complexes within L1210 cells (Fig. 10.4). The EPR signal was not present when L1210 cells were either cultured alone or cocultured with unactivated macrophages, or cocultured with activated macrophages in the presence of NMMA. If coculture (which supposes cell contacts) is necessary for the observed paracrine effects, the way and means of NO intercellular transfer is, however, controversial (see chapter 9).

Further experiments showed that the molecular targets observed by EPR were indeed high molecular weight fractions. On the basis of protein content, the intensity of the signal given by mitochondrial preparations was 50% of that obtained with an equivalent amount of whole cells. Furthermore a large proportion of the remaining nitrosyl-iron-thiolate EPR signal appeared in the cytosolic fraction. The use of membranes of progressively decreasing cut-offs (300 kDa-3 kDa) showed that most (70%) of the EPR signal of the cytosol was carried by a high molecular weight fraction (> 300 kDa) (Table 10.1).[70,72] These data could be related to the suggestion that ferritin (24 subunits of 480 kDa total mass) could be a target for NO (see chapter 6 and below) but have to be reconciled with the above mentioned difference in microwave power saturation behavior (Fig. 10.5).[69]

In a similar paracrine system of coculture of human leukemic cells (target)

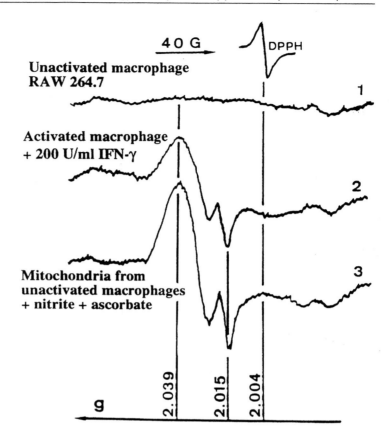

Fig. 10.2. EPR spectra at 77 K from digitonin-permeabilized macrophages' continuous cell line RAW 264.7. Cells were incubated for 24 hr in Dulbecco's modified essential medium alone (1 and 3), or added with 200 U/ml IFNγ (2). 3: an aliquot of digitonin-permeabilized macrophages was incubated with nitrite (5 mmol/l) and ascorbate (5 mmol/l) under helium for a further 30 min. DPPH, 1,1-diphenyl-2-picrylhydrazyl. (Reproduced with permission from: Henry Y, Ducrocq C, Drapier J-C et al. Nitric oxide, a biological effector. Electron paramagnetic resonance detection of nitrosyl-iron-protein complexes in whole cells. Eur Biophys J 1991; 20:1-15.) Copyright © Springer-Verlag GmbH & Co KG, Heidelberg.

and vascular cells (smooth muscle or endothelial NO donor cells), activation by IFNγ and TNFα led to apoptotic death of the leukemic target cells.[73]

OTHER MOLECULAR TARGETS

These data are not easily reconciled either with the results of Vanin et al[74,75] which showed that cultured J774 macrophages treated 5 min with NO presented a g_{av} = 2.03 EPR signal attributed to low molecular weight iron-dinitrosyl complex arising from loosely bound non-heme "free" iron; for these authors this signal could not arise from [FeS] clusters, as their own intrinsic EPR signal intensities at g = 1.94 and 1.92 were not diminished.

More recent concurrent results of Cooper and Brown show that, in rat brain synaptosomes, the large g_{av} = 2.04 EPR signal of the $Fe(SR)_2(NO)_2$ complex (> 30 μmol/l),

obtained by addition of SNP to the brain cells did not correlate at all with the inhibition of synaptosomal respiration.[76] Respiration inhibition is reported to result only from the inhibition of complex IV. The argument is based on the use of SNP, a donor of NO^+ in the dark and of NO in the light, which to us, is not a reliable NO-donor, as it gives rise to release of Fe^{2+}, CN^- and EPR-detectable reduced SNP, interfering with the g_{av} = 2.04 signal. Other experiments with either authentic NO gas, or better using the L-arginine-NOS pathway, are needed to substantiate their criticisms of previous experiments by several research groups, performed on NOS II-containing cellular systems. In fact, Drapier and Hibbs already mentioned in their report that complex IV was also inhibited.[51] It is quite possible however that the Cooper and Brown conclusion that cy-

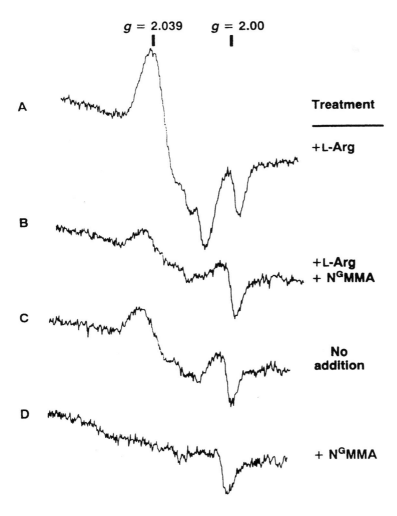

$g = 2.039$ $g = 2.00$

A

Treatment
————————
+L-Arg

B

**+L-Arg
+ N^GMMA**

C

**No
addition**

D

+ N^GMMA

Fig. 10.3. EPR spectra at 77 K of cytotoxic BCG-activated macrophages, after 20 hr incubation with the indicated additions of L-arginine and/or N^G-monomethyl-L-arginine. (Reproduced with permission from the authors: Lancaster JR, Hibbs JB. EPR demonstration of iron-nitrosyl complex formation by cytotoxic activated macrophages. Proc Natl Acad Sci USA 1990; 87:1223-1227.) Copyright © the authors.

tochrome *c* oxidase is the primary target for NO inhibition of brain cell respiration is quite correct (see chapter 14).[76]

In fact, as described in chapter 6, the so-called g_{av} = 2.03 EPR signal seems rather unspecific. It could arise from many different complexes of general formula $[Fe^I(SR)_2(NO)_2]$, generally with [FeS]-containing proteins (Fig. 10.5.).[69,77-87] The most striking instance of proteins forming a $Fe(SR)_2(NO)_2$ complex, although they do not contain any classical [FeS] cluster but only free thiolate groups, is that of nitrosylated Fe^{II}-metallothionein and of one nitrosylated complex of ferritin.[88,89] Metallothionein (6-7 kDa) contains two metal-cysteine domains with adamantane-like

geometry (see chapter 6 and below).[88] Ferritin (480 kDa) forms also a $Fe(SR)_2(NO)_2$ complex at the so-called B-site with residue Cys-130 as one of the iron ligands (see chapter 6 and below).[89] The arguments of enzymatic inhibition upon NO treatment and measurements of the molecular weights of the implicated species observed by EPR are therefore of great importance for correct assignments (Table 10.1).

Two recent reports showed that "reactivated" aconitase of doubtful purity from different sources (*E. coli*, recombinant human cytosolic and porcine heart mitochondrial) reacts with peroxynitrite (~ 10-fold excess with respect to aconitase concentration) and is inactivated by it through the

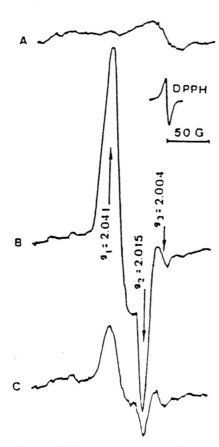

Fig. 10.4. EPR spectra at 77 K of tumor target cells and activated macrophages. A: Control L1210 cells alone or cocultivated with unactivated macrophages. B: L1210 cells detached from IFNγ activated macrophage monolayer. C: IFNγ activated macrophage monolayer alone. DPPH, 1,1-diphenyl-2-picrylhydrazyl. (Reproduced with permission from: Drapier J-C, Pellat C, Henry Y. Generation of EPR-detectable nitrosyl-iron complexes in tumor target cells cocultured with activated macrophages. J Biol Chem 1991; 266:10162-10167.) Copyright © The American Society for Biochemistry & Molecular Biology, Inc., Bethesda, MD, USA.

loss of the labile Fe α atom.[90,91] Both reports stress that "reactivated" aconitase is not inactivated by anaerobic NO itself, contrarily to what is found in whole cells or in mitochondria.

EPR DETECTION OF NITROSYLATED [FeS] CLUSTERS IN OTHER CELL TYPES

Most of the identification work of NO molecular targets was performed on peritoneal murine macrophages, on bone-marrow-derived macrophages or cultured macrophage cell lines.[54,63,65,69,74,75] The functioning of NOS II has been also characterized by EPR spectroscopy in other cell types from mouse or rat as well as from human: rat bone marrow-derived mononuclear phagocytes infected with gram negative bacteria, or LPS-stimulated,[75,92] rat and human pancreatic β-cells in Langerhans islets exposed to IL-1β,[93-96] rat and human hepatocytes exposed to IFNγ, TNFα, IL-1β and LPS, or hepatocytes isolated from *Corynebacterium parvum*-treated rats,[97-99] rat fibroblasts exposed to IFNγ, TNFα and LPS,[100] rat cultured vascular smooth muscle cells exposed to IFNγ or TNFα.[101]

The formation of the $Fe(SR)_2(NO)_2$ complex was also demonstrated in cultured endothelial cells from porcine and bovine aortae when NOS III (eNOS) was stimulated by bradykinin or calcium ionophore A23187.[102]

RIBONUCLEOTIDE REDUCTASE

Blockage of DNA replication in macrophages and in tumor target cells upon induction of the L-arginine-NO pathway is nearly simultaneous with mitochondrial respiration inhibition and not a consequence of it, through a lack of ATP synthesis. The enzyme implicated in the DNA replication impairment is ribonucleotide reductase (RNR) (EC 1.17.4.1), a rate-limiting enzyme in DNA synthesis (see chapter 6).[103-105] RNR catalyzes the reduction of ribonucleotides and regulates the deoxyribonucleotides pool. It is a heterodimer of two homodimeric subunits R1 and R2. Protein R1 binds the substrates and the

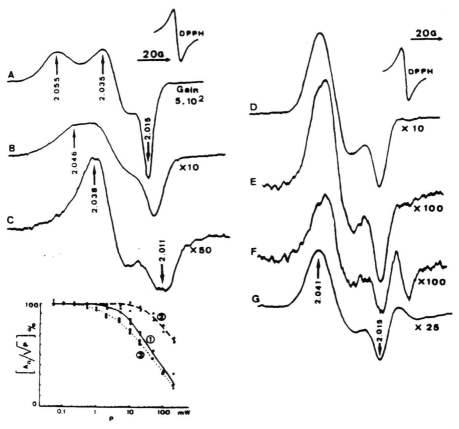

Fig. 10.5. Comparison of EPR spectra at 77 K of activated macrophage-injured L1210 cells and complexes formed by NO with ligands. Protein and cysteine solutions were deoxygenated with helium and treated with nitrite (20 mmol/l) and ascorbate (20 mmol/l) for 20 min at room temperature. A: transferrin; B: BSA and FeSO₄ (1 mmol/l); C: ferritin; D: cysteine and FeSO₄ (1 mmol/l); E: S. platensis ferredoxin; F: P. umbilicalis ferredoxin; G: activated macrophage-injured L1210 cells. Inset: normalized microwave power (P) saturation profiles of EPR signal amplitude (An) from activated macrophage-injured L1210 cells (1), nitrosyl-ferritin (2), and NO-Fe²⁺-cysteine complex (3). DPPH, 1,1-diphenyl-2-picrylhydrazyl. (Reproduced with permission from: Drapier J-C, Pellat C, Henry Y. Generation of EPR-detectable nitrosyl-iron complexes in tumor target cells cocultured with activated macrophages. J Biol Chem 1991; 266:10162-10167.) Copyright © The American Society for Biochemistry & Molecular Biology, Inc., Bethesda, MD, USA.

allosteric effectors, and contains five important cysteine residues per protomer implicated in electron transfer toward the R2 subunit: two CysSH pairs undergoing CysSH/CysSSCys reactions with thioredoxin or glutathione-glutathione reductase, and one thiol residue which might form a thiyl radical. Subunit R2 contains a protein-bound binuclear Fe-Fe center and a stable tyrosyl radical, both necessary for the enzymatic activity of RNR.[103-105] Thus

RNR contains several redox centers susceptible to react with NO: cysteine residues of R1 subunit, the tyrosyl radical and the Fe-Fe group of R2 subunit. This was found to be true for the pure enzyme or its isolated subunits in vitro (see chapter 6).

AUTOCRINE CYTOKINE ACTIVATION OF TA3 TUMOR CELLS

Indeed it was shown that activation of NOS II by IFNγ associated with TNFα

Table 10.1. Estimation of the size of the iron-nitrosyl complex(es) present in the cytosol of injured L1210 cells.

Cut off	Intensity of the EPR signal (arbitrary units) Mean of four experiments
> 300 kDa	12.0
> 100 kDa	3.3
> 30 kDa	1.6
< 30 kDa	Not detectable

(Reproduced with permission from: Drapier J-C, Pellat C, Henry Y. In: Moncada S, Marletta MA, Hibbs JB et al, eds. The Biology of Nitric Oxide. London, UK: Portland Press, 1992: 72-76.) Copyright © The Biochemical Society and Portland Press.

and/or LPS inhibited RNR activity in cellular extracts of murine adenocarcinoma cell line TA3, explaining the inhibition of DNA synthesis determined by [³H]thymidine incorporation.[59] The TA3 M2 subclone, selected for enhanced RNR activity was found to be relatively less sensitive than the wild-type. Similar results were obtained after incubating partially purified RNR from L1210 mouse lymphoma cells with authentic NO.[60] Lysates of activated RAW 264.7 macrophages also inhibited partially purified RNR from L1210 cells. L1210 cell DNA synthesis was completely inhibited by activated macrophages and partially restored by providing deoxyribonucleosides.[60]

EPR experiments carried out on TA3 cytosol or on pure recombinant *E. Coli* or mouse R2 protein brought evidence that the tyrosyl radical disappeared in the presence of NO-releasing agents, sydnonimine, nitroprusside and *S*-nitrosoglutathione, thereby explaining the enzyme inhibition.[106,107]

Further EPR experiments were performed on a hydroxyurea-resistant subclone of TA3 cells overexpressing 40-fold the R2 subunit, called TA3 M2 or later TA3 H2, thus allowing the tyrosyl radical detection by EPR spectroscopy at 77 K in intact cells (40 x 10⁶ cells per EPR sample).[61] This clone and the wild-type cell line can, like murine mammary adenocarcinoma EMT6 cells, express the L-arginine-derived NO pathway after activation by cytokines in the same way as macrophages.[59,108] It is an autocrine model of NO-dependent cytostasis, as compared to the paracrine system of cocultured activated macrophages and L1210 tumor target cells, described above.[64,69] A decrease in the tyrosyl radical EPR signal centered at $g = 2.005$ was observed 4 hours after NOS II induction by IFNγ and TNFα. The extent of the tyrosyl radical loss is directly related to the L-arginine-NOS metabolism as measured by the concentration of L-citrulline produced.[61] It was verified by immunoblot analysis of the R2 subunit that the expression of the protein was not influenced by NOS activity.[61] Simultaneously to the tyrosyl signal quenching, a $g_{av} = 2.04$ EPR signal appeared, previously ascribed (see above) to $Fe(SR)_2(NO)_2$ complexes (Fig. 10.6).[61]

Furthermore, the time course and the extent of the tyrosyl radical loss parallels that of the inhibition of DNA synthesis, as measured by [³H]-thymidine incorporation (Fig. 10.7).[62,109] All phenomena, tyrosyl radical loss, $g_{av} = 2.04$ signal appearance and inhibition of DNA synthesis, are early events detectable 3 to 4 hours after NOS II activation and are fully or partially reversible. On the other hand, DNA synthesis inhibition did not correlate with formation of $Fe(SR)_2(NO)_2$ complexes.[61,62,109]

PARACRINE SYSTEMS IN COCULTURE

Similar phenomena were observed in a paracrine system constituted of cocultures of L1210 cells overexpressing RNR R2 subunits (called L1210-R2, obtained by transfection) and inducible murine macrophages. After 4 hr coculture with macrophages, quenching of the tyrosyl radical was observed by EPR, simultaneously to marked cytostasis. Target cells, withdrawn from activated macrophages, partially recovered from cytostasis. The tyrosyl radical of RNR was restored in parallel within 90 minutes (Fig. 10.8).[62,109]

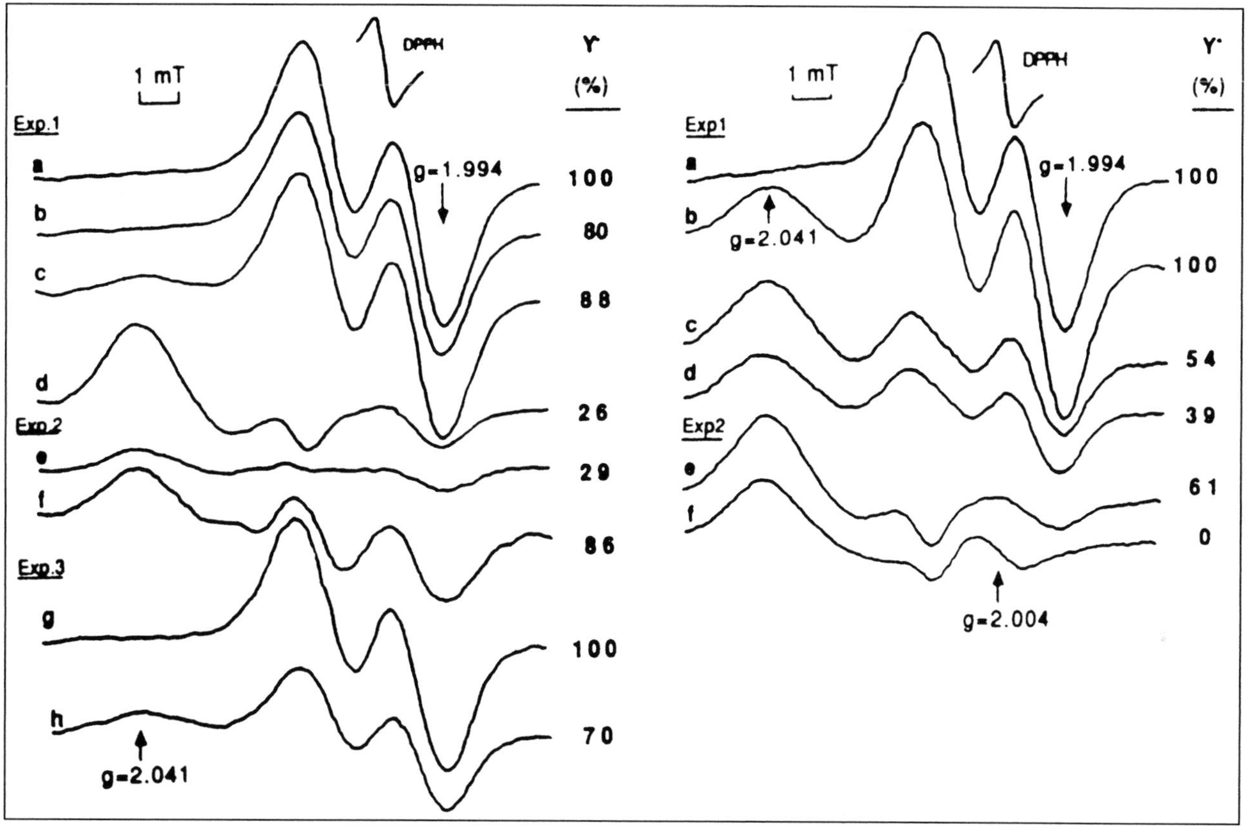

Fig. 10.6. Left panel. Characterization of the NO synthase product inducing a decrease in the tyrosyl EPR signal of TA3 H2 cells. R2-overproducing cells were cultured for 7 hr with N^α-nitro-L-arginine (4.5 mmol/l (Exp. 1), oxyhemoglobin (30 μmol/l) (Exp. 2), and $NaNO_2$ (1 mmol/l), $NaNO_3$ (1 mmol/l), citrulline (1 mmol/l), or S-nitrosoglutathione (2 mmol/l) (Exp. 3). Treated cells referred to as cell cultures incubated with IFNγ (40 U/ml), LPS (1 μg/ml), and TNFα (500 U/ml). Exp. 1: a, untreated control cells; b, untreated cells and N^α-nitro-L-arginine; c, treated cells and N^α-nitro-L-arginine; d, treated cells alone. Exp. 2: e, treated cells alone; f, treated cells and oxyhemoglobin. Exp. 3: g, untreated cells and nitrate, nitrite and citrulline; h, untreated cells and S-nitrosoglutathione. The tyrosyl EPR signal Y: was expressed as the percentage of the g = 1.994 feature of the signal in untreated cells.

Right panel. Kinetics of the decrease in the tyrosyl EPR signal of TA3 H2 cells after NO synthase induction. R2-overproducing cells cultured for 10 hr were treated fro 4 hr (b, e), 7 hr (c, f), or 10 hr (d) with IFNγ (40 U/ml), LPS (1 μg/ml), and TNFα (500 U/ml). Cell cultures contained 3x10⁶ (Exp. 1) or 16x10⁶ (Exp. 2) cells/plate. DPPH, 1,1-diphenyl-2-picrylhydrazyl. (Reproduced with permission from: Lepoivre M, Flaman J-M, Henry Y. Early loss of the tyrosyl radical in ribonucleotide reductase of adenocarcinoma cells producing nitric oxide. J Biol Chem 1992; 267:22994-23000.) Copyright © The American Society for Biochemistry & Molecular Biology, Inc., Bethesda, MD, USA.

Despite the fact that both NO and hydroxyurea were able to quench the R2 tyrosyl free radical, it was shown on human erythroleukemia K-562 cells that these phenomena follow different molecular mechanisms.[62]

POTENTIAL CHEMOTHERAPEUTIC TARGETS IN RIBONUCLEOTIDE REDUCTASE

Finally, as explained in chapter 6, the reversion of the tyrosyl radical loss induced by NO was also observed by EPR spectroscopy of pure ferric R2 protein there-

Fig. 10.7. Quenching of the tyrosyl radical of ribonucleotide reductase by nitric oxide. Relationship between inhibition of DNA synthesis and either decrease in R2 tyrosyl EPR signal (A) or formation of dinitrosyl-iron complexes (B) in TA3-H2 cells. Monolayer of TA3 H2 cells were cultured for 4-10 hr with IFNγ, LPS and TNFα. Concentration of R2 tyrosyl free radical (A) or DNIC (B) were measured by the intensity in arbitrary units of the EPR signal at g-values of 1.994 and 2.041, respectively. Cytostasis was evaluated by a short 90-min thymidine pulse, performed just after cell harvest for EPR analysis. Linear regression analysis excluded the points indicated by an arrow. (Reproduced with permission from: Lepoivre M, Flaman J-M, Bobé P et al. Quenching of the tyrosyl free radical of ribonucleotide reductase by nitric oxide. Relationship to cytostasis induced in tumor cells by cytotoxic macrophages. J Biol Chem 1994; 269:21891-21897.) Copyright © The American Society for Biochemistry & Molecular Biology, Inc., Bethesda, MD, USA.

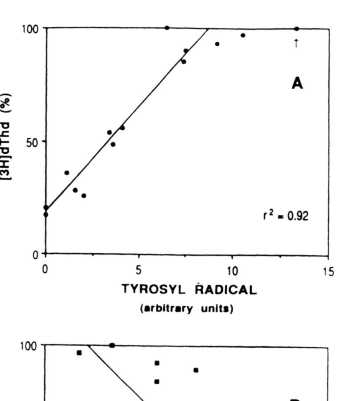

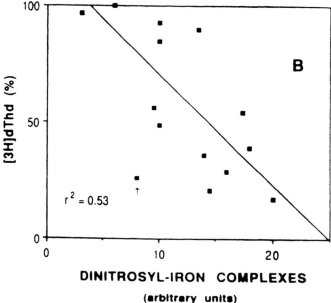

fore characterizing the intrinsic reactivity of R2 tyrosyl radical toward NO.[109,110] Nitric oxide also binds to the diiron(II) center of R2 subunit in the reduced form.[111]

The cellular effects of L-arginine-derived NO described above provide a conceptual basis for research of new NO-producing drugs having RNR as a potential chemotherapeutic target, such as *S*-nitrosothiols[110,112] or benzamidoximes.[113,114]

PROTEINS OF THE IRON METABOLISM

A massive loss of intracellular iron is one of the plausible causes of tumoral cell cytotoxicity by activated macrophages.[46,47,50] Due to the ferrous ion's high reactivity, the iron metabolism is one of the most finely regulated and intricated, so that many different molecular targets can be held simultaneously responsible for the action of NO. It has been under intense study in the last few years.

The unliganded ferric ion is insoluble at physiological pH, while the ferrous ion is highly reactive with many cellular redox components (Haber-Weiss and Fenton reactions) generating highly toxic hydroxyl radicals. To be available for cellular biosynthesis without being damaging, these two ions have to be reversibly "protected" by specific ligations.[115] In a simplified view of iron metabolism, at least six types of proteins cooperate in maintaining iron homeostasis in both redox states Fe^{II} and Fe^{III}: transferrin receptor on the cell surface for iron uptake and release; proteins of the transferrin family for Fe^{III} intra- and extracellular transport; ferritin for Fe^{III} storage; ferroxidases (ceruloplasmin and ferritin H-subunits) for the catalysis of iron oxidation by oxygen, within ferritin itself; metallothionein for Fe^{II} short-term storage and transport; and finally a regulatory protein called IRE-binding protein (IRE for iron response element of messenger RNA) or more recently known as iron regulatory protein (IRP), which regulates mRNA translation of several iron-related proteins (see below) (reviewed in refs. 116-119). Except for the transferrin

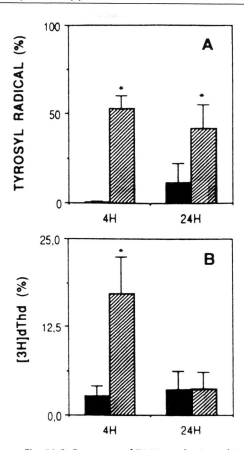

Fig. 10.8. Recovery of DNA synthesis and R2 tyrosyl EPR signal in L1210-R2 cells withdrawn from cytotoxic macrophages. Tumor cells were cultured for 4 or 24 hr with previously activated macrophages. Their relative R2 tyrosyl radical content (A) or their capacity to proliferate (B) were investigated either immediately (closed bars) or after a 90-min recovery period in the absence of macrophages (hatched bars). (Reproduced with permission from: Lepoivre M, Flaman J-M, Bobé P et al. Quenching of the tyrosyl free radical of ribonucleotide reductase by nitric oxide. Relationship to cytostasis induced in tumor cells by cytotoxic macrophages. J Biol Chem 1994; 269:21891-21897.) Copyright © The American Society for Biochemistry & Molecular Biology, Inc., Bethesda, MD, USA.

receptor, all these proteins contain specific iron-binding sites, which could be direct targets for NO, therefore disrupting iron homeostasis. As we shall see below, this scheme is complicated by crossregulations with other metabolisms, such as that of heme, or those of copper and zinc.

EPR spectroscopy has recently brought some controversial evidence of NO binding by three of the above mentioned metalloproteins: metallothionein, transferrin and ferritin, that we shall discuss in turn (see also chapter 6).

METALLOTHIONEIN AND ZINC FINGER PROTEINS

Metallothioneins (MTs) are a class of ubiquitous low molecular weight (6-7 kDa), cysteine-rich (20 out of 61 or 62 amino acid residues), inducible, intracellular proteins. MTs play a role in the homeostasis of transition metals (Zn, Cu, Fe) in the detoxication of heavy nonessential metals (Cd, Hg, Pt, etc.) and in the scavenging of free radicals.[120] Apo-MT can be induced by different classes of inducers: metal ions,[121,122] exogenous organic chemicals such as anticancer drugs,[120,123] hormones, oxidative stress,[124] X-ray irradiation,[125] toxic shock syndrome toxin-1,[126] bacterial endotoxin, various cytokines, TNFα, IL-6,[127,128] etc. There are 14 human MT genes of which 8 are known to be functional, each with a unique expression profile that could be inducer-specific and tissue-specific.[129] Being also regulated along the cell cycle, MT can be used as a marker of cell proliferation.[130,131] The MT induction is regulated mainly at the transcriptional level through DNA motifs, called MRE (metal responsive element) for vertebrate cells.[132] Finally another important role of apo-MT can be the sequestration of zinc, by exchange from the zinc-finger domains of various transcription factors.[133,134]

As we have described in chapter 6, MT binds 7 Fe^{2+} atoms for 20 cysteine groups and forms in vitro ternary complexes $Fe(SR)_2(NO)_2$ giving rise to a g_{av} = 2.04 axial EPR signal, easily detectable at 77 K and indistinguishable from

that of nitrosylated [FeS] cluster-containing proteins.[88]

This property has been used in a cellular system, derived from mouse fibroblasts NIH3T3. These cells were transfected with a plasmid containing mouse metallothionein-I gene (NIH3T3/MT cells) and were found to have a 4-fold increase in intracellular, cytoplasmic MT. These NIH3T3/MT cells are 10-fold more resistant than control (NIH3T3/TM) cells, containing the promoter-free inverted gene, to the powerful lipid peroxidizing agent, *tert*-butyl hydroperoxide.[124] NIH3T3/MT cells were more resistant to the cytotoxic effects of NO released from *S*-nitrosoacetyl-penicillamine (SNAP).[135] This could be correlated to the intensity of the EPR signal at g_{av} = 2.04 characteristic of the $Fe(SR)_2(NO)_2$ complex in the respective NIH3T3/MT cells as compared to NIH3T3/TM cells and NIH3T3 cells. A similar result was obtained when NO was released in coculture of NIH3T3/MT cells with cells infected with a retroviral expressing human NOS II (iNOS) gene, called NIH3T3-DFG-iNOS cells, in the presence of tetrahydrobiopterin.[135] It was demonstrated that MT overexpression in these NIH3T3/MT cells, or MT addition to NIH3T3 cells, protected against DNA damage by NO as measured by single-strand breaks in the isolated nuclei and in whole cells).[135]

Another interesting effect of NO on metallothionein is to release metals from the Zn/Cd-MT complex.[136] NO as a gas or generated from *S*-nitrosocysteine causes a time-dependent and dose-dependent *S*-nitrosylation of thiol groups of MT, followed by disulfide formation. Thus *S*-nitrosylation of MT may have serious consequences by impairing detoxication properties of MT toward toxic heavy metals (Cd, Hg, etc.). The same authors demonstrated that NO also interferes with the DNA-binding activity of a zinc-finger protein, transcription activator LAC9 which has a $[Cys_6Zn(II)_2]$-type zinc-finger.[136] This goes probably through *S*-nitrosylation preventing Zn^{2+} binding, thus inhibiting signal transduction and regulation. When one is reminded that over a thousand of the proteins involved in gene expression are zinc-finger proteins, their *S*-nitrosylation could be of great importance in the regulation of gene expression (reviewed in ref. 185).

Transferrin and Lactoferrin

The transferrins, serum transferrin (sTF), ovotransferrin (oTF) and lactoferrin or lactotransferrin (lTF), are a family of glycoproteins found in the blood and body fluids of many species, which reversibly sequestrate, solubilize and transport the Fe^{3+} ion.[137] lTF is expressed in the lactating mammary gland and in many biological fluids, tears, bronchial secretions, seminal fluids, etc. It is secreted by many endothelial tissues and probably functions as a bactericidal agent. Human lTF has been characterized as an inflammatory response protein synthesized by polymorphonuclear leukocytes.

X-ray crystallographic analysis of human lTF and rabbit sTF has shown that the single chain (~ 80 kDa) forms two globular lobes, each binding one Fe^{3+} ion and one CO_3^{2-} or HCO_3^- anion acting in synergy. The bicarbonate anion can be replaced by several other synergistic anions: containing a carboxylate group, oxalate, malonate, amino acids, etc., or nonsynergistic anions: chloride, pyrophosphate, adenosine triphosphate, etc.[138] Each lobe has the same folding, and each iron binding site has similar protein ligands: two tyrosines (Tyr-92 and Tyr-192 in the N-lobe of human lFT), a histidine (His-253) and an aspartate (Asp-60).[139-142] Ferric iron binding is very tight (K_{app} ~ 10^{19} to 10^{22} mol^{-1} l, depending on the origin) and cooperative for the two atoms; Fe^{2+} binding is comparatively weak (K_{app} ~ 10^5 mol^{-1} l). The redox potential of the Fe^{3+}/Fe^{2+} couple is rather low at -400 mV (versus NHE), making bound Fe^{3+} difficult to reduce.[143] Large conformational changes occur during iron binding or release.[141] Binding is reversible, through the interaction with the transferrin receptor, a transmembrane,

homodimeric glycoprotein (~ 180 kDa), followed by endocytosis and endosomal acidification which allows iron release.[144]

Iron Fe^{3+} can be substituted with the transition elements: Cu^{2+}, Zn^{2+}, Cd^{2+}, Mn^{2+}, Cr^{3+}, Mn^{3+}, Co^{3+} and VO^{2+}, the aluminum column series: Al^{3+}, Sc^{3+}, Ga^{3+}, In^{3+} and Tl^{3+}, the lanthanides: Ce^{3+}, Ce^{4+}, Sm^{3+}, etc., with widely different binding constants.[145] Many spectroscopic methods: EPR and Mössbauer, ^{1}H-NMR, ^{13}C-NMR, and metal-centered NMR for ^{27}Al, ^{45}Sc, ^{113}Cd or ^{205}Tl-bound oTF, etc., have been used to probe the metal binding sites (to quote only recent reports, refs. 145-150). In particular EPR spectroscopy of the ferric transferrins shows that the iron is high-spin (S = 5/2) in a distorted axial symmetry with a zero-field splitting D parameter and a rhombicity component of the crystal field E/D parameter, which are very dependent on the nature of the anion ligands.[149-152]

NO binding to transferrin was proposed on the basis of the EPR spectrum obtained at 77 K after incubation of human sTF with nitrite (20 mmol/l) and ascorbate (20 mmol/l) under anaerobiosis obtained by argon flushing.[69] The EPR signal with g-values at 2.055, 2.035 and 2.015, indicative of a $Fe^{(II)}$-NO complex in a rhombic symmetry, was used as an argument to differentiate it from the signals obtained with [FeS]-NO-containing proteins, with an axial symmetry and g-values at 2.04 and 2.015. The argument was good; however a more careful EPR study has recently shown that the observed $Fe^{(II)}$-NO complex in sTF does not arise from reduced iron bound at the normal iron high-affinity binding site of sTF, but from iron atoms bound at another site (30% of total iron) and not removed by Chelex treatment (Ducastel B & Henry Y, unpublished results). Neither ascorbate + nitrite mixture, nor pure NO gas in solution can reduce Fe^{3+} in sTF or lTF, as ascertained by the bound Fe^{3+} specific high-spin signal near g = 4.3 or by UV-visible spectroscopy. The detected $Fe^{(II)}$-NO signal arises most probably from exogenous nonspecifically bound $Fe^{(II)}$-NO complex, such as $Fe^{(II)}$(NO)-phosphate, $Fe^{(II)}$(NO)-amino acid or $Fe^{(II)}$(NO)-ascorbate complexes (Ducastel B & Henry Y, unpublished results).[78,153,154] A similar nonspecific EPR spectrum is obtained with BSA-$Fe^{(II)}$-NO complex.[69,78,153]

An interaction of NO with lTF was also proposed on the basis of a modification of the high-spin Fe^{3+} g = 4.3 EPR signal and was interpreted to be a Fe^{3+}-NO complex.[155] The same signal was obtained when lTF was added to the supernatant of LPS-activated macrophages. We have some doubt that such a Fe^{3+}-NO complex could give a ferric high-spin EPR signal around g = 4.3, as NO is usually a strong field ligand and the resulting Fe^{3+}-NO complex would be diamagnetic (S = 0) or integer-spin (S = 1), thus EPR-undetectable. We suggest that NO could have been oxidized, through lack of strict anaerobiosis, into NO_2^- which could bind to TF.

We can conclude provisionally that the above evidences of any direct interaction between transferrins and nitricoxide are firstly nonspecific and secondly very superficial. To us it seems altogether to be negative.

In fact, the effect of nitric oxide on transferrin and iron uptake by cells seems to be redox related.[156,157] While a NO^+ donor, sodium nitroprusside (SNP) decreased Fe uptake by human melanoma SK-MEL-28 cells without decreasing the sTF uptake; true NO-donors, 3-morpholino-sydnonimine (SIN-1) and SNAP, also decreased sTF uptake.[156] These effects seem to be indirect and to be correlated to interaction of NO with the IRE-binding protein, which in turn affects the level of transferrin receptor (see below).[157]

A curious result deserves to be mentioned here: transferrin, in its holo-form only, would induce NOS II mRNA in rat cultured aortic smooth muscle cells; a result which should be interesting to confirm.[158]

FERRITIN

We have described the tissue distribution, the functions and the structure of ferritin (FTN) and bacterioferritin (BFR) in chapter 6. We have also described the evidence of direct interaction between nitric oxide and some FTN and BFR mononuclear iron sites called A and B types, brought by EPR spectroscopy.[69,70,89,159,160] Let us recall that both A type and B type EPR signals, assigned respectively to histidine-Fe-NO and to cysteine-Fe-NO ternary complexes within FTN or BFR, are obviously not specific of these proteins or of any non-heme iron-containing protein either. Let us finally recall that the largest part of the EPR signal at g_{av} = 2.04 (type B) found in the cytosol of L1210 cells after coculture with activated macrophages, belongs to a protein selected by a membrane of cut-off superior to 300 kDa, which could well be ferritin (total mass ~ 480 kDa).[70]

Ferritin has appeared to be a most probable target of NO in activated macrophages and cocultured L10 or L1210 cells because the intracellular iron loss is massive after a 4 hr lag, 40% after 10 hr coculture to 70 % after 24 hr.[46,47] The inhibition of non-heme iron-containing enzymes ([FeS]-containing enzymes, RNR, etc.) by NO, which could lead to iron release, could not explain such a massive loss. The interaction of NO with hemoproteins cannot induce any iron release from the heme either. Most of the intracellular non-heme iron (25 % of total iron pool) is being stored in FTN, it was an obvious target for NO, thereby explaining the iron loss following macrophage activation. The question then arises: does the iron release result from direct iron-NO interaction in ferritin or is it indirect?

Ferritin which has a midpoint potential of -190 mV (versus NHE) at pH 7.0, for each of its iron atoms (-310 mV at pH 8.0),[161] can release Fe^{2+} ions upon reaction with many radicals: superoxide anion,[162,163] semiquinone, anthracycline, bipyridyl, nitroaromatic radicals,[164] aqueous extracts of cigarette smoke,[165] etc. All these compounds able to release Fe^{2+} in anaerobiosis have redox potentials lower than that of FTN.[164,166] Would NO be able to release Fe^{2+} ions?

Nitric oxide was found to mediate the release iron from horse spleen FTN.[167] However this result was demonstrated by two groups to be artifactual (Ducastel B and Henry Y, unpublished results).[168] The main problem in Reif and Simmons's experiments was the use of SNP as a NO-donor. The release of NO from SNP occurs only under light or in the presence of a reducer, and becomes rapidly autocatalytic, following the release of Fe^{2+} ions from SNP (see chapter 2). The release of Fe^{2+} ions from FTN, using SNP as a NO-donor, results then from an autocatalytic reduction of FTN by SNP under light. The Fe^{2+} ion chelator used, ferrozine (3-(2-pyridyl)5,6-bis(4-phenylsulphonic)-1,2,4-triazine), is in fact able to mobilize iron from FTN, in the absence of NO or NO-donors (Ducastel B & Henry Y, unpublished results).[168] We were unable to observe any release of Fe^{2+} ions from FTN by authentic NO solutions in anaerobiosis (Ducastel B & Henry Y, unpublished results). Using *S*-nitrosothiols, SNAP and *S*-nitrosocysteamine as NO-donors and measuring iron exchange by ^{59}Fe radioactivity, it was shown firstly that NO does not promote Fe uptake by FTN.[168] It does not promote any iron release from FTN either.[168] Nitric oxide can act sometimes as a reducer of metal cations, but its redox potential ($E^{\prime\circ} = 240$ mV for the NO_2^-/NO couple) is not low enough to reduce FTN.

Thus our conclusion, provisional as in the case of transferrin, is that although NO does interact with some iron atoms of FTN or BRF, it is not this direct interaction which provokes iron release from FTN. Also NO does not directly alter intracellular FTN levels.[169] It is the interaction with the IRE-binding protein (see below) which indirectly induces the unloading of ferritin.[169] In fact treatment of hepatocytes by low doses of NO (derived from SNAP or from NOS II induction) induces an increase in FTN levels, correlated to an increase of the non-heme iron-NO $g_{av} = 2.04$ EPR signal, which imparts an auto-protection of hepatocytes to a second exposure of a higher dose of NO.[170] This could well involve an upregulation of protective proteins, heme oxygenase-1 (hsp32) and ferritin (see chapter 14).[170,171]

IRON-REGULATORY PROTEIN OR CYTOSOLIC ACONITASE

Iron metabolism is auto-regulated through modifications of the available quantities of ferritin, which stores iron, and of transferrin receptor, which controls iron uptake and release. This post-transcriptional regulation is under the control of interactions between the mRNAs of these two proteins and the IRE-binding protein (IRE-BP or IRP), itself dependent of the iron "status". Iron concentration variations have opposite effects on the translation of the two mRNAs; increase of overall iron concentration increases the translation of ferritin subunit mRNAs and decreases the stability of transferrin receptor mRNA. The mRNAs for ferritin H and L chains, for erythroid 5-aminolevulinate synthase (eALAS) and for mitochondrial aconitase contain, at the 5'-untranslated end, a common IRE regulatory noncoding sequence of 26-35 nucleotides, for which IRP has a variable affinity.[172] The IRE sequence of mRNA for the transferrin receptor is located at its 3'-untranslated end. IRP, a 90 kDa cytoplasmic protein, has been found to possess an aconitase activity and to contain a [4Fe-4S] cluster (holoenzyme).[118,119,173] In fact it has a high sequence homology with mitochondrial aconitase and is identical to previously determined cytosolic aconitase. The function of the protein is determined by the presence or the absence of the intact [4Fe-4S] cluster. It functions as an active aconitase when the cluster is present or as mRNA-binding protein when the cluster is absent. The two functions are mutually exclusive, and the interconversion between the two forms is determined by intracellular iron concentrations in a way which depends very much on the tissue under study.[174-180]

Recently Drapier et al[181,182] have demonstrated that NO, derived from NOS II activation in mouse macrophages and several murine cell lines, modulates the capacity of the IRP to bind IRE sequences. The effect of NO probably goes through direct binding to the [4Fe-4S] cluster, thereby leading to the liberation of a labile iron atom of the cluster as in mitochondrial aconitase[68] and dissociation of the whole [FeS] cluster to form the apoenzyme.[182] In this sequence of events, it would confer to the apoenzyme a high affinity for IRE, thus modulating mRNA translation for ferritin subunits and transferrin receptor. Similar results were simultaneously reported by Weiss et al.[183] It was the first time that, through the interaction of NO with a metalloenzyme, *a post-transcriptional regulatory role in genetic expression had been attributed to L-arginine-derived NO*. The modulation of IRP bifunctional properties was further examined on RAW 264.7 cytosolic extracts by measuring the aconitase and the IRE-binding activity.[184] The results show that only NO derived from NONOate or *S*-nitrosothiols or SIN-1 in the presence of SOD, but neither peroxynitrite ONOO⁻ nor $O_2^{-\bullet}$ can convert IRP into its IRE binding form by targeting the [FeS] cluster.[184,185]

Curiously, iron itself regulates NOS activity at the level of nuclear transcription.[186] NOS II activity and mRNA for NOS II were measured in the macrophage cell line J774A.1 activated with LPS/IFNγ, in the presence of added Fe^{3+} ions or in the presence of iron chelator desferrioxamine. The changes in NOS II activity (increase in the presence of desferrioxamine, decrease in the presence of Fe^{3+}) are totally accounted for by the variations of mRNA levels.[186] It seems therefore that there is a regulatory loop between iron metabolism and the L-arginine-NO pathway.[186] Similar results were obtained with NOS II-gene-transfected fibroblasts and following mRNA levels for ferritin and transferrin receptor.[187] Other data indicates that NO-mediated activation of IRP crossregulates alterations of iron hepatic homeostasis occurring in acute liver inflammation.[188] It has also been shown that *N*-methyl-D-aspartate (NMDA), via NO, stimulates the IRE-binding function of IRP, perhaps contributing to glutamate neurotoxicity.[189]

FERROCHELATASE AND OTHER [FeS]-CONTAINING PROTEINS

A most important step in iron metabolism is the termination of the heme biosynthetic pathway catalyzed by ferrochelatase (protoheme ferrolyase, EC 4.99.1.1) a membrane-associated protein. It catalyzes the insertion of Fe^{2+} ion into the protoporphyrin IX macrocycle and is ubiquitous in all cells. It inserts also Co^{2+} and Zn^{2+}, but not Fe^{3+}, and is inhibited by Hg^{2+}, Mn^{2+} etc.[190] EPR, UV-visible, MCD and Mössbauer spectroscopic methods have determined that mammalian, and human ferrochelatase in particular, contain one [2Fe-2S] cluster per molecule (40 kDa) and a minor component of high-spin $Fe^{(III)}$ heme.[191-194] Ferrochelatase binds an IRE present at the 5'- untranslated end of the mRNA for erythroid 5-aminolevulinate synthase (eALAS, EC 2.3.1.37), the first enzyme of the heme biosynthetic pathway.[195] This binding could provide a regulatory mechanism for the heme biosynthesis pathway in differentiating erythroid cells.

Nitric oxide interacts with mammalian ferrochelatase.[196] The induction of NO synthesis by the activation of mouse macrophage cell line RAW 264.7 by LPS and/or IFNγ reduces the ferrochelatase activity in these cells, with no change in the amount of protein. NO donors, SIN-1 and SNAP, have strong inhibitory effects on recombinant human ferrochelatase, but little effect on *E. coli* ferrochelatase which does not contain any [FeS] cluster.[196] Ferrochelatase inhibition by NO is correlated to the disruption of the [FeS] cluster, as estimated by the loss of visible absorbance bands. It forms an EPR-detectable complex characterized by the usual g = 2.04 and g = 2.014 (g_{av} = 2.03) axial signal.[197] Thus the NO-mediated increase in IRE-binding protein activity and NO-mediated decrease in ferrochelatase activity are two aspects of crossregulation of iron and heme metabolisms.[44,196]

Another beautiful example of cellular regulation and defense is that of SoxR protein. Oxidative stress corresponds to an excessive or locally inadequate production of oxygen-derived oxidants, such as superoxide anion $O_2^{-\bullet}$ or H_2O_2. There exists multifunctional defense systems triggered by different signals of oxidative stress. One instance is the redox-sensing SoxR protein of *E. coli*, which is a [2Fe-2S] cluster-containing protein, as revealed by EPR spectroscopic studies.[198,199] SoxR protein is activated by $O_2^{-\bullet}$ and by NO, and is in turn a transcription activator of the soxS gene, whose product activates about 10 other promoters.[199] The cascade activation is completely dependent on the presence of the [2Fe-2S] cluster in the SoxR protein.[198] The activation of SoxR protein by NO could play a role in the bacterial defense against the cytotoxic activated macrophage and may contribute to bacterial virulence, as shown by the study of mixtures of several strains of *E. coli* expressing differently the soxS gene and mouse peritoneal macrophages.[200]

Other parts of the oxidative stress response are the DNA-repair enzymes, endonuclease III,[201] a [4Fe-4S] cluster-containing enzyme; endonuclease IV, more specifically induced by oxidative damage;[200] Mn-containing superoxide dismutase and possibly IRP itself.[202,203]

CONCLUSION

We have focused in this chapter on the cellular effects of nitric oxide, particularly on metalloenzymes, viewed whenever it was possible through the spectral window of EPR spectroscopy.[66,72,204] The molecular targets of NO in the experiments that we have compiled are "spontaneous" and the NO source was usually L-arginine, mostly through the NOS II inducible pathway. Therefore two requirements for the biological relevance of the effects observed were met.

However in several instances, overexpression of the NO target, such as R2 subunit of RNR, or overexpression of NOS II, were used in order to facilitate detection. The cellular effects of the overexpression itself have to be estimated on their own. For instance what would be the effect on a cell of overexpressing the guanylate cyclase gene, if it were possible?

Through the description of many pathways implicating metalloenzymes, we have pointed *to many regulation loops or crossregulations*, which makes analysis more difficult and should render scientists more cautious and apt to reconsider some points of view. A final sequel of chapter 6 and of the present one shall be chapter 14: a discussion of hypothetical or still controversial molecular targets of NO and their crossregulations.

REFERENCES

1. Ignarro LJ. Biosynthesis and metabolism of endothelium-derived nitric oxide. Annu Rev Pharmacol Toxicol 1990; 30:535-560.
2. Ignarro LJ. Haem-dependent activation of guanylate cyclase and cyclic GMP formation by endogenous nitric oxide: a unique transduction mechanism for transcellular signalling. Pharmacol Toxicol 1990; 67:1-7.
3. Ignarro LJ. Signal transduction mechanisms involving nitric oxide. Biochem Pharmacol 1991; 41:485-490.
4. Garbers DL, Lowe DG. Guanylyl cyclase receptors. J Biol Chem 1994; 269:30741-30744.
5. Arnold WP, Mittal CK, Katsuki S et al. Nitric oxide activates guanylate cyclase and increases guanosine 3':5'-cyclic monophosphate levels in various tissue preparations. Proc Natl Acad Sci USA 1977; 74:3203-3207.
6. Miki N, Kawabe Y, Kuriyama K. Activation of cerebral guanylate cyclase by nitric oxide. Biochem Biophys Res Commun 1977; 75:851-856.
7. Mittal C, Murad F. Activation of guanylate cyclase by superoxide dismutase and hydroxyl radical: a physiological regulator of guanosine 3',5'-monophosphate formation. Proc Natl Acad Sci USA 1977; 74:4360-4364.
8. Murad F, Mittal CK, Arnold WP et al. Guanylate cyclase: activation by azide, nitro compounds, nitric oxide, and hydroxyl radical and inhibition by hemoglobin and myoglobin. Adv Cyclic Nucleotide Res

1978; 9:145-157.

9. Gruetter CA, Barry B, McNamara DB et al. Relaxation of bovine coronary artery and activation of coronary arterial guanylate cyclase by nitric oxide, nitroprusside and a carcinogenic nitrosoamine. J Cyclic Nucleotide Res 1979; 5:211-224.

10. Ignarro LJ, Edwards JC, Gruetter DY et al. Possible involvement of *S*-nitrosothiols in the activation of guanylate cyclase by nitroso compounds. FEBS Lett 1980; 110:275-278.

11. Ignarro LJ, Barry BK, Gruetter DY et al. Guanylate cyclase activation by nitroprusside and nitrosoguanidine is related to formation of *S*-nitrosothiol intermediates. Biochem Biophys Res Commun 1980; 94:93-100.

12. Ignarro LJ, Gruetter CA. Requirement of thiols for activation of coronary arterial guanylate cyclase by glyceryl trinitrate and sodium nitrite. Possible involvement of *S*-nitrosothiols. Biochim Biophys Acta 1980; 631:221-231.

13. Feelisch M, Noack EA. Correlation between nitric oxide formation during degradation of organic nitrates and activation of guanylate cyclase. Eur J Pharmacol 1987; 139:19-30.

14. Brüne B, Schmidt K-U, Ullrich V. Activation of soluble guanylate cyclase by carbon monoxide and inhibition by superoxide anion. Eur J Biochem 1990; 192:683-688.

15. Utz J, Ullrich V. Carbon monoxide relaxes ileal smooth muscle through activation of guanylate cyclase. Biochem Pharmacol 1991; 41:1195-1201.

16. Schmidt HHHW. NO, CO and OH. Endogenous soluble guanylyl cyclase-activating factors. FEBS Lett 1992; 307:102-107.

17. Maines MD. Carbon monoxide: an emerging regulator of cGMP in the brain. Mol Cell Neurosci 1993; 4:389-397.

18. Marks GS. Heme oxygenase: the physiological role of one of its metabolites, carbon monoxide and interactions with zinc protoporphyrin, cobalt protoporphyrin and other metalloporphyrins. Cell Mol Biol 1994; 40:863-870.

19. Stone JR, Sands RH, Dunham WR et al. Electron paramagnetic resonance spectral evidence for the formation of a penta-coordinated nitrosyl-heme complex on soluble guanylate cyclase. Biochem Biophys Res Commun 1995; 207:572-577.

20. Stone JR, Marletta MA. Heme stoichiometry of heterodimeric soluble guanylate cyclase. Biochemistry 1995; 34:14668-14674.

21. Stone JR, Marletta MA. The ferrous heme of soluble guanylate cyclase: formation of hexacoordinate complexes with carbon monoxide and nitrosomethane. Biochemistry 1995; 34:16397-16403.

22. Verma A, Hirsch DJ, Glatt CE et al. Carbon monoxide: a putative neural messenger. Science 1993; 259:381-384.

23. Vigne P, Feolde E, Ladoux A et al. Contributions of NO synthase and heme oxygenase to cGMP formation by cytokine and hemin treated brain capillary endothelial cells. Biochem Biophys Res Commun 1995; 214:1-5.

24. Wedel B, Humbert P, Harteneck C et al. Mutation of His-105 in the β_1 subunit yields a nitric oxide-insensitive form of soluble guanylyl cyclase. Proc Natl Acad Sci USA 1994; 91:2592-2596.

25. Gerzer R, Böhme E, Hofmann F et al. Soluble guanylate cyclase purified from bovine lung contains heme and copper. FEBS Lett 1981; 132:71-74.

26. Stone JR, Marletta MA. Soluble guanylate cyclase from bovine lung: activation with nitric oxide and carbon monoxide and spectral characterization of the ferrous and ferric states. Biochemistry 1994; 33:5636-5640.

27. Ignarro LJ, Wood KS, Wolin MS. Activation of purified soluble guanylate cyclase by protoporphyrin IX. Proc Natl Acad Sci USA 1982; 79:2870-2873.

28. Wolin MS, Wood KS, Ignarro LJ. Guanylate cyclase from bovine lung. A kinetic analysis of the regulation of the purified soluble enzyme by protoporphyrin IX, heme, and nitrosyl-heme. J Biol Chem 1982; 257:13312-13320.

29. Ohlstein EH, Wood KS, Ignarro LJ. Purification and properties of heme-deficient hepatic soluble guanylate cyclase: effects of heme and other factors on enzyme activation by NO, NO-heme, and protoporphyrin IX. Arch Biochem Biophys 1982;

218:187-198.

30. Ignarro LJ, Ballot B, Wood KS. Regulation of soluble guanylate cyclase activity by porphyrins and metalloporphyrins. J Biol Chem 1984; 259:6201-6207.

31. Ignarro LJ, Adams JB, Horwitz PM et al. Activation of soluble guanylate cyclase by NO-hemoproteins involves NO-heme exchange. Comparison of heme-containing and heme-deficient enzyme forms. J Biol Chem 1986; 261:4997-5002.

32. Mülsch A, Gerzer R. Purification of heme-containing soluble guanylyl cyclase. Methods Enzymol 1991; 195:377-383.

33. DeRubertis FR, Craven PA, Pratt DW. Electron spin resonance study of the role of nitrosyl-heme in the activation of guanylate cyclase by nitrosoguanidine and related agonists. Biochem Biophys Res Commun 1978; 83:158-167.

34. Craven PA, DeRubertis FR. Restoration of the responsiveness of purified guanylate cyclase to nitrosoguanidine, nitric oxide, and related activators by heme and hemoproteins. Evidence for involvement of the paramagnetic nitrosyl-heme complex in enzyme activation. J Biol Chem 1978; 253:8433-8443.

35. Craven PA, DeRubertis FR, Pratt DW. Electron spin resonance study of the role of NO-catalase in the activation of guanylate cyclase by NaN_3 and NH_2OH. Modulation of enzyme responses by heme proteins and their nitrosyl derivatives. J Biol Chem 1979; 254:8213-8222.

36. Traylor TG, Sharma VS. Why NO? Biochemistry 1992; 31:2847-2849.

37. Traylor TG, Duprat AF, Sharma VS. Nitric oxide-triggered heme-mediated hydrolysis: a possible model for biological reactions of NO. J Am Chem Soc 1993; 115:810-811.

38. Tsai A. How does NO activates hemeproteins? FEBS Lett 1994; 341:141-145.

39. Yu AE, Hu S, Spiro TG et al. Resonance Raman spectroscopy of soluble guanylyl cyclase reveals displacement of distal and proximal heme ligands by NO. J Am Chem Soc 1994; 116:4117-4118.

40. Stone JR, Sands RH, Dunham WR et al. Spectral and ligand-binding properties of an unusual hemoprotein, the ferric form of soluble guanylate cyclase. Biochemistry

1996; 35:3258-3262.

41. Burstyn JN, Yu AE, Dierks EA et al. Studies of the heme coordination and ligand binding properties of soluble guanylyl cyclase (sGC): characterization of Fe(II)sGC and Fe(II)sGC(CO) by electronic absorption and magnetic circular dichroism spectroscopies and failure of CO to activate the enzyme. Biochemistry 1995; 34:5896-5903.

42. Kharitonov VG, Sharma VS, Pilz RB et al. Basis of guanylate cyclase activation by carbon monoxyde. Proc Natl Acad Sci USA 1995; 92:2568-2571.

43. Stone JR, Marletta MA. Spectral and kinetic studies on the activation of soluble guanylate cyclase by nitric oxide. Biochemistry 1996; 35:1093-1099.

44. Kim Y-M, Bergonia HA, Müller C et al. Loss and degradation of enzyme-bound heme induced by cellular nitric oxide synthesis. J Biol Chem 1995; 270:5710-5713.

45. Bonkale WL, Winblad B, Ravid R et al. Reduced nitric oxide responsive soluble guanylyl cyclase in the superior temporal cortex of patients with Alzheimer's disease. Neurosci Lett 1995; 187:5-8.

46. Hibbs JB, Taintor RR, Vavrin Z. Iron depletion: possible cause of tumor cell cytotoxicity induced by activated macrophages. Biochem Biophys Res Commun 1984; 123:716-723.

47. Drapier J-C, Hibbs JB. Murine cytotoxic activated macrophages inhibit aconitase in tumor cells. Inhibition involves the iron-sulfur prosthetic group and is reversible. J Clin Invest 1986; 78:790-797.

48. Stuehr DJ, Marletta MA. Induction of nitrite/nitrate synthesis in murine macrophages by BCG infection, lymphokines, or interferon-γ. J Immunol 1987; 139:518-525.

49. Hibbs JB, Vavrin Z, Taintor RR. L-arginine is required for expression of the activated macrophage effector mechanism causing selective metabolic inhibition in target cells. J Immunol 1987; 138:550-565.

50. Wharton M, Granger DL, Durack DT. Mitochondrial iron loss from leukemia cells injured by macrophages. A possible mechanism for electron transport chain defects. J Immunol 1988; 141:1311-1317.

51. Drapier J-C, Hibbs JB. Differentiation of

murine macrophages to express nonspecific cytotoxicity for tumor cells results in L-arginine-dependent inhibition of mitochondrial iron-sulfur enzymes in the macrophage effector cells. J Immunol 1988; 140:2829-2838.

52. Drapier J-C, Wietzerbin J, Hibbs JB. Interferon gamma and tumor necrosis factor induce the L-arginine-dependent cytotoxic effector mechanism in murine macrophages. Eur J Immunol 1988; 18:1587-1592.

53. Hibbs JB, Taintor RR, Vavrin Z et al. Nitric oxide: a cytotoxic activated macrophage effector molecule. Biochem Biophys Res Commun 1988; 157:87-94.

54. Hibbs JB, Taintor RR, Vavrin Z et al. Synthesis of nitric oxide from a terminal guanidino nitrogen atom of L-arginine: a molecular mechanism regulating cellular proliferation that targets intracellular iron. In: Moncada S and Higgs EA, eds. Nitric Oxide from L-Arginine: a Bioregulatory System. Amsterdam: Elsevier Science Publishers BV, 1990:189-223.

55. Participants in the 39th Forum in Immunology. L-arginine-derived nitric oxide and the cell-mediated immune response. Res Immunol 1991; 142:553-602.

56. Nathan C. Nitric oxide as a secretory product of mammalian cells. FASEB J 1992; 6:3051-3064.

57. Stuehr DJ, Griffith OW. Mammalian nitric oxide synthases. Adv Enzymol 1992; 65:287-346.

58. Hibbs Jr JB. Cytokine induced synthesis of nitric oxide from L-arginine: a cytotoxic mechanism that targets intraculler iron. In: Riederer P and Youdim MBH, eds. Iron in Central Nervous System Disorders. Wien, New York: Springer-Verlag, 1993: 155-171.

59. Lepoivre M, Chenais B, Yapo A et al. Alterations of ribonucleotide reductase activity following induction of the nitrite-generating pathway in adenocarcinoma cells. J Biol Chem 1990; 265:14143-14149.

60. Kwon NS, Stuehr DJ, Nathan CF. Inhibition of tumor cell ribonucleotide reductase by macrophage-derived nitric oxide. J Exp Med 1991; 174:761-767.

61. Lepoivre M, Flaman J-M, Henry Y. Early loss of the tyrosyl radical in ribonucleotide

reductase of adenocarcinoma cells producing nitric oxide. J Biol Chem 1992; 267:22994-23000.

62. Lepoivre M, Flaman J-M, Bobé P et al. Quenching of the tyrosyl free radical of ribonucleotide reductase by nitric oxide. Relationship to cytostasis induced in tumor cells by cytotoxic macrophages. J Biol Chem 1994; 269:21891-21897.

63. Pellat C, Henry Y, Drapier J-C. IFN-γ-activated macrophages: detection by electron paramagnetic resonance of complexes between L-arginine-derived nitric oxide and non-heme iron proteins. Biochem Biophys Res Commun 1990; 166:119-125.

64. Pellat C, Henry Y, Drapier J-C. Detection by electron paramagnetic resonance of a nitrosyl-iron-type signal in interferon-γ-activated macrophages. In: Moncada S and Higgs EA, eds. Nitric Oxide from L-Arginine: a Bioregulatory System. Amsterdam, Elsevier Science Publishers BV, 1990: 281-289.

65. Lancaster JR, Hibbs JB. EPR demonstration of iron-nitrosyl complex formation by cytotoxic activated macrophages. Proc Natl Acad Sci USA 1990; 87:1223-1227.

66. Henry Y, Ducrocq C, Drapier J-C et al. Nitric oxide, a biological effector. Electron paramagnetic resonance detection of nitrosyl-iron-protein complexes in whole cells. Eur Biophys J 1991; 20:1-15.

67. Beinert H, Kennedy MC. Engineering of protein bound iron-sulfur clusters. A tool for the study of protein and cluster chemistry and mechanism of iron-sulfur enzymes. Eur J Biochem 1989; 186:5-15.

68. Beinert H. Recent developments in the field of iron-sulfur proteins. FASEB J 1990; 4:2483-2491.

69. Drapier J-C, Pellat C, Henry Y. Generation of EPR-detectable nitrosyl-iron complexes in tumor target cells cocultured with activated macrophages. J Biol Chem 1991; 266:10162-10167.

70. Drapier J-C, Pellat C, Henry Y. Characterization of the nitrosyl-iron complexes generated in tumour cells after co-culture with activated macrophages. In: Moncada S, Marletta MA, Hibbs JB et al, eds. The Biology of Nitric Oxide. London, UK: Port-

land Press, 1992:72-76.

71. Pellat C, Henry Y, Drapier J-C. Detection of nitrosyl-iron complexes in tumor target cells after coculture with activated macrophages. In: Melzer MS and Mantovani A, eds. Cellular and Cytokine Networks in Tissue Immunity. Wiley-Liss, 1991:229-234.

72. Henry Y, Lepoivre M, Drapier J-C et al. EPR characterisation of molecular targets for NO in mammalian cells and organelles. FASEB J 1993; 7:1124-1134.

73. Geng YJ, Hellstrand K, Wennmalm Å et al. Apoptotic death of human leukemic cells induced by vascular cells expressing nitric oxide synthase in response to γ-interferon and tumor necrosis factor-a. Cancer Res 1996; 56:866-874.

74. Vanin AF, Men'shikov GB, Moroz IA et al. The source of non-heme iron that binds nitric oxide in cultivated macrophages. Biochim Biophys Acta 1992; 1135:275-279.

75. Vanin AF, Mordvintcev PI, Hauschildt S et al. The relationship between L-arginine-dependent nitric oxide synthesis, nitrite release and dinitrosyl-iron complex formation by activated macrophages. Biochim Biophys Acta 1993; 1177:37-42.

76. Cooper CE, Brown GC. The interactions between nitric oxide and brain nerve terminals as studied by electron paramagnetic resonance. Biochem Biophys Res Commun 1995; 212:404-412.

77. McDonald CC, Phillips WD, Mower HF. An electron spin resonance of some complexes of iron, nitric oxide, and anionic ligands. J Am Chem Soc 1965; 87:3319-3326.

78. Woolum JC, Tiezzi E, Commoner B. Electron spin resonance of iron-nitric oxide complexes with amino acids, peptides and proteins. Biochim Biophys Acta 1968; 160:311-320.

79. Woolum J, Commoner B. Isolation and identification of a paramagnetic complex from the livers of carcinogen-treated rats. Biochim Biophys Acta 1970; 201:131-140.

80. Salerno JC, Ohnishi T, Lim J et al. Tetranuclear and binuclear iron-sulfur clusters in succinate dehydrogenase: a method of iron quantitation by formation of paramagnetic complexes. Biochim Biophys Res Commun 1976; 73:833-840.

81. Meyer J. Comparison of carbon monoxide, nitric oxide, and nitrite as inhibitors of the nitrogenase from *Clostridium pasteurianum*. Arch Biochem Biophys 1981; 210:246-256.

82. Reddy D, Lancaster JR, Cornforth DP. Nitrite inhibition of *Clostridium botulinum*: electron spin resonance detection of iron-nitric oxide complexes. Science 1983; 221:769-770.

83. Michalski WP, Nicholas DJD. Inhibition of nitrogenase by nitrite and nitric oxide in *Rhodopseudomonas sphaeroides* f. sp. *denitrificans*. Arch Microbiol 1987; 147:304-308.

84. Butler AR, Glidewell C, Li M-H. Nitrosyl complexes of iron-sulfur clusters. Adv Inorg Chem 1988; 32:335-393.

85. Payne MJ, Glidewell C, Cammack. Interactions of iron-thiol-nitrosyl compounds with the phophoroclastic system of *Clostridium sporogenes*. J Gen Microbiol 1990; 136:2077-2087.

86. Hyman MR, Seefeldt LC, Morgan TV et al. Kinetic and spectroscopic analysis of the inactivating effects of nitric oxide on the individual components of *Azotobacter vinelandii* nitrogenase. Biochemistry 1992; 31:2947-2955.

87. Calmels S, Ohshima H, Henry Y et al. Characterization of bacterial cytochrome cd₁-nitrite reductase as one enzyme responsible for catalysis of nitrosation of secondary amines. Carcinogenesis 1996; 17:533-536.

88. Kennedy MC, Gan T, Antholine WE et al. Metallothionein reacts with Fe^{2+} and NO to form products with g = 2.039 ESR signal. Biochem Biophys Res Commun 1993; 196:632-635.

89. Lee M, Arosio P, Cozzi A et al. Identification of the EPR-active iron-nitrosyl complexes in mammalian ferritins. Biochemistry 1994; 33:3679-3687.

90. Hausladen A, Fridovich I. Superoxide and peroxynitrite inactivate aconitases, but nitric oxide does not. J Biol Chem 1994; 269:29405-29408.

91. Castro L, Rodriguez M, Radi R. Aconitase is readily inactivated by peroxynitrite, but not by its precursor, nitric oxide. J Biol Chem 1994; 269:29409-29415.

92. Keller R, Keist R, Klauser S et al. The

macrophage response to bacteria: flow of L-arginine through the nitric oxide and urea pathways and induction of tumoricidal activity. Biochem Biophys Res Commun 1991; 177:821-827.

93. Corbett JA, Lancaster JR, Sweetland MA et al. Interleukin-1ß-induced formation of EPR-detectable iron-nitrosyl complexes in islets of Langerhans. J Biol Chem 1991; 266:21351-21354.

94. Corbett JA, Wang JL, Hughes JH et al. Nitric oxide and cyclic GMP formation induced by interleukin 1β in islets of Langerhans. Evidence for an effector role of nitric oxide in islet dysfunction. Biochem J 1992; 229-235.

95. Corbett JA, Wang JL, Sweetland MA et al. Interleukin 1β induces the formation of nitric oxide by β-cells purified from rodent islets of Langerhans. Evidence for the β-cell as a source and site of action of nitric oxide. J Clin Invest 1992; 90:2384-2391.

96. Corbett JA, Sweetland MA, Wang JL et al. Nitric oxide mediates cytokine-induced inhibition of insulin secretion by human islets of Langerhans. Proc Natl Acad Sci USA 1993; 90:1731-1735.

97. Nussler AK, Geller DA, Sweetland MA et al. Induction of nitric oxide synthesis and its reactions in cultured human and rat hepatocytes stimulated with cytokines plus LPS. Biochem Biophys Res Commun 1993; 194:826-835.

98. Stadler J, Bergonia HA, Di Silvio M et al. Nonheme iron-nitrosyl complex formation in rat hepatocytes: detection by electron paramagnetic resonance spectroscopy. Arch Biochem Biophys 1993; 302:4-11.

99. Nussler AK, Di Silvio M, Liu Z et al. Further characterization and comparison of inducible nitric oxide synthase in mouse, rat, and human hepatocytes. Hepatology 1995; 21:1552-1560.

100. Lancaster JR, Werner-Felmayer G, Wachter H. Coinduction of nitric oxide synthesis and intracellular nonheme iron-nitrosyl complexes in murine cytokine-treated fibroblasts. Free Rad Biol Med 1994; 16:869-870.

101. Geng YJ, Petersson AS, Wennmalm Å et al. Cytokine-induced expression of nitric

oxide synthase results in nitrosylation of heme and nonheme iron proteins in vascular smooth muscle cells. Exp Cell Res 1994; 214:418-428.

102. Mülsch A, Mordvintcev P, Vanin AF et al. Formation and release of dinitrosyl iron complexes by endothelial cells. Biochem Biophys Res Commun 1993; 196:1303-1308.

103. Stubbe J. Ribonucleotide reductases. Adv Enzymol 1990; 63:349-419.

104. Fontecave M, Nordlund P, Eklund H et al. The redox centers of ribonucleotide reductase of *Escherichia coli.* Adv Enzymol 1992; 65:147-183.

105. Reichard P. From RNA to DNA, why so many ribonucleotide reductases? Science 1993; 260:1773-1777.

106. Lepoivre M, Fieschi F, Coves J et al. Inactivation of ribonucleotide reductase by nitric oxide. Biochem Biophys Res Commun 1991; 179:442-448.

107. Lepoivre M, Fieschi F, Coves J et al. Inhibition of ribonucleotide reductase by nitric oxide donors. In: Moncada S, Marletta MA, Hibbs J et al, eds. The biology of nitric oxide. London, UK: Portland Press, 1992: 95-98.

108. Lepoivre M, Boudbid H, Petit JF. Antiproliferative activity of γ-interferon combined with lipopolysaccharide on murine adenocarcinoma: dependence on an L-arginine metabolism with production of nitrite and citrulline. Cancer Res 1989; 49:1970-1976.

109. Lepoivre M, Bobé P, Henry Y. Reversible alteration of ribonucleotide reductase E.P.R. properties in tumor target cells cocultured with NO-producing macrophages. In: Moncada S, Feelisch M, Busse R, Higgs EA, eds. The Biology of Nitric Oxide, Vol 4. London, UK, Portland Press, 1994:186-189.

110. Roy B, Lepoivre M, Henry Y et al. Inhibition of ribonucleotide reductase by nitric oxide derived from thionitrites: reversible modifications of both subunits. Biochemistry 1995; 34:5411-5418.

111. Haskin CJ, Ravi N, Lynch JB et al. Reaction of NO with the reduced R2 protein of ribonucleotide reductase from *Escherichia coli.* Biochemistry 1995; 34:11090-11098.

112. Roy B, Du Moulinet d'Hardemare A, Fontecave M. New thionitrites: synthesis, stability, and nitric oxide generation. J Org Chem 1994; 59:7019-7026.

113. Szekeres T, Gharehbaghi K, Fritzer M et al. Biochemical and antitumor activity of trimidox, a new inhibitor of ribonucleotide reductase. Cancer Chemother Pharmacol 1994; 34:63-66.

114. Szekeres T, Vielnascher E, Novotny L et al. Iron binding capacity of trimidox (3,4,5-trihydroxybenzamidoxime), a new inhibitor of the enzyme ribonucleotide reductase. Eur J Clin Chem Clin Biochem 1995; 33:785-789.

115. Ponka P, Schulman HM, Woodworth RC. Iron Transport and Storage. Boca Raton, USA: CRC Press, 1990.

116. Crichton RR, Charloteaux-Wauters M. Iron transport and storage. Eur J Biochem 1987; 164:485-506.

117. Crichton RR, Ward RJ. Iron metabolism - New perspectives in view. Biochemistry 1992; 24:11255-11264.

118. Klausner RD, Rouault TA. A double life: cytosolic aconitase as a regulatory RNA binding protein. Mol Biol Cell 1993; 4:1-5.

119. Klausner RD, Rouault TA, Harford JB. Regulating the fate of mRNA: the control of cellular iron metabolism. Cell 1993; 72:19-28.

120. Lazo JS, Pitt BR. Metallothioneins and cell death by anticancer drugs. Annu Rev Pharmacol Toxicol 1995; 35:635-653.

121. Pauwels M, Van Weyenbergh J, Soumillion A et al. Induction by zinc of specific metallothionein isoforms in human monocytes. Eur J Biochem 1994; 220:105-110.

122. Takeda K, Fujita H, Shibahara S. Differential control of the metal-mediated activation of the human heme oxygenase-1 and metallothionein IIA genes. Biochem Biophys Res Commun 1995; 207:160-167.

123. Yu X, Wu Z, Fenselau C. Covalent sequestration of melphalan by metallothionein and selective alkylation of cysteines. Biochemistry 1995; 34:3377-3385.

124. Schwarz MA, Lazo JS, Yalowich JC et al. Cytoplasmic metallothionein overexpression protects NIH 3T3 cells from *tert*-butyl hydroperoxide toxicity. J Biol Chem 1994;

269:15238-15243.

125. Shibuya K, Satoh M, Muraoka M et al. Induction of metallothionein synthesis in transplanted murine tumors by X irradiation. Radiat Res 1995; 143:54-57.

126. Choudhuri S, McKim JM, Klaassen CD. Induction of metallothionein by superantigenic bacterial exotoxin: probable involvement of the immune system. Biochim Biophys Acta 1994; 1225:171-179.

127. Sato M, Sasaki M, Hojo H. Differential induction of metallothionein synthesis by interleukin-6 and tumor necrosis factor-α in rat tissues. Int J Immunopharmac 1994; 16:187-195.

128. Sato M, Sasaki M, Hojo H. Antioxidative roles of metallothionein and manganese superoxide dismutase induced by tumor necrosis factor-a and interleukin-6. Arch Biochem Biophys 1995; 316:738-744.

129. Stennard FA, Holloway AF, Hamilton J et al. Characterization of six additional human metallothionein genes. Biochim Biophys Acta 1994; 1218:357-365.

130. Goulding H, Jasani B, Pereira H et al. Metallothionein expression in human breast cancer. Br J Cancer 1995; 72:968-972.

131. Nagel WW, Vallee BL. Cell cycle regulation of metallothionein in human colonic cancer cells. Proc Natl Acad Sci USA 1995; 92:579-583.

132. Otsuka F, Iwamatsu A, Suzuki K et al. Purification and characterization of a protein that binds to metal responsive elements of the human metallothionein IIA gene. J Biol Chem 1994; 269:23700-23707.

133. Zeng J, Heuchel R, Schaffner W et al. Thionein (apometallothionein) can modulate DNA binding and transcription activation by zinc finger containing factor Sp1. FEBS Lett 1991; 279:310-312.

134. Zeng J, Vallee BL, Kägi JHR. Zinc transfer from transcription factor IIIA fingers to thionein clusters. Proc Natl Acad Sci USA 1991; 88:9984-9988.

135. Schwarz MA, Lazo JS, Yalowich JC et al. Metallothionein protects against the cytotoxic and DNA-damaging effects of nitric oxide. Proc Natl Acad Sci USA 1995; 92:4452-4456.

136. Kröncke K-D, Fehsel K, Schmidt T et al.

Nitric oxide destroys zinc-sulfur clusters inducing zinc release from metallothionein and inhibition of the zinc finger-type yeast transcription activator Lac9. Biochem Biophys Res Commun 1994; 200:1105-1110.

137. Chasteen ND, Woodworth RC. Transferrin and lactoferrin. In: Ponka P, Schulman HM, Woodworth RC, eds. Iron Transport and Storage. Boca Raton, USA: CRC Press, 1990: 67-79.

138. Schlabach MR, Bates GW. The synergistic binding of anions and Fe^{3+} by transferrin. Implications for the interlocking sites hypothesis. J Biol Chem 1975; 250:2182-2188.

139. Bailey S, Evans RW, Garratt RC et al. Molecular structure of serum transferrin at 3.3 Å resolution. Biochemistry 1988; 27:5804-5812.

140. Anderson BF, Baker HM, Norris GE et al. Structure of human lactoferrin: crystallographic structure analysis and refinement at 2.8 Å resolution. J Mol Biol 1989; 209:711-734.

141. Anderson BF, Baker HM, Norris GE et al. Apolactoferrin structure demonstrates ligand-induced conformational change in transferrins. Nature 1990; 344:784-787.

142. Smith CA, Anderson BF, Baker HM et al. Metal substitution in transferrins: the crystal structure of human copper-lactoferrin at 2.1 Å resolution. Biochemistry 1992; 31:4527-4533.

143. Harris DC, Rinehart AL, Hereld D et al. Reduction potential of iron in transferrin. Biochim Biophys Acta 1985; 838:295-301.

144. Bakoy OE, Thorstensen K. The process of cellular uptake of iron from transferrin. A computer simulation program. Eur J Biochem 1994; 222:105-122.

145. Harris WR, Chen Y. Electron paramagnetic resonance and difference ultraviolet studies of Mn^{2+} binding to serum transferrin. J Inorg Biochem 1994; 54:1-19.

146. Kubal G, Mason AB, Patel SU et al. Oxalate- and Ga^{3+}-induced structural changes in human serum transferrin and its recombinant N-lobe. ^{1}H NMR detection of preferential C-lobe Ga^{3+} binding. Biochemistry 1993; 32:3387-3395.

147. Aramini JM, Vogel HJ. A scandium-45 NMR study of ovotransferrin and its half-molecules. J Am Chem Soc 1994; 116:1988-1993.

148. Aramini JM, Krygsman PH, Vogel HJ. Thallium-205 and carbon-13 NMR studies of human sero- and chicken ovotransferrin. Biochemistry 1994; 33:3304-3311.

149. Seidel A, Bill E, Häggström L et al. Complementary Mössbauer and EPR studies of iron(III) in diferric human serum transferrin with oxalate or bicarbonate as synergistic anions. Arch Biochem Biophys 1994; 308:52-63.

150. Grady JK, Mason AB, Woodworth RC et al. The effect of salt and site-directed mutations on the iron(III)-binding site of human serum transferrin as probed by EPR spectroscopy. Biochem J 1995; 309:403-410.

151. Dubach J, Gaffney BJ, More K et al. Effect of the synergistic anion on electron paramagnetic resonance spectra of iron-transferrin anion complexes is consistent with bidendate binding of the anion. Biophys J 1991; 59:1091-1100.

152. Battistuzzi G, Sola M. Fe^{3+} binding to ovotransferrin in the presence of α-amino acids. Biochim Biophys Acta 1992; 1118:313-317.

153. Vanin AF. Identification of divalent iron complexes with cysteine in biological systems by the EPR method. Biokhimia 1967; 32:228-232 (English translation 228-232).

154. Jezowska-Trzebiatowska B, Jezierski A. Electron spin resonance spectroscopy of iron nitrosyl complexes with organic ligands. J Molec Struct 1973; 19:635-640.

155. Carmichael AJ, Steel-Goodwin L, Gray B et al. Nitric oxide interaction with lactoferrin and its production by macrophage cells studied by EPR and spin trapping. Free Rad Res Comms 1993; 19:S201-209.

156. Richardson DR, Neumannova V, Ponka P. Nitrogen monoxide decreases iron uptake from transferrin but does not mobilise iron from prelabelled neoplastic cells. Biochim Biophys Acta 1995; 1266:250-260.

157. Richardson DR, Neumannova V, Nagy E et al. The effect of redox-related species of nitrogen monoxide on transferrin and

iron uptake and cellular proliferation of erythroleukemia (K562) cells. Blood 1995; 86:3211-3219.

158. Takenaka K, Suzuki S, Sakai N et al. Transferrin induces nitric oxide synthase mRNA in rat cultured aortic smooth muscle cells. Biochem Biophys Res Commun 1995; 213:608-615.

159. LeBrun NE, Cheesman RM, Thomson AJ et al. An EPR investigation of non-haem iron sites in *Escherichia coli* bacterioferritin and their interaction with phosphate. A study using nitric oxide as a spin probe. FEBS Lett 1993; 323:261-266.

160. LeBrun NE, Wilson MT, Andrews SC et al. Kinetic and structural characterization of an intermediate in the biomineralization of bacterioferritin. FEBS Lett 1993; 333:197-202.

161. Watt GD, Frankel RB, Papaefthymiou GC. Reduction of mammalian ferritin. Proc Natl Acad Sci USA 1985; 82:3640-3643.

162. Bolann BJ, Ulvik RJ. On the limited ability of superoxide to release iron from ferritin. Eur J Biochem 1990; 193:899-904.

163. Bolann BJ, Ulvik RJ. Decay of superoxide catalyzed by ferritin. FEBS Lett 1993; 318:149-152.

164. Monteiro HP, Vile GF, Winterbourn CC. Release of iron from ferritin by semiquinone, anthracycline, bipyridyl, and nitroaromatic radicals. Free Rad Biol Med 1989; 6:587-591.

165. Moreno JJ, Foroozesh M, Church DF et al. Release of iron from ferritin by aqueous extracts of cigarette smoke. Chem Res Toxicol 1992; 5:116-123.

166. Reif DW. Ferritin as a source of iron for oxidative damage. Free Rad Biol Med 1992; 12:417-427.

167. Reif DW, Simmons RD. Nitric oxide mediates iron release from ferritin. Arch Biochem Biophys 1990; 283:537-541.

168. Laulhère JP, Fontecave M. Nitric oxide does not promote iron release from ferritin. BioMetals 1996; 9:10-14.

169. Oria R, Sanchez L, Houston T et al. Effect of nitric oxide on expression of transferrin receptor and ferritin and on cellular iron metabolism in K562 human eryth–roleukemia cells. Blood 1995; 85:2962-2966.

170. Kim Y-M, Bergonia HA, Lancaster JR. Nitrogen oxide-induced autoprotection in isolated rat hepatocytes. FEBS Lett 1995; 374:228-232.

171. Balla G, Jacob HS, Balla J et al. Ferritin: a cytoprotective antioxidant stratagem of endothelium. J Biol Chem 1992; 267:18148-18153.

172. Theil EC. Iron regulatory elements (IREs): a family of mRNA non-coding sequences. Biochem J 1994; 304:1-11.

173. Melefors Ö, Hentze MW. Iron regulatory factor - the conductor of cellular iron regulation. Blood Reviews 1993; 7:251-258.

174. Haile DJ, Rouault TA, Tang CK et al. Reciprocal control of RNA-binding and aconitase activity in the regulation of the iron-responsive element binding protein: role of the iron-sulfur cluster. Proc Natl Acad Sci USA 1992; 89:7536-7540.

175. Haile DJ, Rouault TA, Harford JB et al. Cellular regulation of the iron-responsive element binding protein: disassembly of the cubane iron-sulfur cluster results in high-affinity RNA binding. Proc Natl Acad Sci USA 1992; 89:11735-11739.

176. Kennedy MC, Mende-Mueller L, Blondin GA et al. Purification and characterization of cytosolic aconitase from beef liver and its relationship to the iron-responsive element binding protein. Proc Natl Acad Sci USA 1992; 89:11730-11734.

177. Basilion JP, Rouault TA, Massinople CM et al. The iron-responsive element-binding protein: localization of the RNA-binding site to the aconitase active-site cleft. Proc Natl Acad Sci USA 1994; 91:574-578.

178. Hirling H, Henderson BR, Kühn LC. Mutational analysis of the [4Fe-4S]-cluster converting iron regulatory factor from its RNA-binding form to cytoplasmic aconitase. EMBO J 1994; 13:453-461.

179. Philpott CC, Klausner RD, Rouault TA. The bifunctional iron-responsive element binding protein/cytosolic aconitase: the role of active-site residues in ligand binding and regulation. Proc Natl Acad Sci USA 1994; 91:7321-7325.

180. Ward RJ, Kühn LC, Kaldy P et al. Control of cellular iron homeostasis by iron-responsive elements *in vivo*. Eur J Biochem 1994;

220:927-931.

181. Drapier J-C, Hirling H, Wietzerbin J et al. Biosynthesis of nitric oxide activates iron regulatory factor in macrophages. EMBO J 1993; 12:3643-3649.

182. Drapier J-C, Hirling H, Bouton C et al. Evidence that nitric oxide modulates IRP activities by targetting its [Fe-S] cluster. J Inorg Chem 1994; 56:44.

183. Weiss G, Goossen B, Doppler W et al. Translational regulation via iron-responsive elements by the nitric oxide/NO-synthase pathway. EMBO J 1993; 12:3651-3657.

184. Bouton C, Raveau M, Drapier J-C. Modulation of iron regulatory protein functions. Further insights into the role of nitrogen- and oxygen-derived reactive species. J Biol Chem 1996; 271:2300-2306.

185. Drapier J-C, Bouton C. Modulation by nitric oxide of metalloprotein regulatory activities. BioEssays 1996, 18:549-556.

186. Weiss G, Werner-Felmayer G, Werner ER et al. Iron regulates nitric oxide synthase activity by controlling nuclear transcription. J Exp Med 1994; 180:969-976.

187. Pantopoulos K, Hentze MW. Nitric oxide signaling to iron-regulatory protein: direct control of ferritin mRNA translation and transferrin receptor mRNA stability in transfected fibroblasts. Proc Natl Acad Sci USA 1995; 92:1267-1271.

188. Cairo G, Pietrangelo A. Nitric oxide-mediated activation of iron-regulatory protein controls hepatic iron metabolism during acute inflammation. Eur J Biochem 1995; 232:358-363.

189. Jaffrey SR, Cohen NA, Rouault TA et al. The iron-responsve element binding protein: a target for synaptic actions of nitric oxide. Proc Natl Acad Sci USA 1994; 91:12994-12998.

190. Ferreira GC, Franco R, Lloyd SG et al. Structure and function of ferrochelatase. J Bioenerg Biomembr 1995; 27:221-229.

191. Dailey HA, Finnegan MG, Johnson MK. Human ferrochelatase is an iron-sulfur protein. Biochemistry 1994; 33:403-407.

192. Ferreira GC. Mammalian ferrochelatase. Overexpression in *Escherichia coli* as a soluble protein, purification and characterization. J Biol Chem 1994; 269:4396-4400.

193. Ferreira GC, Franco R, Lloyd SG et al. Mammalian ferrochelatase, a new addition to the metalloenzyme family. J Biol Chem 1994; 269:7062-7065.

194. Franco R, Moura JJG, Moura I et al. Characterization of the iron-binding site in mammalian ferrochelatase by kinetic and Mössbauer methods. J Biol Chem 1995; 270:26352-26357.

195. Ferreira GC. Ferrochelatase binds the iron-responsive element present in the erythroid 5-aminolevulinate mRNA. Biochem Biophys Res Commun 1995; 214:875-878.

196. Furukawa T, Kohno H, Tokunaga R et al. Nitric oxide-mediated inactivation of mammalian ferrochelatase in vivo and in vitro: possible involvement of the iron-sulphur cluster of the enzyme. Biochem J 1995; 310:533-538.

197. Sellers VM, Johnson MK, Dailey HA. Function of the [2Fe-2S] cluster in mammalian ferrochelatase: a possible role as a nitric oxide sensor. Biochemistry 1996; 35:2699-2704.

198. Hidalgo E, Demple B. An iron-sulfur center essential for transcriptional activation by the redox-sensing SoxR protein. EMBO J 1994; 13:138-146.

199. Hidalgo E, Bollinger JM, Bradley TM et al. Binuclear [2Fe-2S] clusters in the *Escherichia coli* SoxR protein and role of the metal centers in transcription. J Biol Chem 1995; 270:20908-20914.

200. Nunoshiba T, DeRojas-Walker T, Wishnok JS et al. Activation by nitric oxide of an oxidative-stress response that defends *Escherichia coli* against activated macrophages. Proc Natl Acad Sci USA 1993; 90:9993-9997.

201. Cunningham RP, Asahara H, Bank JF et al. Endonuclease III is an iron-sulfur protein. Biochemistry 1989; 28:4450-4455.

202. Pantopoulos K, Hentze MW. Rapid responses to oxidative stress mediated by iron regulatory protein. EMBO J 1995; 14:2917-2924.

203. Juckett MB, Weber M, Balla J et al. Nitric oxide donors modulate ferritin and protect endothelium from oxidative injury. Free Rad Biol Med 1996; 20:63-73.

204. Henry YA, Singel DJ. Metal-nitrosyl interactions in nitric oxide biology probed by electron paramagnetic resonance spectroscopy. In: Feelisch M, Stamler JS, eds. Methods in Nitric Oxide Research. John Wiley & Sons, 1996:357-372.

CHAPTER 11

OVERPRODUCTION OF NITRIC OXIDE IN PHYSIOLOGY AND PATHOPHYSIOLOGY: EPR DETECTION

Yann A. Henry

INTRODUCTION

Stimulating developments in current research on the biological roles of NO are the immediate implications to physiology and pathophysiology, both in animal models and in humans.[1-3] Many diseases have been related to overproduction or underproduction of NO by NOS isoenzymes. The use of NO-donors, such as organic nitrates, nitroprusside or molsidomines, has in fact largely preceded the discovery of EDRF/NO and is quite well monitored (see chapter 12). The use of NO gas and that of NOS inhibitors have been under careful trial for several diseases since 1990. We shall limit the evocation of NO-related pathophysiology in animal models and in humans to a miscellaneous and restricted number of cases in which EPR spectroscopy has made some contribution to the estimation or quantification of NO: cancer, infections and septic shock, autoimmune diseases, organ transplantations and a few other cases in which NOS II induction occurs. Whenever possible we shall focus on human pathologies. Due to the low sensitivity of EPR spectroscopy, most of the cases of detection derive from NO synthesized by inducible NOS in a long-term process which allows NO accumulation on metalloprotein targets. Several selective reviews have been published on this subject.[4-9]

The constitutive and inducible NO synthase isoenzymes, their structure and function, their regulations and their cellular localizations have been introduced in chapter 8. We shall use the nomenclature of Förstermann and Schmidt,[10] reviewed by Nathan and Xie.[11]

Nitric Oxide Research from Chemistry to Biology: EPR Spectroscopy of Nitrosylated Compounds, edited by Yann A. Henry, Annie Guissani and Béatrice Ducastel.
© 1997 R.G. Landes Company.

Constitutive NOS in Human

Let us recall that the two constitutive NOS, the so-called neuronal (nNOS, NOS1 or NOS I)[12] and the endothelial (eNOS, NOS3 or NOS III)[13-17] have been detected in several human cell types: endothelium,[13,14,18] human umbilical vein endothelial cells (HUVEC),[19,20] cerebellum,[21,22] cavernosal smooth muscle,[23] skeletal muscle[24] and placenta.[25-27]

Inducible NOS in Human

The inducible NOS (iNOS, NOS2 or NOS II) first found in macrophages, appears to be ubiquitous,[28,29] including the central nervous system (CNS).[30] NO has been evoked in expired (exhaled) air, arising probably from alveolar macrophages.[31] As for NOS II in human macrophages, we have stated in chapter 8 the existence of a large controversy with respect to its expression and regulation pathways.[32-41]

NOS II is also found in other human cell types: neutrophils,[42] keratinocytes,[43] hepatocytes,[44-48] articular chondrocytes,[49-51] glomerular mesangial cells,[52] astrocytes,[53,54] aortic smooth muscle cells,[47] thyrocytes,[55] pancreatic β cells,[56] lung epithelial cells,[57,58] retinal pigmented epithelial cells,[59] osteoblasts,[60] etc.

Animal Models in Pathophysiology

Many animal models of various pathologies have been studied by use of pig, rabbit, guinea pig, rat and mouse in particular. However the relevance of rodent models to human pathologies has to be ascertained for each pathological case and for each affected organ specifically (see for instance, refs. 48, 61, 62).

CANCER AND INTERLEUKIN-2 IMMUNOTHERAPY

In view of the importance of cancer in human morbidity and mortality, cancer tissues were subjected to EPR spectra analysis by adventurous biophysicists like Vanin or Commoner very early (1965-1970) in the history of this spectroscopic method, without knowing the nature of the results and without any ability of a realistic interpretation of the molecular origin of the nitrosyl group. Actually these scattered results went mostly unnoticed, until the physiological and pathophysiological roles of NO began to be suspected.

Animal Models of Cancer

Ascites hepatoma tumors in rats were subjected to an EPR spectroscopy test and found to exhibit signals typical of nitrosyl hemoglobin (HbNO) (Fig. 11.1).[63] More precisely only the α subunits of the hemoglobin tetramer were NO-bound, and as we have explained in chapter 6 it corresponded to a $\alpha NO\alpha_{deoxy}\beta_{deoxy}\beta_{deoxy}$ or a $\alpha NO\alpha NO\beta_{deoxy}\beta_{deoxy}$ hybrid species in the T "deoxy-like" structure, with the EPR "signature" of a three-line superhyperfine structure. It was shifted to a $\alpha NO\alpha O_2\beta O_2\beta O_2$ or $\alpha NO\alpha NO\beta O_2\beta O_2$ species in the R "oxy-like" structure upon bubbling oxygen to the whole ascites hepatoma.[63]

This experiment together with the similar discoveries of Vanin et al and Commoner et al,[64,65] was the first well characterized EPR evidence for the presence of NO in cancer tissues.

The origin of the induction of NO synthesis by circulating cytokines was only demonstrated 25 years later.[66] Several different murine tumor types were grown subcutaneously, then dissociated from the animals into cell suspension and analyzed by EPR spectroscopy (Fig. 11.2).

Three different EPR signals were detected, arising firstly from $Fe(SR)_2(NO)_2$ complexes, secondly from two different nitrosyl heme species which could not be attributed to hemoglobin. The dependence of the EPR signals of cultured tumor cells on cytokines or on the presence of activated macrophages in coculture showed that the cell-mediated immune response to syngeneic tumors involves the induction of NOS II.[66] However it is not known whether this cytokine-dependent NO production corresponds to a stimulation or an inhibition of tumor cell proliferation, or is just an EPR-detectable epiphenomenon.

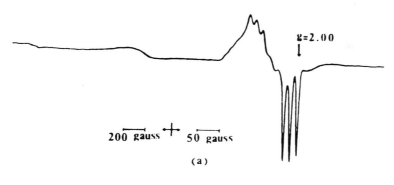

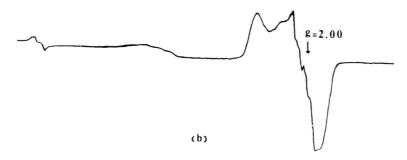

Fig. 11.1. Typical EPR spectra of hepatoma ascites tumors. (a), whole ascites; (b), after bubbling oxygen gas to whole ascites hepatoma. (Reproduced with permission from: Maruyama T, Kataoka N, Nagase S et al. Identification of three-line electron spin resonance signal and its relationship to ascites tumors. Cancer Res 1971; 31:179-184.) Copyright © The American Association for Cancer Research, Inc., Philadelphia, PA, USA.

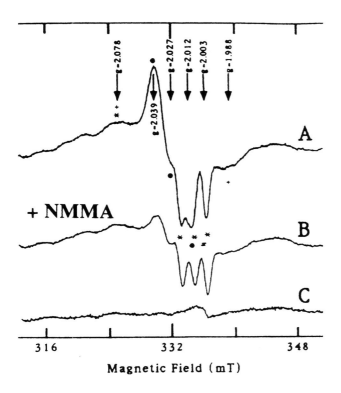

Fig. 11.2. EPR spectra of 3.2x10⁸ freshly dissociated RD-995 tumor cells. A, four separate EPR active species are present in this sample: an Fe(RS)₂(NO)₂ signal (●) with g-values at 2.039 and 2.012, a pentacoordinate heme-nitrosyl signal (*), with a triplet at 2.012 and a broad peak at 2.078, a hexacoordinate heme-nitrosyl signal (+) at g = 1.988 and g = 2.078, and a semiquinone radical signal at g = 2.003 (≠); B, same as A from a mouse treated with NMMA; C, cells grown in the absence of cytokines or activated macrophages. (Reproduced with permission from: Bastian NR, Yim CY, Hibbs JB et al. Induction of iron-derived EPR signals in murine cancers by nitric oxide. Evidence for multiple intracellular targets. J Biol Chem 1994; 269:5127-5131.) Copyright © The American Society for Biochemistry & Molecular Biology, Inc., Bethesda, MD, USA.

Although the role of infiltrating T cells and macrophages in the immune response against cancers remains uncertain, synthesis of NO following activation of macrophages has been invoked as having antiproliferative effects in vitro. In an in vivo model of UV light-induced murine skin cancers, EPR signals were detected in dissociated tumors.[67] NO was again found to be associated to both heme and [FeS] complexes. The same tumor cell line cultured in vitro for 5-9 passages did not exhibit such signals. Other controls included the absence of endotoxin contamination. In the same experiments NO was proven to have correlative antiproliferative effects as measured by [^{3}H]-thymidine incorporation, probably through RNR inhibition.

The exact role of NO as a pro- and antitumor agent was tested on a model of tumor growth.[68] A human colon adenocarcinoma cell line (DLD-1) and an engineered cell line, derived from DLD-1 expressing murine NOS II (clone iNOS-19), were inoculated subcutaneously to nude (immunodepressed athymic) mice and compared in their growth rates. The iNOS-19 clone grew more slowly in vitro than the wild-type, while the reverse was true in nude mice. Furthermore the iNOS-19 tumors were more vascularized. In fact the role of NO in tumor cell growth control could depend on its local concentration. This could be related to the positive correlation between NOS expression and tumor grade found in several human cancers (see below).

Human Cancer

The role of nitrogen oxides of dietary origin or synthesized by macrophages present in a chronic infection locus has been stressed in the etiology of human cancers, particularly in the digestive and urogenital tracts.[69-73] Solid tumors contain infiltrating macrophages, T-lymphocytes, endothelial cells and fibroblasts, all responsive to cytokine activation and regulation, in addition to tumor cells. The NOS activity levels and cellular localization have been investigated in human gynecological cancer.[74] The NOS activity was found to

be confined to malignant tissue and resulted from the expression of constitutive and inducible isoenzymes. It correlated with the differentiation of the tumor. Similar results were obtained in human breast cancer, with a correlation between NOS expression and tumor grade.[75] Human CNS neoplastic tumors were also found to produce high levels of NO through the expression of the two constitutive NOS isotypes, while inducible NOS was less often expressed.[76]

EPR spectroscopy has been used to detect HbNO in human liver, colon and stomach tumor tissues (Fig. 11.3).[77]

Most of the HbNO EPR signal is found in the necrotic central regions but is rare in peripheral tissue. HbNO formation results from the reaction of free NO, derived from "living" cells, with deoxyhemoglobin resulting from hypoxic necrotic and ill-vascularized tissue. Some methemoglobin was also detected, together with a large signal around $g = 4.3$ assigned to tumor-induced transferrin.[77]

Interleukin-2 Immunotherapy

The clinical effects of IL-2 cancer immunotherapy have been tested in two susceptible cancers, renal carcinoma and malignant melanoma. The response has been relatively low. IL-2 therapy induces synthesis of pro-inflammatory cytokines IFNγ, TNFα and IL-1 by immuno-competent cells; cytokines which in turn induce NOS II. EPR spectroscopy, together with other analytical methods, has been recently used in order to test IL-2 immunotherapy in a mouse model skin cancer (Meth A tumor) (Fig. 11.4).[78]

In patients receiving IL-2 immunotherapy, evidence of cytokine-induced NO synthesis was gathered.[79-82] IL-2 administration to patients with renal cell carcinoma or malignant melanoma was followed by increased NO synthesis as reflected by serum nitrate levels (8-fold increase) and 24-hr urine nitrate excretion (9-fold increase).[79] Tracer studies showed that the origin of NO was L-arginine. Similar results were found for patients with the same tumor types, treated with IL-2 and anti-CD3

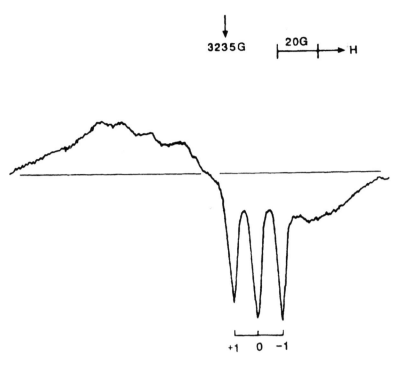

Fig. 11.3. Typical EPR spectrum at 77 K for nitrosyl hemoglobin from the necrotic region of a stomach tumor. (Reproduced with permission from: Symons MCR, Rowland IJ, Deighton N et al. Electron spin resonance studies of nitrosyl haemoglobin in human liver, colon and stomach tumour tissues. Free Rad Res Comms 1994; 21:197-202.) Copyright © International Publishers Distributors, Bâle, Switzerland.

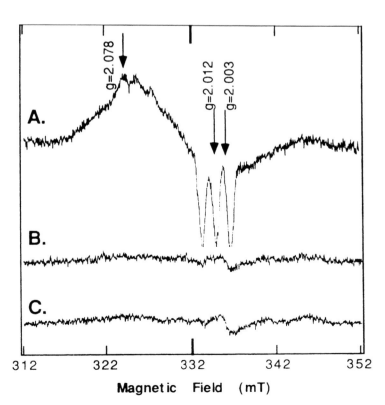

Fig. 11.4. IL-2 therapy-induced heme-nitrosyl EPR signal at 77 K in Meth A ascites tumor. Meth A ascites cells were harvested from mice treated with IL-2 or with concomitant NMMA, NOS inhibitor, administration. Ascites cells were washed from RBC before conditioning as a thick suspension in EPR tube. A, Meth A ascites, contained pentacoordinate (triplet at g = 2.012) and hexacoordinate nitrosyl heme (g = 2.078) signals; B, Ascites cells from control untreated animals; C, administration of NMMA to IL-2-treated mice. (Reproduced with permission from: Yim C-Y, McGregor JR, Kwon O-D et al. Nitric oxide synthesis contributes to IL-2-induced antitumor responses against intraperitoneal Meth A tumor. J Immunol 1995; 155:4382-4390.) Copyright © 1995. The American Association of Immunologists, Bethesda, MD, USA.

monoclonal antibody-activated lymphocytes (T-AK cells).[81] Increased NO synthesis could be correlated to a significant decrease in blood pressure of all patients. Actually hypotension is a dangerous side effect of IL-2 immunotherapy. A relationship between NO synthesis and changes in immunological (IFNγ, TNFα and neopterin) and vascular parameters (systolic blood pressure and pulse rate) was established in patients with metastatic cancer.[82] A model study on mice showed that the effects of IL-2 immunotherapy could be enhanced by supplemental dietary L-arginine.[80]

PARASITES, FUNGI AND BACTERIAL INFECTIONS

Many parasites, fungi and bacterial infections have been found, in animal models and in humans, to involve host defense response by immunocompetent cells through the induction of NOS II by bacterial peptides and by IFNγ, TNFα and IL-1β cytokines. We present here miscellaneous examples, with (in a few cases) a proof of NO binding to intracellular metalloprotein targets detected by EPR spectroscopy of whole tissues.

BORDETELLA PERTUSSIS (WHOOPING COUGH) INFECTION

Bordetella pertussis is a gram-negative bacteria ("bacille de la coqueluche" of Bordet and Gengou). Infection in humans induces violent, debilitating coughing episodes following the release of a specific peptidoglycan of the muramyl peptides family, known as tracheal cytotoxin (TCT). Like other muramyl peptides, such as that of *Neisseria gonorrhoeae*, TCT has been demonstrated to trigger the production of intracellular IL-1 by respiratory epithelial cells.[83] TCT release and IL-1 production result in inhibition of DNA synthesis in epithelial cells and the specific destruction of ciliated cells. The critical intermediary step has been demonstrated to be NO biosynthesis by the cytokine inducible pathway.[83] In hamster trachea epithelial cells treated with TCT, aconitase is inhibited, and the specific EPR signals at $g_\perp = 2.044$

and $g_{//} = 2.015$ ($g_{av} = 2.04$) of the $Fe(SR)_2(NO)_2$ complexes were detected. IL-1 alone produced the same effect. DNA synthesis inhibition was simultaneous to aconitase inhibition. NOS inhibitors, aminoguanidine and NMMA, reversed the TCT and the IL-1 effects. These results explain the epithelial autotoxicity of nitric oxide and its role in the respiratory cytopathology of pertussis.[83]

CORYNEBACTERIUM PARVUM LIVER INFLAMMATION

Rat and human hepatocytes contain the NOS II pathway responsive to IFNγ, TNFα, IL-1β and LPS,[45,48,84,85] like neighboring Kupffer cells.[86-88] EPR spectroscopy analysis of cultured hepatocytes after exposure in vitro to inflammatory stimuli showed the typical $g_{av} = 2.04$ axial signal of the $Fe(SR)_2(NO)_2$ complexes, exclusively associated to the cytosolic fraction of hepatocytes (Fig. 11.5).[45]

The same EPR signals were found in hepatocytes freshly isolated from *Corynebacterium parvum*-treated rats or mice.[45,48,85,89] Notable differences were established in the induction process of NOS II by cytokines in hepatocytes as compared to macrophages,[45] and in the intracellular and intercellular effects of NO. Whole blood of *C. parvum*-treated mice showed also the presence of HbNO, as detected by EPR spectroscopy. Liver tissue of the same animals showed the presence of $Fe(SR)_2(NO)_2$ complexes' $g_{av} = 2.04$ axial signal, together with the P-420-NO complex characteristic signals (see chapter 6) (Fig. 11.6).[45] NOS II was found to be present in the liver cytosolic fraction. Chronic liver inflammation by *C. parvum* in mice seems to reveal a role for hepatic P450's functions.[89]

Similar experiments were performed in vitro with human hepatocytes upon activation by cytokines,[46] particularly IL-1β.[90] To our knowledge, no trial has yet been reported for any human liver disease.

HUMAN DUODENAL ULCER

The pathology of human duodenal ulcer arises probably from the association of

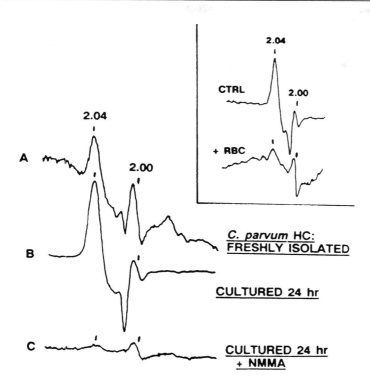

Fig. 11.5. EPR spectra of C. parvum-infected hepatocytes. A, hepatocytes freshly isolated (2x10⁷ cells/ml); B, same as A after 24 hr culture without additional stimuli (6x10⁷ cells/ml); C, same as B in the presence of NMMA. Inset: hepatocytes isolated from C. parvum-treated rat and cultured for 18 hr in the absence (top) or presence (bottom) of rat erythrocytes. (Reproduced with permission from: Stadler J, Bergonia HA, Di Silvio M et al. Nonheme iron-nitrosyl complex formation in rat hepatocytes: detection by electron paramagnetic resonance spectroscopy. Arch Biochem Biophys 1993; 302:4-11.) Copyright © Academic Press Ltd, London, UK.

Helicobacter pylori gastric infection and peptic disease.[91] The NOS II activity was measured in the stomach antrum and fundus of 17 duodenal ulcer patients and was found to be respectively 2- and 1.5-fold higher than in 14 normal subjects, irrespective of the presence or absence of *H. pylori*. This stimulated gastric mucosal NOS II activity may contribute to the pathogenesis of the disease.[91]

CEREBRAL MALARIA

Neuropaludism is an endemic disease to which 40% of the world population is subjected. Two to three hundred million people are infected each year. Cerebral malaria or neuropaludism is one of the major complications of *Plasmodium falciparum* infection and a major cause of infant mortality in Africa. The pathophysiological mechanisms leading to an unrousable coma state and to recovery or death are not elucidated. A cytokine theory of *P. falciparum* human brain infection has

been proposed recently.[92,93] One of the cellular aspects of the disease is that red blood cells (RBC) infected during the parasite cycle, adhere to endothelial cells, leading to mechanical blockage of small blood vessels in the brain and to cerebral hypoxia. This sequestration theory of malaria has been complemented by the cytokine theory.[92,93] A role of circulating inflammatory cytokines, TNFα and IL-1, was proposed, with a correlative role of induced NO synthesis on NMDA receptors and on brain vessel vasodilatation.

The evolution of NO plasma concentrations, measured as the NO_2^-/NO_3^- level, was studied in 28 African children with cerebral malaria.[94] Two groups were compared, a group of children with favorable outcome versus a group presenting neurologic sequelae. Both groups had high levels of plasma NO (44.3 ± 36.5 µmol/l), they were significantly higher in the group with favorable outcome. However there was no statistically significant relationship with

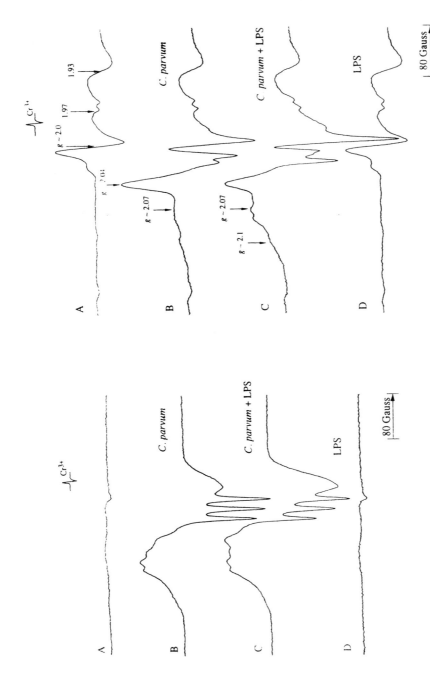

Fig. 11.6. Left panel. EPR spectra at 77 K of the whole blood from, (A), a normal mouse; (B), a mouse infected with C. parvum (7 days before blood drawing); (C), a mouse infected with C. parvum (as in B) and injected with LPS (2 hr before); (D), a mouse injected with LPS (2 hr before blood drawing). Right panel. EPR spectra at 77 K of liver tissue of the same animals as in left panel. (Reproduced with permission from: Chamulitrat W, Jordan SJ, Mason RP et al. Targets of nitric oxide in a mouse model of liver inflammation by Corynebacterium parvum. Arch Biochem Biophys 1995; 316:30-37.) Copyright © Academic Press Ltd, London, UK.

the coma score or clinical measurements. There was an increase of plasma NO with quinine treatment. A deleterious or a protective role of NO has to be fully investigated.[94]

Evidence was recently presented that human RBC infected with *P. falciparum* exhibited a high Ca^{2+}-independent-NOS activity.[95] The NOS apparent molecular mass was < 100 kDa, showing that the enzyme expressed by *P. falciparum* is a very different enzyme from the NOS found in mammalian cells. Furthermore the parasite *P. falciparum* released a soluble factor (> 100 kDa) able to induce NO synthesis in human endothelial cells, but it's very different from the host's cytokines.[95]

Negative evidence for NO production in *P. berghei*-infected CBA/J mice, as compared to resistant BALB/c mice, has been found in cerebral tisues, in RBC or in plasma, as measured by EPR spectroscopy (V Asensio, PB Falanga & Y Henry, unpublished results).[96] No difference was observed either in the expression levels of mRNAs for NOS II (iNOS) and for NOS I (nNOS) in the brain of both resistant and infected mice. NOS inhibitors had no effect on mice survival.[96,97] This is another instance of obvious large differences of behavior in rodent animal models and human, in apparently comparable pathologies.

HIV PATIENTS

HIV patients are subjects to many opportunistic infections, by fungus *Cryptococcus neoformans*, cytomegalovirus (CMV), *Pneumocystis carinii*, etc. The level of nitric oxide synthesis has been detected in several patient groups.

Fungus *C. neoformans* brain infection leads to antifungal drug-resistant meningoencephalitis. In the brain, human astrocytes are known to express the NOS II pathway.[53] It has recently been shown that human astrocytes, when activated by IL-1β and IFNγ, inhibited the *C. neoformans* growth.[98] These results showed that astrocytes which are ubiquitous in the brain participate in the host's defense and that NO has an antimicrobial activity in human cells.[98]

The production of nitric oxide was examined in patients with advanced HIV infection and in intensive care unit control patients.[99] Extrinsic nitrate and nitrite consumption were carefully controlled. Nitric oxide synthesis was found to be within normal range.[99]

Production of NO by peripheral blood mononuclear cells and polymorphonuclear leukocytes of patients with HIV-1 has recently been tested.[40,100] While cells isolated from asymptomatic seropositives had normal NO production, it was 3- to 5-fold higher in patients with disseminated mycobacterial infection, encephalopathy, cerebral toxoplasmosis or pneumonia. These high levels of NO arise from defense mechanisms in the host and might explain disseminated microvascular damage and nervous system disorders in acquired immunodeficiency syndrome (AIDS) patients.[40,100]

A clear demonstration of the expression of NOS II in CMV-infected glial cells of retinas of AIDS patients has been given.[101] Chorioretinis is the most common AIDS-associated ocular infection, causing varying ophthalmic lesions. The expression of NOS II was tested by NADPH-diaphorase staining and NOS immunochemistry, in parallel with immunodetection of glial cells and CMV. NOS II was localized in CMV-infected glial cells. It suggests that high levels of NO produced in the human retina as a result of CMV infection might induce retinitis.[101]

SEPTIC SHOCK

Septic shock is a devastating clinical state, as there are two hundred thousand septic shock patients per year in the USA and more than one hundred thousand in the European community, half of which die as a consequence, often within hours of the shock onset.[102] It complicates all types of infections. It is characterized by persistent hypotension refractory to vasopressors, low systemic resistance and an elevated, but inadequate, cardiac output.[102] Multiple organ failure can develop as a consequence, such as adult respiratory distress syndrome (ARDS), renal or hepatic failure. There exists no real specific therapy except intensive care. Circulating cytokines contributing to this state are TNFα, IL-1 and IL-2, following massive bacterial invasion. Gram-positive bacteria, such as *Streptococcus pyogenes* or *Staphylococcus aureus* produce enterotoxins (for instance TSST-1), which in turn induce IL-1 production by monocytes and TNFα by macrophages. LPS is the main factor inducing septic shock following gram-negative bacterial infection, also followed by release of IL-1, IL-2 and TNFα in the bloodstream.[102] They can have synergistic effects on the activation of circulating macrophages and also on the cardiovascular system or the gastrointestinal tract, through the activation of both endothelial cells and smooth muscle cells inducing the NOS II pathway.[2,103] As mentioned above many other cell types can be induced along this pathway, Kupffer cells and hepatocytes in the liver, for instance. At first, inhibition of macrophage-activating cytokines was thought to be a correct approach to a specific treatment of septic shock.[104] In a more realistic way, measurements of circulating IL-1β, TNFα and IL-6 were used as a marker of the severity of the shock state.[105]

ANIMAL MODELS OF SHOCKS

The involvement of the L-arginine-NO inducible pathway was rapidly tested in animal models, mostly by use of NOS inhibitors.[106,107] NO overproduction has been followed in the time course of rat and mouse endotoxic shock, by sampling of the animal venous or arterial blood and EPR spectroscopy analysis.[108-110] Nitrosylated complex of hemoglobin carrying αNO hybrids and loosely bound heme(Fe^{II})NO species are detected in venous blood 3 hr after LPS injection, are maximal after 4-8 hr, and have disappeared after one day, before the animal dies on day 3 (Fig. 11.7).[108]

Similar results were obtained following hemorrhagic shock. Experiments comparing the effects of LPS with those of recombinant IL-1 and TNFα showed that 5-15 μmol/l HbNO could be detected in

rat blood by treatment with LPS or IL-1 or TNFα alone.[111] Extremely high levels of 30-80 μmol/l HbNO could be reached for combinations of these three NOS II inducers acting in synergy.[111] IL-1 and TNFα also synergistically induce NOS II in rat vascular smooth muscle cells.[112] The synergy in endotoxin-treated rats also exists between IL-1 and TNFα with IFNγ (Fig. 11.8).[111,113] This suggests that the increase in TNFα release by IFNγ plays a key role in NO generation in LPS-treated rats.[111,113] Iron might also potentiate LPS-induced NO production.[114]

Finally a very thorough analysis of the EPR spectra obtained in the complete arteriovenous cycle in cytokine-treated rats has been provided by Kosaka et al.[115] It demonstrates in vivo the formation of "natural" hemoglobin hybrids such as αNOα$_{deoxy}$β$_{deoxy}$β$_{deoxy}$ or αNOαNOβ$_{deoxy}$β$_{deoxy}$ in the "deoxy-like" or "low-affinity" T structure in the venous cycle, with the EPR "signature" of a three-line superhyperfine structure. It was shifted to a αNOαO$_2$βO$_2$βO$_2$ or αNOαNOβO$_2$βO$_2$ species in the "oxy-like" or "high-affinity" R structure in arterial blood (Figs. 11.9 and 11.10).[115]

This beautiful demonstration of the allosteric quaternary structure shift of hemoglobin in vivo amply justifies, 20 years behind, the academic work performed with reconstituted αNOαNOβ$_{deoxy}$β$_{deoxy}$ and αNOαNOβO$_2$βO$_2$ hybrid species, which were then only meant to test the allosteric theory and predictions of Perutz based on X-ray data analysis.[5,116,117] This has been extensively analyzed in chapter 4.

Chemical intoxication can facilitate endotoxic shock by LPS. Such is the case in CCl$_4$-injected mice.[118] Nitric oxide was detected by EPR spectroscopy as HbNO in RBC and P-420-NO in liver tissues.[118]

Another effect of acute LPS endotoxemia in rat, following IL-1 release and NO production, is the largely increased number of endothelial cells and macrophages in rat liver, as compared to animal treated with IL-1 alone.[119] This could well explain the irrepressible character of endotoxic shock.[119]

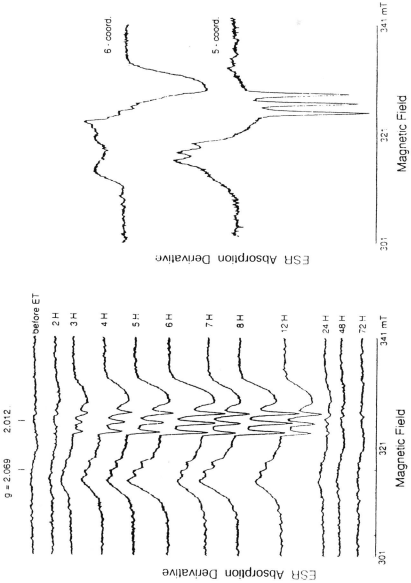

Fig. 11.7. Left panel. Formation of EPR signals in rat blood during endotoxic shock. Endotoxin was administered as a single dose. Blood samples from anesthesized rats were taken by puncture of the jugular vein at various indicated times. Right panel. Demonstration of two different types of EPR signals in rat blood during endotoxic shock. They are interpreted as a pentacoordinate nitrosyl heme complex and hexacoordinate nitrosyl heme complex of the α subunit of hemoglobin. (Reproduced with permission from: Westenberger U, Thanner S, Ruf HH et al. Formation of free radicals and nitric oxide derivative of hemoglobin in rats during shock syndrome. Free Rad Res Comms 1990; 11:167-178.) Copyright © International Publishers Distributors, Bâle, Switzerland.

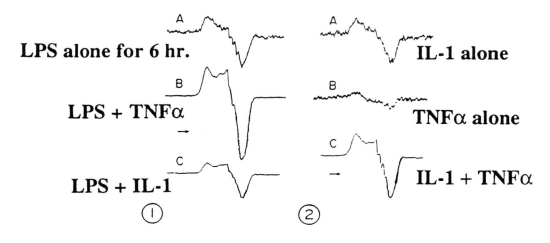

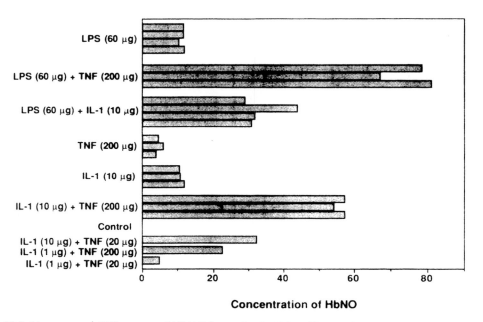

Fig. 11.8. Upper panel. EPR spectra of HbNO formation in LPS and/or cytokine-treated rat blood. Lower panel. Comparison of HbNO generation (expressed in μmol/l) by LPS, IL-1 and/or TNFα treatment of rat. (Reproduced with permission from: Kosaka H, Harada N, Watanabe M, et al. Synergistic stimulation of nitric oxide hemoglobin production in rats by recombinant interleukin 1 and tumor necrosis factor. Biochem Biophys Res Commun 1992; 189:392-397.) Copyright © Academic Press, Inc, Orlando, Florida, USA.

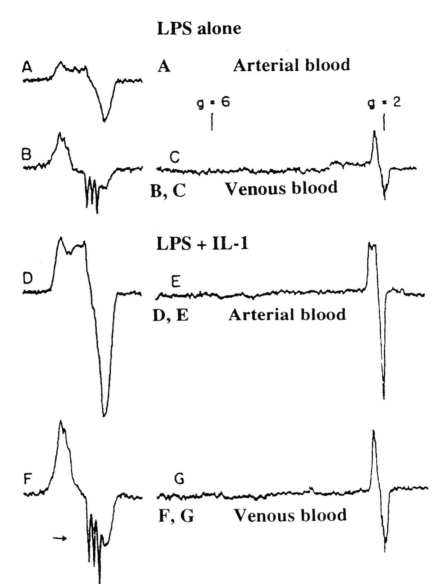

Fig. 11.9. EPR spectra of HbNO in rats treated with LPS or with both LPS and IL-1. After 6 hr, blood samples were taken from abdominal vein and from aorta of the same anesthesized rat, transferred to EPR tubes and frozen in liquid nitrogen. LPS-treatment alone: A, arterial blood; B,C, venous blood. LPS and IL-1-treatment: D,E, arterial blood; F,G, venous blood. Left hand spectra: g = 2 region. Right hand spectra: full-scan showing the absence of methemoglobin EPR signal at g = 6. (Reproduced with permission from: Kosaka H, Sawai Y, Sakaguchi H et al. ESR spectral transition by arteriovenous cycle in nitric oxide hemoglobin of cytokine-treated rats. Am J Physiol 1994; 266:C1400-1405.) Copyright © American Physiological Society, Bethesda, MD, USA.

As to the exact role of NO itself in the endotoxic shock, several results compel one to keep balanced points of view. For instance an increase in serum N^{ω}-hydroxy-L-arginine has been demonstrated in LPS-treated rats.[120] A potent and selective inhibitor of NOS II, *S*-methylisothiourea sulfate, has beneficial effects on arterial pressure and improves survival of LPS-treated rats.[121] An increase in NO exhaled in animals with LPS-induced lung inflammation was found to be an early marker of sepsis development.[122] All three recent results imply a role of NOS II induction in the development of endotoxic shock. On the contrary, endogenously synthesized NO has a protective role in staphylococcal enterotoxin B-induced mice.[123]

Similar balanced or mitigated conclusions can be drawn from septic shock experiments performed on NOS II-lacking mice ("Knock-out" mice).[124,125] NOS II induction is beneficial to defend the host against infection, but might be dangerous for the host itself. Again, comparisons like Dr. Jekyll and Mr. Hyde, friend or foe, double-edge sword, Janus face, etc., show that "further work is needed to understand anything", an undoubtedly useful conclusion!

ENDOTOXIC SHOCK IN HUMANS

Nitric oxide levels were measured in patients as a first test of its role as a mediator in sepsis.[126] A control group of 14 healthy volunteers had an average NO_2^-/NO_3^- level of 28.9 ± 3.6 µmol/l (range 16-62 µmol/l). Patients with clinical sepsis had significantly higher levels (63.1 ± 6.5 µmol/l), while patients with trauma had lower levels (12.8 ± 1.5 µmol/l). Trauma patients which developed sepsis while in the intensive care unit, kept low NO_2^-/NO_3^- levels. This result would explain increased vasodilation observed in sepsis.[126]

Rapidly after the cytokine and NO theory on sepsis development was formulated, a controversy arose in the clinical "arena"[127] as to whether inhibiting NOS might be useful.[128-132] The first problem would be to select inhibitors specific for

NOS II, work which is still in full development (see chapter 8). The second problem is to determine the various mechanisms involving endotoxins, cytokines and NO, and to delineate subsets of septic shock patients in order to define new therapeutic strategies with antibodies, cytokine inhibitors or anti-inflammatory agents.[133-135] There is certainly no clear-cut response. Such a pilot study tried to correlate NO_2^-/NO_3^- serum levels with those of TNFα and IL-6.[136]

To our knowledge, no EPR measurements of NO bound to hemoglobin in RBC of septic shock patients have ever been reported. As to our own experience, out of 22 blood samples of septic shock patients, only 2 presented a weak EPR signal assigned to HbNO. It is not sure that the EPR detection of HbNO would present much interest in prognosis, HbNO being only one of the end products of NO synthesis like NO_2^-/NO_3^- in serum.

AUTOIMMUNE DISEASES

An essential role in autoimmunity has been attributed to nitric oxide or is at least often discussed. The profound causes of autoimmune diseases are not really understood and in general no specific treatment exists. In most autoimmune diseases, both organ-specific and systemic, an inflammatory reaction occurs, and the only therapeutics are the use of nonspecific anti-inflammatory substances.[137] For instance NOS II induction has been well characterized in several human inflammatory bowel diseases: ulcerative colitis, Crohn's disease and toxic megacolon.[138-144] The inducible NOS pathway has been evoked in many diverse autoimmune diseases: psoriasis vulgaris,[145] several CNS diseases,[146] multiple sclerosis,[147,148] schizophrenia,[149] etc. Since human articular chondrocytes are known to express NOS II, following the induction by IL-1, an obvious role for NO was also proposed in the development of rheumatic diseases: osteoarthritis and rheumatoid arthritis, which lead to the inhibition of synthesis of the cartilaginous matrix.[51,150-152] A proper understanding of the mechanisms leading to NO synthesis could offer new therapeutic possibilities.

DIABETES MELLITUS

Diabetes mellitus is another instance in which the discovery of a role for NO has given a better focused view of a disease. In insulin-dependent diabetes IL-1β induces pancreatic β cells of Langerhans islets' modification and eventual destruction.[153] The most obvious effect is a time-dependent inhibition of the glucose-stimulated insulin secretion. In this process NO and cGMP formations have been proven within islets and more precisely in β cells, which have been detected as being both the source of NO and the target for its inhibitory action.[154-159] IL-1β could also have a synergistic effect with IFNγ and TNFα.[160,161] The cellular effects of IL-1β over β-islets are reversible early after the induction of NOS II.[162] By using EPR spectroscopy, evidence was provided for the destruction of [Fe-S] clusters of mitochondrial respiratory enzymes and aconitase, explaining a 40-60% inhibition of mitochondrial oxidation of D-glucose to CO_2, and in turn the inhibition of insulin secretion (Fig. 11.11).[154-156]

As in the rodent model, it has been demonstrated that NO mediates cytokine-induced inhibition of insulin secretion by human islets of Langerhans.[56] Individual effects of cytokines IL-1β, IFNγ and TNFα upon human islets were null in terms of NO production; however a high synergy was observed in combination of IL-1β + IFNγ (5-fold increase in nitrite formation) or of the three cytokines together (8-fold). Nitrite concentration in the isolated human islets culture medium could then reach 5 to 8 µmol/l. EPR spectroscopy brought evidence of binding of NO to mitochondrial [Fe-S]-containing enzymes under conditions which impair their activities (Fig. 11.12).[56]

Several authors have developed the cytokines theory of the destruction of islets' β cells but have contested any role of NO production in insulin-dependent diabetes mellitus in human.[61,62,161,163]

MULTIPLE SCLEROSIS AND EXPERIMENTAL ALLERGIC ENCEPHALOMYELITIS

There are several CNS diseases resulting from autoimmune states, such as multiple sclerosis. The source of NO in neuroimmune disorders is probably astroglial cells which contain both NOS I and NOS II, and produce NOS II inducer cytokines: IL-1, IL-6, TNF and PGE_2.[148,164] Experimental allergic encephalomyelitis (EAE) is a demyelinating autoimmune disorder induced by T lymphocyte and macrophage activation, which is used as an animal model of multiple sclerosis.[165,166] NO was localized in spinal cords of affected mice, as a nitrosylated [Fe-S] protein complex. This result indicates that NO may play a role in EAE and perhaps in multiple sclerosis (Fig. 11.13).[165]

The level of NO present in the CNS of rats infected with Borna disease virus or with rabies virus, or having received myelin basic protein (MBP)-specific T cells, was measured by spin-trapping and EPR spectroscopy. It was found to be much higher (up to 30-fold) than in controls.[166] The spin-trapping technique is explained in chapter 13.

ORGAN TRANSPLANTATIONS

Endothelial cells play a primary role in allograft rejection in their interaction with immunocompetent cells of the host and subsequently as the target of the humoral and cellular response by lymphocytes and macrophages.[167] A better understanding of the induction mechanisms of NOS II by cytokines and of the exact role played by NO as an endogenous immunomodulator in rejection could be of some help in organ transplantation.[168]

ALLOGRAFTS: ANIMAL MODELS

Cytokines involved in macrophage and hepatocyte activation, and in β cell impairment, participate also in the cellular immune response to an alloantigenic stimulus. In this process, the generation of cytotoxic T lymphocytes, which leads to cell proliferation inhibition, is accompanied

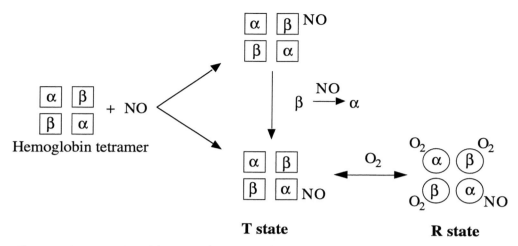

Fig. 11.10. Interpretation of the spectral transition of HbNO observed in figure 11.9, occuring during the arteriovenous cycle, in terms of the allosteric equilibrium and of preferential binding of NO to an α subunit of hemoglobin (see chapter 4 for detailed interpretation). (Reproduced with permission from: Kosaka H, Sawai Y, Sakaguchi H et al. ESR spectral transition by arteriovenous cycle in nitric oxide hemoglobin of cytokine-treated rats. Am J Physiol 1994; 266:C1400-1405.) Copyright © American Physiological Society, Bethesda, MD, USA.

by a large production of NO derived from the NOS II pathway, as revealed by NO_2^-/NO_3^- concentrations in a rat sponge matrix allograft model.[169-172]

This was confirmed by the same group in rat heart allografts followed by EPR spectroscopy of rat blood and tissues.[172] As in septic and hemorrhagic shocks, NO was detected as the αNO complex in hemoglobin tetramers in the rat packed RBC. A g_{av} = 2.04 EPR signal was also detected in grafted heart tissue homogenates at the site of allograft rejection, but in no other tissue (lung, spleen or liver). Both EPR signals were maximal on day 5 after allograft, simultaneously with the onset of allograft rejection. Two relevant negative controls were performed; no EPR signal was detected in blood or in heart tissue in syngeneic grafts nor when allografts were treated with the immunosuppressor FK506 (Fig. 11.14).[172]

N^ω-monomethyl-L-arginine (NMMA), an inhibitor of NOS, administered to mice receiving allogeneic heart transplants, inhibits nitric oxide production, but does not affect graft rejection. High output

NO synthesis was detected by EPR spectroscopy of heme-NO complexes and was inhibited by NMMA. However this NO production or its inhibition was not essential for allograft rejection.[173]

The use of another NOS inhibitor, aminoguanidine, more specific of NOS II, led to somewhat different results. Aminoguanidine treatment prolonged graft survival and reduced the histological grade of rejection.[174]

Detection of mRNA for NOS II and of NOS II itself showed that NO production is induced in infiltrating macrophages and cardiac myocytes of the rejecting allogeneic grafts.[175] Urinary nitrate excretion can be used as an index of acute cardiac allograft rejection preceding external signs of rejection.[176] As mentioned above, administration of NMMA causes only a small increase in graft survival, suggesting that NO would play a minor role as a cytotoxic agent in cardiac graft rejection.[173,177] A systematic appraisal of NO level could also be of help in the test of preservation solutions for transplanted organs.[178,179]

Similar results were obtained in two other experimental models of small bowel

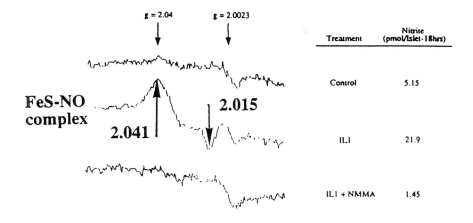

Treatment	Nitrite (pmol/Islet-18hrs)
Control	5.15
ILI	21.9
ILI + NMMA	1.45

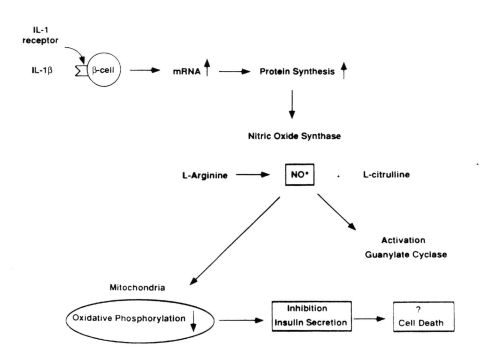

Fig. 11.11. Upper panel. IL-1-induced formation of a g_{av} = 2.04 EPR signal, observed at 77 K, and simultaneous formation of nitrite by islets. Islets were cultured for 18 hr alone, in the presence of IL-1, or in the presence of IL-1 and NMMA. Lower panel. Proposed model for IL-1β-mediated β cell dysfunction and destruction. (Reproduced with permission from: Corbett JA, Lancaster JR, Sweetland MA et al. Interleukin-1β-induced formation of EPR-detectable iron-nitrosyl complexes in islets of Langerhans. J Biol Chem 1991; 266:21351-21354.) Copyright © The American Society for Biochemistry & Molecular Biology, Inc., Bethesda, MD, USA, and from (Corbett JA, Wang JL, Sweetland MA et al. Interleukin 1β induces the formation of nitric oxide by β cells purified from rodent islets of Langerhans. Evidence for the β-cell as a source and site of action of nitric oxide. J Clin Invest 1992; 90:2384-2391). Copyright © The American Society for Clinical Investigation, New York, NY, USA.

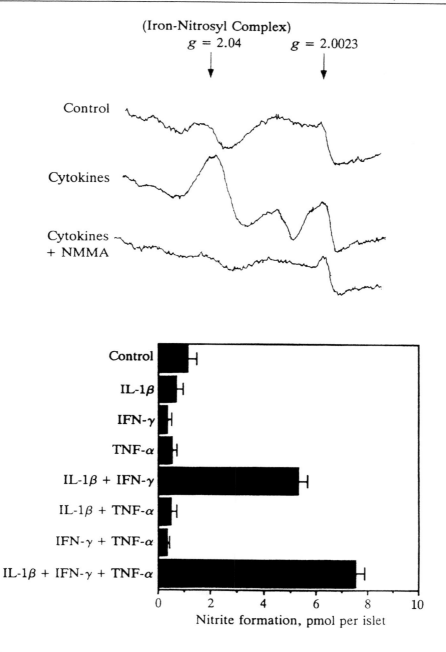

Fig. 11.12. Upper panel. Effects of cytokines on formation of a g_{av} = 2.04 EPR signal, observed at 77 K, by human islets (~10³ islets per ml). Lower panel. Comparison of effects of cytokines (IL-1β at 75 U/ml, 3.5 nmol/l TNFα, IFNγ at 750 U/ml) on nitrite formation by human islets. (Reproduced with permission from: Corbett JA, Sweetland MA, Wang JL et al. Nitric oxide mediates cytokine-induced inhibition of insulin secretion by human islets of Langerhans. Proc Natl Acad Sci USA 1993; 90:1731-1735.) Copyright © The National Academy of Sciences, Washington, DC, USA.

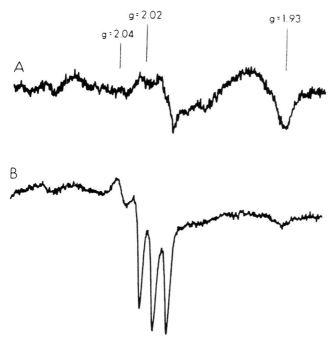

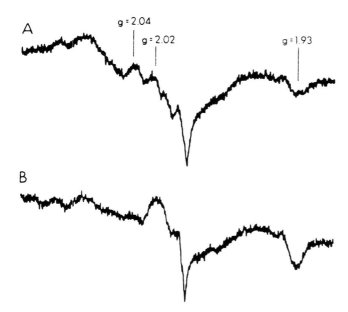

Fig. 11.13. Upper panel. EPR spectra of spinal cords of normal mice at 20 K. A: before exposure to NO; B: After exposure to NO (positive control). Lower panel. EPR spectra at 10 K of EAE-affected cords (A), as compared to control spinal cords (B). Appearance of the $g_{av} = 2.04$ signal of [FeS]-NO complex and disappearance of signals at $g = 2.02$ and 1.93 of [FeS]-proteins. (Reproduced with permission from: Lin RF, Lin T-S, Tilton RG et al. Nitric oxide localized to spinal cords of mice with experimental allergic encephalomyelitis: an electron paramagnetic resonance study. J Exp Med 1993; 178:643-648.) Copyright © Rockefeller University Press, New York, NY, USA.

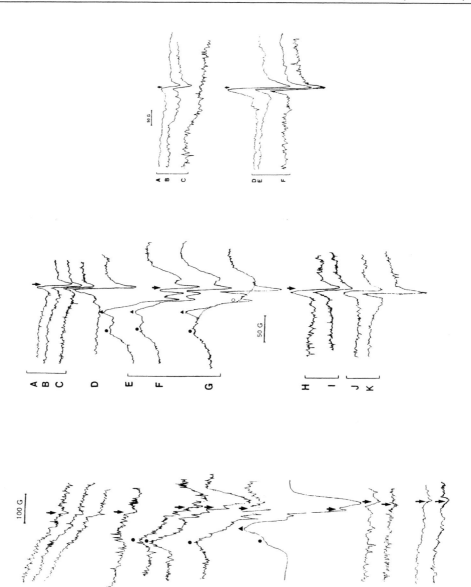

Fig. 11.14. Left panel. EPR spectra of packed rat RBC during rejection of heart allograft. (A-C): three recipients of allogeneic grafts (POD 3); (D): untreated erythrocytes; (E-G): three recipients of allogeneic grafts (POD 5); (H): erythrocytes treated with NO gas; (I, J): two animal recipients of syngeneic heart grafts (POD 3); (K, L): same POD 5. Arrows denote g = 2.00, circle g = 2.078 signal arising from αNO in hemoglobin, triangle, g = 2.04 from βNO component in HbNO. Central panel. EPR spectra of rat heart tissue homogenate. (A-C): three recipients of allogeneic grafts (POD 3); (D): untreated erythrocytes; (E-G): three recipients of allogeneic grafts (POD 5); (H, I): two animal recipients of syngeneic heart grafts (POD 3); (J, K): same POD 5. Right panel. Effect of the immunosuppressant FK506 on iron-nitrosyl complex formation in heart allografts. All spectra are POD 5. (A-C): erythrocytes; (D-F): heart tissue. (Reproduced with permission from: Lancaster JR, Langrehr JM, Bergonia HA et al. EPR detection of heme and nonheme iron-containing protein nitrosylation by nitric oxide during rejection of rat heart allograft. J Biol Chem 1992; 267:10994-10998.) Copyright © The American Society for Biochemistry & Molecular Biology, Inc., Bethesda, MD, USA.

transplantation.[171] Using male Lewis and LBNF$_1$ rats, transplantation led either to graft rejection alone (LBNF$_1$ to Lewis) or to graft-versus-host disease (GVHD) that affects several organs (Lewis to LBNF$_1$). In both kinds of transplantation, easily identified HbNO EPR signals were detected in packed RBC on post-operation day 6 (POD 6) (small signal) which became large on POD 9. No such signal occured in cases of syngeneic (Lewis to Lewis) or immunosuppressant FK 506-treated allogeneic grafts. Measurements of NO$_2^-$/NO$_3^-$ levels and evaluation of circulating HbNO in RBC might prove useful markers to monitor rejection of a small bowel allograft (Fig. 11.15).[171]

EPR spectroscopy was used to detect HbNO in the blood of pancreas transplanted rats and showed that NO generation could be used as a rejection marker.[180] Near-infrared spectroscopy which allows one to follow hemoglobin oxygenation, HbNO and oxidized cytochrome oxidase was used to evaluate simultaneously NO synthesis and tissue oxygenation in rat liver allograft.[181]

Transplantation in Human

The L-arginine/NO pathway role in acute graft response has been tested in several types of organ transplantation in patients. The importance of NO as a possible predictive parameter of acute liver allograft rejection has been compared to several other mediators and markers of rejection: IL-2 receptor (IL-2R)-positive lymphocytes and TNFα.[182] Plasma concentrations of acid-labile nitrosocompounds (NO$_x$) were found to increase during acute rejection, to correlate with rejection severity and to be reduced after a high dose of glucocorticoid administration. Correlations of NO$_x$ with IL-2R-positive lymphocytes and TNFα were demonstrated, showing that the easy monitoring of NO$_x$ may be useful in the detection of acute allograft rejection, at least in human liver transplantation.[182,183] NO formation was also found to be a predictive parameter for acute GVHD after human allogeneic bone marrow trans-

plantation.[184] Various cytokines: IL-2, IL-6, TNFα and IL-2R levels have also been monitored after cardiac transplantation, in order to assess the graft prognosis and their role in allograft dysfunction.[185] Other indirect factors related to NO synthesis have been measured or tested. In liver transplantation, hepatic reperfusion liberates arginase from the implanted graft causing L-arginine deficiency which might affect NO synthesis.[186] Lung transplantation is often followed by pulmonary hypertension associated with transient graft dysfunction, a phenomenon probably not related to cytokine circulation and NOS II. Inhalation of NO gas (80 ppm) over 40-69 hours has been tested positively in the treatment of this transient dysfunction.[187]

VASCULAR DISEASES

The important role of NO on vascular tone is one of its major features in physiology. Probably due to the low level of NO necessary for such effects it has not been detected directly by EPR spectroscopy in circulating blood. However by an ingenious method, detailed in chapter 9, it has been possible to measure nitrite derived from NO in the basal state and following ischemia.[188] In healthy subjects the venous plasma level of nitrite ranged from 0-0.6 μmol/l (0.31 ± 0.10 μmol/l). In a healthy subject plasma nitrite level was significantly increased in response to three different forearm or leg ischemia periods. Similar low NO levels were measured by other methods in HUVEC cells.[19,20]

Endothelial Injury

These NO levels correspond to its production by the constitutive NOS III in endothelial cells. They are relevant to the NO/EDRF functions upon smooth muscle cells and as inhibitor of platelet aggregation. However following endothelial damage of various origins, vascular smooth muscle cells are susceptible to respond to circulating cytokines and LPS,[103,189,190] giving rise to more damaging inflammatory processes.

A de-endothelializing balloon injury to

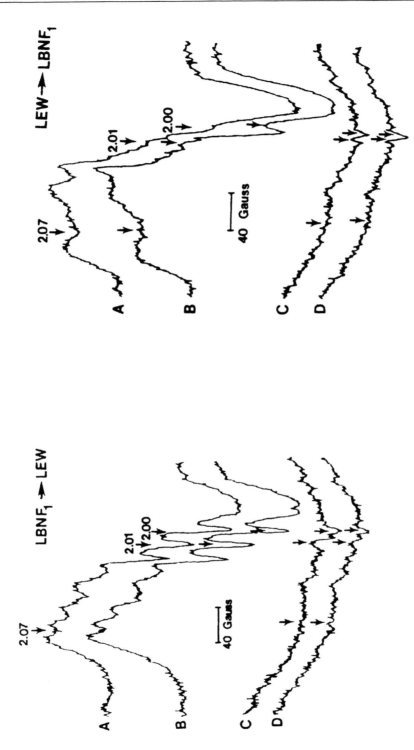

Fig. 11.15. EPR spectra of packed RBC during small bowel allograft rejection. Left panel. A,B, Lewis recipients of LBNF₁ small bowel graft; C,D, FK 506-treated recipients. HbNO signals at g = 2.07 and triplet at g = 2.01 are characterized. Right panel. Graft-versus-host disease from LBNF₁ recipients of Lewis small bowel grafts, with or without FK 506 treatment. The EPR signal is characteristic of the αNO complex of hemoglobin in the R-state. (Reproduced with permission from: Langrehr JM, Müller AR, Bergonia HA et al. Detection of nitric oxide by electron paramagnetic resonance spectroscopy during rejection and graft-versus-host disease after small bowel transplantation in the rat. Surgery 1992; 112:395-402.) Copyright © Mosby-Year Book, Inc., St Louis, MO, USA.

the rat carotid artery provided a good model of endothelial injury. Arterial smooth muscle cells expressed NOS II mRNA, and a systemically detectable level of NO was measured by EPR spectroscopy in RBC (Fig. 11.16).[191]

This result could be important to understand the maintenance of vascular tone and nonthrombogenicity, depending on NO production by vascular smooth muscle of injured arteries.[191] It immediately raises the question of the formation of a neointima after angioplasty.

MYOCARDIAL INFARCTION AND ISCHEMIA

Myocardial infarction has also been associated with inflammatory processes.[192,193] In fact in a model of infarction of rabbits, induction of NOS II was detectable in the infarcted region as compared to noninfarcted region of the heart of the same animal.[194]

However the source of NO produced in ischemic heart has been recently questioned.[195] The direct reduction of nitrite in acidic and reducing conditions could be an enzyme-independent source of NO detected by EPR spectroscopy of its heme adducts (Fig. 11.17).[195]

A similar study showed that the nitrosyl heme compound detected in myocardial ischemia was that of myoglobin MbNO, which presents an EPR spectrum similar to that of HbNO (see chapter 4).[196]

FOCAL CEREBRAL ISCHEMIA

Nitric oxide may also participate in the pathophysiology associated with cerebral ischemia and reperfusion.[197] In an experimental model of focal ischemia followed by reperfusion in the rat, HbNO was detected in the cerebral circulating blood by EPR spectroscopy (Fig. 11.18).[198]

Similar results were obtained in the same animal model of focal ischemia and reperfusion, using NO spin-trapping and EPR spectroscopy (see chapter 13 for the detail of the method),[199-201] and in situ detection of NO by an electrochemical porphyrin microsensor.[202]

Nitric oxide could have neuroprotective and deleterious effects on the evolution of focal cerebral ischemia, depending on its synthesis localization and rate.[203]

HYPERTHERMIA

Heat stroke is characterized by a cascade of events in the cardiovascular system. Hyperthermia stimulates NO formation within the blood circulation shown by the EPR detection of NO-heme in blood. In portal venous blood the EPR signal characteristic of $(\alpha^{2+}NO\beta^{3+})_2$ hybrid molecules was detected together with those of ceruloplasmin, transferrin and a semiquinone radical.[204]

OTHER PHYSIOLOGICAL AND PATHOPHYSIOLOGICAL STATES

A role has been proposed for NO in many, and probably far too many, physiological and pathophysiological states, such as pain,[205] migraine,[206] male impotence,[23] prion encephalopathy,[207] etc. We shall mention only a final case in which EPR spectroscopy has been used.

PREGNANCY

An interesting instance of EPR detection of bound NO in animal blood is that found during pregnancy in rats. Pronounced cardiovascular changes occur during normal pregnancy in women and in other mammals.[25] NO biosynthesis has been proposed to have a role in the specific immune state of pregnancy preventing the rejection of the semi-allograft constituted by the fetus and its placenta. EPR spectra of the red cell fraction showed a typical signal of αNO complex in hemoglobin of pregnant rat versus that of virgin rat (Fig. 11.19).[25]

Constitutive NOS III has been characterized in human placenta.[26,27,208] It has also been characterized in the uterus and seems to play an important role during pregnancy, by inhibiting contractility.[209] An NO donor, glyceryl trinitrate has been successfully used in order to arrest preterm labor and to prolong gestations of only 23-33 weeks, for up to 7-8 more weeks.[210]

CONCLUSION

The present chapter might appear as an indigestible hotchpotch of various diseases, mixed together by a mass of tasteless EPR spectroscopic sauce. In fact most of the diseases mentioned here are related by non-specific immune host defense response, by imbalance of circulating cytokines and finally by NOS II induction. As for the repetitive illustration of EPR spectra, it represents a simple fact: due to the low sensitivity of the EPR spectroscopic method, only relatively concentrated nitrosyl-metalloprotein complexes are detected. So are hemoglobin, myoglobin, P-450 and [FeS]-containing proteins. It does not mean that other metalloenzyme targets are not interacting with NO, as we shall discuss in chapter 14.

Another point has been repeatedly discussed: "Are rodents good models of humans?"[211,212] The question is quite general, as we have seen several instances of different responses in rodent models as compared to human in several pathologies: activation

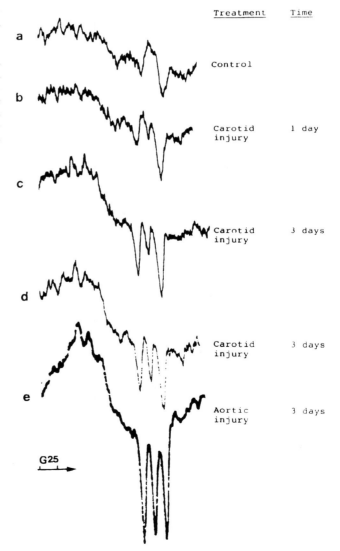

Fig. 11.16. EPR spectra of HbNO in peripheral blood RBC after arterial injury. a, sham-operated control rat; b, c, d, carotid injury effects with time; e, de-endothelializing injury in the thoracic aorta. (Reproduced with permission from: Hansson GK, Geng Y, Holm J et al. Arterial smooth muscle cells express nitric oxide synthase in response to endothelial injury. J Exp Med 1994; 180:733-738.) Copyright © The Rockefeller University Press, New York, NY, USA.

Treatment   Time

a

Control

b

Carotid injury   1 day

c

Carotid injury   3 days

d

Carotid injury   3 days

e

Aortic injury   3 days

G25

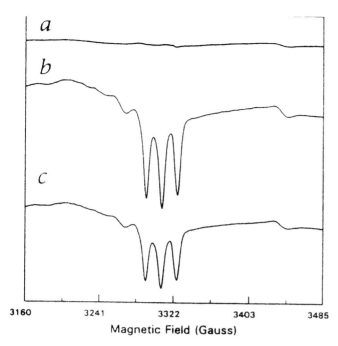

Fig. 11.17. EPR spectra at 77 K of heart tissue subjected to prolonged ischemia. a, control non-ischemic heart tissue; b, after 8 hr ischemia; c, same as b in the presence of NAME. (Reproduced with permission from the authors: Zweier JL, Wang P, Samouilov A et al. Enzyme-independent formation of nitric oxide in biological tissues. Nature Medicine 1995; 1:804-809.) Copyright © Nature Medicine, Washington, DC, USA.

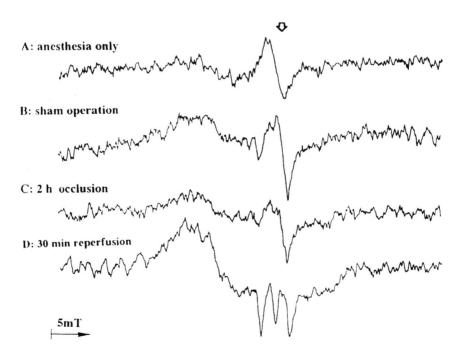

Fig. 11.18. EPR spectra of the jugular blood of rats with middle cerebral artery occlusion. (Reproduced with permission from: Kumura E, Yoshimine T, Tanaka S et al. Nitrosyl hemoglobin production during reperfusion after focal cerebral ischemia in rats. Neurosci Lett 1994; 177:165-167.) Copyright © Elsevier Sci Publishers Ltd, Bay 15K, Shannon Industrial Estate, Co. Clare, Ireland.

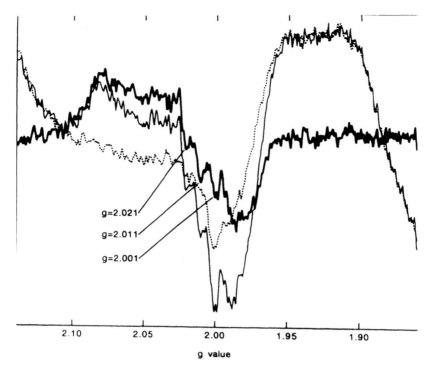

g value

Fig. 11.19. EPR spectra of RBC from the blood of a virgin, control rat (dotted line), a late pregnant rat (thin line), and the difference spectrum (thick line). The triplet centered at g = 2.011 is characteristic of HbNO. (Reproduced with permission from: Conrad KP, Joffe GM, Kruszyna H et al. Identification of increased nitric oxide biosynthesis during pregnancy in rats. FASEB J 1993; 7:566-571.) Copyright © The FASEB Journal, Bethesda, MD, USA.

of monocytes/macrophages,[33,41,212] activation of hepatocytes,[48] cerebral malaria,[96,97] insulin-dependent diabetes mellitus,[61,62,163] etc.

Finally we have made some progress over chapters 6 and 10 as to the biological relevance of NO interaction with metalloproteins, as detected by EPR spectroscopy. NO is certainly produced in quite a few pathophysiological states. But often the question remains: is the detection of NO bound to a given metalloprotein a simple spectroscopic epiphenomenon or is it related in any way to the disease? In fact, rare are the cases in which correlations were established between NO level and disease grade, and none in which a real dual cause-effect relationship could be drawn. It would be naïve to expect it.

REFERENCES

1. Moncada S, Palmer RMJ, Higgs EA. Nitric oxide: physiology, pathophysiology, and pharmacology. Pharmacol Rev 1991; 43:109-142.

2. Nathan C. Nitric oxide as a secretory product of mammalian cells. FASEB J 1992; 6:3051-3064.

3. Moncada S, Higgs EA. Molecular mechanisms and therapeutic strategies related to nitric oxide. FASEB J 1995; 9:1319-1330.

4. Henry Y, Ducrocq C, Drapier J-C et al. Nitric oxide, a biological effector. Electron paramagnetic resonance detection of nitrosyl-iron-protein complexes in whole cells. Eur Biophys J 1991; 20:1-15.

5. Henry Y, Lepoivre M, Drapier J-C et al. EPR characterisation of molecular targets for NO in mammalian cells and organelles. FASEB J 1993; 7:1124-1134.

6. Wilcox DE, Smith RP. Detection and quantification of nitric oxide using electron magnetic resonance spectroscopy. Methods: a companion to Methods in Enzymology 1995; 7:59-70.

7. Henry YA, Singel DJ. Metal-nitrosyl interactions in nitric oxide biology probed by electron paramagnetic resonance spectroscopy. In: Feelisch M, Stamler JS. Methods in Nitric Oxide Research. John Wiley & Sons, 1996:357-372.

8. Singel DJ, Lancaster JR. Electron paramagnetic resonance spectroscopy and nitric oxide biology. In: Feelisch M, Stamler JS. Methods in Nitric Oxide Research. John Wiley & Sons, 1996:341-356.

9. Kosaka H, Shiga T. Detection of nitric oxide by electron paramagnetic resonance using hemoglobin. In: Feelisch M, Stamler JS. Methods in Nitric Oxide Research. John Wiley & Sons, 1996:373-381.

10. Förstermann U, Schmidt HHHW, Pollock JS et al. Isoforms of nitric oxide synthase. Characterization and purification from different cell types. Biochem Pharmacol 1991; 42:1849-1857.

11. Nathan C, Xie Q. Nitric oxide synthases: roles, tolls, and controls. Cell 1994; 78:915-918.

12. Kishimoto J, Spurr N, Liao M et al. Localization of brain nitric oxide synthase to human chromosome 12. Genomics 1992; 14:802-804.

13. Marsden PA, Schappert KT, Chen HS et al. Molecular cloning and characterization of human endothelial nitric oxide synthase. FEBS Lett 1992; 307:287-293.

14. Marsden PA, Heng HHQ, Scherer SW et al. Structure and chromosomal localization of the human constitutive endothelial nitric oxide synthase gene. J Biol Chem 1993; 268:17478-17488.

15. Nadaud S, Bonnardeaux A, Lathrop M et al. Gene structure, polymorphism and mapping of the human endothelial nitric oxide gene. Biochem Biophys Res Commun 1994; 198:1027-1033.

16. Robinson LJ, Weremowicz S, Morton CC et al. Isolation and chromosomal localization of the human endothelial nitric oxide synyhase (NOS3) gene. Genomics 1994; 19:350-357.

17. Miyahara K, Kawamoto T, Sase K et al. Cloning and structural characterization of the human endothelial nitric-oxide-synthase gene. Eur J Biochem 1994; 223:719-726.

18. Janssens SP, Shimouchi A, Quertermous T et al. Cloning and expression of the cDNA encoding human endothelium-derived relaxing factor/nitric oxide synthase. J Biol Chem 1992; 267:14519-14522.

19. Tsukahara H, Gordienko DV, Goligorsky MS. Continuous monitoring of nitric oxide release from human umbilical vein endothelial cells. Biochem Biophys Res Commun 1993; 193:722-729.

20. Sato I, Morita I, Kaji K et al. Reduction of nitric oxide producing activity associated with in vitro aging in cultured human umbilical vein endothelial cell. Biochem Biophys Res Commun 1993; 195:1070-1076.

21. Schmidt HHHW, Murad F. Purification and characterization of a human NO synthase. Biochem Biophys Res Commun 1991; 181:1372-1377.

22. Klatt P, Heinzel B, Mayer B et al. Stimulation of human nitric oxide synthase by tetrahydrobiopterin and selective binding of the cofactor. FEBS Lett 1992; 305:160-162.

23. Pickard RS, Powell PH, Zar MA. The effect of inhibitors of nitric oxide biosynthesis and cyclic GMP formation on nerve-evoked relaxation of human cavernosal smooth muscle. Br J Pharmacol 1991; 104:755-759.

24. Nakane M, Schmidt HHHW, Pollock JS et al. Cloned human brain nitric oxide synthase is highly expressed in skeletal muscle. FEBS Lett 1993; 316:175-180.

25. Conrad KP, Joffe GM, Kruszyna H et al. Identification of increased nitric oxide biosynthesis during pregnancy in rats. FASEB J 1993; 7:566-571.

26. Conrad KP, Vill M, McGuire PG et al. Expression of nitric oxide synthase by syncytiotrophoblast in human placental villi. FASEB J 1993; 7:1269-1276.

27. Garvey EP, Tuttle JV, Covington K et al. Purification and characterization of the constitutive nitric oxide synthase from human placenta. Arch Biochem Biophys 1994; 311:235-241.

28. Chartrain NA, Geller DA, Koty PP et al. Molecular cloning, structure, and chromosomal localization of the human inducible nitric oxide synthase gene. J Biol Chem 1994; 269:6765-6772.

29. Marsden PA, Heng HHQ, Duff CL et al. Localization of the human gene for inducible nitric oxide synthase (NOS2) to chromosome 17q11.2-q12. Genomics 1994; 19:183-185.

30. Sparrow JR. Inducible nitric oxide synthase in the central nervous system. J Molec Neurosci 1995; 5:219-229.

31. Gustafsson LE, Leone AM, Persson MG et al. Endogenous nitric oxide is present in the exhaled air of rabbits, guinea pigs and humans. Biochem Biophys Res Commun 1991; 181:852-857.

32. Denis M. Tumor necrosis factor and granulocyte macrophage-colony stimulating factor stimulate human macrophages to restrict growth of virulent *Mycobacterium avium* and to kill avirulent *M. avium*; killing effector mechanism depends on the generation of reactive nitrogen intermediates. J Leuk Biol 1991; 49:380-387.

33. Denis M. Human monocytes/macrophages: NO or no NO? J Leuk Biol 1994; 55:682-684.

34. Kolb JP, Paul-Eugène N, Damais C et al. Interleukin-4 stimulates cGMP production by IFN-γ-activated human monocytes. Involvement of the nitric oxide synthase pathway. J Biol Chem 1994; 269:9811-9816.

35. Dumarey CH, Labrousse V, Rastogi N et al. Selective *Mycobacterium avium*-induced production of nitric oxide by human monocytes-derived macrophages. J Leuk Biol 1994; 56:36-40.

36. Zembala M, Siedlar M, Marcinkiewicz J et al. Human monocytes are stimulated for nitric oxide release *in vitro* by some tumor cells but not by cytokines and lipopolysaccharide. Eur J Immunol 1994; 24:435-439.

37. Paul-Eugène N, Mossalayi D, Sarfati M et al. Evidence for a role of FcεRII/CD23 in the IL-4-induced nitric oxide production by normal human mononuclear phagocytes. Cell Immunol 1995; 163:314-318.

38. Dugas B, Mossalayi MD, Damais C et al. Nitric oxide production by human mono-cytes: evidence for a role of CD23. Immunol Today 1995; 16:574-580.

39. Weinberg JB, Misukonis MA, Shami PJ et al. Human mononuclear phagocyte inducible nitric oxide synthase (iNOS): analysis of iNOS mRNA, iNOS protein, biopterin, and nitric oxide production by blood monocytes and peritoneal macrophages. Blood 1995; 86:1184-1195.

40. Torre D, Ferrario G, Bonetta G et al. Production of nitric oxide from peripheral blood mononuclear cells and polymorphonuclear leukocytes of patients with HIV-1. AIDS 1995; 9:979-980.

41. Albina JE. On the expression of nitric oxide synthase by human macrophages. Why no NO? J Leuk Biol 1995; 58:643-649.

42. Bryant JL, Mehta P, Von der Porten A et al. Co-purification of 130 kD nitric oxide synthase and a 22 kD link protein from human neutrophils. Biochem Biophys Res Commun 1992; 189:558-564.

43. Heck DE, Laskin DL, Gardner CR et al. Epidermal growth factor suppresses nitric oxide and peroxide production by keratinocytes. Potential role for nitric oxide in the regulation of wound healing. J Biol Chem 1992; 267:21277-21280.

44. Nussler AK, Di Silvio M, Billiar TR et al. Stimulation of the nitric oxide synthase pathway in human hepatocytes by cytokines and endotoxin. J Exp Med 1992; 176:261-264.

45. Stadler J, Bergonia HA, Di Silvio M et al. Nonheme iron-nitrosyl complex formation in rat hepatocytes: detection by electron paramagnetic resonance spectroscopy. Arch Biochem Biophys 1993; 302:4-11.

46. Nussler AK, Geller DA, Sweetland MA et al. Induction of nitric oxide synthesis and its reactions in cultured human and rat hepatocytes stimulated with cytokines plus LPS. Biochem Biophys Res Commun 1993; 194:826-835.

47. Geller DA, Lowenstein CJ, Shapiro RA et al. Molecular cloning and expression of inducible nitric oxide synthase from human hepatocytes. Proc Natl Acad Sci USA 1993; 90:3491-3495.

48. Nussler AK, Di Silvio M, Liu Z-Z et al. Further characterization and comparison of

inducible nitric oxide synthase in mouse, rat, and human hepatocytes. Hepatology 1995; 21:1552-1560.

49. Stadler J, Stefanovic-Racic M, Billiar TR et al. Articular chondrocytes synthesize nitric oxide in response to cytokines and lipopolysaccharide. J Immunol 1991; 147:3915-3920.

50. Palmer RMJ, Hickery MS, Charles IG et al. Induction of nitric oxide synthase in human chondrocytes. Biochem Biophys Res Commun 1993; 193:398-405.

51. Häuselmann HJ, Oppliger L, Michel BA et al. Nitric oxide and proteoglycan biosynthesis by human articular chondrocytes in alginate culture. FEBS Lett 1994; 352:361-364.

52. Nicolson AG, Haites NE, McKay NG et al. Induction of nitric oxide synthase in human mesangial cells. Biochem Biophys Res Commun 1993; 193:1269-1274.

53. Mollace V, Colasanti M, Rodino P et al. Cytokine-induced nitric oxide generation by cultured astrocytoma cells involve a Ca^{++}-calmodulin-independent NO-synthase. Biochem Biophys Res Commun 1993; 191:327-334.

54. Hu S, Sheng WS, Peterson PK et al. Differential regulation by cytokines of human astrocyte nitric oxide production. Glia 1995; 15:491-494.

55. Kasai K, Hattori Y, Nakanishi N et al. Regulation of inducible nitric oxide production by cytokines in human thyrocytes in culture. Endocrinol 1995; 136:4261-4270.

56. Corbett JA, Sweetland MA, Wang JL et al. Nitric oxide mediates cytokine-induced inhibition of insulin secretion by human islets of Langerhans. Proc Natl Acad Sci USA 1993; 90:1731-1735.

57. Adcock IM, Brown CR, Kwon O et al. Oxidative stress induces NFκB DNA binding and inducible NOS mRNA in human epithelial cells. Biochem Biophys Res Commun 1994; 199:1518-1524.

58. Robbins RA, Barnes PJ, Springall DR et al. Expression of inducible nitric oxide in human lung epithelial cells. Biochem Biophys Res Commun 1994; 203:209-218.

59. Goureau O, Hicks D, Courtois Y. Human retinal pigmented epithelial cells produce nitric oxide in response to cytokines. Biochem Biophys Res Commun 1994; 198:120-126.

60. Ralston SH, Todd D, Helfrich M et al. Human osteoblast-like cells produce nitric oxide and express inducible nitric oxide synthase. Endocrinology 1994; 135:330-336.

61. Rabinovitch A, Suarez-Pinzon W, Strynadka K et al. Human pancreatic islet β-cell destruction by cytokines is independent of nitric oxide production. J Clin Endocrinol Metab 1994; 79:1058-1062.

62. Welsh N, Eizirik DL, Sandler S. Nitric oxide and pancreatic β-cell destruction in insulin dependent diabetes mellitus: don't take NO for an answer. Autoimmunity 1994; 18:285-290.

63. Maruyama T, Kataoka N, Nagase S et al. Identification of three-line electron spin resonance signal and its relationship to ascites tumors. Cancer Res 1971; 31:179-184.

64. Vithayathil AJ, Ternberg JL, Commoner B. Changes in electron spin resonance signals of rat liver during chemical carcinogenesis. Nature 1965; 207:1246-1249.

65. Vanin AF, Vakhnina LV, Chetverikov AG. Nature of the EPR signals of a new type found in cancer tissues. Biofizika 1970; 15:1044-1051 (English translation 1082-1089).

66. Bastian NR, Yim CY, Hibbs JB et al. Induction of iron-derived EPR signals in murine cancers by nitric oxide. Evidence for multiple intracellular targets. J Biol Chem 1994; 269:5127-5131.

67. Yim C-Y, Bastian NR, Smith JC et al. Macrophage nitric oxide synthesis delays progression of ultraviolet light-induced murine skin cancers. Cancer Res 1993; 53:5507-5511.

68. Jenkins DC, Charles IG, Thomsen LL et al. Roles of nitric oxide in tumor growth. Proc Natl Acad Sci USA 1995; 92:4392-4396.

69. Miwa M, Stuehr DJ, Marletta et al. Nitrosation of amines by stimulated macrophages. Carcinogenesis 1987; 7:955-958.

70. Correa P. A human model of gastric carcinogenesis. Cancer Res 1988; 48:3554-3560.

71. Bartsch H, Ohshima H, Pignatelli B. Inhibitors of endogenous nitrosation mechanisms and implications in human cancer prevention. Mutation Res 1988; 202:307-324.

72. Ohshima H, Tsuda M, Adachi H et al. L-arginine-dependent formation of N-nitrosamines by the cytosol of macrophages activated with lipopolysaccharide and interferon-γ. Carcinogenesis 1991; 12:1217-1220.

73. Ohshima H, Bartsch H. Chronic infections and inflammatory processes as cancer risk factors: possible role of nitric oxide in carcinogenesis. Mutation Res 1994; 305:253-264.

74. Thomsen LL, Lawton FG, Knowles RG et al. Nitric oxide synthase activity in human gynecological cancer. Cancer Res 1994; 54:1352-1354.

75. Thomsen LL, Miles DW, Happerfield L et al. Nitric oxide synthase activity in human breast cancer. Br J Cancer 1995; 72:41-44.

76. Cobbs CS, Brenman JE, Aldape KD et al. Expression of nitric oxide synthase in human central nervous system tumors. Cancer Res 1995; 55:727-730.

77. Symons MCR, Rowland IJ, Deighton N et al. Electron spin resonance studies of nitrosyl haemoglobin in human liver, colon and stomach tumour tissues. Free Rad Res Comms 1994; 21:197-202.

78. Yim C-Y, McGregor JR, Kwon O-D et al. Nitric oxide synthesis contributes to IL-2-induced antitumor responses against intraperitoneal Meth A tumor. J Immunol 1995; 155:4382-4390.

79. Hibbs JB, Westenfelder C, Taintor R et al. Evidence for cytokine-inducible nitric oxide synthesis from L-arginine in patients receiving interleukin-2 therapy. J Clin Invest 1992; 89:867-877.

80. Lieberman MD, Nishioka K, Redmond P et al. Enhancement of interleukin-2 immunotherapy with L-arginine. Ann Surgery 1992; 215:157-165.

81. Ochoa JB, Curti B, Peitzman AB et al. Increased circulating nitrogen oxides after human tumor immunotherapy: correlation with toxic hemodynamic changes. J Natl Cancer Inst 1992; 84:864-867.

82. Miles D, Thomsen L, Balkwill F et al. Association between biosynthesis of nitric oxide and changes in immunological and vascular parameters in patients treated with interleukin-2. Eur J Clin Invest 1994; 24:287-290.

83. Heiss LN, Lancaster JR, Corbett JA et al. Epithelial autotoxicity of nitric oxide: role in respiratory cytopathology of pertussis. Proc Natl Acad Sci USA 1994; 91:267-270.

84. Nussler AK, Billiar TR. Inflammation, immunoregulation, and inducible nitric oxide synthase. J Leuk Biol 1993; 54:171-178.

85. Nussler AK, Liu Z-Z, Di Silvio M et al. Hepatocyte inducible nitric oxide synthase is influenced by cell density. Am J Physiol 1994; 266: C394-C401.

86. Billiar TR, Curran RD, Stuehr DJ et al. An L-arginine-dependent mechanism mediates Kupffer cell inhibition of hepatocyte protein synthesis in vitro. J Exp Med 1989; 169:1467-1472.

87. Curran RD, Billiar TR, Stuehr DJ et al. Hepatocytes produce nitrogen oxides from L-arginine in response to inflammatory products of Kupffer cells. J Exp Med 1989; 170:1769-1774.

88. Suzuki H, Menegazzi M, Carcereri de Prati A et al. Nitric oxide in the liver: physiopathological roles. Adv Neuroimmunol 1995; 5:379-410.

89. Chamulitrat W, Jordan SJ, Mason RP et al. Targets of nitric oxide in a mouse model of liver inflammation by *Corynebacterium parvum*. Arch Biochem Biophys 1995; 316:30-37.

90. Geller DA, de Vera ME, Russell DA et al. A central role for IL-1β in the in vitro and in vivo regulation of hepatic inducible nitric oxide synthase. IL-1β induces hepatic nitric oxide synthesis. J Immunol 1995; 155:4890-4898.

91. Rachmilewitz D, Karmeli F, Eliakim R et al. Enhanced gastric nitric oxide activity in duodenal ulcer patients. Gut 1994; 35:1394-1397.

92. Clark IA, Rockett KA, Cowden WB. Proposed link between cytokines, nitric oxide and human cerebral malaria. Parasitol Today 1991; 7:205-207.

93. Clark IA, Rockett KA. The cytokine theory of human cerebral malaria. Parasitol Today 1994; 10:410-412.

94. Cot S, Ringwald P, Mulder B et al. Nitric oxide in cerebral malaria. J Infect Diseases 1994; 169:1417-1418.

95. Ghigo D, Todde R, Ginsburg H et al. Erythrocyte stages of *Plasmodium falciparum* exhibit a high nitric oxide synthase (NOS) activity and release an NOS-inducing soluble factor. J Exp Med 1995; 183:677-688.

96. Asensio V. Neuropaludisme et monoxyde d'azote. PhD doctorate thesis. Université Paris XII-Val de Marne. 1995.

97. Asensio VC, Ohshima H, Falanga PB. *Plasmodium berghei*: is nitric oxide involved in the pathogenesis of mouse cerebral malaria? Exp Parasitol 1993; 77:111-117.

98. Lee SC, Dickson DW, Brosnan CF et al. Human astrocytes inhibit *Cryptococcus neoformans* growth by a nitric oxide-mediated mechanism. J Exp Med 1994; 180:365-369.

99. Evans TG, Rasmussen K, Wiebke G et al. Nitric oxide synthesis in patients with advanced HIV infection. Clin Exp Immunol 1994; 97:83-86.

100. Bukrinsky MI, Nottet HSLM, Schmidt-mayerova H et al. Regulation of nitric oxide synthase activity in human immunodeficiency virus type 1 (HIV-1)-infected monocytes: implications for HIV-associated neurological disease. J Exp Med 1995; 181:735-745.

101. Dighiero P, Reux I, Hauw J-J et al. Expression of inducible nitric oxide synthase in cytomegalovirus-infected glial cells of retinas from AIDS patients. Neurosci Lett 1994; 166:31-34.

102. Dal Nogare AR. Southwestern Internal Medicine Conference: Septic shock. Am J Med Sci 1991; 302:50-65.

103. Stoclet J-C, Fleming I, Gray G et al. Nitric oxide and endotoxemia. Circulation 1993; 87:V77-V80.

104. Redmond HP, Chavin KD, Bromberg JS et al. Inhibition of macrophage-activating cytokines is beneficial in the acute septic response. Ann Surg 1991; 214:502-509.

105. Damas P, Ledoux D, Nys M et al. Cytokine serum level during severe sepsis in human IL-6 as a marker of severity. Ann Surg 1992; 215:356-362.

106. Kilbourn RG, Jubran A, Gross SS et al. Reversal of endotoxin-mediated shock by N^G-methyl-L-arginine, an inhibitor of nitric oxide synthesis. Biochem Biophys Res Commun 1990; 172:1132-1138.

107. Hutcheson IR, Whittle BJR, Boughton-Smith NK. Role of nitric oxide in maintaining vascular integrity in endotoxin-induced acute intestinal damage in the rat. Br J Pharmacol 1990; 101:815-820.

108. Westenberger U, Thanner S, Ruf HH et al. Formation of free radicals and nitric oxide derivative of hemoglobin in rats during shock syndrome. Free Rad Res Comms 1990; 11:167-178.

109. Wang Q, Jacobs J, DeLeo J et al. Nitric oxide hemoglobin in mice and rats in endotoxic shock. Life Sci 1991; PL55-PL60.

110. Kosaka H, Watanabe M, Yoshihara H et al. Detection of nitric oxide production in lipopolysaccharide-treated rats by ESR using carbon monoxide hemoglobin. Biochem Biophys Res Commun 1992; 184:1119-1124.

111. Kosaka H, Harada N, Watanabe M et al. Synergistic stimulation of nitric oxide hemoglobin production in rats by recombinant interleukin 1 and tumor necrosis factor. Biochem Biophys Res Commun 1992; 189:392-397.

112. Beasley D, Eldridge M. Interleukin-1β and tumor necrosis factor-α synergistically induce NO synthase in rat vascular smooth muscle cells. Am J Physiol 1994; 266:R1197-R1203.

113. Kosaka H, Sakaguchi H, Sawai Y et al. Effect of interferon-γ on nitric oxide hemoglobin production in endotoxin-treated rats and its synergism with interleukin 1 or tumor necrosis factor. Life Sci 1994; 54:1523-1529.

114. Kubrina LN, Mikoyan VD, Mordvintcev PI et al. Iron potentiates bacterial lipopolysaccharide-induced nitric oxide formation in animal organs. Biochim Biophys Acta 1993; 1176:240-244.

115. Kosaka H, Sawai Y, Sakaguchi H et al. ESR spectral transition by arteriovenous cycle in nitric oxide hemoglobin of cytokine-treated rats. Am J Physiol 1994; 266:C1400-C1405.

116. Henry Y, Banerjee R. Electron paramagnetic studies of nitric oxide haemoglobin derivatives: isolated subunits and nitric oxide hybrids. J Mol Biol 1973; 73:469-482.

117. Henry Y, Cassoly R. Chain non-equivalence in nitric oxide binding to hemoglobin. Biochem Biophys Res Commun 1973; 51:659-665.

118. Chamulitrat W, Jordan SJ, Mason RP. Nitric oxide production during endotoxic shock in carbon tetrachloride-treated rats. Mol Pharmacol 1994; 46:391-397.

119. Feder LS, Laskin DL. Regulation of hepatic endothelial cell and macrophage proliferation and nitric oxide production by GM-CSF, M-CSF, and IL-1β following acute endotoxemia. J Leuk Biol 1994; 55:507-513.

120. Hecker M, Schott C, Bucher B et al. Increase in serum N^G-hydroxy-L-arginine in rats treated with bacterial lipoplysaccharide. Eur J Pharmacol 1995; 275:R1-R3.

121. Szabo C, Southan GJ, Thiemermann C. Beneficial effects and improved survival in rodent models of septic shock with S-methylisothiurea sulfate, a potent and selective inhibitor of inducible nitric oxide synthase. Proc Natl Acad Sci USA 1994; 91:12472-12476.

122. Stewart TE, Valenza F, Ribeiro SP et al. Increased nitric oxide in exhaled gas as an early marker of lung inflammation in a model of sepsis. Am J Respir Crit Care Med 1995; 151:713-718.

123. Florquin S, Amraoui Z, Dubois C et al. The protective role of endogenously synthesized nitric oxide in staphyloccal enterotoxin B-induced shock in mice. J Exp Med 1994; 180:1153-1158.

124. MacMicking JD, Nathan C, Hom G et al. Altered responses to bacterial infection and endotoxic shock in mice lacking inducible nitric oxide synthase. Cell 1995; 81:641-650.

125. Laubach VE, Shesely EG, Smithies O et al. Mice lacking inducible nitric oxide synthase are not resistant to lipopolysaccharide-induced death. Proc Natl Acad Sci USA 1995; 92:10688-10692.

126. Ochoa JB, Udekwu AO, Billiar TR et al. Nitrogen oxide levels in patients after trauma and during sepsis. Ann Surg 1991; 214:621-626.

127. Editorial. Nitric oxide in the clinical arena. Lancet 1991; 338:1560-1562.

128. Nava E, Palmer RMJ, Moncada S. Inhibition of nitric oxide synthesis in septic shock: how much is beneficial? Lancet 1991; 338:1555-1557.

129. Petros A, Bennett D, Vallance P. Effect of nitric oxide synthase inhibitors on hypotension in patients with septic shock. Lancet 1991; 338:1557-1558.

130. Hotchkiss RS, Karl IE, Parker JL et al. Inhibition of NO synthesis in septic shock. Lancet 1992; 339:434-435.

131. Geroulanos S, Schilling J, Cakmakci M et al. Inhibition of NO synthesis in septic shock. Lancet 1992; 339:435.

132. Cohen J, Silva A. NO inhibitors and septic shock. Lancet 1992; 339:751.

133. Berrazueta JR, Salas E, Amado JA et al. Induction of nitric oxide synthase in human mammary arteries in vitro. Eur J Pharmacol 1994; 251:303-305.

134. Glauser MP, Heumann D, Baumgartner JD et al. Pathogenesis and potential strategies for prevention and treatment of septic shock: an update. Clin Infect Dis 1994; 18:S205-S216.

135. Natanson C, Hoffman WD, Suffredini AF et al. Selected treatment strategies for septic shock based on proposed mechanisms of pathogenesis. Ann Intern Med 1994; 120:771-783.

136. Barthlen W, Stadler J, Lehn NL et al. Serum levels of end products of nitric oxide synthesis correlate positively with tumor necrosis factor α and negatively with body temperature in patients with postoperative abdominal sepsis. Shock 1994; 2:398-401.

137. Vladutiu AO. Role of nitric oxide in autoimmunity. Clinic Immunol Immunopathol 1995; 76:1-11.

138. Boughton-Smith NK, Evans SM, Hawkey CJ et al. Nitric oxide activity in ulcerative colitis and Crohn's disease. Lancet 1993; 342:338-340

139. Tran DD, Visser JJ, Pool MO et al. Enhanced systemic nitric oxide production in inflammatory bowel disease. Lancet 1993; 341:1150.

140. Boughton-Smith NK. Pathological and therapeutic implications for nitric oxide in inflammatory bowel disease. J Royal Soc Med 1994; 87:312-313.

141. Lundberg JON, Hellström PM, Lundberg JM et al. Greatly increased luminal nitric oxide in ulcerative colitis. Lancet 1994; 344:1673-1674.

142. Lundberg JON, Hellström PM, Lundberg JM et al. Nitric oxide in ulcerative colitis. Lancet 1995; 345:449.

143. Mourelle M, Casellas F, Guarner F et al. Induction of nitric oxide synthase in colonic smooth muscle from patients with toxic megacolon. Gastroenterol 1995; 109:1497-1502.

144. Rachmilewitz D, Stamler JS, Bachwich D et al. Enhanced colonic nitric oxide generation and nitric oxide synthase activity in ulcerative colitis and Crohn's disease. Gut 1995; 36:718-723.

145. Kolb-Bachofen V, Fehsel K, Michel G et al. Epidermal keratinocyte expression of inducible nitric oxide synthase in skin lesions of psoriasis vulgaris. Lancet 1994; 344:139.

146. Lipton SA, Choi Y-B, Pan Z-H et al. A redox-based mechanism for the neuroprotective and neurodestructive effects of nitric oxide and related nitroso-compounds. Nature 1993; 364:626-632.

147. Johnson AW, Land JM, Thomson EJ et al. Evidence for increased nitric oxide production in multiple sclerosis. J Neurol 1995; 58:107-115.

148. Bagasra O, Michaels FH, Zheng YM et al. Activation of the inducible form of nitric oxide synthase in the brains of patients with multiple sclerosis. Proc Natl Acad Sci USA 1995; 92:12041-12045.

149. Smith RS, Maes M. The macrophage-T-lymphocyte theory of schizophrenia: additional evidence. Med Hypotheses 1995; 45:135-141.

150. Farrell AJ, Blake DR, Palmer RMJ et al. Increased concentrations of nitrite in synovial fluid and serum samples suggest increased nitric oxide synthesis in rheumatic diseases. Ann Rheum Dis 1992; 51:1219-1222.

151. Stefanovic-Racic M, Stadler J, Evans CH. Nitric oxide and arthritis. Arthritis Rheum 1993; 36:1036-1044.

152. McCartney-Francis N, Allen JB, Mizel DE et al. Suppression of arthritis by an inhibitor of nitric oxide synthase. J Exp Med 1993; 178:749-754.

153. Southern C, Schulster D, Green IC. Inhibition of insulin secretion by interleukin-1β and tumour necrosis factor-α via an L-arginine-dependent nitric oxide generating mechanism. FEBS Lett 1990; 276:42-44.

154. Corbett JA, Lancaster JR, Sweetland MA et al. Interleukin-1β-induced formation of EPR-detectable iron-nitrosyl complexes in islets of Langerhans. J Biol Chem 1991; 266:21351-21354.

155. Corbett JA, Wang JL, Hughes JH et al. Nitric oxide and cyclic GMP formation induced by interleukin 1β in islets of Langerhans. Evidence for an effector role of nitric oxide in islet dysfunction. Biochem J 1992; 229-235.

156. Corbett JA, Wang JL, Sweetland MA et al. Interleukin 1β induces the formation of nitric oxide by β-cells purified from rodent islets of Langerhans. Evidence for the β-cell as a source and site of action of nitric oxide. J Clin Invest 1992; 90:2384-2391.

157. Bergmann L, Kröncke KD, Suschek C et al. Cytotoxic action of IL-1β against pancreatic islets is mediated via nitric oxide formation and is inhibited by N^G-monomethyl-L-arginine. FEBS Lett 1992; 299:103-106.

158. Eizirik DL, Cagliero E, Björklund A et al. Interleukin-1β induces the expression of an isoform of nitric oxide synthase in insulin-producing cells, which is similar to that observed in activated macrophages. FEBS Lett 1992; 308:249-252.

159. Corbett JA, Kwon G, Misko TP et al. Tyrosine kinase involvement in IL-1β-induced expression of iNOS by β-cells purified from islets of Langerhans. Am J Physiol 1994; 267:C48-54.

160. Yamada K, Otabe S, Inada C et al. Nitric oxide and nitric oxide synthase mRNA induction in mouse islet cells by interferon-γ plus tumor necrosis factor-α. Biochem Biophys Res Commun 1993; 197:22-27.

161. Sternesjö J, Bendtzen K, Sandler S. Effects of prolonged exposure in vitro to interferon-

γ and tumor necrosis factor-α on nitric oxide and insulin production of rat pancreatic islets. Autoimmunity 1995; 20:185-190.

162. Corbett JA, McDaniel ML. Reversibility of interleukin-1β-induced islet destruction and dysfunction by the inhibition of nitric oxide synthase. Biochem J 1994; 299:719-724.

163. Eizirik DL, Sandler S, Welsh N et al. Cytokines suppress human islet function irrespective of their effects on nitric oxide generation. J Clin Invest 1994; 93:1968-1974.

164. Mollace V, Nistico G. Release of nitric oxide from astroglial cells: a key mechanism in neuroimmune disorders. Adv Neuroimmunol 1995; 5:421-430.

165. Lin RF, Lin T-S, Tilton RG et al. Nitric oxide localized to spinal cords of mice with experimental allergic encephalomyelitis: an electron paramagnetic resonance study. J Exp Med 1993; 178:643-648.

166. Hooper DC, Ohnishi ST, Kean R et al. Local nitric oxide production in viral and autoimmune diseases of the central nervous system. Proc Natl Acad Sci USA 1995; 92:5312-5316.

167. Scoazec JY, Lesèche G. Immunologie des cellules endothéliales et rejet de greffe. Médecine/Science 1993; 9:1094-1101.

168. Langrehr JM, Hoffman RA, Lancaster JR et al. Nitric oxide - a new endogenous immunomodulator. Transplantation 1993; 55:1205-1212.

169. Langrehr JM, Hoffman RA, Billiar TR et al. Nitric oxide synthesis in the in vivo allograft response: a possible regulatory mechanism. Surgery 1991; 110:335-342.

170. Langrehr JM, Dull KE, Ochoa JB et al. Evidence that nitric oxide production by in vivo allosensitized cells inhibits the development of allospecific CTL. Transplantation 1992; 53:632-640.

171. Langrehr JM, Müller AR, Bergonia HA et al. Detection of nitric oxide by electron paramagnetic resonance spectroscopy during rejection and graft-versus-host disease after small-bowel transplantation in the rat. Surgery 1992; 112:395-402.

172. Lancaster JR, Langrehr JM, Bergonia HA et al. EPR detection of heme and nonheme iron-containing protein nitrosylation by nitric oxide during rejection of rat heart

allograft. J Biol Chem 1992; 267:10994-10998.

173. Bastian NR, Xu SR, Shao XL et al. Nω-monomethyl-L-arginine inhibits nitric oxide production in murine cardiac allografts but does not affect graft rejection. Biochim Biophys Acta 1994; 1226:225-231.

174. Worrall NK, Lazenby WD, Misko TP et al. Modulation of in vivo alloreactivity by inhibition of inducible nitric oxide synthase. J Exp Med 1995; 181:63-70.

175. Yang X, Chowdhury N, Cai B et al. Induction of myocardial nitric oxide synthase by cardiac allograft rejection. J Clin Invest 1994; 94:714-721.

176. Winlaw DS, Schyvens CG, Smythe GA et al. Urinary nitrate excretion is a noninvasive indicator of acute cardiac allograft rejection and nitric oxide production in the rat. Transplantation 1994; 58:1031-1036.

177. Winlaw DS, Schyvens CG, Smythe GA et al. Selective inhibition of nitric oxide production during cardiac allograft rejection causes a small increase in graft survival. Transplantation 1995; 60:77-82.

178. Oz MC, Pinsky DJ, Koga S et al. Novel preservation solution permits 24-hour preservation in rat and baboon cardiac transplant models. Circulation 1993; 88:291-297.

179. Pinsky DJ, Naka Y, Chowdhury NC et al. The nitric oxide/cyclic GMP pathway in organ transplantation: critical role in successful lung preservation. Proc Natl Acad Sci USA 1994; 91:12086-12090.

180. Tanaka S, Kamiike W, Ito T et al. Generation of nitric oxide as a rejection marker in rat pancreas transplantation. Transplantation 1995; 60:713-717.

181. Ohdan H, Fukuda Y, Suzuki S et al. Simultaneous evaluation of nitric oxide synthesis and tissue oxygenation in rat liver allograft rejection using near-infrared spectroscopy. Transplantation 1995; 60:530-535.

182. Devlin J, Palmer RMJ, Gonde CE et al. Nitric oxide generation. A predictive parameter of acute allograft rejection. Transplantation 1994; 58:592-595.

183. Ioannidis I, Hellinger A, Dehmlow C et al. Evidence for increased nitric oxide production after liver transplantation in humans. Transplantation 1995; 59:1293-1297.

184. Weiss G, Schwaighoffer H, Herold M et al. Nitric oxide formation as predictive parameter for acute graft-versus-host disease after human allogeneic bone marrow transplantation. Transplantation 1995; 60:1239-1244.

185. Deng MC, Erren M, Kammerling L et al. The relation of interleukin-6, tumor necrosis factor-α, IL-2, and IL-2 receptor levels to cellular rejection, allograft dysfunction, and clinical events early after cardiac transplantation. Transplantation 1995; 60:1118-1124.

186. Roth E, Steininger R, Winkler S et al. L-arginine deficiency after liver transplantation as an effect of arginase efflux from the graft. Transplantation 1994; 57:665-669.

187. Adatia I, Lillehei C, Arnold JH et al. Inhaled nitric oxide in the treatment of postoperative graft dysfunction after lung transplantation. Ann Thorac Surg 1994; 57:1311-1318.

188. Wennmalm Å, Lanne B, Petersson A-S. Detection of endothelium-derived factor in human plasma in the basal state and following ischemia using electron paramagnetic resonance spectrometry. Anal Biochem 1990; 187:359-363.

189. Geng Y, Petersson A-S, Wennmalm Å et al. Cytokine-induced expression of nitric oxide synthase results in nitrosylation of heme and nonheme iron proteins in vascular smooth muscle cells. Exp Cell Res 1994; 214:418-428.

190. Schulz R, Triggle CR. Role of NO in vascular smooth muscle and cardiac muscle function. Trends in Pharmacol Sci 1994; 15:255-259.

191. Hansson GK, Geng Y, Holm J et al. Arterial smooth muscle cells express nitric oxide synthase in response to endothelial injury. J Exp Med 1994; 180:733-738.

192. Winlaw DS, Smythe GA, Keogh AM et al. Increased nitric oxide production in heart failure. Lancet 1994; 344:373-374.

193. Dusting GJ, Macdonald PS. Endogenous nitric oxide in cardiovascular disease and transplantation. Ann Med 1995; 27:395-406.

194. Wildhirt SM, Dudek RR, Suzuki H et al. Involvement of inducible nitric oxide synthase in the inflammatory process of myocardial infarction. Int J Cardiol 1995; 50:253-261.

195. Zweier JL, Wang P, Samouilov A et al. Enzyme-independent formation of nitric oxide in biological tissues. Nature Medicine 1995; 1:804-809.

196. Konorev EA, Joseph J, Kalyanaraman B. S-nitrosoglutathione induces formation of nitrosylmyoglobin in isolated hearts during cardioplegic ischemia. An electron spin resonance study. FEBS Lett 1996; 378:111-114.

197. Iadecola C, Pelligrino DA, Moskowitz MA et al. Nitric oxide synthase inhibition and cerebrovascular regulation. J Cereb Blood Flow Metab 1994; 14:175-192.

198. Kumura E, Yoshimine T, Tanaka S et al. Nitrosyl hemoglobin production during reperfusion after focal cerebral ischemia in rats. Neurosci Lett 1994; 177:165-167.

199. Dugan LL, Lin TS, He YY et al. Detection of free radicals by microdialysis/spin trapping EPR following focal cerebral ischemia-reperfusion and a cautionary note on the stability of 5,5-dimethyl-1-pyrroline N-oxide (DMPO). Free Rad Res 1995; 23:27-32.

200. Sato S, Tominaga T, Ohnishi T et al. ESR spin-trapping study of nitric oxide formation during bilateral carotid occlusion in the rat. Biochim Biophys Acta 1993; 195:197.

201. Sato S, Tominaga T, Ohnishi T et al. Electron paramagnetic resonance study on nitric oxide production during brain focal ischemia and reperfusion in the rat. Brain Res 1994; 647:91-96.

202. Zhang ZG, Chopp M, Bailey F et al. Nitric oxide changes in the rat brain after transient middle cerebral artery occlusion. J Neurol Sci 1995; 128:22-27.

203. Verrechia C, Boulu RG, Plotkine M. Neuroprotective and deleterious effects of nitric oxide on focal cerebral ischemia-induced neurone death. Adv Neuroimmunol 1995; 5:359-378.

204. Hall DM, Buettner GR, Matthes RD et al. Hyperthermia stimulates nitric oxide formation: electron paramagnetic resonance detection of ·NO-heme in blood. J Appl Physiol 1994; 77:548-553.

205. Holthusen H, Arndt JO. Nitric oxide evokes pain in humans on intracutaneous injection. Neurosci Lett 1994; 165:71-74.

206. Olesen J, Thomsen LL, Iversen H. Nitric oxide is a key molecule in migraine and other vascular headaches. Trends in Pharmacol Sci 1994;15:149-153.

207. Brown DR, Schmidt B, Kretzschmar HA. Role of microglia and host prion protein in neurotoxicity of a prion protein fragment. Nature 1996; 380:345-347.

208. McLaughlin MK, Conrad KP. Nitric oxide biosynthesis during pregnancy: implications for circulatory changes. Clin Exp Pharmacol Physiol 1995; 22:164-171.

209. Yallampalli C, Izumi H, Byam-Smith M et al. An L-arginine-nitric oxide-cyclic guanosine monophosphate system exists in the uterus and inhibits contractility during pregnancy. Am J Obstet Gynecol 1994; 170:175-185.

210. Lees C, Campbell S, Jauniaux E et al. Arrest of preterm labour and prolongation of gestation with glyceryl trinitrate, a nitric oxide donor. Lancet 1994; 343:1325-1326.

211. Participants in the 39th Forum in Immunology. L-arginine-derived nitric oxide and the cell-mediated immune response. Res Immunol 1991; 142:553-602.

212. Granger DL. Macrophage production of nitrogen oxides in host defence against microorganisms. Res Immunol 1991; 142:570-572.

══════════ CHAPTER 12 ══════════

Palliatives to Underproduction of Nitric Oxide as Assayed by EPR Spectroscopy

Claire Ducrocq and Annie Guissani

INTRODUCTION

Nitric oxide has been shown to be the endogenous stimulator of soluble guanylate cyclase (sGC) and as such to be the endogenous vasodilator of the smooth muscle in both the cardiovascular and cerebrovascular systems, and the digestive system. It is also an effector molecule released by macrophages, hepatocytes and other cell types after cytokine activation.[1-4]

What are the effects of an underproduction of NO on given organic systems and what are the palliatives to such situations? At the end of the 70s, it was already known that inorganic nitrate was biosynthesized in humans.[5] On the other hand it was demonstrated that NO, various nitrosylated compounds and derivatives capable of forming NO in incubations, activated sGC; this enzyme then increased the production of guanosine 3',5'-cyclic monophosphate (cGMP) which in turn allowed the vascular smooth muscle cell relaxation. This explained the high activity of sGC in some tissues exposed to environmental nitrosylated compounds.[6] It was then shown that the formation of the paramagnetic nitrosyl heme complex is a common and essential step in the process by which NO or NO-forming compounds activate sGC (see chapter 10, for a full development on the activation of sGC by NO).[7-9] Among these nitrosylated compounds, glyceryl trinitrate (GTN) and amyl nitrite are powerful drugs commonly used for more than a century in the treatment of cardio-

Nitric Oxide Research from Chemistry to Biology: EPR Spectroscopy of Nitrosylated Compounds, edited by Yann A. Henry, Annie Guissani and Béatrice Ducastel.

vascular diseases such as angina pectoris.[10-12] Just a century ago, one famous patient was Alfred Nobel, the inventor of dynamite (a clever riddle: which scientists are going to share a Nobel Price on the subject?).

L-ARGININE AND N^{ω}-HYDROXY-L-ARGININE

Direct NO formation from its endogenous precursor L-arginine was demonstrated in animal tissues in vivo, by the use of a specific NO-trap (Fe^{2+}-diethyldithiocarbamate), leading to a stable paramagnetic mononitrosyl iron complex, characterized by an EPR signal at g = 2.035 and 2.020, and a triplet hyperfine structure (see chapter 13 for a description of the methodology). The formation of [^{15}NO-Fe-diethyldithiocarbamate] from L-(^{15}N-guanidineimino)-arginine administered to animals was monitored in mouse liver after lipopolysaccharide (LPS) injection and demonstrated that NO forms in vivo and that NO arises from the guanidino nitrogens of L-arginine.[13]

Intravenous administration of L-arginine caused vasodilation in normotensive subjects and antihypertensive effects in patients with essential or secondary hypertension. L-arginine increases biological indicators of NO release in vivo, and modulates the release of neuroendocrine hormones.[14]

N^{ω}-hydroxy-L-arginine (NOHA) is an intermediate in the biosynthesis of NO from arginine; it can be liberated from NO-synthase (NOS), released into cell culture medium, and may be an alternate substrate for the enzymes responsible for NO generation; induced relaxation is indeed consistently greater with NOHA than with L-arginine.[15] However, it requires an enzymatic reduction process that both NO-synthase and cytochromes P-450 are able to realize (Fig. 12.1).[16-18]

Among the commonly known NO-releasing compounds, *S*-nitrosothiols, the molsidomine metabolite, SIN-1, adducts with heme (nitrosyl heme) or amines (NONOates) can spontaneously release NO in aqueous solutions, while sodium nitro-

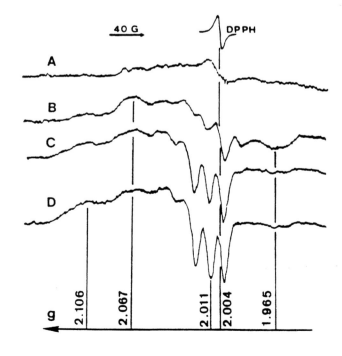

Fig. 12.1. Oxidation of N^{ω}-hydroxy-L-arginine catalyzed by cytochrome P-450 from dexamethasone-treated rat liver microsomes. EPR spectra at 77K of rat liver microsomes incubated with NOHA and NADPH. A: anaerobic microsomes (90 μmol/l P-450) + NADPH (5 mmol/l) + NOHA (250 μmol/l); B: introduction of O_2 into the EPR tube after warming at 20°C and immediate freezing; C: same as B after 5 min warming at 20°C; D: same as B after 20 min warming. (Reproduced with kind permission from: Boucher J-L, Genet A, Vadon S et al. Cytochrome P450 catalyzes the oxidation of N^{ω}-hydroxy-L-arginine by NADPH and O_2 to nitric oxide and citrulline. Biochem Biophys Res Commun 1992; 187:880-886.) Copyright © Academic Press, Inc., Orlando, Florida, USA.

prusside (SNP), sydnonymines, inorganic and organic nitrites and organic nitrates are activated under reductive conditions or by illumination. The only drug requiring an oxidative conversion to give NO is hydroxylamine NH_2OH. Nitric oxide release can be directly measured in the drug incubation medium, but is generally monitored by the marked elevation of cGMP levels in various organs and/or the guanylate cyclase activation in cell-free systems in the presence of the NO-donor. However, the vasodilator agents produced dilation of the systemic vasculature, leading to arterial hypotension, right ventricular ischemia and consequently heart failure at high frequency.

METAL COMPLEXES AS NITRIC OXIDE DONORS

NITROPRUSSIDE

Sodium nitroprusside (SNP; disodium pentacyanonitrosylferrate(2-) dihydrate, $Na_2Fe(CN)_5NO$, $2H_2O$) is composed of a ferric ion center complexed with five cyanide moieties and a nitrosyl group (Fig. 12.2). SNP is a stable and EPR-silent, diamagnetic salt. Its use in humans was first reported in 1928, as reviewed by Friederich and Butterworth,[19] and since then SNP has remained commonly in use as a vasodilating drug, due to its ability to release NO towards the vascular wall.

Nitroprusside is known to release NO under normal daylight or a visible lamp. A NO level of about 450 nmol/l was found in a buffered 1 mmol/l SNP solution kept at room temperature.[20] Another evaluation for the continuous measurement of NO during visible irradiation used EPR methodology to detect the transformation of nitrosyl nitroxides by NO into imino nitroxides (see chapter 13). Increasing SNP concentrations from 0.25 mmol/l to 2 mmol/l led to a linear increase in the rate of NO production, from 7-40 pmol/s. The maximum rate of NO production achieved was 45 pmol/s observed at SNP concentrations above 2 mmol/l.[21] The NO release can be activated through one electron reduction of SNP by hemoproteins such as the microsomal P-450/P-450 reductase system, or thiols.[22] In the course of SNP activation, reduced intermediates containing NO, the one-electron reduced form $[Fe(CN)_5NO]^{3-}$ (pentacyano complex) and its decomposition product $[Fe(CN)_4NO]^{2-}$

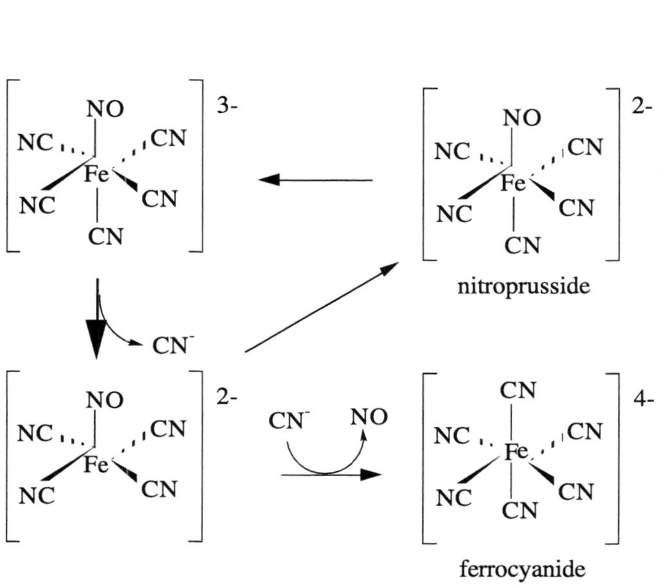

Fig. 12.2. Transformation of nitroprusside and release of NO under illumination.

(tetracyano complex), are produced, which are paramagnetic and EPR detectable.[22,23] Loss of the *trans* cyanide ligand from the pentacoordinated complex is required for the dissociation of NO, which could lead to cyanide poisoning.[24,25]

In SNP-treated tissues and homogenates, the pentacyanide complex is the only EPR-detectable species, together with the $Fe(NO)_2(SR)_2$ complexes spontaneously encountered in cells where endogenous NO production occurs (see chapters 10 and 11).[22,23,26]

Reactions of SNP with both heme(Fe^{II}) and sulfhydryl groups liberate NO. Oxygen is susceptible first to shift the penta-tetracyanide equilibrium in the direction of the tetracyanide form and also to transfer NO more probably to oxyheme(Fe^{II}) than to heme(Fe^{II}).[23] NO is formed during the following reductive metabolism (Fig. 12.3).

These results obtained in a model hemoglobin system were confirmed in a study of SNP metabolization by intact porcine endothelial cells (Fig. 12.4).[27]

Several authors have discussed cyanide toxicity induced by SNP infusions.[28] Friederich and Butterworth recently reviewed SNP current indications, contraindications and toxic side effects, in medical use.[19] For instance intracavernous use of SNP seems inappropriate as a treatment of male impotence.[29]

NITROSYL HEMES

A reversible binding of NO to hemoproteins is generally attributed to an interaction of NO with a heme(Fe^{III}), while interaction with heme(Fe^{II}) is considered very tight. Such an example is located in the salivary glands of the blood-sucking insect *Rhodnius prolixus*,[30,31] the vector of Chagas' disease. This cherry red colored gland yielded four major nitrosyl heme(Fe^{III}) proteins, or nitrophorins, which are found naturally loaded with NO arising from the L-arginine/NOS metabolism within the gland. Upon dilution these nitrosyl hemoproteins release NO, which inhibits platelet function and induces vasodilation. In addition to dilution, another factor such as histamine could further increase NO dissociation from the nitrophorins. Indeed histamine can bind with a high affinity to the heme protein and displace native NO (see chapter 14).[32] This hemoprotein NO-carrier found in insects, could provide a model and some new ideas in the pharmacology of NO-donors.

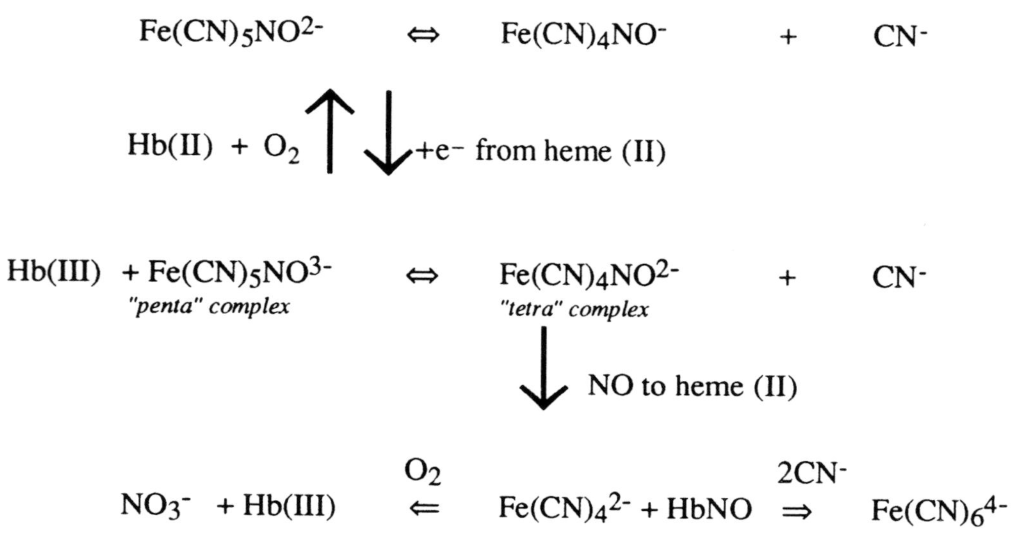

$$Fe(CN)_5NO^{2-} \Leftrightarrow Fe(CN)_4NO^- + CN^-$$

$$Hb(II) + O_2 \uparrow \downarrow +e^- \text{ from heme (II)}$$

$$Hb(III) + Fe(CN)_5NO^{3-} \Leftrightarrow Fe(CN)_4NO^{2-} + CN^-$$
"penta" complex   *"tetra" complex*

$$\downarrow \text{ NO to heme (II)}$$

$$NO_3^- + Hb(III) \overset{O_2}{\Leftarrow} Fe(CN)_4^{2-} + HbNO \overset{2CN^-}{\Rightarrow} Fe(CN)_6^{4-}$$

Fig. 12.3. Mechanism of cyanide poisoning by nitroprusside in the presence of hemoglobin.

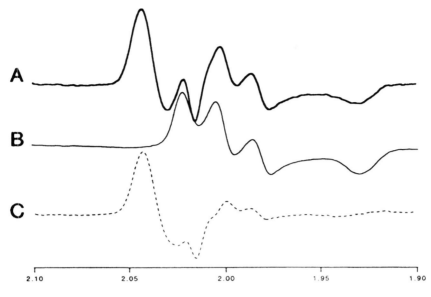

Fig. 12.4. EPR spectrum (A) at -138°C of porcine endothelial cells (2x10⁶) incubated with 2 mmol/l nitroprusside for 30 min at 37°C under anaerobic conditions. The spectrum can be separated into two components: spectrum B of the reduced form of SNP ($g_\perp = 2.003$, $A_\perp(^{14}N) = 30$ G, $g_{//} = 1.930$), spectrum C of the $Fe(NO)_2(SR)_2$ complex ($g_\perp = 2.033$ and $g_{//} = 2.015$). (Reproduced with kind permission from: Rochelle LG, Kruszyna H, Kruszyna R et al. Bioactivation of nitroprusside by porcine endothelial cells. Toxicol Appl Pharmacol 1994; 128:123-128.) Copyright © Academic Press, Ltd, London, UK.

NITROSYL COMPLEXES OF IRON-SULFUR CLUSTERS

The recent upsurge of interest in iron-sulfur nitrosyl complexes has been stimulated by the understanding of naturally occurring [2Fe-2S] and [4Fe-4S] clusters found in proteins and their ability to form nitrosyl complexes with NO or nitrite under physiological conditions (see chapters 6 and 10). Several types of iron-sulfur-nitrosyl complexes have been investigated as potent vasodilators, antiplatelet aggregation agents and bacteriostatic agents.[33-35] The complex $[Fe_4S_3(NO)_7]^-$, Roussin's black salt, was shown to inhibit several very toxic *Clostridium* and *Salmonella* species, and was extracted from nitrite-treated proteins.[36,37] It is chemically and spectroscopically very similar to paramagnetic mononuclear dinitrosyl iron complexes. The transfer of nitrosyl groups from these complexes to non-iron atoms is problematical, as reported by Butler et al.[36,37]

Iron-centered paramagnetic complexes, formed by reactions between iron salts and nitric oxide in the presence of anions and characterized by their EPR spectra at $g_{av} = 2.03$, were first reported over 20 years ago. The identification of the $g_{av} = 2.03$ species as being of the general type $[Fe(NO)_2(SR)_2]$, where SR may be either a free cysteine or a cysteine-containing protein at the iron binding site, was far from definitive. These paramagnetic mono-iron complexes are both readily formed in vivo and associated with cancerous states in experimental animals (see chapters 10 and 11). Recently it was found that the activation of NO synthesis in macrophages and endothelial cells resulted in the formation of dinitrosyl iron complexes in these cells, characterized by the $g_{av} = 2.03$ EPR signal. Either a loosely bound non-heme iron, or iron release due to destruction of iron-sulfur centers by NO oxidation products or from ferritin, are the possible sources of iron for the formation of the $g_{av} = 2.03$ complexes.[38-40]

NITRITE, ORGANIC NITRITES AND NITRATES

NITROUS ACID AND NITRITE

Concentrated nitrous acid is a NO-donor through a decomposition giving rise to nitric acid; in dilute solution it slowly dismutates into $NO^{\bullet}$ and $NO_2^{\bullet}$:

$$2HNO_2 \leftrightarrow NO^{\bullet} + NO_2^{\bullet} + H_2O$$

Rate constants of this equilibrium favor nitrous acid formation, and this explains why nitrous acid induces a dilation of aortic ring strips and this more specifically when in more acidic solutions. This is the basis of the seminal experiments of Furchgott, leading to the identification of EDRF.[41]

However, inorganic nitrite is thought to require conversion to NO in vascular tissue for vasodilating activity. A millimolar nitrite level is a minimum amount required to produce vasodilation of aortic rings. Mineral nitrite can be reduced to iron-nitrosyl adduct by hemoproteins such as myoglobin, hemoglobin, cytochrome P-450 and model porphyrins.[42,43] These complexes might insert oxygen between C-H, S-H, $C=O$, $N=O$.[44]

ORGANIC NITRITES

Organic nitrites containing the -C-O-N=O function, such as amyl and *n*-butyl nitrite, are commonly used as recreational drugs by homosexuals. While they react readily with thiols to form *S*-nitrosothiols, they also undergo metabolic conversion to NO.[45,46] This conversion is located in the cytosol of smooth muscle cells and requires the presence of thiols. EC_{50} values of 10^{-5} mol/l for the relaxation of aortic rings have been reported for both nitrites.[47] Recently, it has been shown that the glutathione *S*-transferases stimulated the formation of the *S*-nitrosoglutathione (GSNO) from organic nitrites. Like NO, *S*-nitrosothiols may cause cytotoxicity in hepato-

cytes and cytostasis in activated T lymphocytes.[48]

ORGANIC NITRATES

Organic nitrates, such as nitroglycerin (or glyceryl trinitrate, GTN) and isosorbide nitrate, contain the $-C-O-NO_2$ group. They found their major industrial application as explosives. Shortly after the first successful synthesis of nitroglycerin in the mid-nineteenth century, nitrates were found to be effective for the treatment of heart disease. These compounds are potent vasodilators and have enjoyed extensive use in cardiovascular therapy. The duration and strength of their effects vary according to their structures and pharmacokinetics. GTN, the most potent one, is generally thought to be a smooth muscle relaxant. Organic nitrates' pharmacokinetics appear related to both accumulation of cGMP and vascular smooth muscle relaxation.[49-51] The common pathway followed by GTN and EDRF, both leading to formation of *S*-nitrosothiols and subsequent activation of guanylate cyclase was demonstrated.[45,52,53] The identification of NO with EDRF was enabled because it appeared to be formed by bioconversion of organic nitrovasodilators in intact blood vessel and in cultured cells.[54-56]

Glyceryl trinitrate, denoted here as $RONO_2$, can be reduced with hemoglobin to glyceryl dinitrates, written as ROH, generating also nitrosyl hemoglobin hybrids (see chapter 4):[57-60]

$$RONO_2 + 3H^+ + 3\ Hb(Fe^{II}) \rightarrow ROH + NO^{\bullet} + H_2O + 3\ Hb(Fe^{III})$$

Similarly HbNO can be detected after injection into rat veins or mixing human blood with isosorbide dinitrate, a long-acting type of nitrovasodilator (Fig. 12.5).[61]

The enzymatic processes able to produce EDRF from organic nitrates is still under study. The first enzymatic system to be well characterized was the glutathione-dependent organic nitrate reductase, also called glutathione *S*-transferase, which generates nitrite.[62] Nitrite ions may not be

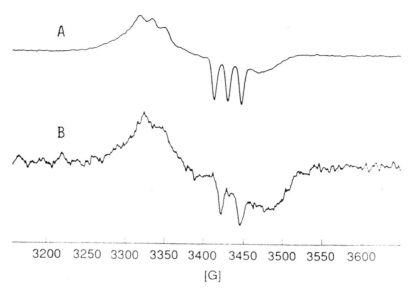

Fig. 12.5. EPR spectra at 110 K of HbNO (A) and ^{15}N-HbNO (B) production in the venous blood of rats 30 min after injection of isosorbide dinitrate solution and the ^{15}N-compound, respectively. (Reproduced with permission from: Kosaka H, Tanaka S, Yoshii T et al. Direct proof of nitric oxide formation from a nitrovasodilator metabolized by erythrocytes. Biochem Biophys Res Commun 1994; 204:1055-1060.) Copyright © Academic Press, Inc, Orlando, Florida, USA.

directly involved to account for the vasodilation derived from GTN (EC_{50} value of 10^{-9} mol/l for the relaxation of contracted aortic ring).[46] On the other hand, it appears that the NO-generating enzymes for the organic nitrites are distinct from those for the organic nitrates. These dissimilar metabolic activation mechanisms for organic nitrates and nitrites are consistent with a small cross-tolerance exhibited towards nitroglycerin-tolerant vessels.[63]

However, an enzymatic system converting GTN to NO, as measured by chemiluminescence, is also located in the cellular plasma membrane in the bovine coronary artery smooth muscle cell.[64] In aortic strips a possible S-nitrosothiol intermediate could be involved in GTN metabolism.[65] Direct demonstration of NO generation from organic nitrate can be assessed by the 2-line superfine structure in the EPR spectra of Hb-^{15}NO in the venous blood of rats treated by organic nitrate.[66]

In patients receiving GTN for coronary diseases, nitrosyl hemoglobin was detected in whole blood by EPR spectroscopy at 77 K, several hours after intravenous GTN injection (Fig. 12.6).[67] However not all subjects presented an EPR signal within the detection limit of the method. The individual differences could perhaps be accounted for by temperature-dependent instability of HbNO in the blood collection.

The biotransformation of glyceryl trinitrate into dinitrates and mononitrates occurs in most microorganism species.[68,69] By measuring nitrite formation and scanning the EPR spectra of entire cells of an eukaryotic fungus *Phanerochaete chrysosporium*, the correlation between the nitrate conversion and the appearance of nitrite and nitrosyl complexes indicated several distinct metabolic pathways. The metabolic process which led to NO implied the formation of a nitrosyl non-heme iron-sulfur complex, via NO binding to a reduced

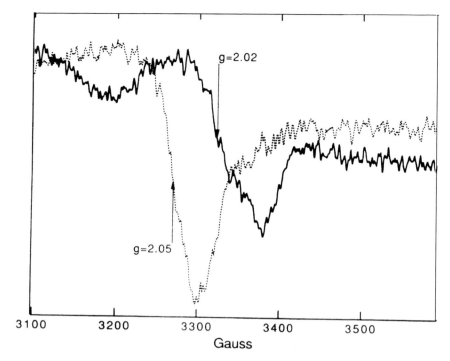

Fig. 12.6. EPR spectra of whole blood of a patient. The spectrum of blood sample before
nitroglycerin treatment with a signal attributed to ceruloplasmin at g = 2.05 (broken line)
was substracted from the spectrum of a sample obtained after the patient had been
receiving nitroglycerin for several hours to obtain the HbNO signal at 2.02 (solid line).
EPR spectra were obtained at 77 K and resulted from averaging of 100 scans.
(Reproduced with kind permission from: Cantilena LR, Smith RP, Frasur S et al. Nitric
oxide hemoglobin in patients receiving nitroglycerin as detected by electron paramagnetic resonance spectroscopy. J Lab Clin Med 1992; 120:902-907.) Copyright © Mosby-
Year Book Inc, St Louis, MO, USA.

heme at the beginning of the conversion (Fig. 12.7).[70,71]

It has been suggested that the microbial metabolism of organic nitrates is closely linked to assimilatory and/or dissimilatory reduction of nitrate by microorganisms.[72] A similar reductive enzymatic pathway able to convert GTN to NO was obvious in rat liver where the system cytochrome P-450 plus cytochrome P-450 reductase reduced the nitrate function (Fig. 12.8).[73,74]

This process is greatly inhibited by molecular oxygen and depended particularly on the presence of the isoenzyme P-450 CYP3A.[75] The effects of oxygen and cytochrome P-450 inhibitors on the biotransformation of GTN by rat aortic microsomes has also been assessed.[76]

Different organic nitrates have variable specificity in NO formation among animal vascular tissues: aorta, femoral artery, vena cava, etc., and organs: heart, kidney, spleen, liver, etc.[77] As prodrugs they can exert their medicinal properties in nonvascular tissues. A particular nitrate, 18-nitro-oxyandrostenedione (18-ONO$_2$A), which is a potent inhibitor of aldosterone biosynthesis, is now recognized to release NO during its P-450 catalyzed reductive metabolism.[78] Formation of NO was followed by detection of P-450 and P-420-Fe$^{(II)}$-NO complexes by EPR spectroscopy.[78,79] As this P-450-dependent oxidation is inhibited to a great extent upon addition of SOD, O$_2$$^{•-}$, could be involved in this transformation.

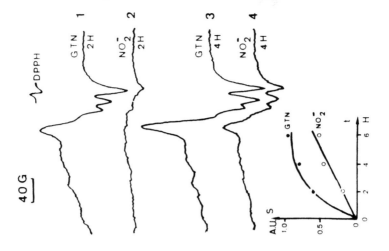

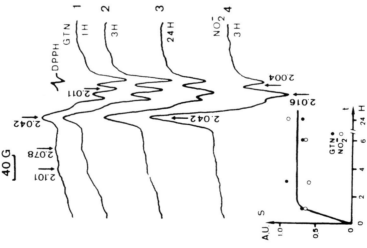

Fig. 12.7. Nitroglycerin metabolization by Phanerochaete chrysosporium. EPR spectra at 77 K of P. chrysosporium sampled during GTN and nitrite incubations at 37°C. Left panel. The fungus pellets are sampled and frozen at indicated times from 2 mmol/l GTN (spectra 1-3) and 4 mmol/l nitrite (spectrum 4) incubations with a mycelium suspension. Inset: variation with time of the area S of the integrated spectra, in arbitrary units. Right panel. The fungus was incubated during 2 and 4 hr with either (spectra 1 and 3) 2 mmol/l GTN or (spectra 2 and 4) nitrite added continuously at the concentration calculated from the measured concentrations of glyceryl mono- and dinitrates. Inset: Area S of the integrated spectra, as a function of time. (Reproduced with kind permission from: Servent D, Ducrocq C, Henry Y et al. Nitroglycerin metabolism by Phanerochaete chrysosporium: evidence for nitric oxide and nitrite formation. Biochim Biophys Acta 1991; 1074:320-325.) Copyright © Elsevier Science, Sara Burgerhartstraat 25, 1055 KV Amsterdam, The Netherlands.

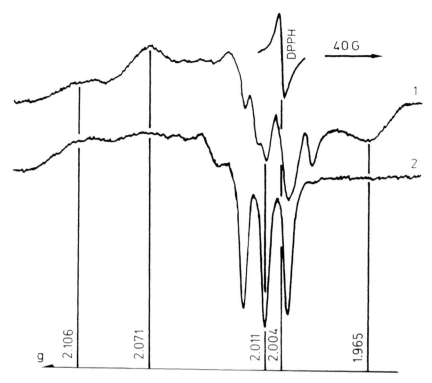

Fig. 12.8. Nitroglycerin metabolization by liver microsomes of dexamethasone-treated rats. EPR spectra at 77 K of cytochrome P-450-NO and P-420-NO. 1: after incubation of microsomes (48 µmol/l) under argon with nitrite (9 mmol/l) and ascorbate (9 mmol/l); 2: after an hour incubation with GTN (1.5 mmol/l) and NADPH (15 mmol/l). Resonances at $g_x = 2.071$, $g_z = 2.004$ ($A_z = 20$ G) and $g_y = 1.965$ are assigned to P-450-NO, those at $g_x = 2.106$ and $g_z = 2.01$ ($A_z = 16$ G) to P-420-NO. (Reproduced with kind permission from: Henry Y, Ducrocq C, Drapier J-C et al. Nitric oxide, a biological effector. Electron paramagnetic resonance detection of nitrosyl-iron-protein complexes in whole cells. Eur Biophys J 1991; 20:1-15.) Copyright © Springer Verlag Tiergartenstasse 17, D-69121 Heidelberg, Germany.

It has recently been reported that organic nitrates and nitrites have other NO-related physiological effects. They are able to modulate adrenergic neuroeffector transmission.[80] They produce detectable concentrations of NO in exhaled air in vivo.[81] Decreased synthesis of NO in the uterus is associated with initiation of labor in animals, and uterine muscle quiescence is induced in vitro by an increase in cGMP.[82] GTN can be successfully used, through patch diffusion, to stop preterm labor and to increase the gestation period by up to 8 weeks.[82]

OTHER DERIVATIVES

HYDROXYLAMINE

Hydroxylamine acts as a vasorelaxant on rat aortic strips or in a whole lung rat perfusion model, but its properties are attributed to the release of NO in biological systems.[61,62,83] It does not directly act on guanylate cyclase but rather requires a three-electron oxidation to exert its vasorelaxant effect (EC_{50} for relaxation of norepinephrine-contracted aortic rings is 10^{-6} mol/l). Catalase is known to catalyze this reaction according to the following scheme:

$$E(Fe^{III}) + H_2O_2 \rightarrow E(Fe^V=O) + H_2O$$

$$E(Fe^V=O) + NH_2OH \rightarrow H_2O +$$
$$E(Fe^{III}HNO) \rightarrow E(Fe^{II}NO) + H^+$$

$$E(Fe^{II}NO) \leftrightarrow E(Fe^{II}) + NO^\bullet$$

If L-arginine is converted to NO through a hydroxylamine intermediate, this last step must occur regardless of whether the oxidation is catalyzed by catalase or another oxidative pathway, such as cytochrome *c* oxidase (see chapter 6).[83]

S-NITROSOTHIOLS

Because they are formed in vivo or in vitro under physiological conditions by reactions between cysteine, glutathione, peptides or proteins containing cysteine residues, and oxidized derivatives of authentic NO or EDRF, *S*-nitrosothiols are considered as the principal reservoir of NO in vivo. An evaluation of *S*-nitrosothiols in mammalian plasma pointed out a concentration 3 orders magnitude greater than that of free NO. The predominant species is the *S*-nitroso-serum albumin.[84] All *S*-nitrosothiols exhibit antiplatelet and vasodilatory properties mediated by cGMP and they prolong the EDRF effects according to their structure. These compounds decomposed spontaneously to NO and the corresponding disulfides. The half-life of *S*-nitrosothiols in buffered solutions ranged from a few minutes (for *S*-nitroso-*N*-acetyl-penicillamine (SNAP) and *S*-nitroso-cysteine) to several hours (for *S*-nitroso-glutathione and *S*-nitroso-serum albumin).[48,84,85] *S*-nitrosohemoglobin could also function as another highly concentrated NO reservoir in circulating red blood cells.[86]

As the spontaneous liberation of NO cannot account for relaxation in vitro by *S*-nitrosothiol, various mechanisms have been proposed to explain this discrepancy: chemical or enzymatic ways could be explored. First, the homolytic rupture releasing NO from *S*-nitrosothiols is dependent upon many chemical and physical factors and for example can be excited photochemically or catalyzed by Cu^{2+}.[87] The decomposition of *S*-nitrosocysteine is also catalyzed by traces of iron ions by the intermediate formation of dinitrosyl-iron complexes described by Vanin et al.[88] Such catalytical effects to increase NO release have been found associated with membrane components of the vascular smooth muscle cell.[89] Moreover, there is evidence that heterolytic decompositions leading to NO^+ (nitrosyl) or NO^- (nitroxyl) release are more rapid than $NO^\bullet$ release at physiological pH.[90,91] Involvement of thiolate ion and NO^+ species can be attributed to transnitrosation reactions, rather than nitrosation by way of an oxydative NO derivative; thus, the half-lives of SNAP or *S*-nitrosoalbumin are close to that of *S*-nitrosocysteine in the presence of increasing amounts of cysteine.[92] These reactions constitute the significant pathways of *S*-nitrosothiol metabolism.

NITRIC OXIDE ADDUCTS WITH NUCLEOPHILES, DIAZENIUMDIOLATES, NONOATES

NO adducts with polyamines or secondary amines produced stable solid salts of general formula $XN(O^-)N=O$, where X is a nucleophile residue.[93-95] A wide variety of diazeniumdiolate derivatives (Drago's salts, NONOates, discovered in the early 60s) can be synthesized simply by exposing nucleophilic compounds to NO. Use of dilute solutions of diamines favored formation of the intramolecular zwitterions: $RR'N-NO-NO^- + H^+ \rightarrow RR'NH + 2NO$.

These dissolved compounds undergo slow decomposition which regenerates NO with first order rates varying in a predictable way with structure. For example the half-life of $Et_2N[N(O)NO]Na$, diethylamine-NO adduct (DEA-NO) and that of spermine adduct are respectively 2 min and 40 min at pH 7.4 and 37°C.[96]

NO release from these complexes is not catalyzed by exogenous thiol or albumin; thus they can be used as agents for controlled biological release of NO because their potency as vasorelaxants correlates closely with the rates of spontaneous reversion to NO in simple aqueous buffer.[95,97]

NONOates act as potent vasodilators and have an antiplatelet effect.[93,97-99] They were shown to inhibit proliferation of human melanoma cells in vitro and of vascular smooth muscular cells. The most effective NO donors for antiproliferative activity were long-lived NO donors with $IC_{50} = 180$ µmol/l for spermine-NO adduct and 40 µmol/l for diethylenetriamine-NO adduct (DETA-NO) whose $\tau_{1/2}$ is 20 hr (see chapter 2). This antiproliferative effect was related to the NO release and corresponded to the inhibition of the DNA synthesis.[96] Other NONOate molecules are currently synthesized and assayed.[100]

The NONOates' properties can be conveniently altered by changing the identity of the nucleophilic residue. The current trends consists in the choice of the nucleophilic entity in order to target the organ where the drug is intended to deliver NO. Another example consists of masking NO with a photo-labile 2-nitrobenzyl group to give a completely stable NO donor until photolysis ("caged" nitric oxide), whereupon it releases NO.[101] Other NO donors with quite different structures, such as furoxan derivatives, are currently under in vitro trials.[102]

MOLSIDOMINE AND SYDNONIMINES

Molsidomine (N-ethoxycarbonyl-3-morpholinosydnonymine) is a vasodilator drug clinically used in Europe. It first undergoes an enzymatic hydrolysis to the 3-morpholinosydnonimine (SIN-1) mediated by liver esterases.[103] Subsequently, in aqueous solution, the ring opens in a pH-dependent manner to yield N-morpholino-iminoacetonitrile (SIN-1A) which spontaneously releases NO. The bioactivation process requires oxidation by oxygen, and releases continuously $O_2^{•-}$, $NO^{•}$, NO_x and other potent oxidants such as peroxynitrite and hydroxyl radical.[104]

SIN-1 is often used as a model for continuous release of peroxynitrite. Cytotoxic effects of SIN-1 have been demonstrated in a large variety of cell lines. Assays were performed using SOD or catalase or $OH^{•}$ trapping to propose the structure of the

toxic agent. If peroxynitrite is the cause of cytotoxicity, SOD will be protective since dismutation of $O_2^{•-}$ will prevent formation of $ONOO^-$. If NO is the toxic agent, SOD can potentiate NO by preventing its reaction with $O_2^{•-}$. Actually, effects of SOD on the SIN-1 toxicity are very different according to the cell type. In the case of human hepatoma, the potentiation of SIN-1 cytotoxicity by both Cu,Zn-SOD and Mn-SOD which is abolished by catalase is linked to the formation of H_2O_2 by SOD's catalyzed dismutation of $O_2^{•-}$.[105]

MISCELLANEOUS DRUGS

Liver cytochromes P-450 and NOS have been found to catalyze the oxidative cleavage of the C=NOH bond, not only of the endogenous substrate N-hydroxy-L-arginine, but also of many xenobiotic compounds containing aldoxime, ketoxime, amidoxime and N-hydroxyguanidine functions, by the following mechanism:[106-109]

$$RR'\ C = N\text{-}OH \rightarrow RR'\ C = O + NO^{•},\ NO_2^-,\ NO_3^-.$$

FK409 {(E)-4-ethyl-2-[(E)-hydroxy-imino]-5-nitro-3-hexene-amide}, isolated from the fermentation broth of *Streptomyces griseosporeus*, exhibits vasorelaxant and antiplatelet properties due to spontaneous NO-release.[110] The rate limiting step in the degradation of FK409 might be the hydroxyl ion-dependent subtraction of the H atom at the 5-position (Fig. 12.10).[111,112]

Nitric oxide is assumed to be generated during rapid FK409 degradation (rate constant 2.2 10^{-2} min^{-1} at pH 7). FK144420, a synthetic analogue of FK409, releases NO more slowly and shows a longer duration of biological effects.[113] This natural compound could be a useful drug for study of the physiological role of NO.

INHALATION OF NITRIC OXIDE: NO GAS, A SUPPLETIVE DRUG

Quantification of NO in exhaled air from animals and humans by chemiluminescence was performed on healthy subjects. A NO concentration of 6 parts per billion (ppb) in the exhaled air during normal breathing peaked to 100-300 ppb after 5-60 s breathholding.[114] During exercise and hyperventilation, the amount of NO excreted increased.

The phenomenon of acute or chronic hypoxic pulmonary vasoconstriction is known to increase the pulmonary vascular resistance in patients with parenchymal lung disease, pneumonia or adult respiratory distress syndrome (ARDS). In the case of ARDS, acute vasoconstriction and widespread occlusion of the pulmonary microvasculature should be improved by selective pulmonary vasodilators without systemic effects.[115] But most drug trials have been complicated by decreased systemic vascular resistance and peripheral hypotension and consequent heart failure

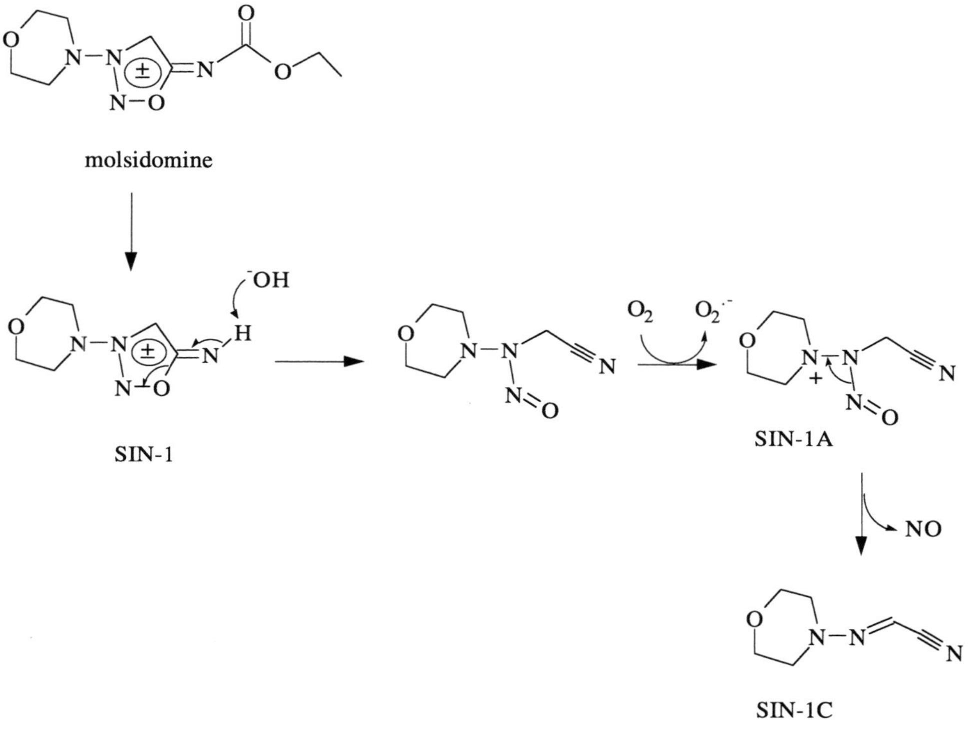

molsidomine

Fig. 12.9. Decomposition of molsidomine into SIN-1, SIN-1A, O_2^{-}, NO and SIN-1C.

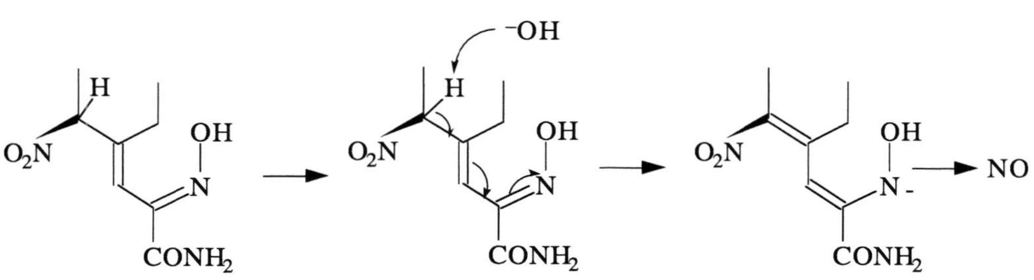

Fig. 12.10. Proposed kinetic scheme of NO release from FK 409.

due to the lack of pharmacological specificity.[116-119]

For this reason, inhalation of NO gas in a concentration of 2-80 parts per million (ppm) was assayed to dilate selectively the pulmonary circulation and improve arterial oxygenation.[120-122] The onset of pulmonary vasodilation occurred within seconds after beginning NO inhalation, and the vasodilator effect was maximal within 3 minutes. EC_{50} for improvement of oxygenation and that for reduction of mean pulmonary artery pressure were respectively about 100 ppb, a concentration similar to that of a very polluted urban atmosphere (see chapter 1), and 2-3 ppm.[123,124] A small but significant methemoglobinemia occurred.[125]

As higher oxides such as NO_2 are highly toxic in doses above 5 ppm, it is of vital importance to ensure a safe and precise delivery of NO (see chapter 2).[126-129] The formation of NO_2 from NO is dependent on time and on the initial concentration of oxygen. So 50% oxidation rate of NO to NO_2 in air is 72 hr, 7 hr or 40 min when the NO concentration varied from 1 ppm to 10 or 100 ppm.[130] The therapeutic use of 2-20 ppm of NO was applied to patients with severe ARDS and newborn patients.[131-136] Inhaled NO has therapeutic potential in postoperative pulmonary hypertension for congenital heart disease,[137-139] and in patients undergoing cardiac surgery.[140-142] Continuous inhalation is utilized during the perioperative period in a premature neonate with a severe congenital diaphragmatic hernia[143] and congenital heart disease.[144-146]

Cigarette smoke contains concentrations of 400-1000 ppm of NO,[147] which could act as a bronchial vasodilator and as an inhibitor of constitutive endothelial NO synthase (NOS III) (see chapters 8 and 14). Cigarette smoking appears to decrease to 4-5 ppb the level of expired NO, which could be correlated to the increased risk for hypertension in smokers,[148] characterized by the often reported early-morning cardiac infarction of the heavy smoker before his first cigarette.

CONCLUSION

The search for effective palliatives to underproduction of nitric oxide is certainly the most promising aspect of all the research in the domain of nitric oxide. This goes with the rapid unraveling of the functional mechanisms of the NOS isoenzymes and of their cellular localization. In view of the ubiquity of nitric oxide and of its functional polyvalence, the development of a pharmacology of NO donors requires a very high organ specificity of the drugs. The case of bronchial inhalation, now in widespread use less than ten years after the discovery of NO biosynthesis in mammals, is, in the field of drug development, absolutely unique. In fact, NO is a "natural" gas and should be considered as an "honorary" drug, except that its potentially detrimental manipulation in air requires extremely precise safety rules.

Developing the pharmacology of NO-releasing drugs also requires a sound knowledge of their chemistry and of their organ-specific cellular metabolisms. The role of EPR spectroscopy in these pharmacological studies can only be very modest, except by use of NO-specific-spin trapping methods. As we shall overview in the next chapter several promising methods are under trial, which potentially allow observation of the kinetics of free NO appearance and decay, and perhaps NO imaging in organs.

REFERENCES

1. Moncada S, Palmer RMJ, Higgs EA. Nitric oxide: physiology, pathophysiology, and pharmacology. Pharmacol Rev 1991; 43:109-142.
2. Stuehr DJ, Griffith OW. Mammalian nitric oxide synthases. Adv Enzymol 1992; 65:287-346.
3. Nathan C. Nitric oxide as a secretory product of mammalian cells. FASEB J 1992; 6:3051-3064.
4. Moncada S, Higgs EA. Molecular mechanisms and therapeutic strategies related to nitric oxide. FASEB J 1995; 9:1319-1330.
5. Green LC, Ruiz de Luzuriaja K, Wagner DA et al. Nitrate biosynthesis in man. Proc Natl Acad Sci USA 1981; 78:7764-7768.

6. Arnold WP, Mittal CK, Katsuki S et al. Nitric oxide activates guanylate cyclase and increases guanosine 3':5'-cyclic monophosphate levels in various tissue preparations. Proc Natl Acad Sci USA 1977; 74:3203-3207.

7. Craven PA, DeRubertis FR. Restoration of the responsiveness of purified guanylate cyclase to nitrosoguanidine, nitric oxide, and related activators by heme and hemoproteins. Evidence for involvement of the paramagnetic nitrosyl-heme complex in enzyme activation. J Biol Chem 1978; 253:8433-8443.

8. Craven PA, DeRubertis FR, Pratt DW. Electron spin resonance study of the role of NO· catalase in the activation of guanylate cyclase by NaN_3 and NH_2OH. Modulation of enzyme responses by heme proteins and their nitrosyl derivatives. J Biol Chem 1979; 254:8213-8222.

9. Ignarro LJ, Adams JB, Horwitz PM et al. Activation of soluble guanylate cyclase by NO-hemoproteins involves NO-heme exchange. Comparison of heme-containing and heme-deficient enzyme forms. J Biol Chem 1986; 261:4997-5002.

10. Harrisson DG, Bates JN. The nitrovasodilators. New ideas about old drugs. Circulation 1993; 87:1461-1467.

11. Bennett BM, McDonald BJ, Nigam R et al. Biotransformation of organic nitrates and vascular smooth muscle cell function. Trends in Pharmacol Sci 1994; 15:245-249.

12. Reeves JT. Brunton's use of amyl nitrite in angina pectoris: an historic root of nitric oxide research. News in Physiol Sci 1995; 10:141-144.

13. Kubrina LN, Caldwell WS, Mordvintcev PI et al. EPR evidence for nitric oxide production from guanidino nitrogens of L-arginine in animal tissues in vivo. Biochim Biophys Acta 1992; 1099:233-237.

14. Hishikawa K, Nakaki Y, Suzuki H et al. Role of L-arginine-nitric oxide pathway in hypertension. J Hypertens 1993; 11:639-645.

15. Wallace GC, Gulati P, Fukuto JM. N^ω-hydroxy-L-arginine: a novel arginine analog capable of causing vasorelaxation in bovine intrapulmonary artery. Biochem Biophys Res Commun 1991; 176:528-534.

16. Boucher J-L, Genet A, Vadon S et al. Cytochrome P450 catalyzes the oxidation of N^ω-hydroxy-L-arginine by NADPH and O_2 to nitric oxide and citrulline. Biochem Biophys Res Commun 1992; 187:880-886.

17. Renaud J-P, Boucher J-L, Vadon S et al. Particular ability of liver P450s3A to catalyze the oxidation of N^ω-hydroxyarginine to citrulline and nitrogen oxides and occurrence in NO synthases of a sequence very similar to the heme-binding sequence of P450s. Biochem Biophys Res Commun 1993; 192:53-60.

18. Mansuy D, Boucher J-L, Clement B. On the mechanism of nitric oxide formation upon oxidative cleavage of C=N(OH) bonds by NO-synthases and cytochromes P450. Biochimie 1995; 77:661-667.

19. Friederich JA, Butterworth JF. Sodium nitroprusside: twenty years and counting. Anesth Analg 1995; 81:152-162.

20. Shibuki K. An electrochemical microprobe for detecting nitric oxide release in brain tissue. Neurosci Res 1990; 9:69-76.

21. Singh RJ, Hogg N, Neese F et al. Trapping of nitric oxide formed during photolysis of sodium nitroprusside in aqueous and lipid phases: an electron spin resonance study. Photochem Photobiol 1995; 61:325-330.

22. Rao DNR, Elguindi S, O'Brien PJ. Reductive metabolism of nitroprusside in rat hepatocytes and human erythrocytes. Arch Biochem Biophys 1991; 286:30-37.

23. Kruszyna H, Kruszyna R, Rochelle LG et al. Effects of temperature, oxygen, heme ligands and sufhydryl alkylation on the reactions of nitroprusside and nitroglycerin with hemoglobin. Biochem Pharmacol 1993; 46:95-102.

24. Smith RP, Kruszyna H. Nitroprusside produces cyanide poisoning via a reaction with hemoglobin. J Pharm Exp Ther 1974; 191:557-563.

25. Wilcox DE, Kruszyna H, Kruszyna R et al. Effect of cyanide on the reaction of nitroprusside with hemoglobin: relevance to cyanide interference with the biological activity of nitroprusside. Chem Res Toxicol 1990; 3:71-76.

26. Kowaluk EA, Seth P, Fung H-L. Metabolic activation of sodium nitroprusside to nitric

oxide in vascular smooth muscle. J Pharmacol Exp Ther 1992; 262:916-922.

27. Rochelle LG, Kruszyna H, Kruszyna R et al. Bioactivation of nitroprusside by porcine endothelial cells. Toxicol Appl Pharmacol 1994; 128:123-128.

28. Zerbe NF, Wagner BK. Use of vitamine B12 in the treatment and prevention of nitroprusside-induced cyanide toxicity. Crit Care Med 1993; 21:465-467.

29. Brock G, Breza J, Lue TF. Intracavernous sodium nitroprusside: inappropriate impotence treatment. J Urol 1993; 150:864-867.

30. Ribeiro JMC, Hazzard JMH, Nussenzveig RH et al. Reversible binding of nitric oxide by a salivary heme protein from a blood-sucking insect. Science 1993; 260:539-541.

31. Ribeiro JMC, Walker FA. High affinity histamine-binding and antihistaminic activity of the salivary nitric oxide-carrying heme protein (nitrophorin) of *Rhodnius prolixus*. J Exp Med 1994; 180:2251-2257.

32. Champagne DE, Nussenzveig RH, Ribeiro JMC. Purification, partial characterization and cloning of nitric oxide-carrying heme proteins (nitrophorins) from salivary glands of the blood-sucking insect *Rhodnius prolixus*. J Biol Chem 1995; 270:8691-8695.

33. Flitney FW, Megson IL, Clough T et al. Nitrosylated iron-sulphur clusters, a novel class of nitrovasodilator: studies on the rat isolated tail artery. J Physiol 1990; 430:42P-42P.

34. Flitney FW, Megson IL, Flitney DE et al. Iron-sulphur cluster nitrosyls, a novel class of nitric oxide generator: mechanism of vasodilator action on rat isolated tail artery. Br J Pharmacol 1992; 107:842-848.

35. Ludbrook SB, Scrutton MC, Joannou CL et al. Inhibition of platelet aggregation by Roussin's black salt, sodium nitroprusside and other metal nitrosyl complexes. Platelets 1995; 6:209-212.

36. Butler AR, Glidewell C, Li M-H. Nitrosyl complexes of iron-sulfur clusters. Adv Inorg Chem 1988; 32:335-393.

37. Butler AR, Williams DLH. The physiological role of nitric oxide. Chem Soc Rev 1993; 1993:233-241.

38. Mülsch A, Mordvintcev P, Vanin AF et al. The potent vasodilating and guanylyl cy-

clase activating dinitrosyl-iron(II) complex is stored in a protein-bound form in vascular tissue and is released by thiols. FEBS Lett 1991; 294:252-256.

39. Vanin AF, Men'shikov GB, Moroz IA et al. The source of non-heme iron that binds nitric oxide in cultivated macrophages. Biochim Biophys Acta 1992; 1135:275-279.

40. Mülsch A, Mordvintcev P, Vanin AF et al. Formation and release of dinitrosyl iron complexes by endothelial cells. Biochem Biophys Res Commun 1993; 196:1303-1308.

41. Furchgott RF. Studies on relaxation of rabbit aorta by sodium nitrite: the basis for the proposal that the acid-activatable inhibitory factor from bovine retractor penis is inorganic nitrite and the endothelium-derived relaxing factor is nitric oxide. In: Vanhoutte PM, ed. Vasodilatation: Vascular Smooth Muscle, Peptides, Autonomic Nerves, and Endothelium. Raven Press, Ltd., New York. 1988:401-414.

42. Bonnett R, Chandra S, Charalambides AA et al. Nitrosation and nitrosylation of haemoproteins and related compounds. Part 4. Pentaco-ordinate nitrosylprotohaem as the pigment of cooked cured meat. Direct evidence from ESR spectroscopy. J Chem Soc Perkin I 1980; 1706-1710.

43. Kruszyna R, Kruszyna H, Smith RP et al. Nitrite conversion to nitric oxide in red cells and its stabilization as a nitrosylated valency hybrid of hemoglobin. J Pharmacol Exp Therap 1987; 241:307-313.

44. Castro C, O'Shea SK. Activation of nitrite ion by iron(III) porphyrins. Stoichiometric oxygen transfer to carbon, nitrogen, phosphorus, and sulfur. J Org Chem 1995; 60:1922-1923.

45. Ignarro LJ. Endothelium-derived nitric oxide: pharmacology and relationship to the actions of organic nitrate esters. Pharmaceut Res 1989; 6:651-659.

46. Kowaluk EA, Chung S-J, Fung H-L. Nitrite ion is not an active intermediate in the vascular metabolism of organic nitrates and organic nitrites to nitric oxide. Drug Metab Dispos 1993; 21:967-968.

47. Kowaluk EA, Fung H-L. Vascular nitric oxide generating activities for organic ni-

trites and nitrates are distinct. J Pharmacol Exp Ther 1991; 259:519-525.

48. Meyer DJ, Kramer H, Özer N et al. Kinetics and equilibria of *S*-nitrosothiol-thiol exchange between glutathione, cysteine, penicillamines and serum albumin. FEBS Lett 1994; 345:177-180.

49. Brien JF, McLaughin E, Breedon TH et al. Biotransformation of glyceryl trinitrate occurs concurrently with relaxation of rabbit aorta. J Pharmacol Exp Ther 1986; 237:609-614.

50. Feelisch M, Noack EA. Correlation between nitric oxide formation during degradation of organic nitrates and activation of guanylate cyclase. Eur J Pharmacol 1987; 139:19-30.

51. Bult H, Bosmans JM, Vrints CJM et al. Isosorbidedinitrate and SIN-1 as dilators of human coronary arteries and platelet inhibitors. J Cardiovasc Pharmacol 1995; 25:572-578.

52. Ignarro LJ. Biosynthesis and metabolism of endothelium-derived nitric oxide. Annu Rev Pharmacol Toxicol 1990; 30:535-560.

53. Ignarro LJ. Signal transduction mechanisms involving nitric oxide. Biochem Pharmacol 1991; 41:485-490.

54. Schrör K, Förster S, Woditsch I. On-line measurement of nitric oxide release from organic nitrates in the intact coronary circulation. Naunyn-Schmiedeberg's Arch Pharmacol 1991; 344:240-246.

55. Salvemini D, Mollace V, Pistelli A et al. Metabolism of glyceryl trinitrate to nitric oxide by endothelial cells and smooth muscle cells and its induction by *Escherichia coli* lipopolysaccharide. Proc Natl Acad Sci USA 1992; 89:982-986.

56. Feelisch M, Brands F, Kelm M. Human endothelial cells bioactivate organic nitrates to nitric oxide: Implications for the reinforcement of endothelial defence mechanisms. Eur J Clin Invest 1995; 25:737-745.

57. Bennett BM, Brien JF, Nakatsu K et al. Role of hemoglobin in the differential biotransformation of glyceryl trinitrates and isosorbide dinitrate by human erythrocytes. J Pharmacol Exp Ther 1985; 234:228-232.

58. Bennett BM, Kobus SM, Brien JF et al. Requirement for reduced, unliganded he-moprotein for the hemoglobin- and myoglobin-mediated biotransformation of glyceryl trinitrate. J Pharmacol Exp Ther 1986; 237:629-635.

59. Kruszyna H, Kruszyna R, Smith RP et al. Red blood cell generate nitric oxide from directly acting, nitrogenous vasodilators. Toxicol Applied Pharmacol 1987; 91:429-438.

60. Kruszyna R, Kruszyna H, Smith RP et al. Generation of valency hybrids and nitrosylated species of hemoglobin in mice by nitric oxide vasodilators. Toxicol Applied Pharmacol 1988; 94:458-465.

61. Kosaka H, Tanaka S, Yoshii T et al. Direct proof of nitric oxide formation from a nitrovasodilator metabolised by erythrocytes. Biochem Biophys Res Commun 1994; 204:1055-1060.

62. Lau DT-W, Benet LZ. Nitroglyceryl metabolism in subcellular fractions of rabbit liver. Dose dependency of glyceryl dinitrate formation and possible involement of multiple isozymes of glutathione *S*-transferases. Drug Metab Dispos 1990; 18:292-297.

63. Schrör K, Woditsch I, Förster S. Generation of nitric oxide from organic vasodilators during passage through the coronary vascular bed and its role in coronary vasodilation and nitrate tolerance. Blood Vessels 1991; 28:62-66.

64. Chung S-J, Fung H-L. Identification of the subcellular site for nitroglycerin metabolism to nitric oxide in bovine coronary smooth muscle cells. J Pharmacol Exp Ther 1990; 253:614-619.

65. Marks GS, McLaughlin BE, Jimmo SL et al. Time-dependent increase in nitric oxide formation concurrent with vasodilation induced by sodium nitroprusside, 3-morpholinosydnonimine, and *S*-nitroso-*N*-acetylpenicillamine but not by glyceryl trinitrate. Drug Metab Dispos 1995; 23:1248-1252.

66. Kohno M, Masumizu T, Mori A. ESR demonstration of nitric oxide production from nitroglycerin and sodium nitrite in the blood of rats. Free Rad Biol Med 1995; 18:451-457.

67. Cantilena LR, Smith RP, Frasur S et al. Nitric oxide hemoglobin in patients receiving nitroglycerin as detected by electron

paramagnetic resonance spectroscopy. J Lab Clin Med 1992; 120:902-907.

68. Ducrocq C, Servy C, Lenfant M. Bioconversion of glyceryl trinitrate into mononitrates by *Geotrichum candidum*. FEMS Microbiol Letters 1989; 65:219-222.

69. Ducrocq C, Servy C, Lenfant M. Formation of glyceryl 2-mononitrate by regioselective bioconversion of glyceryl trinitrate: efficiency of the filamentous fungus *Phanerochaete chrysosporium*. Biotechnol Appl Biochem 1990; 12:325-330.

70. Servent D, Ducrocq C, Henry Y et al. Nitroglycerin metabolism by *Phanerochaete chrysosporium*: evidence for nitric oxide and nitrite formation. Biochim Biophys Acta 1991; 1074:320-325.

71. Servent D, Ducrocq C, Henry Y et al. Multiple enzymatic pathways involved in the metabolism of glyceryl trinitrate in *Phanerochaete chrysosporium*. Biotechnol Applied Biochem 1992; 15:257-266.

72. White GF, Snape JR. Microbial cleavage of nitrate esters: defusing the environment. J Gen Microbiol 1993; 139:1947-1957.

73. Servent D, Delaforge M, Ducrocq C et al. Nitric oxide formation during microsomal hepatic denitration of glyceryl trinitrate: involvement of cytochrome P-450. Biochem Biophys Res Commun 1989; 163:1210-1216.

74. Henry Y, Ducrocq C, Drapier J-C et al. Nitric oxide, a biological effector. Electron paramagnetic resonance detection of nitrosyl-iron-protein complexes in whole cells. Eur Biophys J 1991; 20:1-15.

75. Delaforge M, Servent D, Wirsta P et al. Particular ability of cytochrome P-450 CYP3A to reduce glyceryl trinitrate in rat liver microsomes: subsequent formation of nitric oxide. Chem Biol Interactions 1993; 86:103-117.

76. McDonald BJ, Bennett BM. Cytochrome P-450 mediated biotransformation of organic nitrates. Can J Physiol Pharmacol 1990; 68:1552-1557.

77. Mülsch A, Bara A, Mordvintcev P et al. Specificity of different organic nitrates to elicit NO formation in rabbit vascular tissues and organs *in vivo*. Br J Pharmacol 1995; 116:2743-2749.

78. Delaforge M, Piffeteau A, Boucher J-L et al. Nitric oxide formation during the cytochrome P-450 dependent reductive metabolism of 18-nitro-oxyandrostenedione. J Pharmacol Exp Ther 1995; 274:634-640.

79. Jousserandot A, Boucher J-L, Desseaux C et al. Formation of nitrogen oxides including NO from oxidative cleavage of C=NOH bonds: a general cytochrome P450-dependent reaction. Bioorg Med Chem Lett 1995; 5:423-426.

80. Cederqvist B, Persson MG, Gustafsson LE. Direct demonstration of NO formation *in vivo* from organic nitrites and nitrates and correlation to effects on blood pressure and to *in vitro* effects. Biochem Pharmacol 1994; 47:1047-1053.

81. Persson MG, Agvald P, Gustafsson LE. Detection of nitric oxide in exhaled air during administration of nitroglycerin *in vivo*. Br J Pharmacol 1994; 111:825-828.

82. Lees C, Campbell S, Jauniaux E et al. Arrest of preterm labour and prolongation of gestation with glyceryl trinitrate, a nitric oxide donor. Lancet 1994; 343:1325-1326.

83. DeMaster EG, Raij L, Archer SL et al. Hydroxylamine is a vasorelaxant and a possible intermediate in the oxidative conversion of L-arginine to nitric oxide. Biochem Biophys Res Commun 1989; 163:527-533.

84. Stamler JS, Jaraki O, Osborne J et al. Nitric oxide circulates in mammalian plasma primarily as an *S*-nitroso adduct of serum albumin. Proc Natl Acad Sci USA 1992; 89:7674-7677.

85. Stamler JS, Simon DI, Osborne JA et al. *S*-Nitrosylation of proteins with nitric oxide: synthesis and characterization of biologically active compounds. Proc Natl Acad Sci USA 1992; 89:444-448.

86. Jia L, Bonaventura C, Bonaventura J et al. *S*-nitrosohaemoglobin: a dynamic activity of blood involved in vascular control. Nature 1996; 380:221-226.

87. McAninly J, Williams DLH, Askew SC et al. Metal ion catalysis in nitrosothiols (RSNO) decomposition. J Chem Soc Chem Comm 1993; 23:1758-1759.

88. Vanin AF. Roles of iron ions and cysteine in formation and decomposition of *S*-

nitrosocysteine and *S*-nitrosoglutathione. Biochemistry (Moscow) 1995; 60:441-447.

89. Kowaluk EA, Fung H-L. Spontaneous liberation of nitric oxide cannot account for *in vitro* relaxation by *S*-nitrosothiols. J Pharmacol Exp Ther 1990; 255:1256-1264.

90. Stamler JS, Singel DJ, Loscalzo J. Biochemistry of nitric oxide and its redox-activated forms. Science 1992; 258:1898-1902.

91. Lipton SA, Choi Y-B, Pan Z-H et al. A redox-based mechanism for the neuroprotective and neurodestructive effects of nitric oxide and related nitroso-compounds. Nature 1993; 364:626-632.

92. Arnelle DR, Stamler JS. NO$^+$, NO$^•$, and NO$^-$ donation by *S*-nitrosothiols: implications for regulation of physiological functions by *S*-nitrosylation and acceleration of disulfide formation. Arch Biochem Biophys 1995; 318:279-285.

93. Maragos CM, Morley D, Wink DA et al. Complexes of NO with nucleophiles as agents for the controlled biological release of nitric oxide. Vasorelaxant effects. J Med Chem 1991; 34:3242-3247.

94. Saavedra JE, Dunams TM, Flippen-Anderson JL et al. Secondary amine /nitric oxide complex ions, $R_2N[N(O)NO]^-$. O-functionalization chemistry. J Org Chem 1992; 57:6134-6138.

95. Hrabie JA, Klose JR, Wink DA et al. New nitric oxide-releasing zwitterions derived from polyamines. J Org Chem 1993; 58:1472-1476.

96. Mooradian DL, Hutsell TC, Keefer LK. Nitric oxide (NO) donor molecules: effect of NO release rate on vascular smooth muscle cell proliferation in vitro. J Cardiovasc Pharmacol 1995; 25:674-678.

97. Morley D, Keefer LK. Nitric oxide/ nucleophile complexes: a unique class of nitric oxide-based vasodilators. J Cardiovasc Pharmacol 1993; 22(Suppl 7):S3-S9.

98. Diodati JG, Quyyumi AA, Hussain N et al. Complexes of nitric oxide with nucleophiles as agents for the controlled biological release of nitric oxide: antiplatelet effect. Thrombosis and Haemostasis 1993; 70:654-658.

99. Vanderford PA, Wong J, Chang R et al. Diethylamine/nitric oxide (NO) adduct, an NO donor, produces potent pulmonary and systemic vasodilation in intact newborn lambs. J Cardiovasc Pharmacol 1994; 23:113-119.

100. Shimaoka M, Iida T, Ohara A et al. NOC, a nitric-oxide-releasing compound, induces dose dependent apoptosis in macrophages. Biochem Biophys Res Commun 1995; 209:519-526.

101. Makings LR, Tsien RY. Caged nitric oxide. Stable organic molecules from which nitric oxide can be photoreleased. J Biol Chem 1994; 269:6282-6285.

102. Ferioli R, Folco GC, Ferretti C et al. A new class of furoxan derivatives as NO donors: mechanism of action and biological activity. Br J Pharmacol 1995; 114:816-820.

103. Tanayama S, Nakai Y, Fujita T et al. Biotransformation of molsidomine (*N*-ethoxycarbonyl-morpholinosydnonimine), a new anti-anginal agent, in rats. Xenobiotica 1974; 4:175-191.

104. Feelisch M, Ostrowski J, Noack E. On the mechanism of NO release from sydnonimines. J Cardiovasc Pharmacol 1989; 14(Suppl. 11):S13-S22.

105. Gergel D, Misik V, Ondrias K et al. Increased cytotoxicity of 3-morpholinosydnonimine to HepG2 cells in the presence of superoxide dismutase. J Biol Chem 1995; 270:20922-20929.

106. Andronik-Lion V, Boucher J-L, Delaforge M et al. Formation of nitric oxide by cytochrome P450-catalyzed oxidation of aromatic amidoximes. Biochem Biophys Res Commun 1992; 185:452-458.

107. Clement B, Jung F. *N*-hydroxylation of the antiprotozoal drug pentamidine catalyzed by rabbit liver cytochrome P-450 2C3 or human liver microsomes, microsomal retroreduction, and further oxidative transformation of the formed amidoximes. Possible relationship to the biological oxidation of arginine to N^G-hydroxyarginine, citrulline, and nitric oxide. Drug Metab Dispos 1994; 22:486-497.

108. Szekeres T, Gharehbaghi K, Fritzer M et al. Biochemical and antitumor activity of trimidox, a new inhibitor of ribonucleotide reductase. Cancer Chemother Pharmacol 1994; 34:63-66.

109. Sennequier N, Boucher J-L, Battioni P et al. Superoxide anion efficiently performs the oxidative cleavage of C=NOH bonds of amidoximes and N-hydroxyguanidines with formation of nitrogen oxides. Tetrahedron Lett 1995; 36:6059-6062.

110. Kita Y, Osaki R, Sakai S et al. Antianginal effects of FK409, a new spontaneous NO releaser. Br J Pharmacol 1994; 113:1137-1140.

111. Fukuyama S, Kita Y, Hirasawa Y et al. A new nitric oxide (NO) releaser: spontaneous NO release from FK409. Free Rad Res 1995; 23:443-452.

112. Fukuyama S, Hirasawa Y, Cox D et al. Acceleration of nitric oxide (NO) release from FK409, a spontaneous NO releaser, in the presence of sulfhydryl-bearing compounds. Pharmaceut Res 1995; 12:1948-1952.

113. Kita Y, Ohkubo K, Hirasawa Y et al. FR144420, a novel, slow, nitric oxide-releasing agent. Eur J Pharmacol 1995; 275:125-130.

114. Persson MG, Wiklund NP, Gustafsson LE. Endogenous nitric oxide in single exhalations and the change during exercise. Am Rev Respir Dis 1993; 148:1210-1214.

115. Rimar S, Gillis CN. Selective pulmonary vasodilation by inhaled nitric oxide is due to hemoglobin inactivation. Circulation 1993; 88:2884-2887.

116. Higenbottam T. Inhaled nitric oxide: a magic bullet? Quarterly J Med 1993; 86:555-558.

117. Pearl MD. Inhaled nitric oxide. The past, the present, and the future. Anesthesiol 1993; 78:413-416.

118. Goldman AP, Rees PG, Macrae DJ. Is it time to consider domiciliary nitric oxide? Lancet 1995; 345:199-200.

119. Lunn RJ. Inhaled nitric oxide therapy. Mayo Clin Proc 1995; 70:247-255.

120. Frostell C, Fratacci M-D, Wain JC et al. Inhaled nitric oxide. A selective pulmonary vasodilator reversing hypoxic pulmonary vasoconstriction. Circulation 1991; 83:2038-2047.

121. Frostell CG, Blomqvist H, Hedenstierna G et al. Inhaled nitric oxide selectively reverses human hypoxic pulmonary vasoconstriction without causing systemic vasodilation. Anesthesiol 1993; 78:427-435.

122. Rossaint R, Falke KJ, López F et al. Inhaled nitric oxide for the adult respiratory distress syndrome. N Eng J Med 1993; 328:399-405.

123. Gerlach H, Rossaint R, Pappert D et al. Time-course and dose-response of nitric oxide inhalation for systemic oxygenation and pulmonary hypertension in patients with adult respiratory distress syndrome. Eur J Clin Invest 1993; 23:499-502.

124. Gerlach H, Pappert D, Lewandowski K et al. Long-term inhalation with evaluated low doses of nitric oxide for selective improvement of oxygenation in patients with adult respiratory distress syndrome. Intensive Care Med 1993; 19:443-449.

125. Kinsella JP, Abman SH. Methaemoglobin during nitric oxide therapy with high-frequency ventilation. Lancet 1993; 342:615-615.

126. Foubert L, Fleming B, Latimer R et al. Safety guidelines for use of nitric oxide. Lancet 1992; 339:1615-1616.

127. Bouchet M, Renaudin M-H, Raveau C et al. Safety requirement for use of inhaled nitric oxide in neonates. Lancet 1993; 341:968-969.

128. Laguenie G. Measurement of nitric dioxide formation from nitric oxide by chemiluminescence in ventilated children. Lancet 1993; 341:969-969.

129. Miller OI, Celermajer DS, Deanfield JE et al. Guidelines for the safe administration of inhaled nitric oxide. Arch Dis Child 1994; 70:F47-F49.

130. Stenqvist O, Kjelltoft B, Lundin S. Evaluation of a new system for ventilatory administration of nitric oxide. Acta Anaesthesiol Scand 1993; 37:687-691.

131. Kinsella JP, Neish SR, Shaffer E et al. Low-dose inhalational nitric oxide in persistent pulmonary hypertension of the newborn. Lancet 1992; 340:819-820.

132. Kinsella JP, Neish SR, Dunbar D et al. Clinical responses to prolonged treatment of persistent pulmonary hypertension of the newborn with low doses of inhaled nitric oxide. J Pediatr 1993; 123:103-108.

133. Adnot S, Kouyoumdjian C, Defouilloy C et

al. Hemodynamic and gas exchange responses to infusion of acetylcholine and inhalation of nitric oxide in patients with chronic obstructive lung disease and pulmonary hypertension. Am Rev Resp Dis 1993; 148:310-316.

134. Bigatello LM, Hurford WE, Kacmarek RM et al. Prolonged inhalation of low concentrations of nitric oxide in patients with severe adult respiratory distress syndrome. Effects on pulmonary hemodynamics and oxygenation. Anesthesiol 1994; 80:761-770.

135. Finer NN, Etches PC, Kamstra B et al. Inhaled nitric oxide in infants referred for extracorporeal membrane oxygenation: dose response. J Pediatr 1994; 124:302-308.

136. Young JD, Brampton WJ, Knighton JD et al. Inhaled nitric oxide in acute respiratory failure in adults. Br J Anaesth 1994; 73:499-502.

137. Girard C, Lehot J-J, Pannetier J-C et al. Inhaled nitric oxide after mitral valve replacement in patients with chronic pulmonary artery hypertension. Anesthesiol 1992; 77:880-883.

138. Haydar A, Mauriat P, Pouard P et al. Inhaled nitric oxide for post operative pulmonary hypertension in patients with congenital heart defects. Lancet 1992; 340:1545-1545.

139. Journois D, Pouard P, Mauriat P et al. Inhaled nitric oxide as a therapy for pulmonary hypertension after operations for congenital heart defects. J Thorac Cardiovasc Surg 1994; 107:1129-1135.

140. Rich GF, Murphy GD, Roos CM et al. Inhaled nitric oxide. Selective pulmonary vasodilation in cardiac surgical patients. Anesthesiol 1993; 78:1028-1035.

141. Rich GF, Lowson SM, Johns RA et al. Inhaled nitric oxide selectively decreases pulmonary vascular resistance without impairing oxygenation during one-lung ventilation in patients undergoing cardiac surgery. Anesthesiol 1994; 80:57-62.

142. Snow DJ, Gray SJ, Ghosh S et al. Inhaled nitric oxide in patients with normal and increased pulmonary vascular resistance after cardiac surgery. Br J Anaesth 1994; 72:185-189.

143. Lévêque C, Hamza J, Berg AE et al. Successful repair of a severe left congenital diaphragmatic hernia during continuous inhalation of nitric oxide. Anesthesiol 1994; 80:1171-1175.

144. Roberts JD, Polaner DM, Todres ID et al. Inhaled nitric oxide (NO): a selective pulmonary vasodilator for the treatment of persistent pulmonary hypertension of the newborn. Circulation 1991; 84:A1279-A1279.

145. Roberts JD, Polaner DM, Lang P et al. Inhaled nitric oxide in persistent pulmonary hypertension of the newborn. Lancet 1992; 340:818-819.

146. Roberts JD, Lang P, Bigatello LM et al. Inhaled nitric oxide in congenital heart disease. Circulation 1993; 87:447-453.

147. Norman V, Keith CH. Nitrogen oxides in tobacco smoke. Nature 1965; 205:915-916.

148. Schilling J, Holzer P, Guggenbach M et al. Reduced endogenous nitric oxide in the exhaled air of smokers and hypertensives. Eur Respir J 1994; 7:467-471.

CHAPTER 13

NITRIC OXIDE-SPECIFIC SPIN-TRAPPING EPR METHODS

Yann A. Henry

U p to this point we have dealt with NO's own paramagnetism and that of its spontaneous complexes with metalloproteins into whole cells and organelles. Direct detection of NO by EPR spectroscopy being unpracticable, spin-trapping methods have been used as in the case of unstable oxygen-derived free radicals. We will give an overview of the spin-trapping method in solution, and shall focus on the now widely used Vanin, Mordvintcev and Mülsch method of formation of an EPR-detectable ternary complex of the type FeL$_2$NO.

USE OF HEMOGLOBIN AS A NITRIC OXIDE SPIN-TRAP

Due to the rapid formation of a relatively stable paramagnetic nitrosylated complex, deoxyhemoglobin seems a natural spin-trap to choose. However for experiments on organs, cells, or organelles requiring properly oxygenated media, the reaction to consider is that of oxidation by NO of oxyhemoglobin into methemoglobin and possibly other interfering reactions, rather than HbNO formation. In both cases the absorbance method is sensitive, as hemoglobin has high extinction coefficients (100-160 mmol^{-1} cm^{-1}) in the Soret band (390-450 nm). We have detailed in chapter 4 all the reactions between deoxyHb, or HbO$_2$ and nitric oxide and other nitrogen oxides, which are to be taken into account when using hemoglobin as a spin-trap detected by UV-visible spectrophotometry or EPR spectroscopy.[1-5]

We have also detailed in chapter 11 the use of agarose-bound hemoglobin for the EPR spectroscopy detection of HbNO at 77 K.[6,7] Finally we have described in chapters 5 and 11 the spontaneous binding of NO to hemoglobin within red blood cells, using NO gas or in pathophysiological conditions (reviewed in refs. 8 and 9).

Nitric Oxide Research from Chemistry to Biology: EPR Spectroscopy of Nitrosylated Compounds, edited by Yann A. Henry, Annie Guissani and Béatrice Ducastel.
© 1997 R.G. Landes Company.

CONVENTIONAL SPIN-TRAPPING

Nitrones and nitroso spin-traps conventionally used for spin-trapping of oxygen-derived free radicals, 3,5-dibromo-4-nitrosobenzene sulphonate (DBNBS), 2-methyl-2-nitrosopropane (MNP) and 5,5-dimethyl-1-pyrroline *N*-oxide (DMPO), have been used to detect free NO in cultured neuronal cells,[10] collagen-activated human platelets,[11] LPS-activated macrophages,[12] and human monocyte-derived macrophages stimulated by gp120 HIV envelope glycoprotein,[13] to give a few examples. Caution must be exercised as spin-adducts have a complex chemistry.[14-17] Artifacts are even more likely than for the formation of spin-adducts with $O_2^{\bullet-}$, $OH^{\bullet}$, and C-centered free radicals; many control experiments are needed in order to avoid such pitfalls.

Phenyl *N-tert*-butylnitrone (PBN), a commonly used spin-trap, can decompose under light or by a $Fe^{(III)}$-catalyzed hydrolysis to form NO.[18,19] PBN has also been found to inhibit the induction of NOS II in a septic shock model in mice.[16,20] The use of PBN or other nitrones in animal models may thus alter in several ways the biological response of the animals.

These spin-traps have also been used in order to test the decomposition pathways of peroxynitrite, with conflicting results with respect to the production of $OH^{\bullet}$.[21-23]

PROSPECTS ON NITRIC OXIDE-SPECIFIC SPIN-TRAPS

The development of NO-specific spin-traps would be adequate to quantify small amounts of NO in the presence of oxygen-derived free radicals. Due to the high sensitivity of EPR spectroscopy at detecting spin-adducts' narrow lines at room temperature, the spin-trap methods are certain to see a great development provided that a strict specificity can be obtained. Several attempts have been made. One makes use of tetraalkyl-*ortho*-quinodimethane derivatives as a nitric oxide cheletropic trap (NOCT) for both $NO^{\bullet}$ and $NO_2^{\bullet}$;[24] the method was tested with Kupffer cells, producing both $NO^{\bullet}$ and $O_2^{\bullet-}$. A whole series of compounds of this type have been synthesized, including water soluble derivatives.[25,26]

A more complicated method has been developed in which aminoxyl radicals of the type $R^1N(O^{\bullet})R^2$ are formed.[27] By a similar method, phenolic antioxidants react with NO to form covalent paramagnetic spin-adducts.[28,29] Additions of NO to long chain -dienes to form a paramagnetic ring adduct and to nitromethane have also been proposed.[30,31] The chemistry of these different reactions is probably too complicated to be used in biological systems.

SCAVENGING OF NITRIC OXIDE BY REACTION WITH STABLE FREE RADICALS

We mentioned in chapter 2 that NO reacts readily with other free radicals. This was taken advantage of in its reaction with some stable free radicals of the 2-phenyl-4,4,5,5-tetramethylimidazoline-1-oxyl 3-oxide (PTIO) family, water soluble carboyxy-PTIO, and carboxymethoxy-PTIO.[32,33] The imidazolineoxyl *N*-oxides react with NO in a stoichiometric (1:1) manner in a neutral solution with rate constants on the order of 10^4 mol^{-1} l s^{-1}, to yield the imidazoline-1-oxyl derivative (PTI) and $NO_2^{\bullet}$, and finally nitrite and nitrate.[32,34] In solution at room temperature PTIO derivatives give typical EPR spectra with five hyperfine lines, with 1:2:3:2:1 ratios, and $a_N^{1,3}$ = 0.82 mT resulting from two equivalent *N* atoms. PTI derivatives give EPR spectra with six hyperfine lines, with a_N^1 = 0.98 mT and a_N^3 = 0.44 mT, from two inequivalent *N* atoms. The method was tested by measuring the inhibiton of aorta strips' relaxation by PTIO derivatives.[32] Bradikinin-stimulated endothelial cells gave similar effects.[33]

The scavenging properties of PTIO and carboxy-PTIO have been used in many experimental models, solid tumor,[35] endothelial-dependent vascular relaxation,[36] and activated macrophage cell line J774.1.[37] Through its NO-scavenging property,

PTIO was found to have therapeutic effects in the course of endotoxin septic shock in rats.[38] Carboxy-PTIO was also used to study the chemical exchange of NO between *S*-nitrosothiols and thiols.[39]

Another method is based on the reaction of NO with nitronyl nitroxides to produce imino nitroxides.[34,40,41] Nitronyl nitroxides and imino nitroxides have characteristic and distinct EPR spectra. The conversion has a 1:1 stoichiometry and is specific of NO as neither oxygen-derived free radicals nor $NO_2^\bullet$ can catalyze it. The method has been applied for measurement of NO synthesized in rat cerebellum[42] and of NO released from NO-donors, SNP and SNAP.[41,43] Finally a $NO_2^\bullet$ specific trapping by dimethylsulfoxide (DMSO) has been proposed.[44]

These methods are probably the simplest EPR spin-trapping methods to put into routine practice and do not seem to induce artifacts.

LINE BROADENING OF FUSINITE AND OTHER FREE RADICALS BY NITRIC OXIDE

Some free radicals do not react with nitric oxide by addition or electron transfer. Their interaction is physical and results in line broadening (Heisenberg spin exchange) of the EPR resonance lines of the free radicals by NO, similar to the effect of paramagnetic triplet-state molecular oxygen. This technique has been applied to fusinite, a paramagnetic material derived from woody plant tissue found in coal. Fusinite is injected into animals, and these are subjected to low-frequency EPR spectroscopy (1.2 GHz; L-band, see chapter 3 for basic EPR methodology) and magnetic resonance imaging.[45] This was used for oximetry of tumor tissue.[45] The method was also tested with Chinese hamster ovary cells with fusinite internalized by phagocytosis. The appearance of NO released from NO-donors and its disappearance within mitochondria could be followed by this method.[46]

A similar line broadening effect has been observed by the interaction of NO

with stable nitroxide, 3-carbamoyl-2,2,5,5-tetramethyl-3-pyrroline-1-yloxy (CTPO), or lipid soluble spin label derived from 12-doxylstearic acid.[47]

The method could be extremely useful for examining the physical properties of NO within biological membranes and for NO imaging within whole organs.[45-47]

USE OF Fe-(DETC)$_2$ AND SIMILAR COMPLEXES AS NITRIC OXIDE SPIN-TRAPS

The complex of Fe^{2+} with diethyldithiocarbamate (DETC), Fe^{2+}-$(DETC)_2$ forms with NO a stable ternary complex which can be detected by EPR spectroscopy in liquid aqueous phase or at 77 K.[48] Due to the poor solubility of Fe^{2+}-$(DETC)_2$ complex in water the spin-trap is "dissolved" in yeast cell membranes. The spectral feature of the ternary complex NO-Fe^{2+}-$(DETC)_2$ is an axial EPR signal at 37°C with g-values of $g_\perp = 2.04$ and $g_{//} = 2.02$ ($g_{av} = 2.03$) with an easily recognized SHF triplet at $g_\perp$. The EPR signal could be quantified, and the method is applicable to oxygenated media. Kinetics of NO release from SIN-1, for instance, could be followed (Fig. 13.1).[48-50]

The ternary complex was formed from endogenous iron and from DETC administered i.p. 30 min before the experiment; it was found in animal tissues, and was demonstrated by isotopic substitution to arise from guanidino nitrogens of L-arginine.[50,51] Very often iron sulfate or citrate is also exogenously injected (Fig. 13.2).

The method may suffer from certain drawbacks. DETC is, like other dithiocarbamates, a known inhibitor of the nuclear transcription factor kappa B (NFκB), which is a component in the regulation of NOS induction (see chapter 8). In fact DETC has been found to inhibit induction of murine bone marrow-derived macrophages' NOS by LPS.[52] Iron has an opposite effect. Iron citrate administration potentiates LPS induction of NOS in mice.[53] So these two components of the formation of the ternary complex NO-Fe^{2+}-$(DETC)_2$ have adverse effects on the NO formation by the inducible NOS pathway.

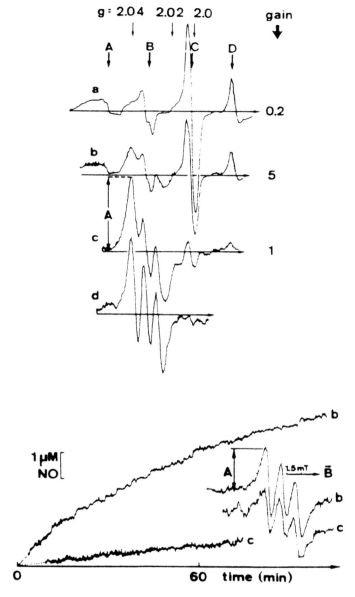

Fig. 13.1. Upper panel. EPR spectra of NO-Fe^{2+}-(DETC)$_2$ complex. (a) EPR spectrum at 77 K of Cu^{2+}-(DETC)$_2$; (b, c, d) Yeast cells loaded with DETC and incubated with SIN-1 (c at 77 K, d at 310 K) or without SIN-1 (b at 77 K). A, B, C, D are quartet hyperfine lines of the Cu^{2+}-(DETC)$_2$ complex. The arrow A is used to measure NO-Fe^{2+}-(DETC)$_2$ complex concentration. Lower panel. Kinetics at 37°C of NO release from SIN-1 in the presence of heat-killed yeast loaded with DETC. (b) pH 7.5, (c) pH 7.0. (Reproduced with kind permission from: Mordvintcev P, Mülsch A, Busse R et al. On-line detection of nitric oxide formation in liquid aqueous phase by electron paramagnetic resonance spectroscopy. Anal Biochem 1991; 199:142-146.) Copyright © Academic Press, Inc., Orlando, Florida, USA.

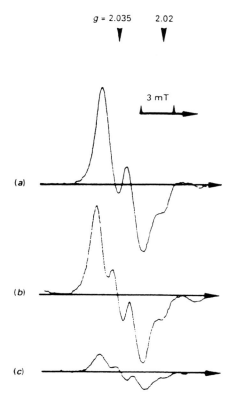

g = 2.035   2.02

3 mT

(a)

(b)

(c)

Fig. 13.2. Upper panel. EPR spectra at 77 K of heat-killed DETC-loaded yeast incubated 1 hr at 37°C with cytosol from LPS-activated murine bone-marrow-derived macrophages, containing [^{15}N]N^G-L-arginine (a), or [^{14}N]N^G-L-arginine (b), or [^{14}N]N^G-L-arginine and N^G-nitro-L-arginine (c). Lower panel. On-line EPR registration at 37°C of NO formation by endothelial cytosol and by SIN-1. (Reproduced with kind permission from: Mülsch A, Vanin A, Mordvintcev P et al. NO accounts completely for the oxygenated nitrogen species generated by enzymatic L-arginine oxygenation. Biochem J 1992; 288:597-603.) Copyright © The Biochemical Society & Portland Press, London, UK.

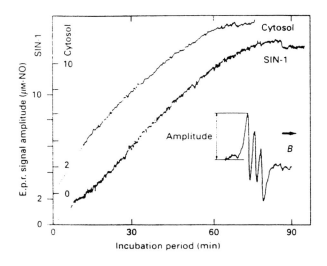

Vanin's method has received numerous applications in stress and shock pathologies (Fig. 13.3).[54-59]

The NO-Fe^{2+}-(DETC)$_2$ complex can be detected in most organs, liver, stomach, and in ascites tumors.[60-62] It has been used to detect NO in many CNS diseases,[63] is-

chemia-reperfusion,[64,65] kaïnate-induced seizure,[66] disease following viral infection,[67] HIV-associated disease,[68] etc. (Fig. 13.4).

A three-dimensional imaging method has been developed using this NO-Fe^{2+}-(DETC)$_2$ complex in rat brain subjected to ischemia-hypoxia.[69] Finally the method has

Fig. 13.3. Upper panel. EPR spectra at 77 K of rat brain cortex obtained from (a) a control rat and (b) a rat killed 2 hr after injection of kaïnate. A, B, C, D, are quartet features of the Cu^{2+}-(DETC)$_2$ complex. The triplet shown in the difference spectrum b-a, arises from the NO-Fe^{2+}-(DETC)$_2$ complex (c). Lower panel. Time course of kaïnate-elicited NO formation in different regions of the rat brain. (Reproduced with kind permission from: Mülsch A, Busse R, Mordvintcev PI et al. Nitric oxide promotes seizure activity in kainate-treated rats. Neuro Reports 1994; 5:2325-2328.) Copyright © International Thomson Publishing Journals, 2-6 Boundary Row, London SE1 8HN, UK.

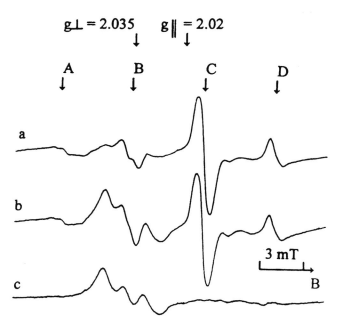

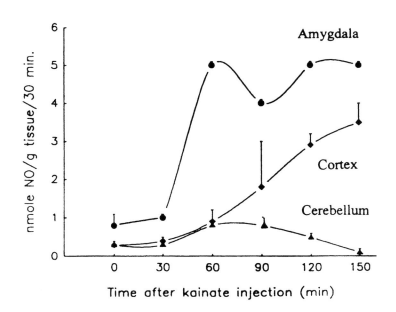

been used to follow the metabolism of some drugs.[70,71]

In a similar method, using the water soluble complex of *N*-methyl-D-glucamine dithiocarbamate (MGD) with Fe^{2+} to trap NO and form a ternary complex $NO-Fe^{2+}-(MGD)_2$, Komarov et al were for the first time able to detect NO in a living animal. $Fe^{2+}-(MGD)_2$ was in-

jected into the mouse lateral vein of the tail, and EPR detection of NO in blood circulating in the mouse tail was performed at the S-band (3.5 GHz) with a 4 mm loop.[72,73] This interesting experiment requires a home-built or largely modified commercial EPR spectrometer and thus is not yet used routinely (Fig. 13.5).

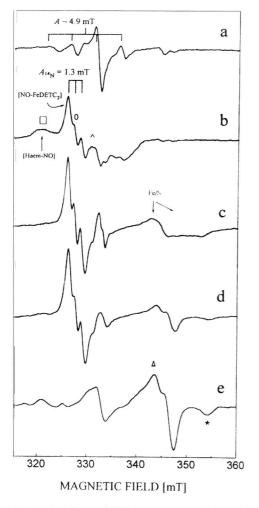

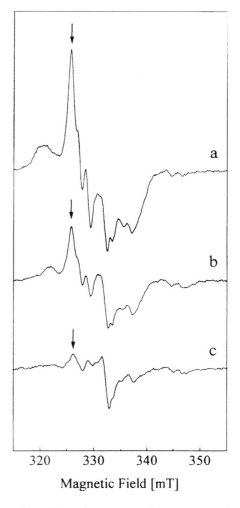

Fig. 13.4. Left panel. EPR spectra at 15 K from endotoxin-treated rats. Spleen homogenate from rats treated with DETC and Fe-citrate alone 30 min before sacrifice (a); and from spleen (b), kidney (c), and liver homogenate (d) from rats administered LPS (5 hr before sacrifice) and DETC plus Fe-citrate (30 min before sacrifice); (e) liver homogenate of untreated rat. g-values are g = 2.08 (□), 2.035 (o), 2.02 (^), 1.93 (Δ), and 1.88 (*). Right panel. Effects of arginase and NMMA on the endotoxin-stimulated $NO-Fe^{2+}-(DETC)_2$ EPR signal. The animals were injected with LPS (5 hr before sacrifice) and saline (a), arginase (b) or NMMA (c). (Reproduced with kind permission from: Bune AJ, Shergill JK, Cammack R et al. L-arginine depletion by arginase induces nitric oxide production in endotoxic shock: an electron paramagnetic resonance study. FEBS Lett 1995; 366:127-13.) Copyright © Elsevier Science, Sara Burgerhartstraat 25, 1055 KV Amsterdam, The Netherlands.

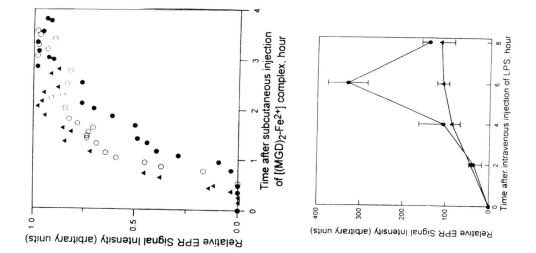

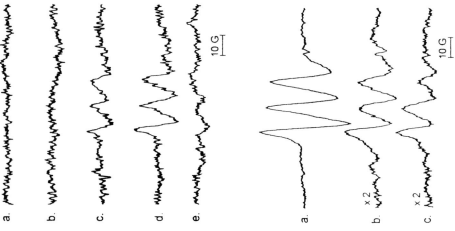

Fig. 13.5. Left upper panel. 3.5 GHz EPR spectra (average of nine 30 sec scans) of the NO-Fe²⁺-(MGD)₂ complex in the circulation of the mouse tail at various times after LPS administration through the lateral tail vein (a) time 0, (b) 2 hr, (c) 4 hr, (d) 6 hr, (e) 6 hr as in (d) with NMMA. Right upper panel. Time course of in vivo EPR signal intensities of the NO-Fe²⁺-(MGD)₂ complex in mouse tail circulation during septic shock. Six hr after i.v. injection of LPS, an aliquot of Fe²⁺-(MGD)₂ complex in water was injected subcutaneously, and the EPR signal was followed for a 4 hr period (three separate experiments). Left lower panel. 3.5 GHz EPR spectra of the NO-Fe²⁺-(MGD)₂ complex in various organs of Fe²⁺-(MGD)₂ and LPS-treated mice, (a) liver, (b) kidney, (c) heparinized whole blood. Right lower panel. Time course of EPR signal intensities of the NO-Fe²⁺-(MGD)₂ complex in liver (circle) and kidney (triangle) tissues of LPS-treated mice. (Reproduced with kind permission from: Lai C-S, Komarov AM. Spin trapping of nitric oxide produced in vivo in septic-shock mice. FEBS Lett 1994; 345:120-124.) Copyright © Elsevier Science, Sara Burgerhartstraat 25, 1055 KV Amsterdam, The Netherlands.

The method has been applied at the normal X-band (9-10 GHz) to several animal models of pathologies, septic shock,[73,74] ischemic heart,[75] influenza virus-induced pneumonia,[76] etc., or in simple macrophage cultures.[77]

CONCLUSION

This chapter is only an overview of spin-trapping methods, pointing out their main advantages over frozen solution EPR spectroscopy: improved sensitivity, potential specificity and time-resolution. Several specific reviews have been published in which different analytical methods have been compared.[8,9,26,48,49,78-83] See the book of methods edited by Martin Feelisch and Jonathan S. Stamler.[84]

REFERENCES

1. Feelisch M, Noack EA. Correlation between nitric oxide formation during degradation of organic nitrates and activation of guanylate cyclase. Eur J Pharmacol 1987; 139:19-30.
2. Kelm M, Feelisch M, Spahr R et al. Quantitative and kinetic characterisation of nitric oxide and EDRF released from cultured endothelial cells. Biochem Biophys Res Commun 1988; 154:236-244.
3. Kosaka H, Watanabe M, Yoshihara H et al. Detection of nitric oxide production in lipopolysaccharide-treated rats by ESR using carbon monoxide hemoglobin. Biochem Biophys Res Commun 1992; 184:1119-1124.
4. Sonoda M, Hashimoto T, Satomi A et al. Diazotization reaction of nitric oxide trapped by hemoglobin. Life Sci 1994; 55:PL 199-204.
5. Murphy ME, Noack E. Nitric oxide assay using hemoglobin method. Methods Enzymol 1994; 233:240-250.
6. Greenberg SS, Wilcox DE, Rubanyi GM. Endothelium-derived relaxing factor released from canine femoral artery by acethycholine cannot be identified as free nitric oxide by electron paramagnetic resonance spectroscopy. Circ Res 1990; 67:1446-1452.
7. Wennmalm Å, Lanne B, Petersson A-S. Detection of endothelium-derived factor in human plasma in the basal state and following ischemia using electron paramagnetic resonance spectrometry. Anal Biochem 1990; 187:359-363.
8. Wilcox DE, Smith RP. Detection and quantification of nitric oxide using electron magnetic resonance spectroscopy. Methods Enzymol 1995; 7:59-70.
9. Kosaka H, Shiga T. Detection of nitric oxide by electron paramagnetic resonance using hemoglobin. In: Feelisch M, Stamler JS. Methods in Nitric Oxide Research. John Wiley & Sons, 1996:373-381.
10. Arroyo CM, Forray C, El-Fakahany EE et al. Receptor-mediated generation of an EDRF-like intermediate in a neuronal cell line detected by spin trapping techniques. Biochem Biophys Res Commun 1990; 170:1177-1183.
11. Prönai L, Ichimori K, Nozaki H et al. Investigation of the existence and biological role of L-arginine/nitric oxide pathway in human platelets by spin-trapping/EPR studies. Eur J Biochem 1991; 202:923-930.
12. Carmichael AJ, Steel-Goodwin L, Gray B et al. Nitric oxide interaction with lactoferrin and its production by macrophage cells studied by EPR and spin trapping. Free Rad Res Comms 1993; 19:S201-S209.
13. Pietraforte D, Tritarelli E, Testa U et al. gp120 HIV envelope glycoprotein increases the production of nitric oxide in human monocyte-derived macrophages. J Leuk Biol 1994; 55:175-182.
14. Arroyo CM, Kohno M. Difficulties encountered in the detection of nitric oxide (NO) by spin trapping techniques. A cautionary note. Free Rad Res Comms 1991; 14:145-155.
15. Carmichael AJ, Steel-Goodwin L, Gray B et al. Reactions of active oxygen and nitrogen species studied by EPR and spin trapping. Free Rad Res Comms 1993; 19:S1-S16.
16. Pou S, Keaton L, Surichamorn W et al. Can nitric oxide be spin trapped by nitrone and nitroso compounds? Biochim Biophys Acta 1994; 1201:118-124.
17. Dugan LL, Lin T-S, He Y-Y et al. Detection of free radicals by microdialysis/spin trapping EPR following focal cerebral is-

chemia-reperfusion and a cautionary note on the stability of 5,5-dimethyl-1-pyrroline *N*-oxide (DMPO). Free Rad Res 1995; 23:27-32.

18. Chamulitrat W, Jordan SJ, Mason RP et al. Nitric oxide formation during light-induced decomposition of phenyl *N-tert*-butylnitrone. J Biol Chem 1993; 268:11520-11527.

19. Chamulitrat W, Parker CE, Tomer KB et al. Phenyl *N-tert*-butyl nitrone forms nitric oxide as a result of its Fe(III)-catalyzed hydrolysis or radical adduct formation. Free Rad Res 1995; 23:1-14.

20. Miyajima T, Kotake Y. Spin trapping agent, phenyl *N-tert*-butyl nitrone, inhibits induction of nitric oxide synthase in endotoxin-induced shock in mice. Biochem Biophys Res Commun 1995; 215:114-121.

21. Yang G, Candy TEG, Boaro M et al. Free radical yields from the homolysis of peroxynitrous acid. Free Rad Biol Med 1992; 12:327-330.

22. Crow JP, Spruell C, Chen J et al. On the pH-dependent yield of hydroxyl radical products from peroxynitrite. Free Rad Biol Med 1994; 16:331-338.

23. Lemercier J-N, Squadrito GL, Pryor WA. Spin trap studies on the decomposition of peroxynitrite. Arch Biochem Biophys 1995; 321:31-39.

24. Korth H-G, Ingold KU, Sustmann R et al. Tetramethyl-ortho-quinodimethane, first member of a family of custom-tailored cheletropic spin traps for nitric oxide. Angew Chem Int Ed Engl 1992; 31:891-893.

25. Korth H-G, Sustmann R, Lommes P et al. Nitric oxide cheletropic traps (NOCTs) with improved thermal stability and water solubility. J Am Chem Soc 1994; 116:2767-2777.

26. Korth H-G, Weber H. Detection of nitric oxide with nitric oxide-trapping reagents. In: Feelisch M, Stamler JS. Methods in Nitric Oxide Research. John Wiley & Sons, 1996:383-391.

27. Lagercrantz C. Spin trapping of nitric oxide (NO•) as aminoxyl radicals by its reaction with two species of short-lived radicals derived from azo compounds such as 2,2'-azobisisobutyronitrile and some aliphatic alcohols. Free Rad Res Comms 1993;

19:387-395.

28. Janzen EG, Wilcox AL, Manoharan V. Reactions of nitric oxide with phenolic antioxidants and phenoxyl radicals. J Org Chem 1993; 58:3597-3599.

29. Wilcox AL, Janzen EG. Nitric oxide reactions with antioxidants in model systems: sterically hindered phenols and α-tocopherol in sodium dodecyl sulphate (SDS) micelles. J Chem Soc Chem Comm 1993; 18:1377-1379.

30. Reszka KJ, Chignell CF, Bilski P. Spin trapping of nitric oxide (•NO) by *aci*-nitromethane in aqueous solutions. J Am Chem Soc 1994; 116:4119-4120.

31. Rockenbauer A, Korecz L. Comment on conversion of nitric oxide into a nitroxide radical using 2,3-dimethylbutadiene and 2,5-dimethylhexadiene. J Chem Soc Chem Comm 1994; 145.

32. Akaike T, Yoshida M, Miyamoto Y et al. Antagonistic action of imidazolineoxyl *N*-oxides against endothelium-derived relaxing factor /•NO through a radical reaction. Biochemistry 1993; 32:827-832.

33. Az-ma T, Fujii K, Yuge O. Reaction between imidazolineoxil *N*-oxide (carboxy-PTIO) and nitric oxide released from cultured endothelial cells: quantitative measurement of nitric oxide by ESR spectrometry. Life Sci 1994; 54:PL185-PL190.

34. Hogg N, Singh RJ, Joseph J et al. Reactions of nitric oxide with nitronyl nitroxides and oxygen: prediction of nitrite and nitrate formation by kinetic simulation. Free Rad Res 1995; 22:47-56.

35. Maeda H, Noguchi Y, Sato K et al. Enhanced vascular permeability in solid tumor is mediated by nitric oxide and inhibited by both new nitric oxide scavenger and nitric oxide synthase inhibitor. Jpn J Cancer Res 1994; 85:331-334.

36. Maeda H, Akaike T, Yoshida M et al. Multiple functions of nitric oxide in pathophysiology and microbiology: analysis by a new nitric oxide scavenger. J Leuk Biol 1994; 56:588-592.

37. Amano F, Noda T. Improved detection of nitric oxide (NO) production in an activated macrophage culture with a radical scavenger, carboxy PTIO, and Griess re-

agent. FEBS Lett 1995; 368:425-428.

38. Yoshida M, Akaike T, Wada Y et al. Therapeutic effects of imidazolineoxyl N-oxide against endotoxin shock through its direct nitric oxide-scavenging activity. Biochem Biophys Res Commun 1994; 202:923-930.

39. Pietraforte D, Mallozzi C, Scorza G et al. Role of thiols in the targeting of S-nitroso thiols to red blood cells. Biochemistry 1995; 34:7177-7185.

40. Joseph J, Kalyanaraman B, Hyde JS. Trapping of nitric oxide by nitroxyl nitroxides: an electron spin resonance investigation. Biochem Biophys Res Commun 1993; 192:926-934.

41. Konorev EA, Tarpey MM, Joseph J et al. Nitronyl nitroxides as probes to study the mechanism of vasodilatory action of nitrovasodilators, nitrone spin traps, and nitroxides: role of NO. Free Rad Biol Med 1995; 18:169-177.

42. Woldman YY, Khramtsov VV, Grigor'ev IA et al. Spin trapping of nitric oxide by nitronylnitroxides: measurement of the activity of NO synthase from rat cerebellum. Biochem Biophys Res Commun 1994; 202:195-203.

43. Singh RJ, Hogg N, Neese F et al. Trapping of nitric oxide formed during photolysis of sodium nitroprusside in aqueous and lipid phases: an electron spin resonance study. Photochem Photobiol 1995; 61:325-330.

44. Pace MD, Kalyanaraman B. Spin trapping of nitrogen dioxide radical from photolytic decomposition of nitramines. Free Rad Biol Med 1993; 15:337-342.

45. O'Hara JA, Goda F, Liu KJ et al. The pO$_2$ in a murine tumor after irradiation: an *in vivo* electron paramagnetic resonance oximetry study. Radiat Res 1995; 144:222-229.

46. Clarkson RB, Norby SW, Smirnov A et al. Direct measurement of the accumulation and mitochondrial conversion of nitric oxide within Chinese hamster ovary cells using an intracellular electron paramagnetic resonance technique. Biochim Biophys Acta 1995; 1243:496-502.

47. Singh RJ, Hogg N, Mchaourab HS et al. Physical and chemical interactions between nitric oxide and nitroxides. Biochim Biophys Acta 1994; 1201:437-441.

48. Mordvintcev P, Mülsch A, Busse R et al. On-line detection of nitric oxide formation in liquid aqueous phase by electron paramagnetic resonance spectroscopy. Anal Biochem 1991; 199:142-146.

49. Mülsch A, Mordvintcev P, Vanin A. Quantification of nitric oxide in biological samples by electron spin resonance spectroscopy. Neuroprotocols 1992; 1:165-173.

50. Mülsch A, Vanin A, Mordvintcev P et al. NO accounts completely for the oxygenated nitrogen species generated by enzymatic L-arginine oxygenation. Biochem J 1992; 288:597-603.

51. Kubrina LN, Caldwell WS, Mordvintcev PI et al. EPR evidence for nitric oxide production from guanido nitrogens of L-arginine in animal tissues in vivo. Biochim Biophys Acta 1992; 1099:233-237.

52. Mülsch A, Schray-Utz B, Mordvintcev PI et al. Diethyldithiocarbamate inhibits induction of macrophage NO synthase. FEBS Lett 1993; 321:215-218.

53. Kubrina LN, Miloyan VD, Mordvintcev PI et al. Iron potentiates bacterial lipopolysaccharide-induced nitric oxide formation in animal organs. Biochim Biophys Acta 1993; 1176:240-244.

54. Voevodskaya NV, Vanin AF. Gamma-irradiation potentiates L-arginine-dependent nitric oxide formation in mice. Biochem Biophys Res Commun 1992; 186:1423-1428.

55. Meerson FZ, Lapshin AV, Mordvintcev PI et al. Increased generation of nitric oxide in tissues of rats following their adaptation to short-term stress (an EPR study). Bull Exp Biol Med 1994; 117:243-245.

56. Wizemann TM, Gardner CR, Laskin JD et al. Production of nitric oxide and peroxynitrite in the lung during acute endotoxemia. J Leuk Biol 1994; 56:759-768.

57. Bune AJ, Shergill JK, Cammack R et al. L-arginine depletion by arginase induces nitric oxide production in endotoxic shock: an electron paramagnetic resonance study. FEBS Lett 1995; 366:127-130.

58. Laskin DL, Rodriguez del Valle M, Heck DE et al. Hepatic nitric oxide production following acute endotoxemia in rats is mediated by increased inducible nitric oxide synthase gene expression. Hepatology 1995;

22:223-234.

59. Malyshev IY, Manukhina EB, Mikoyan VD et al. Nitric oxide is involved in heat-induced HSP70 accumulation. FEBS Lett 1995; 370:159-162.

60. Mikoyan VD, Kubrina LN, Vanin AF. EPR evidence for nitric oxide formation via L-arginine-dependent way in stomach of mice in vivo. Biochem Mol Biol Internat 1994; 32:1157-1160.

61. Obolenskaya MY, Vanin AF, Mordvintcev PI et al. EPR evidence of nitric oxide production by the regenerating rat liver. Biochem Biophys Res Commun 1994; 202:571-576.

62. Kleschyov AL, Sedov KR, Mordvintcev PI et al. Biotransformation of sodium nitroprusside into dinitrosyl iron complexes in tissue of ascites tumors of mice. Biochem Biophys Res Commun 1994; 202:168-173.

63. Tsapin AI, Stepanichev MY, Libe ML et al. Determination of NO-synthase activity in the brain: a new method. Bull Exp Biol Med 1994; 117:39-41.

64. Sato S, Tominaga T, Ohnishi T et al. ESR spin-trapping study of nitric oxide formation during bilateral carotid occlusion in the rat. Biochim Biophys Acta 1993; 195:197.

65. Sato S, Tominaga T, Ohnishi T et al. Electron paramagnetic resonance study on nitric oxide production during brain focal ischemia and reperfusion in the rat. Brain Res 1994; 647:91-96.

66. Mülsch A, Busse R, Mordvintcev PI et al. Nitric oxide promotes seizure activity in kainate-treated rats. NeuroReports 1994; 5:2325-2328.

67. Hooper DC, Ohnishi ST, Kean R et al. Local nitric oxide production in viral and autoimmune diseases of the central nervous system. Proc Natl Acad Sci USA 1995; 92:5312-5316.

68. Bukrinsky MI, Nottet HSLM, Schmidtmayerova H et al. Regulation of nitric oxide synthase activity in human immunodeficiency virus type 1 (HIV-1)-infected monocytes: implications for HIV-associated neurological disease. J Exp Med 1995; 181:735-745.

69. Kuppusamy P, Ohnishi ST, Numagami Y et al. Three-dimensional imaging of nitric oxide production in the rat brain subjected to ischemia-hypoxia. J Cerebral Blood Flow Metab 1995; 15:899-903.

70. Rao DNR, Cederbaum AI. Production of nitric oxide and other iron-containing metabolites during the reductive metabolism of nitroprusside by microsomes and by thiols. Arch Biochem Biophys 1995; 321:363-371.

71. Décout J-L, Roy B, Fontecave M et al. Decomposition of FK 409, a new vasodilator: identification of nitric oxide as a metabolite. Bioorg Med Chem Lett 1995; 5:973-978.

72. Komarov A, Mattson D, Jones MM et al. In vivo spin trapping of nitric oxide in mice. Biochem Biophys Res Commun 1993; 195:1191-1198.

73. Lai C-S, Komarov AM. Spin trapping of nitric oxide produced in vivo in septic-shock mice. FEBS Lett 1994; 345:120-124.

74. Komarov AM, Lai C-S. Detection of nitric oxide production in mice by spin-trapping electron paramagnetic resonance spectroscopy. Biochim Biophys Acta 1995; 1272:29-36.

75. Zweier JL, Wang P, Kuppusamy P. Direct measurement of nitric oxide generation in the ischemic heart using electron paramagnetic resonance spectroscopy. J Biol Chem 1995; 270:304-307.

76. Akaike T, Noguchi Y, Ijiri S et al. Pathogenesis of influenza virus-induced pneumonia: involvement of both nitric oxide and oxygen radicals. Proc Natl Acad Sci USA 1996; 93:2448-2453.

77. Kotake Y, Tanigawa T, Tanigawa M et al. Spin trapping isotopically-labelled nitric oxide produced from [^{15}N]L-arginine and [^{17}O]dioxygen by activated macrophages using a water soluble Fe^{++}-dithiocarbamate spin trap. Free Rad Res 1995; 23:287-295.

78. Henry Y, Ducrocq C, Drapier J-C et al. Nitric oxide, a biological effector. Electron paramagnetic resonance detection of nitrosyl-iron-protein complexes in whole cells. Eur Biophys J 1991; 20:1-15.

79. Archer S. Measurement of nitric oxide in biological models. FASEB J 1993; 7:349-360.

80. Henry Y, Lepoivre M, Drapier J-C et al.

EPR characterisation of molecular targets for NO in mammalian cells and organelles. FASEB J 1993; 7:1124-1134.

81. Henry YA, Singel DJ. Metal-nitrosyl interactions in nitric oxide biology probed by electron paramagnetic resonance spectroscopy. In: Feelisch M, Stamler JS. Methods in Nitric Oxide Research. John Wiley & Sons, 1996:357-372.

82. Singel DJ, Lancaster JR. Electron paramagnetic resonance spectroscopy and nitric oxide biology. In: Feelisch M, Stamler JS. Methods in Nitric Oxide Research. John Wiley & Sons, 1996:341-356.

83. Malinski T, Czuchajowski L. Nitric oxide measurements by electrochemical methods. In: Feelisch M, Stamler JS. Methods in Nitric Oxide Research. John Wiley & Sons, 1996:319-339.

84. Feelisch M, Stamler JS. Methods in Nitric Oxide Research. John Wiley and Sons, Chichester, UK. 1996, 712 pages.

GENERAL DISCUSSION: CROSSREGULATIONS OF METALLOENZYMES TRIGGERED BY NITRIC OXIDE

Yann A. Henry

This final chapter is meant to be a discussion of current results. The divisions into chapters 6, 10, 11 and the present one are in many respects arbitrary. In chapter 6 we have presented EPR results obtained on pure metalloenzymes interacting anaerobically with authentic NO gas in the test tube. Chapter 10 dealt mostly with results obtained with cells either in culture or separated from organ, and activated in vitro. There were the notable exceptions of pure soluble guanylate cyclase and of the isolated proteins of iron metabolism, which could have been treated in chapter 6. In chapter 11, mostly pathological cases were explained, on animal models and on patients whenever possible. In the present chapter, we have gathered recent results, often contradictory at first sight, which hint at several regulation loops triggered by NO synthesis (Table 14.1). All the metalloproteins mentioned here form NO complexes, as proven by EPR spectroscopy (chapter 6), but no useful correlation has yet been made between EPR spectra and cellular functions. So these "left-over" results will be served with almost no EPR "sauce", which shall be a relief to some nauseated readers reaching this last chapter!

DOWNREGULATION OF NITRIC OXIDE SYNTHASE BY NITRIC OXIDE

The basic knowledge on nitric oxide synthase (NOS) has been summarized by our colleagues Sandrine Vadon-Le Goff and Jean-Pierre Tenu in chapter 8 (Table 8.1 and Fig. 14.1). These isoenzymes which have in common all the attributes of a complete integrated soluble P-450 + P-450

Nitric Oxide Research from Chemistry to Biology: EPR Spectroscopy of Nitrosylated Compounds, edited by Yann A. Henry, Annie Guissani and Béatrice Ducastel.
© 1997 R.G. Landes Company.

Table 14.1. Crossregulations triggered by NO

Metalloenzyme	Inhibitor/Activator	Induction/Repression	References
Guanylate cyclase	NO	–	83-86
NO-synthases	NO	+ (NOS II)	8, 9, 13-15, 18-20
P-450 cytochromes	NO	+	39-43
Arginase	NOHA	+	52-54
Argininosuccinate synthase	n.d.	+	57
Argininosuccinate lyase		+	
Cytochrome c oxidase	NO	–	60-66
Indoleamine 2,3-dioxygenase	NO	+	72-79
Heme oxygenase	NO?	+	82, 87-90
Prostaglandin H synthase (cyclooxygenase)	NO	+	99-113
Lipoxygenase	NO	n.d.	101, 114-116
Xanthine oxidase	NO	+	118, 119

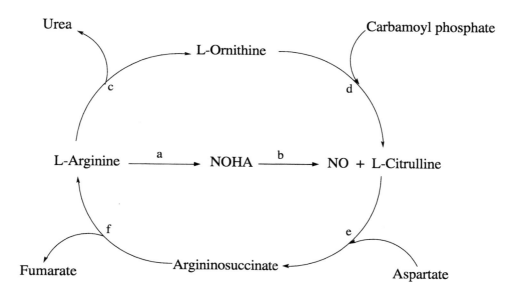

Fig. 14.1. The urea cycle of Hans Krebs and Kurt Henseleit (1932), with the L-arginine-NO pathway and the N^ω-hydroxy-L-arginine (NOHA) intermediate. (a): Nitric oxide synthase (NOS; EC 1.14.13.39), (b): NOS, P-450 enzymes, peroxidases, (c): L-arginine amidinohydrolase or arginase (EC 3.5.3.1), (d): L-ornithine transcarbamoylase (EC 2.1.3.3), (e): Argininosuccinate synthase (EC 6.3.4.5), (f): Argininosuccinate lyase (EC 4.3.2.1).

reductase system, are amazing as they have O_2 as one prime substrate, NO as one prime product and in some instances $O_2^{\bullet-}$ and H_2O_2 as secondary products, depending on NADPH oxidase activity, Ca^{2+} concentration, the absence of substrate L-arginine and the presence of tetrahydrobiopterin (BH_4 or H_4BPT) as a NOS cofactor.[1-3] How can these chemicals, which would love to react together, rapidly and irreversibly (see chapter 2), avoid each other in the vicinity of heme and not rapidly and irreversibly inhibit the enzyme? In view of the affinity of NO for reduced and oxidized heme, which can vary by many orders of magnitude among hemoproteins, one wonders whether NO could inhibit NOS.

CONSTITUTIVE NITRIC OXIDE SYNTHASES

Some answers was given immediately (1992) when NOS were demonstrated to be a soluble P-450 system.[4-7] Addition of dissolved NO gas was found to reversibly inhibit the constitutive NOS I crudely extracted from rat cerebellum, as measured by the production of L-citrulline.[8] NO-donors, SNAP and *N*-methyl-*N'*-nitro-*N*-nitrosoguanidine had the same effect. This inhibition was potentiated by superoxide dismutase (SOD) but was abolished by addition of HbO_2 which reacts rapidly with released NO.[8] Similar results were obtained with a crude extract from bovine cerebellum.[9] The K_m for L-arginine was not significantly changed by NO (200 µmol/l), from (6.4 ± 0.8) µmol/l to (10.6 ± 1.6) µmol/l, whereas the V_{max} was decreased from 80 pmol mg^{-1} min^{-1} to 45 pmol mg^{-1} min^{-1}.[9] The superoxide anion generated by the hypoxanthine-xanthine oxidase system inhibits the rat cerebellum enzyme, an effect partly reversed by SOD and catalase.[10]

Neuronal rat brain NOS (NOS I or nNOS) heme cofactor and its neighboring air-stable flavin semiquinone radical have been well characterized by EPR spectroscopy.[11,12] Nitric oxide inhibits NOS I by interacting with the heme prosthetic group.[13-15] This was proven on pure enzyme prepared from rat cerebellum. The rate of

L-citrulline production was found to be nonlinear even in the presence of excess substrate. It was inhibited by 30% by SOD addition (200 units per ml) and rendered even less linear, while HbO_2 addition (30 µmol/l) increased NOS activity by 2-fold and made the reaction rate linear.[13]

It was first suggested that NO inhibition of NOS I occurs through binding to the ferric heme. Inhibition was reversed by BH_4 which increases NOS activity by diminishing the inhibitory action of NO.[13] Analyses of UV-visible spectrophotometry and resonance Raman (RR) spectroscopy showed that NO binds to the heme in both ferrous and ferric forms, together with substrate L-arginine, in a manner similar to its binding to P-450.[14] This analysis was further developed during steady-state NO synthesis by NOS I.[15] When the three substrates, NADPH, L-arginine and O_2 were present, the majority of the enzyme (70-90%) was in the ferrous-nitrosyl form. The build-up of this complex is rapid (< 2 s) and faster than its decay. Addition of NO scavengers, HbO_2 and $O_2^{\bullet-}$ did not prevent the ferrous-nitrosyl complex formation and correlative enzyme inhibition. Thus a majority of the enzyme is converted to a catalytically inactive ferrous-nitrosyl form whose formation downregulates NO synthesis by NOS I.[15]

The role of BH_4 in protecting rat brain NOS I against NO inhibition has been further examined.[16] Both BH_4 and a BH_4-regeneration system (dihydrobiopterin reductase/NADH) are protective, probably not through a chemical scavenging interaction between BH_4 and NO or its oxidation product NO_2, as NO_2 is not a good inhibitor of NOS I. Nitrite and nitrate are not inhibitors at all. BH_4 would rather act on the enzyme through a redox cycle or as an allosteric modulator.[13,16] Another suggestion is that $O_2^{\bullet-}$ formed by autoxidation of BH_4 by O_2 inactivates NO to peroxynitrite.[3] In fact, rat brain NOS I preparations in the absence of BH_4 are very unstable and rapidly form a P-420 analogue, as characterized by UV-visible and RR spectra of the CO-bound ferrous form.[17]

The normal thiolate heme iron ligand of active NOS I could be replaced by a histidine residue, which would lead to a partially denatured and thus inactive enzyme.[17] The analogy between NOS I and P-450 heme sites seems therefore to be very strong (see below).

Endothelial NOS (NOS III or eNOS) is also reversibly inhibited by exogenous NO and NO-donors.[18,19] The K_m (4.9 µmol/l) was unaffected by NO (50 µmol/l), while V_{max} decreased slightly from 784 pmol mg^{-1} min^{-1} to 633 pmol mg^{-1} min^{-1}.[19] Conversely, exposure of porcine coronary resistance arteriolar endothelial cells (REAC) to hypoxia (10 ppm O_2) for up to 2 hr resulted in increased NO production through an activation of NOS III.[20]

INDUCIBLE NITRIC OXIDE SYNTHASE

Inducible NO synthase (NOS II or iNOS) is inhibited by NO in rat macrophage cell line culture, in alveolar macrophage and in vitro.[21-23] As in the case of NOS I, it was demonstrated that both the ferric and the ferrous forms of NOS II heme bind NO.[24] During NOS II turnover a transient mixture of the two nitrosylated complexes can be detected spectrophotometrically, which disappears to the ferric form in the presence of oxygen. The affinity of NO for ferric heme decreases upon L-arginine binding, but not for the ferrous heme.[24]

NOS II is induced at the transcriptional level (see chapter 8).[25] It is also regulated, as we have seen, by negative NO feedback, by limitation of substrate, of cofactor BH$_4$ and by phosphorylation.[25] The transcriptional induction of NOS II seems to be modulated by NO itself.[26,27] The induction of NOS II in astroglial cells by IL-1β and IFNγ was monitored by the determination of the NOS II mRNA level. NOS inhibitors or HbO$_2$ amplified mRNA expression, while a NO donor (spermine NONOate) reversed this effect.[26] A similar result was obtained in rat neutrophils.[27] These effects could be relevant in selection of drugs specific for one isotype of NOS (Fig. 14.2).[28]

INHIBITION OF MICROSOMAL CYTOCHROMES P-450

Microsomal cytochromes P-450 (CYP) are, like NOS isoenzymes, able to metabolize N^ω-hydroxy-L-arginine (NOHA), the first intermediate in the NOS pathway and aromatic amidoximes to yield NO.[29-31] EPR spectroscopy has been useful to ascertain the NO effective and direct production from these chemicals.[29-31] Other hemoproteins with peroxidase activity: horseradish peroxidase, catalase and hemoglobin, catalyze the oxidation of NOHA by H_2O_2 to NO and L-citrulline.[31] There appears to be common mechanisms of catalysis of oxidative cleavage of C=N(OH) bonds by NO synthases and cytochromes P-450.[32] Cytochromes P-450 CYP3A, induced by dexamethasone, are among the class of P-450—the most active subfamily in this catalysis.[33] Incidently, P-450s, particularly CYP3A isoforms, participate in metabolizing several other NO-donors, notably glyceryl trinitrate, 18-nitro-oxyandrostenedione, etc. (see chapter 12).[34-38]

We have mentioned in chapter 6 the interaction of P-450 monooxygenases with NO, as detected by EPR spectroscopy. Both P-450 and P-420 forms have been characterized in vitro.[39-41]

Nitric oxide inhibits cytochromes P-450 CYP1A1 and CYP2B1, in two phases: first through a reversible binding to heme(Fe^{II}) of the P-450 form, then irreversibly to the P-420 form and by nitrosation of protein residues.[42]

Inhibition by endogenous NO of other rat and human isoforms of CYP1A1 and CYP1A2 has been observed in genetically engineered V79 Chinese hamster cells.[43] This NO concentration-dependent inhibition could be a general and important means of NO synthesis regulation. A similar dependence of cytokine-mediated NO synthesis on P-450 CYP3A, CYP1A and CYP2E activity in cultured rat hepatocytes occurs at the post-translational level.[44] Cytokines, like IL-6, IL-1α and TNFα, strongly repress the inducibility of CYP1A1, CYP1A2 and CYP3A4 in human hepatocytes in primary culture.[45] Moreover

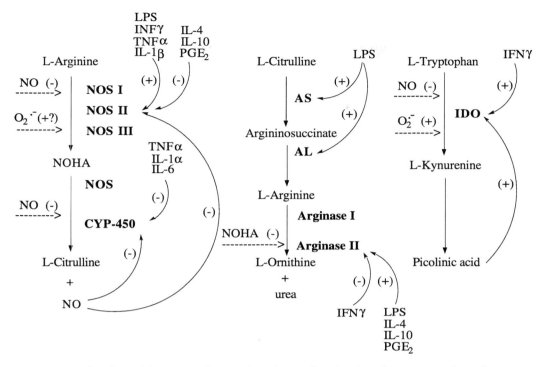

Fig. 14.2. Examples of possible crossregulations of cytokine-induced and cytokine-repressed metalloenzymes of the metabolisms of L-arginine (nitric oxide synthase (NOS), arginase, argininosuccinate synthase (AS), argininosuccinate lyase (AL)) and of L-tryptophan (L-tryptophan-2,3-dioxygenase and indoleamine 2,3-dioxygenase (IDO)). Full arrows indicate induction/repression and dashed arrows indicate enzyme inhibition/activation.

a NO-dependent inhibition of CYP expression at a transcriptional level was also demonstrated.[43] Thus systemic inflammatory response syndromes in the liver might severely affect hepatic detoxication (Fig. 14.2).[43]

Another important cellular effect of NO synthesis in hepatocytes could be a direct but reversible loss of total microsomal heme, affecting P-450 together with catalase, a decreased heme synthesis, and an increased degradation by heme oxygenase.[46]

THE UREA CYCLE

Arginase (L-arginine amidinohydrolase, EC 3.5.3.1) catalyzes the hydrolysis of L-arginine to form L-ornithine and urea. It is part of the urea cycle, of which NOS provides a shortcut between L-arginine and L-citrulline. Argininosuccinate synthase or L-citrulline:L-aspartate ligase (AS; EC 6.3.4.5) and argininosuccinate

lyase (AL; EC 4.3.2.1) together synthesize L-arginine from L-citrulline (Fig. 14.1).

ARGINASE

Found in bacteria, yeast and mammal cells,[47] arginase from rat liver contains a binuclear $Mn^{(II)}$ center as evidenced by EPR spectroscopy.[48] This binuclear $Mn^{(II)}$ site presents analogies with those of other hydrolytic enzymes (enolase and leucineaminopeptidase) and of Mn-catalase. Its function in the urea cycle, the role of arginase in the immune response and the presence of this metal center offer several possibilities of crossregulation with the L-arginine/NOS pathway.[49]

Arginase and the three NOS isoenzymes present different K_m for L-arginine (2-30 μmol/l for NOS and 1-2 mmol/l for arginase) and different inhibitors' specificity.[50,51] NOHA, the stable intermediate in the production of L-citrulline and NO from

L-arginine by NOS is a very strong competitive inhibitor of rat liver arginase, with a K_i (42 µmol/l) 25- to 50-fold lower than its K_m (1-2 mmol/l) for L-arginine.[52] A similar result was obtained on bovine liver arginase (K_i = 150 ± 50 µmol/l), while L-ornithine, L-citrulline, urea and hydroxyurea, products of the hydrolysis of L-arginine and NOHA, were not inhibitors.[53,54] N^ω-hydroxylamino-α-aminoacids formed a class of strong inhibitors of arginase.[54] The inhibitor effect of NOHA was also detected in rat liver homogenates or in rat peritoneal macrophages. Nitric oxide itself was found to be a poor inhibitor (40% at 1 mmol/l concentration).[53]

Arginase inhibition by NOHA was detected in rabbit or rat alveolar macrophages, together with an attenuated L-arginine transport, with respective lower-limit $IC_{50} \geq 15$ µmol/l and ≥ 150 µmol/l.[55] As large concentrations of NOHA are found in conditioned medium of LPS-activated macrophages, the effects of endogenous NOHA on arginase and on L-arginine transport could have important implications for the regulation of NO production in liver or macrophages, which contain both inducible NOS II and arginase.[55]

A crossregulation of arginase and NOS II occurs also at the induction level. Arginase II (the inducible isoenzyme) and NOS II are co-induced in murine macrophages (RAW 264.7) activated by LPS, by distinct transcriptional mechanisms.[56] IFNγ, which causes NOS II induction, prevents arginase II induction. Suppressors of NOS II induction: IL-4, IL-10 and PGE_2, are inducers of arginase in murine bone marrow-derived macrophages.[57] Thus the regulation of NO synthesis by the arginase pathway appears complex at several cellular levels (Fig. 14.2). The expression of arginase in human macrophages, like that of NOS II, is questioned.[58]

ARGININOSUCCINATE SYNTHASE AND ARGININOSUCCINATE LYASE

Two other enzymes in the urea cycle, AS and AL are co-induced with NOS in LPS-treated rats.[59] The level of mRNA for the three proteins was followed in several rat organs: lung, heart, liver and spleen. The time course of induction was found to be somewhat dependent on the organ and cell type. The citrulline to arginine recycling seems to be important in NO synthesis following NOS induction.[59]

INHIBITION OR REGULATION OF CYTOCHROME *C* OXIDASE

We have described in chapter 6 the multiple interactions between cytochrome *c* oxidase (cytochrome aa_3) (CcO) and nitric oxide, giving rise to a variety of EPR spectra, and the reaction cycles catalyzed by CcO, producing nitrous oxide N_2O from NO.

Results obtained in the last two years have shown that in various cell types mitochondrial respiration is inhibited reversibly at the distal end of the respiratory chain, CcO. Incubation of rat skeletal muscle mitochondria with *S*-nitrosoglutathione (GSNO) and dithiothreitol (DTT) reversibly inhibits a) CcO, determined polarographically on intact mitochondria, but not b) complexes I, II and III, determined spectrophotometrically on freeze-thawed mitochondria.[60] Cleeter et al suggested that the inhibiton of complexes I and II described by previous authors are dependent on differences in methodology, particularly in mitochondrial preparations, and could be long-term effects due to NO exposure as compared to the reversible inhibition of CcO.[60] Similar conclusions were drawn by Brown and Cooper, who suggested that CcO inhibition is the primary event of brain cell respiration and not the inhibition of [FeS]-containing proteins of complexes I and II (see chapter 10).[61] Another effect of NO on mitochondria is the reversible de-energization at low oxygen tension which is paralleled by release and re-uptake of mitochondrial Ca^{2+}.[62]

Similar experiments were carried out on rat brain synaptosomes, by following NO and O_2 concentrations with two independent Clark-type electrodes.[63] CcO was found to be inhibited by NO gas or released from SNP under light, both in synaptosomes or isolated. Inhibition was reversible and apparently competitive with oxygen, with half-inhibition NO concentration of 270 nmol/l at O_2 concentration around 145 µmol/l (arterial O_2 concentration), and of 60 nmol/l at O_2 of 30 µmol/l (around tissue O_2 concentration). These levels of NO are within plausible physiological and pathological ranges, suggesting that NO inhibition of CcO and the competition with oxygen may occur in vivo, as discussed by Brown.[63,64]

NO produced by the inducible NOS II pathway caused a similar reversible inhibition of cellular respiration in activated astrocytes.[65] The two electrode methods measuring simultaneously NO and O_2 concentrations, showed that the activated NOS II produced a steady-state level up to 1 µmol/l NO which caused respiration inhibition within the same time scale. The inhibition of respiration was rapidly reversed by addition of a NOS inhibitor or HbO_2.[65] A similar inhibition was observed when astrocytes, not activated by LPS and IFNγ, were incubated with pure exogenous NO. These effects on cellular respiration could perhaps be part of an explanation of neuropathological disorders, such as encephalitis, ischemia or multiple sclerosis (see chapter 11).[65]

Binding of NO to isolated CcO during turnover has been investigated by static and kinetic spectroscopic methods.[66] This study shows that inhibition by NO occurs during turnover by binding to ferrous cytochrome a_3, and is dependent on the oxygen concentration. It means that NO at low physiological concentrations can act as an effective regulator of the mitochondrial respiratory chain (Fig. 14.3).[66]

The competition reactions of NO and O_2 for CcO are further complicated by the above mentioned anaerobic production of nitrous oxide N_2O by reduction of NO, which has been confirmed recently by several spectroscopic methods.[67,68] This occurs at low levels of O_2 and NO. Higher levels of O_2 will favor NO_2 formation, and higher levels of NO would inhibit CcO and NO-reductase activities.[68] Production of N_2O by

reduction of NO is of interest since N_2O interacts with CcO, inducing conformational changes which modify its turnover rate.[69] Do we produce our own endogenous anesthetic laughing gas N_2O from our endogenous NO?

In two recent articles, evidence was presented for the localization of a mitochondrial NOS (mtNOS) in rat nonsynaptic brain, liver, heart, skeletal muscle and kidney mitochondria.[70,71] Immunochemical localization was detected by a mouse anti-eNOS (NOS III) monoclonal antibody. From the multiple evidences of reversible inhibition of mitochondrial respiration, the presence of mtNOS could mean a role for NO as a regulator of oxidative phosphorylation.[71]

CROSSREGULATIONS OF OTHER INDUCIBLE METALLOPROTEINS

As for arginase and NOS, several other inducible proteins are crossregulated by cytokines and by their enzymatic products, both at the transcriptional level and at the enzymatic activity level.

INDOLEAMINE 2,3-DIOXYGENASE

The antimicrobial and antitumoral activities of monocytes and macrophages and derived from their activation by IFNγ produced by T-cells. Together with the induction of NOS II and of the enzymes of the biosynthesis of BH_4 (one NOS cofactor) such as GTP cyclohydrolase, the first and rate-limiting enzyme of the L-tryptophan degradation pathway in human macrophages and cell lines is also induced by IFNγ.[72,73] This first step is catalyzed by indoleamine 2,3-dioxygenase (IDO, EC 1.13.11.17) and by L-tryptophan 2,3-dioxygenase (TPO, EC 1.13.11.11) (Fig. 14.4).

Contrary to L-tryptophan 2,3-dioxygenase, which has only L-tryptophan for substrate (see chapter 6), IDO has a low substrate specificity and plays an important role in the metabolism of serotonin, melatonin and other indoleamine derivatives.[74-76] We have described in chapter 6,

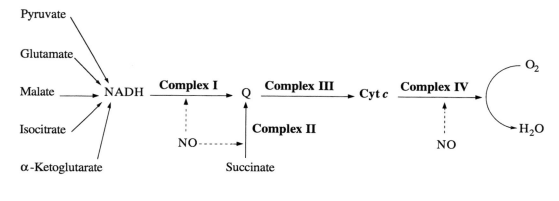

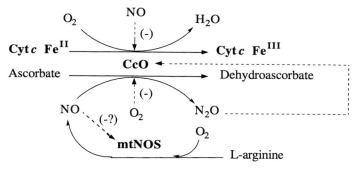

Fig. 14.3. Possible cross-interactions of the four complexes of the respiratory chain, cytochrome c oxidase (CcO), (cytochrome aa₃) and mitochondrial nitric oxide synthase (mtNOS) with O₂, NO and N₂O.

Fig. 14.4. L-tryptophan degradation pathway catalyzed in the first step by indoleamine 2,3-dioxygenase (IDO), to the three final products.

L-Tryptophan

IDO

L-Kynurenine

Quinolinic acid

Picolinic acid

Anthranilic acid

their structure and function, and the EPR spectra of the NO derivatives of both enzymes.[74]

IDO, which catalyzes the incorporation of molecular oxygen into the pyrrole ring, is activated by $O_2^{-\bullet}$.[75,76] On the other hand IDO is inhibited by NO.[77] Thus the metabolisms of L-arginine and L-tryptophan in IFNγ-activated mononuclear phagocytes are functionally regulated and contribute together to their antimicrobial and antiproliferative properties.

Another regulation occurs at the transcriptional level. Picolinic acid, a final product in the catabolism of L-tryptophan initiated by IDO, used in vitro in the millimolar range, is with IFNγ, a co-inducer of NOS II and activates the transcription of NOS II gene through an interaction with a hypoxia-responsive element (Fig. 14.2).[78,79]

HEME OXYGENASE

Carbon monoxide has been offered a putative role as a cGMP regulator in the brain.[80-82] This is firstly based on the activation of soluble guanylate cyclase (sGC) by CO (see chapter 10). However recent experiments with the purest sGC available show that this activation is only to a small extent (1.4- to 4-fold) as compared to that by NO.[83-86]

Carbon monoxide is with biliverdin, a product of the heme catabolism catalyzed by heme oxygenase (HO; EC 1.14.13.39). There are two isoforms of HO: HO-1 which is inducible and highly expressed in detoxifying tissues (liver and spleen) for destruction of red blood cells, and HO-2 which is constitutive and expressed everywhere in the body, particularly in the brain. The two forms differ in molecular mass (33 kDa for HO-1 and 36 kDa for HO-2) and are produced by distinct genes.

HO binds heme in a 1:1 ratio, forming a transitory hemoprotein. HO in the ferrous form binds its product CO. Zinc protoporphyrin and other metalloprotoporphyrins are inhibitors of both HO isoenzymes and also of sGC, while cobalt protoporphyrin and hemin are HO-1 in-

ducers.[82,87] The HO-1 enzyme has a hydrophobic C-terminal sequence involved in membrane binding. A truncated enzyme lacking this sequence is water soluble and remains active. This truncated HO-1 forms an EPR-detectable hexacoordinate nitrosyl heme complex, characterized by a rhombic symmetry (g_x = 2.086, g_y = 1.986, g_z = 2.008) with a well resolved superhyperfine structure, similar to those found for peroxidases (see chapter 6).[88] Recombinant human microsomal HO-2, expressed in *E. coli,* was digested by trypsin to yield a soluble fraction, which retained HO activity.[89] It also forms an EPR-detectable hexacoordinate nitrosyl heme complex.[89] From site-directed mutagenesis experiments, residue His-25 has been assigned as the substrate (heme) ligand. In these two recent reports, it has not been determined whether NO which binds to the heme is an inhibitor or an activator of its degradation in vitro.

Contradictory results were reported in two studies of the effect of NO on heme oxygenase activity in tissues. In rat brain and spleen homogenates, HO activity is decreased by NO derived from NOS or from SNP.[90] On the contrary in rat hepatocytes, NO synthesis resulted in massive loss of catalase, cytochrome P-450 and total microsomal heme, and in an increased heme oxygenase activity.[46]

The relative contribution of HO isoenzymes, as compared to that of NOS isoenzymes, to the activation of sGC mentioned above, has been measured in several tissues. By in situ hybridization in brain slices, a co-localization of mRNA for HO-2 and for sGC was demonstrated, which was a first hint at a possible role of CO as a neuronal messenger.[80] This was confirmed by immunohistochemistry localization of HO-2 to endothelial cells and neurons.[91] Following treatment of rats with buthionine-SR-sulfoximine (BSO), a selective depletor of glutathione, the brain HO-1 induction was found to be accompanied by a decrease in NOS activity, with a constant cGMP level.[92] Thus HO-1 seems to compensate for NOS. The balance of the two enzymes was also tested for in brain capillary endothelial cells.[93] There was no effect of cytokines IL-1, TNF or LPS on the expression of HO-1, which was enhanced by hemin. In vascular smooth muscle cells both isoforms of HO are expressed (Fig. 14.5).[94]

Finally HO-1 is a heat-shock protein (HSP32) and altogether an oxidative-stress protein.[95,96] Heat-shock pretreatment might induce a protection against the effects of damaging oxidants, of cytotoxic NO, in sepsis, etc. Another possible crossregulation can be found in the case of tobacco smoke, which contains NO and oxygen-activated derivatives, and induces the synthesis of classical HSPs (HSP70, 90, 110) and of HO. Circulating arterial HbCO has been used as an index of endogenous CO production in critical illness of patients in intensive care units.[97]

REGULATION BY NITRIC OXIDE OF OTHER METALLOENZYMES OF THE INFLAMMATORY RESPONSE

Two classes of enzymes, having arachidonic acid (5,8,11,14-eicosatetraenoic acid) for substrate, are components of the inflammatory response: prostaglandin H synthase and lipoxygenases. Prostaglandin H synthase or cyclooxygenase (PGHS; EC 1.14.99.1) is the rate-limiting enzyme in the biosynthesis of prostaglandins (PGs), thromboxane A_2 and prostacyclin (PGI_2). It catalyzes the first two steps of prostaglandin biosynthesis and contributes to inflammatory response (nephrosis, sepsis and arthritis). Lipoxygenases (EC 1.13.11.34) play a role in the synthesis of leukotrienes and lipoxins from arachidonic acid. In view of these general functions, their possible interference with the different components of the L-arginine-NOS-NO pathway has recently been examined.

PROSTAGLANDIN H SYNTHASE (CYCLOOXYGENASE)

We have summarized in chapter 6 PGHS structure and function, and its interaction with NO, as detected in vitro by EPR spectroscopy. PGHS is a hemoprotein, producing tyrosyl free radicals, similar to that of ribonucleotide reductase (RNR), during the enzyme turnover.[98] PGHS binds NO in vitro to form two different complexes, one species with a rather unspecific pentacoordinate heme(Fe^{II})-NO EPR spectrum, and a minority hexacoordinate species.[99,100] The affinity of NO for the ferric resting form of the enzyme is relatively weak (K_d = 0.9 mmol/l).[100] Thus, there is no real significance in this direct interaction at concentrations of NO encountered in vivo.[100]

Nitric oxide however inhibits lipid oxidation catalyzed by cyclooxygenase-PGHS, and also that catalyzed by lipoxygenase or hemoglobin.[101] This inhibition could occur via the reduction of PGHS ferric heme by NO to the inactive ferrous form, or by competiton to the heme site, or through a reaction with the tyrosyl radical produced during the turnover.[101,102]

PGHS and NOS enzymes present great similitudes in the existence of two, constitutive and inducible, isoforms and in their activation as well as their transcriptional regulation. The constitutive form PGHS-1 (or COX-1) is ubiquitous, while PGHS-2 (or COX-2) is found in endothelial cells, in fibroblasts and macrophages after treatment with IL-1β and LPS (Fig. 14.6). The co-induction of PGHS-2 and NOS II has been studied by several research groups. In a model of granulomatous inflammation, PGHS-2 activity and protein increased in the acute first 24 hr, and continued to increase during the chronic stage with a peak at day 14, then decreased. PGHS-1 was constant throughout. NOS II activity followed a contrasted course: it increased rapidly in the first 24 hr, continued to increase to day 7, decreased to day 14, followed by a further increase.[103] The two enzyme activities and regulations follow different courses during the inflammatory response. Another instance of differential activation has been found in human smooth muscle cells, in which IL-1β activates preferentially PGHS-2 rather than NOS II pathway.[104] Experimental autoimmune encephalomyelitis (EAE) is an inflammatory

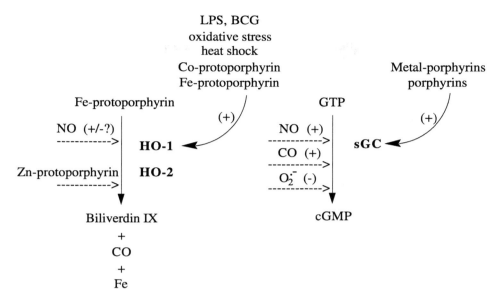

Fig. 14.5. Possible crossregulations of heme oxygenase (HO) and soluble guanylate cyclase (sGC) by porphyrins, metalloporphyrins, CO and NO.

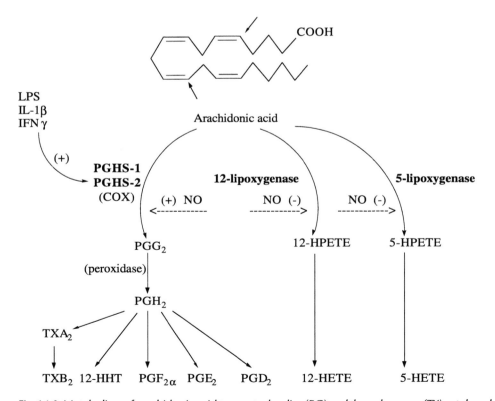

Fig. 14.6. Metabolism of arachidonic acid to prostaglandins (PG) and thromboxanes (TX) catalyzed by prostaglandin H synthase (PGHS), and to leukotrienes catalyzed by lipoxygenases (12-HHT, 12-hydroxyheptadecatrienoic acid; HPETE, hydroperoxyeicosatetraenoic acid; HETE, hydroxyeicosatetraenoic acid).

disorder mediated by IFNγ and IL–1β (see chapter 11). It has been demonstrated, in culture of macrophages in conditioned medium of activated EAE-inducer cells, that IFNγ and IL–1β induce both NOS II and PGHS-2.[105] Finally if IL–1β induces PGHS-2, its product prostaglandin E2 (PGE$_2$) seems to downregulate the induction of NOS II by IL–1β.[106] There are many instances that nitric oxide activates PGHS enzymes in various tissues and cell types: rat hypothalamus,[107] rat uterus,[108] LPS-activated mouse macrophages RAW 264.7,[109] IL-1β-activated human fetal fibroblasts,[109] IL-1β-activated rat Langerhans islets,[110] LPS-activated mouse astrocytes,[111] endothelial cells stimulated by the calcium ionophore A23187,[112] etc. A contradictory result was obtained in rat Kupffer cells, in which endogenous NO inhibits PGE$_2$ and IL-6 productions.[113]

LIPOXYGENASE

We have summarized in chapter 6 the functions of lipoxygenases, dioxygenases which have arachidonic acid for substrate, their structure and their interaction with NO, as detected by EPR spectroscopy.

Some evidence has been provided that NO inhibits lipoxygenase. Microsomal lipoxygenase from bovine seminal vesicles was inhibited by NO gas derived from the reaction of ascorbate with nitrite and transported by a nitrogen stream.[101] In human platelets the 12-lipoxygenase pathway was found to be selectively inhibited over that of PGHS (Fig. 14.6).[114] The same was found by use of NO-donors, SNP and SNAP, over human platelets and human neuroblastoma cells.[115] IL-1 was found to increase the arachidonic acid concentration available for 12-lipoxygenase through a NO-dependent mechanism.[116]

OTHER TARGETS

Many other metalloprotein targets for NO could be discussed, such as catalase,[117] SOD and other enzymes involved in the metabolism of oxygen. We shall restrict discussion to two final examples.

XANTHINE OXIDASE

Xanthine dehydrogenase (XDH) and its proteolytic derivative xanthine oxidase (XO), are regulated by inflammatory mediators in endothelial cells and liver tissue.[118] It has been shown that NO, derived from IFNγ activation of NOS in rat macrophages (bone marrow, alveolar or RAW cell line macrophages), inactivates both XDH and XO, while there was an increase of XDH mRNA. The effect of IFNγ occurs at the enzymatic level by a NO dose-dependent inhibition.[118] As XO produces O$_2$$^{\bullet-}$ and H$_2$O$_2$, its inactivation by NO could be a "protective" regulatory role for NO at the inflammation sites.[118] The inhibition of XO enzymatic activity in vitro by NO is reversible and dose-dependent, does not depend on the presence of the substrate, and thus does not result from the formation of peroxynitrite by the reaction of O$_2$$^{\bullet-}$ with NO.[119] In fact addition of peroxynitrite up to 2 μmol/l does not affect XO activity. The reversible inhibition seems to occur through an interaction with the flavin prosthetic group and not the [FeS] groups.[119]

NITROPHORINS: HOW BUGS TAKE ADVANTAGE OF US

To finish with a humorous note, we evoke the magistral work on nitrophorins by Ribeiro and colleagues, who by a beautiful demonstration have explained how blood-sucking bugs manage to make the most of one meal.[120-126] The blood-sucking insect *Rhodnius proxilus* possesses in its salivary gland a good NO carrier which can be injected into the skin of the bug's victim through its sting. As NO has vasodilatory and antiplatelet activities, the victim bleeds for the great interest of the bug. The NO carrier or nitrophorin is a hemoprotein in the ferric state, as demonstrated by UV-visible and EPR spectroscopy (Fig. 14.7).

The NO binding is reversible and pH-dependent, which makes the NO carrier very efficient.[120] NO is produced by a NOS (185 kDa) very similar to vertebrate constitutive NOS with NADPH, FAD, BH_4, calmodulin and Ca^{2+} as cofactors, dependent on L-arginine and producing L-citrulline. This NOS is also located in the salivary gland.[121] The nitrophorin has a high affinity for histamine, by exchange with NO, which is then released to counteract the host (victim) hemostatic response.[121]

Infection of *Rhodnius proxilus* with *Trypanosoma rangeli* seems to decrease the antihemostatic properties of its salivary gland products.[123] The parasite wins over the bug, which takes more time to draw blood, thus increasing the possibility of in-

oculation of parasites into the mammalian host, which is the final and double looser in the chain: it looses its blood and worse, gets a dangerous parasite infection.[123]

Nitrophorins have been studied at several levels, at the molecular and genetic levels,[124] and at the cellular and organ levels.[125] In *Rhodnius proxilus* there are four isotypes of nitrophorins carrying approximately one heme per molecular mass of around 20 kDa.[124] The heme is probably bound through a histidine residue. A similar nitrophorin has been found in the bedbug *Cimex lectularius*.[126]

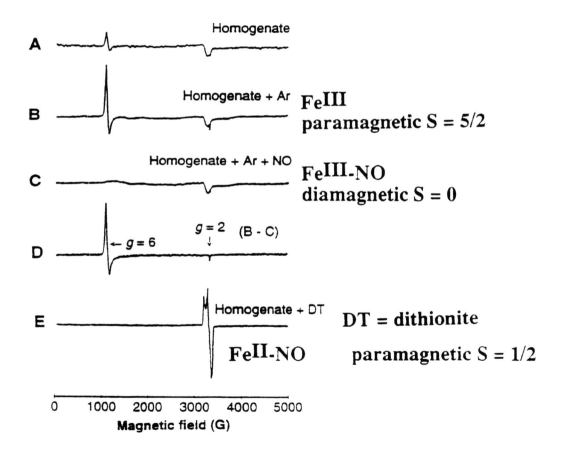

Fig. 14.7. EPR spectra of 100 pairs of Rhodnius proxilus *salivary glands in 125 μl of PBS (A) before argon equilibration; (B) after equilibration under argon for 4 hr; (C) after equilibration of (B) with NO for 2 min; (D) difference spectrum (B-C); (E) homogenate as (A) treated with dithionite. Reproduced with permission from the authors from: Ribeiro JMC, Hazzard JMH, Nussenzveig RH et al. (Reversible binding of nitric oxide by a salivary heme protein from a bloodsucking insect. Science 1993; 260:539-541.) Copyright © The American Association for the Advancement of Science, 1333 H Street, NW, Washington, DC 20005, USA.*

GENERAL CONCLUSION

We have discussed specific problems related to the use of EPR spectroscopy in nitric oxide research all along these fourteen chapters.

SPECIFICITY OF EPR SPECTROSCOPY

EPR spectroscopy has proven to be a useful instrument in metalloprotein studies and more recently in nitric oxide biology. The great interest of the method is certainly its specificity; a given EPR signal is usually a "signature" of a specific complex. A flat note however should be added to this general assertion: a given complex may be found in rather different proteins. The quality of the signature is limited to a given metal atom and its nearest neighboring atoms. Let us recall two such examples of approximate specificity. One is the nitrosylated complex in aconitase compared to that of metallothionein, for which apparently identical EPR spectra arise from fairly different iron-thiolate complexes (see chapters 6 and 10). The second instance is that of loosely bound heme-NO found in P-420, PGHS or T-state hemoglobin, giving similar signals whatever the protein. Care should therefore be taken to further characterize the molecular species responsible for EPR signals by independent methods, such as enzymatic activity measurements, molecular mass determination, immunoreactivity, etc.

SENSITIVITY: COMPARISONS WITH OTHER ANALYTICAL METHODS

As for other spectroscopic methods such as NMR or RR, EPR has the inherent defect of a lack of sensitivity. For nitrosylated hemoproteins an optimistic limit of detection is 1 µmol/l. A similar limit was reached in the detection of nitrite in plasma after dithionite reduction in the presence of agarose-bound hemoglobin.[127] This is to be compared to a threshold concentration of 1 nmol/l for the spectrophotometric evaluation of NO by hemoglobin,[128] to the lower limit of the porphyrin electrode at 10 nmol/l[129-136] and to a lower limit of 1 µmol/l for other

electrodes.[137] However by the use of Hb-spin-trap and an arduous EPR spectrometry technique a similar lower limit of 1 nmol/l of NO could be reached in biological effluents.[138] By use of gas chromatography/mass spectrometry (GC/MS) a lower limit of detection for nitrate is 0.1 µmol/l.[139] There are many other analytical methods, such as electrochemical spectrometry,[140,141] which we shall not attempt to summarize, as a whole book has been devoted to the subject (edited by Martin Feelisch and Jonathan S Stamler, ref. 142). Each of these methods has limitations and interferences. For instance peroxynitrite interferes in the determination of NO by the oxyhemoglobin reaction.[143] Similarly nitrite and nitrate determinations in plasma by the Griess reaction should follow a precise protocol and be carefully evaluated.[144]

BIOLOGICAL RELEVANCE

This specificity *versus* sensitivity balance being admitted, there remains the problem of biological relevance. If one detects EPR signals corresponding to a rather high concentration of 10 µmol/l nitrosylated complex in cells, that is 10-fold the detection limit for this method, is it really relevant to these cells' biological state of cytostasis, apoptosis or lysis? In another instance, in order to easily detect by EPR the RNR tyrosyl radical, modified cells have been built-up into which the RNR R2 subunit is expressed 40-fold over the normal level.[145-147] Is the observed correlation between the tyrosyl radical quenching and the inhibition of DNA synthesis relevant to the life and death of an ordinary cell? This problem is not specific to the EPR method but is general for all analytical methods. It requires many control experiments and much careful work to be taken into account.

EPR spectroscopy specificity remains a bonus in all cases, but it is not a panacea! Other spectroscopic methods such as resonance Raman, when properly interpreted, could be far more powerful.[14,15,17]

This book reflects only the intrinsic paramagnetism of NO and of some of its complexes with metalloproteins. A mass of primordial data, with respect to the reactivity of NO with other cellular compounds, for instance thiol or amine groups of proteins, has deliberately been omitted. To cite only a few instances of enzymes interacting with NO, out of several hundred articles: glyceraldehyde-3-phosphate dehydrogenase, poly(ADP-ribose) synthetase, tumor suppressor gene product p53, etc.,[148-153] would require another whole book.

We have tried in the present one to gather and make a synthesis of data with biological relevance as close as possible to what interests us most—human biology. We have often found ourselves in situations where basic biochemical and biophysical data, from which one can make one's opinion as to a biological relevance, were lacking. Data such as IC_{50}, K_m, V_{max}, K_i, k_{on}, k_{off}, etc., besides structural and spectroscopic data, are necessary to be gathered before anyone can make any guess as to the biological relevance of any given biological observation. We feel that biochemists should once more set to work on getting these "raw" data, not trying too much to spot "meaningful" subjects; this is a priori subjective. The profound biological meaning shall perhaps come as an unexpected reward, as it is presently the case for nitric oxide research.

REFERENCES

1. Pou S, Pou WS, Bredt DS et al. Generation of superoxide by purified brain nitric oxide synthase. J Biol Chem 1992; 267:24173-24176.

2. Hobbs AJ, Fukuto JN, Ignarro LJ. Formation of free nitric oxide from L-arginine by nitric oxide synthase: direct enhancement of generation by superoxide dismutase. Proc Natl Acad Sci USA 1994; 91:10992-10996.

3. Mayer B, Klatt P, Werner ER et al. Kinetics and mechanism of tetrahydrobiopterin-induced oxidation of nitric oxide. J Biol Chem 1995; 270:655-659.

4. White KA, Marletta MA. Nitric oxide synthase is a cytochrome P-450 type hemeprotein. Biochemistry 1992; 31:6627-6631.

5. Stuehr DJ, Ikeda-Saito M. Spectral characterization of brain and macrophage nitric oxide synthases. Cytochrome P-450-like hemeproteins that contain a flavin semiquinone radical. J Biol Chem 1992; 267:20547-2055.

6. McMillan K, Bredt DS, Hirsch DJ et al. Cloned, expressed rat cerebellar nitric oxide synthase contains stoichiometric amounts of heme, which binds carbon monoxide. Proc Natl Acad Sci USA 1992; 89:11141-11145.

7. Klatt P, Schmidt K, Mayer B. Brain nitric oxide synthase is a haemoprotein. Biochem J 1992; 288:15-17.

8. Rogers NE, Ignarro LJ. Constitutive nitric oxide synthase from cerebellum is reversibly inhibited by nitric oxide formed from L-arginine. Biochem Biophys Res Commun 1992; 189:242-249.

9. Rengasamy A, Johns RA. Regulation of nitric oxide synthase by nitric oxide. Mol Pharmacol 1993; 44:124-128.

10. Rengasamy A, Johns RA. Inhibition of nitric oxide synthase by a superoxide generating system. J Pharmacol Exp Ther 1993; 267:1024-1027.

11. Salerno JC, Frey C, McMillan K et al. Characterization by electron paramagnetic resonance of the interactions of L-arginine and L-thiocitrulline with the heme cofactor region of nitric oxide synthase. J Biol Chem 1995; 270:27423-27428.

12. Galli C, MacArthur R, Abu-Soud HM et al. EPR spectroscopic characterization of neuronal NO synthase. Biochemistry 1996; 35:2804-2810.

13. Griscavage JM, Fukuto JM, Komori Y et al. Nitric oxide inhibits neuronal nitric oxide synthase by interacting with the heme prosthetic group. Role of tetrahydrobiopterin in modulating the inhibitory action of nitric oxide. J Biol Chem 1994; 269:21644-21649.

14. Wang J, Rousseau DL, Abu-Soud HM et al. Heme coordination of NO in NO synthase. Proc Natl Acad Sci USA 1994; 91:10512-10516.

15. Abu-Soud HM, Wang J, Rousseau DL et al. Neuronal nitric oxide synthase self-inactivates by forming a ferrous-nitrosyl complex during aerobic catalysis. J Biol Chem 1995; 270:22997-23006.

16. Hyun J, Komori Y, Chaudhuri G et al. The protective effect of tetrahydrobiopterin on the nitric oxide-mediated inhibition of purified nitric oxide synthase. Biochem Biophys Res Commun 1995; 206:380-386.

17. Wang J, Stuehr DJ, Rousseau DL. Tetrahydrobiopterin-deficient nitric oxide synthase has a modified heme environment and forms a cytochrome P-420 analogue. Biochemistry 1995; 34:7080-7087.

18. Buga H, Griscavage JM, Rogers NE et al. Negative feedback regulation of endothelial cell function by nitric oxide. Cir Res 1993; 73:808-812.

19. Ravichandran LV, Fohns RA, Rengasamy A. Direct and reversible inhibition of endothelial nitric oxide synthase by nitric oxide. Am J Physiol 1995; 268:H2216-H2223.

20. Xu X-P, Pollock JS, Tanner MA et al. Hypoxia activates nitric oxide synthase and stimulates nitric oxide production in porcine coronary resistanec arteriolar endothelial cells. Cardiovasc Res 1995; 30:841-847.

21. Assreuy J, Cunha FQ, Liew FY et al. Feedback inhibition of nitric oxide synthase activity by nitric oxide. Br J Pharmacol 1993; 108:833-837.

22. Griscavage JM, Rogers NE, Sherman MP et al. Inducible nitric oxide synthase from a rat alveolar macrophage cell line is inhibited by nitric oxide. J Immunol 1993; 151:6329-6337.

23. Morin C, Fessi H, Devissaguet JP et al. Factors influencing macrophage activation by muramyl peptides: inhibition of NO synthase activity by high levels of NO. Biochim Biophys Acta 1994; 1224:427-432.

24. Hurshman AR, Marletta MA. Nitric oxide complexes of inducible nitric oxide synthase: spectral characterization and effect on catalytic activity. Biochemistry 1995; 34:5627-5634.

25. Xie Q, Nathan C. The high-output nitric oxide pathway: role and regulation. J Leuk Biol 1994; 56:576-582.

26. Park SK, Lin HL, Murphy S. Nitric oxide limits transcriptional induction of nitric oxide synthase in CNS glial cells. Biochem Biophys Res Commun 1994; 201:762-768.

27. Mariotto S, Cuzzolin L, Adami A et al. Inhibition by sodium nitroprusside of the expression of inducible nitric oxide synthase in rat neutrophils. Br J Pharmacol 1995; 114:1105-1106.

28. Amin AR, Vyas P, Attur M et al. The mode of action of aspirin-like drugs: effect on inducible nitric oxide synthase. Proc Natl Acad Sci USA 1995; 92:7926-7930.

29. Andronik-Lion V, Boucher J-L, Delaforge M et al. Formation of nitric oxide by cytochrome P450-catalyzed oxidation of aromatic amidoximes. Biochem Biophys Res Commun 1992; 185:452-458.

30. Boucher J-L, Genet A, Vadon S et al. Formation of nitrogen oxides and citrulline upon oxidation of N^ω-hydroxy-L-arginine by hemeproteins. Biochem Biophys Res Commun 1992; 184:1158-1164.

31. Boucher J-L, Genet A, Vadon S et al. Cytochrome P450 catalyzes the oxidation of N^ω-hydroxy-L-arginine by NADPH and O_2 to nitric oxide and citrulline. Biochem Biophys Res Commun 1992; 187:880-886.

32. Mansuy D, Boucher J-L, Clement B. On the mechanism of nitric oxide formation upon oxidative cleavage of C=N(OH) bonds by NO-synthases and cytochromes P450. Biochimie 1995; 77:661-667.

33. Renaud JP, Boucher J-L, Vadon S et al. Particular ability of liver P450s3A to catalyze the oxidation of N^ω-hydroxyarginine to citrulline and nitrogen oxides and occurrence in NO synthases of a sequence very similar to the heme-binding sequence of P450s. Biochem Biophys Res Commun 1993; 192:53-60.

34. Servent D, Delaforge M, Ducrocq C et al. Nitric oxide formation during microsomal hepatic denitration of glyceryl trinitrate: involvement of cytochrome P-450. Biochem Biophys Res Commun 1989; 163:1210-1216.

35. Servent D, Ducrocq C, Henry Y et al. Nitroglycerin metabolism by *Phanerochaete chrysosporium*: evidence for nitric oxide and nitrite formation. Biochim Biophys Acta 1991; 1074:320-325.

36. Servent D, Ducrocq C, Henry Y et al. Multiple enzymatic pathways involved in the metabolism of glyceryl trinitrate in *Phanerochaete chrysosporium*. Biotechnol Applied Biochem 1992; 15:257-266.

37. Delaforge M, Servent D, Wirsta P et al. Particular ability of cytochrome P-450 CYP3A to reduce glyceryl trinitrate in rat liver microsomes: subsequent formation of nitric oxide. Chem Biol Interactions 1993; 86:103-117.

38. Delaforge M, Piffeteau A, Boucher J-L et al. Nitric oxide formation during the cytochrome P-450-dependent reductive metabolism of 18-nitro-oxyandrostenedione. J Pharmacol Exp Ther 1995; 274:634-640.

39. Ebel RE, O'Keeffe DH, Peterson JA. Nitric oxide complexes of cytochrome P-450. FEBS Lett 1975; 55:198-201.

40. O'Keeffe DH, Ebel RE, Peterson JA. Studies of the oxygen binding site of cytochrome P-450. Nitric oxide as a spin-label probe. J Biol Chem 1978; 253:3509-3516.

41. Tsubaki M, Hiwatashi A, Ichikawa Y et al. Electron paramagnetic resonance study of ferrous cytochrome $P-450_{scc}$-nitric oxide complexes: effects of cholesterol and its analogue. Biochemistry 1987; 26:4527-4534.

42. Wink DA, Osawa Y, Darbyshire JF et al. Inhibition of cytochromes P450 by nitric oxide and a nitric oxide-releasing agent. Arch Biochem Biophys 1993; 300:115-123.

43. Stadler J, Trockfeld J, Schmalix WA et al. Inhibition of cytochromes P4501A by nitric oxide. Proc Natl Acad Sci USA 1994; 91:3559-3563.

44. Kuo PC, Abe KY. Cytokine-mediated production of nitric oxide in isolated rat hepatocytes is dependent on cytochrome P-450III activity. FEBS Lett 1995; 360:10-14.

45. Muntané-Relat J, Ourlin J-C, Domergue J et al. Differential effects of cytokines on the inducible expression of CYP1A1, CYP1A2, and CYP3A4 in human hepatocytes in primary culture. Hepatology 1995; 22:1143-1153.

46. Kim YM, Bergonia HA, Müller C et al. Loss and degradation of enzyme-bound heme induced by cellular nitric oxide synthesis. J Biol Chem 1995; 270:5710-5713.

47. Bond JS, Failla ML, Unger DF. Elevated manganese concentration and arginase activity in livers of streptozotocin-induced diabetic rats. J Biol Chem 1983; 258:8004-8009.

48. Reczkowski RS, Ash DE. EPR evidence for binuclear Mn(II) centers in rat liver arginase. J Am Chem Soc 1992; 114:10992-10994.

49. Schneider E, Dy M. The role of arginase in the immune response. Immunol Today 1985; 6:136-140.

50. Robertson CA, Green BG, Niedzwiecki L et al. Effect of nitric oxide synthase substrate analog inhibitors on rat liver arginase. Biochem Biophys Res Commun 1993; 197:523-528.

51. Hrabak A, Bajor T, Temesi A. Comparison of substrate and inhibitor specificity of arginase and nitric oxide (NO) synthase for arginine analogues and related compounds in murine and rat macrophages. Biochem Biophys Res Commun 1994; 198:206-212.

52. Daghigh F, Fukuto JM, Ash DE. Inhibition of rat liver arginase by an intermediate in NO biosynthesis, N^G-hydroxy-L-arginine: implications for the regulation of nitric oxide biosynthesis by arginase. Biochem Biophys Res Commun 1994; 202:174-180.

53. Boucher J-L, Custot J, Vadon S et al. N^ω-hydroxy-L-arginine, an intermediate in the L-arginine to nitric oxide pathway, is a strong inhibitor of liver and macrophage arginase. Biochem Biophys Res Commun 1994; 203:1614-1621.

54. Custot J, Boucher J-L, Vadon S et al. N^ω-Hydroxylamino-α-aminoacids as a new class of very strong inhibitors of arginases. J Biol Inorg Chem 1996; 1:73-82.

55. Hecker M, Nematollahi H, Hey C et al. Inhibition of arginase by N^G-hydroxy-L-arginine in alveolar macrophages: implications for the utilization of L-arginine for nitric oxide synthesis. FEBS Lett 1995; 359:251-254.

56. Wang WW, Jenkinson CP, Griscavage JM et al. Co-induction of arginase and nitric oxide synthase in murine macrophages activated by lipopolysaccharide. Biochem Biophys Res Commun 1995; 210:1009-1016.

57. Corraliza IM, Soler G, Eichmann K et al. Arginase induction by suppressors of nitric oxide synthesis (IL-4, IL-10 and PGE_2) in murine bone-marrow-derived macrophages. Biochem Biophys Res Commun 1995; 206:667-673.

58. Albina JE. On the expression of nitric oxide synthase by human macrophages. Why no NO? J Leuk Biol 1995;58:643-649.

59. Nagasaki A, Gotoh T, Takeya M et al. Coinduction of nitric oxide synthase, argininosuccinate synthetase, and argininosuccinate lyase in lipopolysaccharide-treated rats. RNA blots, immunoblots, and immunohistochemical analyses. J Biol Chem 1996; 271:2658-2662.

60. Cleeter MWJ, Cooper JM, Darley-Usmar VM et al. Reversible inhibition of cytochrome *c* oxidase, the terminal enzyme of the mitochondrial respiratory chain, by nitric oxide. Implications for neurodegenerative diseases. FEBS Lett 1994; 345:50-54.

61. Cooper CE, Brown GC. The interactions between nitric oxide and brain nerve terminals as studied by electron paramagnetic resonance. Biochem Biophys Res Commun 1995; 212:404-412.

62. Schweizer M, Richter C. Nitric oxide potently and reversibly deenergizes mitochondria at low oxygen tension. Biochem Biophys Res Commun 1994; 204:169-175.

63. Brown GC, Cooper CE. Nanomolar concentrations of nitric oxide reversibly inhibit synaptosomal respiration by competing with oxygen at cytochrome oxidase. FEBS Lett 1994; 356:295-298.

64. Brown GC. Nitric oxide regulates mitochondrial respiration and cell functions by inhibiting cytochrome oxidase. FEBS Lett 1995; 369:136-139.

65. Brown GC, Bolanos JP, Heales SJR et al. Nitric oxide produced by activated astrocytes rapidly and reversibly inhibits cellular respiration. Neurosci Lett 1995; 193:201-204.

66. Torres J, Darley-Usmar V, Wilson MT. Inhibition of cytochrome *c* oxidase in turnover by nitric oxide: mechanism and implications for control of respiration. Biochem J 1995; 312:169-173.

67. Zhao XJ, Sampath V, Caughey WS. Infrared characterization of nitric oxide bonding to bovine cytochrome *c* oxidase and myoglobin. Biochem Biophys Res Commun 1994; 204:537-543.

68. Zhao XJ, Sampath V, Caughey WS. Cytochrome *c* oxidase catalysis of the reduction of nitric oxide to nitrous oxide. Biochem Biophys Res Commun 1995; 212:1054-1060.

69. Dong A, Huang P, Zhao XJ et al. Characterization of sites occupied by the anesthetic nitrous oxide within proteins by infrared spectroscopy. J Biol Chem 1994; 269:23911-23917.

70. Bates TE, Loesch A, Burnstock G et al. Immunocytochemical evidence for a mitochondrially located nitric oxide synthase in brain and liver. Biochem Biophys Res Commun 1995; 213:896-900.

71. Bates TE, Loesch A, Burnstock G et al. Mitochondrial nitric oxide synthase: a ubiquitous regulator of oxidative phosphorylation? Biochem Biophys Res Commun 1996; 218:40-44.

72. Werner ER, Bitterlich G, Fuchs D et al. Human macrophages degrade tryptophan upon induction by interferon-gamma. Life Sci 1987; 41:273-280.

73. Werner ER, Werner-Felmayer G, Fuchs D et al. Parallel induction of tetrahydrobiopterin biosynthesis and indoleamine 2,3-dioxygenase activity in human celles and cell lines by interferon-γ. Biochem J 1989; 262:861-866.

74. Shimizu T, Nomiyama S, Hirata F et al. Indoleamine 2,3-dioxygenase. Purification and some properties. J Biol Chem 1978; 253;4700-4706.

75. Hirata F, Hayaishi O. Studies on indoleamine 2,3-dioxygenase. I. Superoxide anion as substrate. J Biol Chem 1975; 250:5960-5966.

76. Taniguchi T, Hirata F, Hayaishi O. Intracellular utilization of superoxide anion by indoleamine 2,3-dioxygenase of rabbit enterocytes. J Biol Chem 1977; 252:2774-2776.

77. Thomas SR, Mohr D, Stocker R. Nitric oxide inhibits indoleamine 2,3-dioxygenase activity in interferon-γ primed mononuclear phagocytes. J Biol Chem 1994; 269:14457-14464.

78. Melillo G, Cox GW, Biragyn A et al. Regulation of nitric oxide synthase mRNA expression by interferon-γ and picolinic acid. J Biol Chem 1994; 269:8128-8133.

79. Melillo G, Musso T, Sica A et al. A hypoxia-responsive element mediates a novel pathway of activation of the inducible nitric oxide synthase promoter. J Exp Med 1995; 182:1683-1693.

80. Verma A, Hirsch DJ, Glatt CE et al. Carbon monoxide: a putative neural messenger. Science 1993; 259:381-384.

81. Maines MD. Carbon monoxide: an emerging regulator of cGMP in the brain. Mol Cell Neurosci 1993; 4:389-397.

82. Marks GS. Heme oxygenase: the physiological role of one of its metabolites, carbon monoxide and interactions with zinc protoporphyrin, cobalt protoporphyrin and other metalloporphyrins. Cell Mol Biol 1994; 40:863-870.

83. Stone JR, Marletta MA. Soluble guanylate cyclase from bovine lung: activation with nitric oxide and carbon monoxide and spectral characterization of the ferrous and ferric states. Biochemistry 1994; 33:5636-5640.

84. Stone JR, Sands RH, Dunham WR et al. Electron paramagnetic resonance spectral evidence for the formation of a penta-coordinated nitrosyl-heme complex on soluble guanylate cyclase. Biochem Biophys Res Commun 1995; 207:572-577.

85. Stone JR, Marletta MA. Heme stoichiometry of heterodimeric soluble guanylate cyclase. Biochemistry 1995; 34:14668-14674.

86. Stone JR, Marletta MA. The ferrous heme of soluble guanylate cyclase: formation of hexacoordinate complexes with carbon monoxide and nitrosomethane. Biochemistry 1995; 34:16397-16403.

87. Meffert MK, Haley JE, Schuman EM et al. Inhibition of hippocampal heme oxygenase, nitric oxide synthase, and long-term potentiation by metalloporphyrins. Neuron 1994; 13:1225-1233.

88. Ito-Maki M, Ishikawa K, Matera KM et al. Demonstration that histidine 25, but not 132, is the axial heme ligand in rat heme oxygenase-1. Arch Biochem Biophys 1995; 317:253-258.

89. Ishikawa K, Takeuchi N, Takahashi S et al. Heme oxygenase-2. Properties of the heme complex of the purified tryptic fragment of recombinant human heme oxygenase-2. J Biol Chem 1995; 270:6345-6350.

90. Willis D, Tomlinson A, Frederick R et al. Modulation of heme oxygenase activity in rat brain and spleen by inhibitors and donors of nitric oxide. Biochem Biophys Res Commun 1995; 214:1152-1156.

91. Zakhary R, Gaine SP, Dinerman JL et al. Heme oxygenase-2: endothelial and neuronal localization and role in endothelium-dependent relaxation. Proc Natl Acad Sci USA 1996; 93:795-798.

92. Maines MD, Mark JA, Ewing JF. Heme oxygenase, a likely regulator of cGMP production in the brain: induction *in vivo* of HO-1 compensates for depression in NO synthase activity. Mol Cell Neurosci 1993; 4:398-405.

93. Vigne P, Feolde E, Ladoux A et al. Contributions of NO synthase and heme oxygenase to cGMP formation by cytokine and hemin treated brain capillary endothelial cells. Biochem Biophys Res Commun 1995; 214:1-5.

94. Christodoulides N, Durante W, Kroll MH et al. Vascular smooth muscle cell heme oxygenases generate guanylyl cyclase-stimulatory carbon monoxide. Circulation 1995; 91:2306-2309.

95. Ewing JF, Raju VS, Maines MD. Induction of heart heme oxygenase-1 (HSP32) by hyperthermia: possible role in stress-mediated elevation of cyclic 3'-5'-guanosine monophosphate. J Pharmacol Exp Ther 1994; 271:408-414.

96. Motterlini R, Foresti R, Vandegriff K et al. Oxidative-stress response in vascular endothelial cells exposed to acellular hemoglobin solutions. Am J Physiol 1995; 269:H648-H655.

97. Hunter K, Mascia M, Eudaric P et al. Evidence that carbon monoxide is a mediator of critical illness. Cell Mol Biol 1994; 40:507-510.

98. Tsai A, Hsi LC, Kulmacz RJ et al. Characterization of the tyrosyl radicals in ovine prostaglandin H synthase-1 by isotope replacement and site-directed mutagenesis. J Biol Chem 1994; 269:5085-5091.

99. Karthein R, Nastainczyk W, Ruf HH. EPR study of ferric prostaglandin H synthase and its ferrous NO derivative. Eur J Biochem 1987; 166:173-180.

100. Tsai A, Wei C, Kulmacz RJ. Interaction between nitric oxide and prostaglandin H synthase. Arch Biochem Biophys 1994; 313:367-372.

101. Kanner J, Harel S, Granit R. Nitric oxide, an inhibitor of lipid oxidation by lipoxygenase, cyclooxygenase and hemoglobin. Lipids 1992; 27:46-49.

102. Lassmann G, Odenwaller R, Curtis JF et al. Electron spin resonance investigation of tyrosyl radicals of prostaglandin H synthase. Relation to enzyme catalysis. J Biol Chem 1991; 266:20045-20055.

103. Vane JR, Mitchell JA, Appleton I et al. Inducible isoforms of cyclooxygenase and nitric-oxide synthase in inflammation. Proc Natl Acad Sci USA 1994: 91:2046-2050.

104. Lonchampt MO, Schulz J, Mabille K et al. Interleukin-1 activates preferentially cyclooxygenase rather than NO synthase pathway in human smooth muscle cells. Agents Actions 1994; 41:C164-C165.

105. Misko TP, Trotter JL, Cross AH. Mediation of inflammation by encephalitogenic cells: interferon-γ induction of nitric oxide synthase and cyclooxygenase-2. J Neuroimmunol 1995; 61:195-204.

106. Tetsuka T, Daphna-Iken D, Srivastava SK et al. Cross-talk between cyclooxygenase and nitric oxide pathways: prostaglandin E_2 negatively modulates induction of nitric oxide synthase by interleukin 1. Proc Natl Acad Sci USA 1994; 91:12168-12172.

107. Rettori V, Gimeno M, Lyson K et al. Nitric oxide mediates norepinephrine-induced prostaglandin E_2 release from the hypothalamus. Proc Natl Acad Sci USA 1992; 89:11543-11546.

108. Franchi A, Chaud M, Rettori V et al. Role of nitric oxide in eicosanoid synthesis and uterine mobility in estrogen-treated rat uteri. Proc Natl Acad Sci USA 1994; 91:539-543.

109. Salvemini D, Misko TP, Masferrer JL et al. Nitric oxide activates cyclooxygenase enzymes. Proc Natl Acad Sci USA 1993; 90:7240-7244.

110. Corbett JA, Kwon G, Turk J et al. IL-1β induces the coexpression of both nitric oxide synthase and cyclooxygenase by islets of Langerhans: activation of cyclooxygenase by nitric oxide. Biochemistry 1993; 32:13767-13770.

111. Molina-Hidalgo F, Lledo A, Guaza C. Evidence for cyclooxygenase activation by nitric oxide in astrocytes. Glia 1995; 15:167-172.

112. Davidge ST, Baker PN, McLaughlin MK et al. Nitric oxide produced by endothelial cells increases production of eicosanoids through activation of prostaglandin H synthase. Circ Res 1995; 77:274-283.

113. Stadler J, Harbrecht BG, DiSilvio M et al. Endogenous nitric oxide inhibits the synthesis of cyclooxygenase products and interleukin-6 by rat Kupffer cells. J Leuk Biol 1993; 53:165-172.

114. Nakatsuka M, Osawa Y. Selective inhibition of the 12-lipoxygenase pathway of arachidonic acid metabolism by L-arginine or sodium nitroprusside in intact human platelets. Biochem Biophys Res Commun 1994; 200:1630-1634.

115. Maccarrone M, Corasaniti MT, Guerrieri P et al. Nitric oxide-donor compounds inhibit lipoxygenase activity. Biochem Biophys Res Commun 1996; 219:128-133.

116. Ma Z, Ramanadham S, Corbett JA et al. Interleukin-1 enhances pancreatic islet arachidonic acid 12-lipoxygenase product generation by increasing substrate availability through a nitric oxide-dependent mechanism. J Biol Chem 1996; 271:1029-1042.

117. Brown GC. Reversible binding and inhibition of catalase by nitric oxide. Eur J Biochem 1995; 232:188-191.

118. Rinaldo JE, Clark M, Parinello J et al. Nitric oxide inactivates xanthine dehydrogenase and xanthine oxidase in interferon-γ-stimulated macrophages. Am J Respir Cell Mol Biol 1994; 11:625-630.

119. Fukahori M, Ichimori K, Ishida H et al. Nitric oxide reversibly suppresses xanthine oxidase activity. Free Rad Res 1994; 21:203-212.

120. Ribeiro JMC, Hazzard JMH, Nussenzveig RH et al. Reversible binding of nitric oxide by a salivary heme protein from a bloodsucking insect. Science 1993; 260:539-541.

121. Ribeiro JMC, Nussenzveig RH. Nitric oxide synthase activity from a hematophagous insect salivary gland. FEBS Lett 1993; 330:165-168.

122. Ribeiro JMC, Walker FA. High affinity histamine-binding and antihistaminic activity of the salivary nitric oxide-carrying heme protein (nitrophorin) of *Rhodnius proxilus*. J Exp Med 1994; 180:2251-2257.

123. Garcia ES, Mello CB, Azambuja P et al. *Rhodnius proxilus*: salivary antihemostatic components decrease with *Trypanosoma rangeli* infection. Exp Parasitol 1994; 78:287-293.

124. Champagne DE, Nussenzveig RH, Ribeiro JMC. Purification, partial characterization, and cloning of nitric oxide-carrying heme proteins (nitrophorins) from salivary glands of the blood-sucking insect *Rhodnius proxilus*. J Biol Chem 1995; 270:8691-8695.

125. Nussenzveig RH, Bentley DL, Ribeiro JMC. Nitric oxide loading of the salivary nitric oxide-carrying hemoproteins (nitrophorins) in the blood-sucking bug *Rhodnius proxilus*. J Exp Biol 1995; 198:1093-1098.

126. Valenzuela JG, Walker FA, Ribeiro JMC. A salivary nitrophorin (nitric oxide-carrying hemoprotein) in the bedbug *Cimex lectularius*. J Exp Biol 1995; 198:1519-1526.

127. Wennmalm Å, Lanne B, Petersson AS. Detection of endothelial-derived relaxing factor in human plasma in the basal state and following ischemia using electron paramagnetic resonance spectrometry. Anal Biochem 1990; 187:359-363.

128. Kelm M, Feelisch M, Spahr R et al. Quantitative and kinetic characterisation of nitric oxide and EDRF released from cultured endothelial cells. Biochem Biophys Res Commun 1988; 154:236-244.

129. Malinski T, Taha Z, Grunfeld S et al. Diffusion of nitric oxide in the aorta wall monitored in situ by porphyrinic microsensors. Biochem Biophys Res Commun 1993; 193:1076-1082.

130. Kiechle FL, Malinski T. Nitric oxide. Biochemistry, pathophysiology, and detection. Am J Clin Pathol 1993; 100:567-575.

131. Kanai AJ, Strauss HC, Truskey GA et al. Shear stress induces ATP-independent transient nitric oxide release from vascular endothelial cells, measured directly with a porphyrinic microsensor. Circ Res 1995; 77:284-293.

132. Lantoine F, Brunet A, Bedioui F et al. Direct measurement of nitric oxide production in platelets: relationship with cytosolic Ca^{2+} concentration. Biochem Biophys Res Commun 1995; 215:842-848.

133. Vallance P, Patton S, Baghat K et al. Direct measurement of nitric oxide in human beings. Lancet 1995; 346:153-154.

134. Malinski T, Czuchajowski L. Nitric oxide measurements by electrochemical methods. In: Feelisch M, Stamler JS. Methods in Nitric Oxide research. John Wiley & Sons, 1996:319-339.

135. Cespuglio R, Burlet S, Marinesco S et al. NO voltammetric detection in the rat brain. Variations of the signal throughout the sleep-waking cycle. C R Acad Sci Paris 1996; 319:191-200.

136. Buguet A, Burlet S, Auzelle F et al. Dual intervention of NO in experimental African trypanosomiasis. C R Acad Sci Paris 1996; 319:201-207.

137. Shibuki K. An electrochemical microprobe for detecting nitric oxide release in brain tissue. Neurosci Res 1990; 9:69-76.

138. Greenberg SS, Wilcox DE, Rubanyi GM. Endothelium-derived relaxing factor released from canine femoral artery by acetylcholine cannot be identified as free nitric oxide by electron paramagnetic resonance spectroscopy. Circ Res 1990; 67:1446-1452.

139. Wennmalm Å, Benthin G, Edlund A et al. Metabolism and excretion of nitric oxide in humans. An experimental and clinical study. Circ Res 1993; 73:1121-1127.

140. Hayon J, Ozer D, Rishpon J et al. Spectroscopic and electrochemical response to nitrogen monoxide of a cationic iron porphyrin immobilized in nafion-coated electrodes or membranes. J Chem Soc, Chem Commun 1994; 619-620.

141. Blyth DJ, Aylott JW, Richardson DJ et al. Sol-gel encapsulation of metalloproteins for the development of optical biosensors for nitrogen monoxide and carbon monoxide. Analyst 1995; 120:2725-2730.

142. Feelisch M, Stamler JS. Methods in Nitric Oxide Research. John Wiley and Sons, Chichester, UK. 1996, 712 pages.

143. Schmidt K, Klatt P, Mayer B. Reaction of peroxynitrite with oxyhemoglobin: interfer-

ence with photochemical determination of nitric oxide. Biochem J 1994; 301:645-647.

144. Moshage H, Kok B, Huizenga JR et al. Nitrite and nitrate determinations in plasma: a critical evaluation. Clin Chem 1995; 41:892-896.

145. Lepoivre M, Chenais B, Yapo A et al. Alterations of ribonucleotide reductase activity following induction of the nitrite-generating pathway in adenocarcinoma cells. J Biol Chem 1990; 265:14143-14149.

146. Lepoivre M, Flaman J-M, Henry Y. Early loss of the tyrosyl radical in ribonucleotide reductase of adenocarcinoma cells producing nitric oxide. J Biol Chem 1992; 267:22994-23000.

147. Lepoivre M, Flaman J-M, Bobé P et al. Quenching of the tyrosyl free radical of ribonucleotide reductase by nitric oxide. Relationship to cytostasis induced in tumor cells by cytotoxic macrophages. J Biol Chem 1994; 269:21891-21897.

148. Zhang J, Snyder SH. Nitric oxide stimulates auto-ADP-ribosylation of glyceraldehyde-3-phosphate dehydrogenase. Proc Natl Acad Sci USA 1992; 89:9382-9385.

149. Molina y Vedia L, McDonald B, Reep B et al. Nitric oxide-induced S-nitrosylation of glyceraldehyde-3-phosphate dehydrogenase inhibits enzymatic activity and increases endogenous ADP-ribosylation. J Biol Chem 1992; 267:24929-24935.

150. McDonald B, Reep B, Lapetina EG et al. Glyceraldehyde-3-phosphate dehydrogenase is required for the transport of nitric oxide in platelets. Proc Natl Acad Sci USA 1993; 90:11122-11126.

151. Zhang J, Dawson VL, Dawson TM et al. Nitric oxide activation of poly(ADP-ribose) synthetase in neurotoxicity. Science 1994; 263:687-689.

152. Michetti M, Salamino F, Melloni E et al. Reversible inactivation of calpain isoforms by nitric oxide. Biochem Biophys Res Commun 1995; 207:1009-1014.

153. Forrester K, Ambs S, Lupold SE et al. Nitric oxide-induced p53 accumulation and regulation of inducible nitric oxide synthase expression by wild-type p53. Proc Natl Acad Sci USA 1996; 93:2442-2447.

INDEX

A

AABS: A-activator binding site, 183
AIDS: acquired immunodeficiency syndrome, 243
AL: argininosuccinate lyase, 311-312
AO: ascorbate oxydase, 114-115
ARDS: adult respiratory distress syndrome, 93, 243, 283-284
AS: argininosuccinate synthase, 311-312
AscH2: ascorbic acid, 22-23
Av1 and Av2: subunits of nitrogenase from *Azotobacter vinelandii*, 119-120

B

BAL: bronchoalveolar cell pellet, 89, 91
BCG: Bacillus Calmette-Guérin (*Myobacterium bovis*), 52, 198, 208-209, 211
BFR: bacterioferritin, 126-127, 220
BH_4: tetrahydrobiopterin, 176, 179, 182, 309-310, 313, 319
BSA: bovine serum albumin, 198, 213, 220
BSO: buthionine-SR-sulfoxime, 315
BZF: bezafibrate, 72

C

CaM: calmodulin, 176, 179-180
CAT: catalase, 103-104
CcO: cytochrome aa_3 (cyt aa_3), cytochrome c oxidase, 108-114, 116, 118, 123, 312-314
CcP: cytochrome c peroxidase, 101-102, 129
2,3-CDO: catechol 2,3-dioxygenase, 117
CFA: clofibirc acid, 72
cGMP: cyclic guanosine 3',5'-monophophate, 205-206, 208, 249, 271, 273, 276, 280-281, 315
CMV: cytomegalovirus, 243
CNS: central nervous system, 236, 238, 248-249, 298
COX: cyclooxygenase or cytochrome oxidase, 105, 107, 165
Cp1 and Cp2: subunits of nitrogenase from *Clostridium pasteurianum*, 119-120
CPN: ceruloplasmin, 113-115
CPO: chloroperoxidase, 101, 103
CTPO: 3-carbamoyl-2,2,5,5-tetramethyl-3-pyrroline-1-yloxy, 295
CW-EPR: continuous wave EPR, 47
CYP: cytochrome P-450, 310-311

D

DBNBS: 3,5-dibromo-4-nitrosobenzene sulphonate, 294
DETC: diethyldithiocarbamate, 56, 295-299
DMSO: dimethylsulfoxide, 64, 295
DNIC: dinitrosyl-iron complex, 194-195, 198, 216
2,3-DPG: 2,3-diphosphoglycerate, 68, 72, 75, 87, 90
DPPH: 1,1-diphenyl-2-picrylhydrazyl stable free radical, 69, 210, 212-213, 215
DTT: dithiothreitol, 312

E

EAE: experimental allergic encephalomyelitis, 249, 253, 316, 318
eALAS: erythroid 5-aminolevulinate synthase, 221-222
EDRF: endothelium-derived relaxing factor, 5-6, 23, 25, 30-31, 33, 67, 79-80, 88, 90, 93-94, 175-176, 193-199, 206, 235, 255, 276, 281
ENDORelectron nuclear double resonance, 76, 109-110, 158
ESR: electron paramagnetic resonance, 20, 47, 250
EXAFS: extended X-ray absorption fine structure, 57, 99-100, 109, 117, 165

F

FAD: flavin adenine dinucleotide,, 118, 152, 159, 175-176, 179, 319
Fd: ferredoxin, 119
FK409: (E)-4-ethyl-2-[(E)-hydroxyimino]-5-nitro-3-hexenamide, 282
Fld: Flavodoxin, 119
FMLP: peptide formylMet-Leu-Phe, 27
FMN: flavin mononucleotide, 118, 152-153, 175-176, 179
FT: Fourier transform, 111
FTN: ferritin, 125-128, 220-221

G

G: Gauss, 48, 50, 53-55, 58, 100-104, 107-111, 113, 198, 275, 280
GC, sGC: guanylate cyclase, soluble GC, 205-208, 271, 315, 317
GC/MS: gas chromatography/mass spectrometry, 320
GHz: gigahertz, 20, 47, 53, 55, 62, 124, 161, 165, 295, 299-301
GTN: glyceryl trinitrate, 91, 271, 276-280
GTP: guanosine 5'-triphosphate, 182, 206-208, 313
GVHD: graft-versus-host disease, 255

H

Hb: hemoglobin, 50-51, 54, 62, 68, 73-75, 77-78, 80,
    87, 89, 91, 93-94, 102, 110, 129, 195-197,
    276-277, 320
HbNO: nitrosyl hemoglobin, 51, 58, 61-62, 68-70,
    72-73, 75-76, 87, 90, 92-94, 100-101, 106-107,
    195, 197, 207, 236, 238, 240, 244, 246, 248,
    250, 254-258, 260, 276-278, 293
HbO2: oxyhemoglobin, 61, 76, 78-79, 87, 90-91, 94,
    293, 309-310, 313
Hc: hemocyanin, 115-116, 127
HETE: hydroxyeicosatetraenoic acid, 117, 317
12-HHT: 12-hydroxyheptadecatrinoic acid, 117, 317
HIV: human immunodefiency virus, 243–244, 294,
    298
HO: heme oxygenase, 22, 26, 315-317
Hp: haptoglobin, 73
5-HPETE: 5-hydroperoxyeicosatertraenoic acid, 116
Hr: hemerythrin, 122-123, 125
HRP: horseradish peroxidase, 53-54, 66, 101
HSP: heat shock protein, 316
HUVEC: human umbilical vein endothelial cells, 236,
    255

I

i.p.: intraperitoneally, 295
IC$_{50}$: half-inhibitory concentration, 282, 312, 321
IDN: isosorbide dinitrate, 91
IDO: indoleamine 2,3-dioxygenase, 104-105, 311,
    313-315
IFNγ: γ-interferon, 57, 176, 182-184, 208-210,
    212-216, 222, 238, 240, 243-244, 249, 252,
    310, 312-313, 315, 318
IHP: inositolhexaphosphate, 68, 72, 74-75, 90
IL-1β: interleukin-1β, 8, 240, 243-244, 249, 251-252,
    310, 316, 318
IL-2R: interleukin-2 receptor, 255
IL-6: interleukin 6, 183-184, 218, 248-249, 255, 310,
    318
IRE: iron responsive element, 117, 183, 217, 220-222
γ-IRE: IFNγ response element, 183
IRF-1: interferon-response factor 1, 183
IRP: iron regulatory protein or IRE-binding protein,
    117, 217, 221-223

J

J: Joule, 20-22, 48, 243

K

K: Kelvin, 21, 33, 50-51, 53-58, 62-63, 65-66, 68-71,
    73-74, 76-77, 92-93, 100, 102-103, 105, 108,
    111, 116, 118, 119, 121-127, 154-156, 158,
    160-162, 164-166, 195-196, 207, 209-214, 216,
    218-219, 239, 242, 251-254, 259, 277-280, 293,
    295-299
kHz: kilohertz, 53

L

L-1: lipoxygenase 1, 116-117
LegHb: leghemoglobin, 101
LPO: lactoperoxidase, 101-102
LPS: lipopolysaccharide, 8, 57, 176, 182-185,
    208-209, 212, 214-216, 220, 222, 240, 242,
    244, 246-248, 255, 272, 294-295, 297, 299,
    300, 312-313, 316, 318
LTB4: leukotriene B4, 27
lTF: lactoferrin, 219-220

M

Mb, metMb: myoglobin and ferric myoglobin, 65,
    100-102, 110, 129
MbNO: nitrosylmyoglobin, 5, 100-101, 257
MBP: myelin basic protein, 249
MCD: magnetic circular dichroism, 106, 165, 208,
    222
MDP: muramyl dipeptide, 182
MDRF: macrophage-derived relaxing factor, 194
MetHb: methemoglobin, 77-79, 87, 90, 93-94
MGD: *N*-methyl-D-glucamine dithiocarbamate,
    299-300
MHz: megahertz, 20, 50, 68
MNP: 2-methyl-2-nitrosopropane, 294
MPO: myeloperoxidase, 101, 103
MRE: metal responsive element, 218
MT: metallothionein, 121-122, 218
mtNOS: mitochondrial NOS, 313-314
mW: milliwatt, 56, 58

N

N$_2$OR: nitrous oxide reductase, 146, 164-166
NADPH: nicotinamide adenine dinucleotide
    phosphate, 25, 108, 161, 175-176, 178-180,
    243, 272, 280, 309, 319
NAME: *N*$^\omega$-nitro-L-arginine methyl ester, 259
NaR: nitrate reductase, 146
NF-IL-6: nuclear factor-interleukin-6, 183
NFκB: nuclear factor-κB, 295, 183
NHE: normal hydrogen electrode, 219-220
NI: 7-nitroindazole, 182
NiR: nitrite reductase, 146, 149, 152-155, 158-159,
    161, 163, 166
NMDA: *N*-methyl-D-aspartate, 33, 222, 241
NMMA: *N*$^\omega$-monomethyl-L-arginine, 180, 185, 209,
    237, 239-241, 250-251, 299-300
NMR: nuclear magnetic resonance, 47, 53, 57, 99,
    111, 121, 219, 320
NOCT: nitric oxide cheletropic trap, 294
NOHA: *N*$^\omega$-hydroxy-L-arginine or *N*^{G}hydroxy-L-
    arginine, 178-179, 185, 272, 308, 310-312
NONOate: diazeniumdiolate, RR'N[(NO)-N=O]$^-$,
    29, 222, 282, 310
NoR: nitric oxide reductase, 146, 149, 158-161, 163,
    166

NOS: nitric oxide synthase, 8, 101, 105, 146, 167,
    176, 178-180, 182, 185, 194, 210, 214, 222,
    235-236, 238-240, 243-244, 248, 250, 272, 274,
    282, 284, 295, 307-313, 315-316, 318-319
NOS I (nNOS): neuronal constitutive NOS, 175-176,
    179-182, 236, 243, 249, 309-310
NOS II (iNOS): inducible NOS, 22, 176, 178-184,
    194, 198, 205, 208-209, 212-214, 218, 220-223,
    235-236, 238, 240-241, 243-244, 248-250, 255,
    257-258, 294, 310, 312-313, 315-316, 318
NOS III (eNOS): endothelial constitutive NOS, 176,
    179-180, 182, 184, 198, 212, 236, 255, 257,
    284, 310

O

Oe: Oersted, 53
18-ONO₂A: 18-nitro-oxyandrostenedione, 278
oTF: ovotransferrin, 219

P

P-450: P-450 monooxygenases, P-450$_{cam}$ from
    Pseudomonas putida, P-450$_{scc}$ from bovine adrenal
    cortex, 89, 101, 103, 107–108, 161, 176, 178,
    180, 194, 272-273, 276, 278, 280, 282,
    307-311, 315
PBN: phenyl *N-tert*-burylnitrone, 294
3,4-PDO: protocatechuate 3,4-dioxygenase, 117
4,5-PDO: protocatechuate 4,5-dioxygenase, 117
PGE2: [15Z,11α,13E,15S]-11,15-dihydroxy-9-
    oxoprosta-5,13-dienoic acid, 249, 312, 318
PGG2: 15-hydroperoxy-9,11-peroxidoprosta-5,13-
    dienoic acid, 105
PGH2: 15-hydroxy-9,11-peroxidoprosta-5,13-dienoic
    acid, 105-106
PGHS: prostaglandin H (prostaglandin endoperoxide)
    synthase, 105-107, 207, 316-318, 320
PMA: phorbol 12-myristate 13-acetate, 27
PMN: polymorphonuclear leukocyte, neutrophil, 27,
    78
PMS: phenazine methosulfate, 153-154, 156-158, 162
POD: post-operation day, 254-255
ppb: part per billion, 7, 282, 284
PPHN: persistent pulmonary hypertension of the
    newborn, 93-94
ppm: part per million, 7, 25, 89-91, 93-94, 255, 284,
    310
PTI: 2-phenyl-4,4,5,5-tetramethylimidazoline-1-oxyl,
    294
PTIO: 2-phenyl-4,4,5,5-tetramethylimidazoline-1-
    oxyl 3-oxide, 294-295

R

R1 and R2: subunits of ribonucleotide reductase, 27,
    33, 52, 54, 57, 106, 122–126, 212-217, 223,
    294, 320
RBC: red blood cell, 57, 73, 79-80, 87-91, 93-94,
    239, 241, 243-244, 248, 250, 254-258, 260

REAC: resistance arteriolar endothelial cells, 310
RNR: ribonucleotide reductase, 54, 123-126,
    212-214, 217, 220, 223, 238, 316, 320
RR: resonance Raman, 100, 106, 109, 111, 117, 126,
    207-208, 309, 320

S

SDH: succinate dehydrogenase, 118-119
SDS: sodium dodecylsulfate, 107, 207
sGC: soluble guanylate cyclase, 315
SHF: superhyperfine (structure), 62-64, 68, 72, 76,
    101-102, 104, 106-107, 109-110, 207, 295
SIN-1: 3-morpholinosydnonimine *N*-ethylcarbamide,
    31, 91, 220, 222, 272, 282-283, 295-297
SIN-1A: N-morpholinoiminoacetonitrate, 282-283
SNAP: *S*-nitroso-*N*-acetylpenicillamine, 30, 218,
    220-222, 281, 295, 309, 318
SNP: sodium nitroprusside, 30, 91, 210, 220-221,
    272-275, 295, 312, 315, 318
SOD: superoxide dismutase, 23, 34, 222, 278, 282,
    309, 318
SR: thiol residue, 30, 50, 54, 118-122, 125, 128,
    195–199, 210-212, 214, 218, 236, 240,
    274-275, 315
sTF: serum transferrin, 219-220

T

T: Tesla, 20, 47-48, 51, 54, 68-69, 71-75, 79, 195,
    207, 236, 244
t-BuOOH: *tert*-butylhydroperoxide, 78
TCT: tracheal cytotoxin, 240
TGFb: transforming growth factor β, 182-183
TNF: tumor necrosis factor, 183, 249, 316
TNFα: tumor necrosis factor α, 57, 182-184, 208,
    210, 212-216, 218, 238, 240-241, 244, 246,
    248-249, 252, 255, 310
TPO: L-tryptophan 2,3-dioxygenase, 104-105, 313
TSST-1: toxic shock syndrome toxin 1, 244

U

U: unit, 8, 206, 210, 215, 252

X

XDH: xanthine dehydrogenase, 318
XO: xanthine oxidase, 318

DATE DUE

JUN 0 5 1998	
FEB 1 9 2001	
SEP 0 3 2004	